MONOGRAPHS ON STATISTICS AND AP

General Editors

D.R. Cox, V. Isham, N. Keiding, N. Reid, H. Tong, and T. Louis

1 Stochastic Population Models in Ecology and Epidemiology *M.S. Barlett* (1960)
2 Queues D.R. *Cox and W.L. Smith* (1961)
3 Monte Carlo Methods *J.M. Hammersley and D.C. Handscomb* (1964)
4 The Statistical Analysis of Series of Events *D.R. Cox and P.A.W. Lewis* (1966)
5 Population Genetics *W.J. Ewens* (1969)
6 Probability, Statistics and Time *M.S. Barlett* (1975)
7 Statistical Inference *S.D. Silvey* (1975)
8 The Analysis of Contingency Tables *B.S. Everitt* (1977)
9 Multivariate Analysis in Behavioural Research *A.E. Maxwell* (1977)
10 Stochastic Abundance Models *S. Engen* (1978)
11 Some Basic Theory for Statistical Inference *E.J.G. Pitman* (1979)
12 Point Processes *D.R. Cox and V. Isham* (1980)
13 Identification of Outliers *D.M. Hawkins* (1980)
14 Optimal Design *S.D. Silvey* (1980)
15 Finite Mixture Distributions *B.S. Everitt and D.J. Hand* (1981)
16 Classification *A.D. Gordon* (1981)
17 Distribution-free Statistical Methods, 2nd edition *J.S. Maritz* (1995)
18 Residuals and Influence in Regression R.D. *Cook and S. Weisberg* (1982)
19 Applications of Queueing Theory, 2nd edition *G.F. Newell* (1982)
20 Risk Theory, 3rd edition *R.E. Beard, T. Pentikäinen and E. Pesonen* (1984)
21 Analysis of Survival Data *D.R. Cox and D. Oakes* (1984)
22 An Introduction to Latent Variable Models *B.S. Everitt* (1984)
23 Bandit Problems *D.A. Berry and B. Fristedt* (1985)
24 Stochastic Modelling and Control *M.H.A. Davis and R. Vinter* (1985)
25 The Statistical Analysis of Composition Data *J. Aitchison* (1986)
26 Density Estimation for Statistics and Data Analysis *B.W. Silverman* (1986)
27 Regression Analysis with Applications *G.B. Wetherill* (1986)
28 Sequential Methods in Statistics, 3rd edition
G.B. Wetherill and K.D. Glazebrook (1986)
29 Tensor Methods in Statistics *P. McCullagh* (1987)
30 Transformation and Weighting in Regression
R.J. Carrol and D. Ruppert (1988)
31 Asymptotic Techniques of Use in Statistics
O.E. Bardorff-Nielsen and D.R. Cox (1989)
32 Analysis of Binary Data, 2nd edition *D.R. Cox and E.J. Snell* (1989)

33 Analysis of Infectious Disease Data *N.G. Becker* (1989)

34 Design and Analysis of Cross-Over Trials *B. Jones and M.G. Kenward* (1989)

35 Empirical Bayes Methods, 2nd edition *J.S. Maritz and T. Lwin* (1989)

36 Symmetric Multivariate and Related Distributions
K.-T Fang, S. Kotz and K.W. Ng (1990)

37 Generalized Linear Models, 2nd edition *P. McCullagh and J.A. Nelder* (1989)

38 Cyclic and Computer Generated Designs, 2nd edition
J.A. John and E.R. Williams (1995)

39 Analog Estimation Methods in Econometrics *C.F. Manski* (1988)

40 Subset Selection in Regression *A.J. Miller* (1990)

41 Analysis of Repeated Measures *M.J. Crowder and D.J. Hand* (1990)

42 Statistical Reasoning with Imprecise Probabilities *P. Walley* (1991)

43 Generalized Additive Models *T.J. Hastie and R.J. Tibshirani* (1990)

44 Inspection Errors for Attributes in Quality Control
N.L. Johnson, S. Kotz and X, Wu (1991)

45 The Analysis of Contingency Tables, 2nd edition *B.S. Everitt* (1992)

46 The Analysis of Quantal Response Data *B.J.T. Morgan*

47 Longitudinal Data with Serial Correlation: A State-space Approach
R.H. Jones (1993)

48 Differential Geometry and Statistics *M.K. Murray and J.W. Rice* (1993)

49 Markov Models and Optimization *M.H.A. Davis* (1993)

50 Networks and Chaos—Statistical and Probabilistic Aspects
O.E. Barndorff-Nielsen, J.L. Jensen and W.S. Kendall (1993)

51 Number-theoretic Methods in Statistics *K.-T. Fang and Y. Wang* (1994)

52 Inference and Asymptotics *O.E. Barndorff-Nielsen and D.R. Cox* (1994)

53 Practical Risk Theory for Actuaries
C.D. Daykin, T. Pentikäinen and M. Pesonen (1994)

54 Biplots *J.C. Gower and D.J. Hand* (1996)

55 Predictive Inference—An Introduction *S. Geisser* (1993)

56 Model-Free Curve Estimation *M.E. Tarter and M.D. Lock* (1993)

57 An Introduction to the Bootstrap *B. Efron and R.J. Tibshirani* (1993)

58 Nonparametric Regression and Generalized Linear Models
P.J. Green and B.W. Silverman (1994)

59 Multidimensional Scaling *T.F. Cox and M.A.A. Cox* (1994)

60 Kernel Smoothing *M.P. Wand and M.C. Jones* (1995)

61 Statistics for Long Memory Processes *J. Beran* (1995)

62 Nonlinear Models for Repeated Measurement Data
M. Davidian and D.M. Giltinan (1995)

63 Measurement Error in Nonlinear Models
R.J. Carroll, D. Rupert and L.A. Stefanski (1995)

64 Analyzing and Modeling Rank Data *J.J. Marden* (1995)

65 Time Series Models – In econometrics, finance and other fields
D.R. Cox, D.V. Hinkley and O.E. Barndorff-Nielsen (1996)

66 Local Polynomial Modeling and its Application *J. Fan and I. Gijbels* (1996)

67 Multivariate Dependencies—Models, analysis and interpretation *D.R. Cox and N. Wermuth* (1996)

68 Statistical Inference—Based on the likelihood *A. Azzalini* (1996)

69 Bayes and Empirical Bayes Methods for Data Analysis *B.P. Carlin and T.A Louis* (1996)

70 Hidden Markov and Other Models for Discrete-valued Time Series *I.L. Macdonald and W. Zucchini* (1997)

71 Statistical Evidence—A likelihood paradigm *R. Royall* (1997)

72 Analysis of Incomplete Multivariate Data *J.L. Schafer* (1997)

73 Multivariate Models and Dependence Concepts *H. Joe* (1997)

74 Theory of Sample Surveys *M.E. Thompson* (1997)

75 Retrial Queues *G. Falin and J.G.C. Templeton* (1997)

76 Theory of Dispersion Models *B. Jørgensen* (1997)

77 Mixed Poisson Processes *J. Grandell* (1997)

78 Variance Components Estimation—Mixed models, methodologies and applications *P.S.R.S. Rao* (1997)

79 Bayesian Methods for Finite Population Sampling *G. Meeden and M. Ghosh* (1997)

80 Stochastic Geometry: Likelihood and Computation *O.E. Barndorff-Nielsen, W.S. Kendall and M.N.M. van Lieshout*

81 Computer-Assisted Analysis of Mixtures and Applications *D. Böhning* (1999)

82 Classification, 2nd edition *A.D. Gordon* (1999)

83 Semimartingales and their Statistical Inference, *B.L.S. Prakasa Rao* (1999)

Semimartingales and their Statistical Inference

B.L.S. Prakasa Rao

Indian Statistical Institute
New Delhi, India

CRC Press
Taylor & Francis Group
Boca Raton London New York

CRC Press is an imprint of the
Taylor & Francis Group, an **informa** business
A CHAPMAN & HALL BOOK

CRC Press
Taylor & Francis Group
6000 Broken Sound Parkway NW, Suite 300
Boca Raton, FL 33487-2742

First issued in paperback 2019

ISBN-13: 978-1-58488-008-0 (hbk)
ISBN-13: 978-0-367-39975-7 (pbk)

Contents

Preface

Decision making in all spheres of activity involves uncertainity. If rational decisions have to be made, they have to be based on the past observations on the phenomenon. Data collection, model building and inference from the data collected, validation of the model, and refinement of the model are the key steps or building blocks involved in any rational decision-making process. Stochastic processes are widely used for model building in social, physical, engineering, and life sciences. A more recent application is to the area of financial economics. Statistical inference for stochastic processes is of great importance from the theoretical as well as the applications point of view in model building. During the past twenty years, there has been a large amount of progress in the study of inferential aspects for continuous as well as discrete time stochastic processes. Semimartingales are a large class of stochastic processes that include diffusion-type processes, point processes, and diffusion-type processes with jumps that are widely used for stochastic modeling. Our aim in this book is to study the asymptotic theory of statistical inference for semimartingale models in a unified manner. The special case of diffusion-type processes is discussed in a separate book entitled *Statistical Inference for Diffusion Type Processes,* bringing together several methods of estimation of parameters involved when the processes are observed continuously over a period of time or when a sampled data is available as is generally feasible.

It is a great pleasure to thank the Indian Statistical Institute for its support during the preparation of this book. The idea for this book arose when the author gave a course on Semimartingales and Their Statistical Inference at the University of California, Davis during 1986–87 at the invitation of Professor George Roussas. Thanks are due to him and several other colleagues who have sent their preprints and reprints to me over the years, the contents of which have been extensively used in the preparation of this book. Mr. V. P. Sharma and Ms. Simmi Marwah of Indian Statistical Institute have ably assisted me in preparing the Latex version. Thanks are due to Mr. Robert Stern, Executive Editor, Mathematics and Engineering, of CRC Press LLC for readily accepting my book proposal. Finally, it is a great pleasure for me to thank my wife Vasanta for her understanding, endurance, and patience with the endless number of hours I was away in the office working on this book and other books over the years.

B. L. S. Prakasa Rao

To my wife Vasanta

Chapter 1

Semimartingales

1.1 Introduction

Decision making in all spheres of activity involves uncertainty. If rational decisions have to be made, they have to be based on the past observations on the phenomenon. Data collection, model building and inference from the data collected, validation of the model, and refinement of the model are the key steps involved in any rational decision-making process. Stochastic processes are widely used for model building, and the subject of inference for stochastic processes is of importance both from the theoretical point of view as well as from the application aspect. Several books have been published during the last twenty years, starting with Basawa and Prakasa Rao [5], Prakasa Rao [52, 53], Grenander [18], Kutoyants [38], and more recently by Karr [35] and Fleming and Harrington [15] on inference for stochastic processes. Prakasa Rao [54], Prakasa Rao and Bhat [56], Prabhu [50], Prabhu and Basawa [51], and Basawa and Prabhu [4] give comprehensive accounts of recent developments in the area of inference for stochastic processes. In a recent book, Statistical Inference for Diffusion Type Processes (cf. [55]), we have discussed various methods of estimation of the parameters involved in the process when the process is continuously observed over time or when a sampled data on the process is available as is generally feasible. The class of diffusion-type processes is a subclass of the class of semimartingales. Our discussion here concerns the general asymptotic theory of semimartingales and their statistical inference.

The notion of a semimartingale has been found to be of major interest in stochastic modeling, as it includes several types of processes such as point processes, diffusion processes, diffusion processes with jumps, etc., that are widely used for model building. A semimartingale is essentially the sum of a local martingale and a process that is of bounded variation. A brief introduction for such processes is given in [53]. As the concept of a semimartingale and its properties are not widely discussed in the books on statistical inference and are not widely known to the statisticians and modelers, we give a review of some results from the theory of semimartingales in this chapter, giving proofs occasionally. The books by Elliott [12], Bremaud [6], Liptser and Shiryayev [44, 45], and Kallianpur [33], for instance, deal extensively with this topic, and our review in this chapter is based on these as well as a few other related books.

1.2 Stochastic Processes

Filtration and Stopping Times

Let $(\Omega, \mathcal{F})$ be a measurable space and τ be an index set which is either $[0, \infty)$ or $N = \{0, 1, 2, \ldots\}$.

DEFINITION 1.1 *A filtration (flow, filter) $\{\mathcal{F}_t\}$ of $(\Omega, \mathcal{F})$ is a family of sub-σ-algebras $\mathcal{F}_t$, $t \in \tau$ of $\mathcal{F}$ such that*

$$s \leq t \Rightarrow \mathcal{F}_s \subset \mathcal{F}_t .$$

Notation

(i) $\mathcal{F}_\infty = \vee_t \mathcal{F}_t$ is the smallest σ-algebra generated by $\{\mathcal{F}_t, \ t \in \tau\}$.

(ii) $\mathcal{F}_{t+} = \cap_{s>t} \mathcal{F}_s, \ \ t \geq 0.$

(iii) $\mathcal{F}_{t-} = \vee_{s<t} \mathcal{F}_s, \ \ t > 0.$

DEFINITION 1.2 *A filtration is said to be right-continuous if $\mathcal{F}_t = \mathcal{F}_{t+}$. If $\tau = \{0, 1, 2, \ldots\}$, then $\mathcal{F}_{n+} = \mathcal{F}_{n+1}$ and $\mathcal{F}_{n-} = \mathcal{F}_{n-1}$.*

DEFINITION 1.3 *A random variable $T : \Omega \to \tau \cup \{\infty\}$ is said to be a stopping time if $[T \leq t] = \{\omega : T(\omega) \leq t\} \in \mathcal{F}_t$ for every $t \in \tau$.*

Properties of Stopping Times

(i) A constant random variable $T \equiv t \in \tau$ is a stopping time.

(ii) T stopping time, $s \in \tau \Rightarrow T + s$ is a stopping time.

(iii) S, T stopping times $\Rightarrow S \vee T$ and $S \wedge T$ are stopping times.

(iv) $T_n, n \geq 1$ stopping times $\Rightarrow \wedge_n T_n$ and $\vee_n T_n$ are stopping times where $\wedge$ denotes the infinimum and $\vee$ denotes the supremum.

DEFINITION 1.4 *Let T be a stopping time. Define*

$$\mathcal{F}_T = \{A : A \cap [T \leq t] \in \mathcal{F}_t \textit{ for every } t\} \ .$$

Then $\mathcal{F}_T$ is a σ-algebra and $\mathcal{F}_T$ is called the stopped σ-algebra at the stopping time T.

(v) Let S and T be stopping times. Then
(a) $S \leq T \Rightarrow \mathcal{F}_S \subset \mathcal{F}_T$, and
(b) $\mathcal{F}_{S \wedge T} = \mathcal{F}_S \cap \mathcal{F}_T$.

REMARK 1.1 Suppose $\tau = N$. If $X_n, n \geq 0$ is a sequence of random variables such that X_n is $\mathcal{F}_n$-measurable, then $X_T(\omega) = X_{T(\omega)}(\omega)$ is $\mathcal{F}_T$-measurable whenever T is a stopping time. ▮

REMARK 1.2 Let $(\Omega, \mathcal{F}, P)$ be a probability space. A stochastic process $\{X_t, \ t \in \tau\}$ is a family of random variables defined on the probability space $(\Omega, \mathcal{F}, P)$. For most of our discussion, the index set $\tau = [0, \infty)$ or $\tau = \{0, 1, 2, \ldots\}$. ▮

DEFINITION 1.5 *A stochastic process defined on $(\Omega, \mathcal{F}, P)$ is said to be RCLL if the process is continuous on the right and has limits from the left.*

DEFINITION 1.6 *A process $\{Y_t\}$ is said be a modification of another process $\{X_t\}$ if $Y_t = X_t$ a.s. for every $t \in \tau$ (here the null set might depend on the choice of t).*

REMARK 1.3 It is possible that $\{Y_t\}$ is a modification of $\{X_t\}$ and yet the sample paths of X might be continuous and those of Y might be discontinuous.

For instance, let $X_t(\omega) = 0$ for all ω, t and define

$$Y_t(\omega) = 0 \text{ if } t - [t] \neq \omega$$

$$= 1 \text{ if } t - [t] = \omega$$

where $[t]$ denotes the largest integral less than or equal to t. Then $t \to X_t(\omega)$ is continuous for any fixed ω where as $t \to Y_t(\omega)$ is discontinuous. But $X_t = Y_t$ a.s. on $(\Omega, \mathcal{F}, P)$ where $\Omega = [0, 1]$, $\mathcal{F}$ is the Borel σ-algebra and P is the Lebesgue measure on $[0, 1]$. ∎

DEFINITION 1.7 *Two processes $\{X_t\}$ and $\{Y_t\}$ are said to be indistinguishable if for almost every $\omega \in \Omega$, $X_t(\omega) = Y_t(\omega)$ for every $t \in \tau$ (here the null set does not depend on t).*

If $\{X_t\}$ is a modification of $\{Y_t\}$, $\tau = [0, \infty]$ or $\tau = [0, \infty)$, and if both the processes are right-continuous, then $\{X_t\}$ and $\{Y_t\}$ are indistinguishable.

DEFINITION 1.8 *Let $A \subset \tau \times \Omega$. Consider the process I_A defined by*

$$I_A(t, \omega) = 1 \quad \textit{if } (t, \omega) \in A$$
$$= 0 \quad \textit{otherwise}.$$

The set A is said to be evanescent if the process I_A is indistinguishable from the zero process.

REMARK 1.4 A set A is evanescent if and only if the projection of A on Ω is a set of measure zero. ∎

DEFINITION 1.9 *A process $\{X_t\}$ is said to be adapted to a filtration $\{\mathcal{F}_t\}$ if X_t is $\mathcal{F}_t$-measurable for every $t \in \tau$.*

REMARK 1.5 If $\{X_t\}$ is adapted to $\{\mathcal{F}_t\}$ and $\{Y_t\}$ is a modification of $\{X_t\}$, then $\{Y_t\}$ is also adapted to $\{\mathcal{F}_t\}$ if each $\mathcal{F}_t$ contains all null sets of $\mathcal{F}$. ∎

DEFINITION 1.10 *Let $(\Omega, \mathcal{F}, P)$ be a probability space with a filtration*

$\{\mathcal{F}_t\}$. Let $\mathcal{F}^P$ denote the completion of $\mathcal{F}$ under P (that is, it is the smallest σ-algebra generated by $\mathcal{F}$ and the set $\mathcal{N}^P$ of all P-null sets; if $A \in \mathcal{F}$ and $P(A) = 0$, then, for any $B \subset A$, $B \in \mathcal{N}^P$). Let $\mathcal{F}_t^P$, $t \in \tau$ be the σ-algebra generated by $\mathcal{F}_t$ and the P-null sets of $\mathcal{F}^P$. Then $\{\mathcal{F}_t^P\}$ is called the completion of the filtration $\{\mathcal{F}_t\}$.

REMARK 1.6 A filtration is *complete* if $\mathcal{F}$ is complete and each $\mathcal{F}_t$ contains all the P-null sets of $\mathcal{F}$ (here after we assume that the filtration is always complete). ▮

DEFINITION 1.11 *Suppose $\tau = [0, \infty)$. A process $\{X_t\}$ adapted to a filtration $\{\mathcal{F}_t\}$ and taking values in E is said to be progressively measurable (progressive) if for every $t \in \tau$, the map*

$$(s, \omega) \to X_s(\omega) \text{ from } [0, t] \times \Omega \to (E, \mathcal{E})$$

is measurable when $[0, t] \times \Omega$ is given the product σ-algebra $\mathcal{B}([0, t]) \otimes \mathcal{F}_t$ (here E is a metric space with the associated Borel σ-algebra $\mathcal{E}$ and $\mathcal{B}([0, t])$ denotes the Borel σ-algebra on $[0, t]$).

REMARK 1.7 A progressively measurable process is adapted, but the converse is not true in general. ▮

REMARK 1.8 If $\{X_t\}$ is adapted and right-continuous, then it is progressively measurable. This can be seen as follows.

For fixed $t \in [0, \infty)$, consider a partition of $[0, t]$ into 2^n equal intervals. For $s \in [(k-1)2^{-n}t,\ k2^{-n}t)$, $1 \le k \le 2^n$, define

$$X_s^{(n)}(\omega) = X_{k2^{-n}t}(\omega)$$

and define

$$X_s^{(n)}(\omega) = X_t(\omega) \text{ for } s = t\ .$$

Then $X^{(n)}$ is a map of $[0, t] \times \Omega$ into E and it is progressively measurable. Further $X^{(n)} \to X$ as $n \to \infty$ by the right continuity for every $\omega \in \Omega$. Hence, X is a progressively measurable process.

Similarly if $\{X_t\}$ is adapted and left-continuous, then it is progressively measurable. If the process has continuous sample paths and adapted, then it is progressively measurable.

Suppose a process $\{X_t\}$ is progressively measurable on $(\Omega, \mathcal{F})$ with respect to a filtration $\{\mathcal{F}_t\}$. If S is an $\{\mathcal{F}_t\}$-stopping time, then $X_S = X_{S(\omega)}(\omega)$ is $\mathcal{F}_S$-measurable and the stopped process $X_t^S = X_{t\wedge S}$ is progressively measurable. ∎

Predictable σ-Algebra

Let $(\Omega, \mathcal{F}, P)$ be a complete probability space and $\{\mathcal{F}_t\}$ be a right-continuous complete filtration, $t \in [0, \infty)$, of $\mathcal{F}$. Let $\mathcal{F}_\infty = \vee_t \mathcal{F}_t$. Recall that a random variable T defined on $(\Omega, \mathcal{F}, P)$ with values in $[0, \infty) \cup \{\infty\}$ such that $[T \leq t] \in \mathcal{F}_t$, $t \in [0, \infty)$ is called a *stopping time.*

DEFINITION 1.12 *A stopping time T defined on $(\Omega, \mathcal{F}, P)$ is said to be predictable if there exists $T_n \uparrow T$ a.s. where T_n is a stopping time taking values in $[0, \infty)$ and on the set $[T > 0]$, $T_n < T$ a.s. for all n.*

REMARK 1.9

(i) If T is a stopping time, then $T + r$ is predictable, since $T_n = T + r - 1/n \uparrow T + r$ a.s.

(ii) Let $r_n > 0$ and $r_n \downarrow 0$. Since $T = \lim_{n\to\infty} T + r_n$ a.s., any stopping time T is the limit of a decreasing sequence of predictable stopping times.

(iii) If T is a stopping time, $T - r$ need not be a stopping time in general.

∎

DEFINITION 1.13 *Suppose S, T are stopping times. The stochastic interval $[S, T)$ is the set $\{(t, \omega) : S(\omega) \leq t < T(\omega)\}$ and $[T, T] = \{(t, \omega) : T(\omega) = t\} \equiv [T]$ is called the graph of the stopping time T. Note that $[T] = \{(T(\omega), \omega) : \omega \in \Omega\}$.*

Notation: For any stopping time T, define

$$\mathcal{F}_{T-} = \mathcal{F}_0 \vee [A \cap \{T > t\};\ t \in [0, \infty) \text{ and } A \in \mathcal{F}_t]$$

(intuitively, $\mathcal{F}_{T-}$ deals with the events that occurred strictly prior to the stopping time T).

DEFINITION 1.14 *A predictable σ-algebra Σ_p on $[0, \infty) \times \Omega$ is the σ-algebra generated by the evanescent sets and all the stochastic intervals of the form $[T, \infty)$ for T arbitrary predictable stopping time.*

REMARK 1.10 T is a predictable stopping time if and only if $[T] = \{(\omega, T(\omega) : \omega \in \Omega,\ T(\omega) < \infty\} \in \Sigma_p$. ∎

DEFINITION 1.15 *A stochastic process $\{X_t\}$ defined on $(\Omega, \mathcal{F})$ and adapted to a filtration $\{\mathcal{F}_t\}$ with values in a measurable space $(E, \mathcal{E})$ is said to be predictable if the map $X : [0, \infty) \times \Omega \to E$ is measurable when $[0, \infty) \times \Omega$ is given the predictable σ-algebra Σ_p.*

REMARK 1.11 If $\{X_t\}$ and $\{Y_t\}$ are predictable, then $\{X_t\}$ and $\{Y_t\}$ are indistinguishable if and only if $X_T = Y_T$ a.s. for every predictable stopping time T. ∎

Example 1.1
Suppose S and T are stopping times and Z is an $\mathcal{F}_S$-measurable random variable. Then the process

$$X_t(\omega) = Z(\omega) I_{[S,T]}(t, \omega)$$

is predictable. □

REMARK 1.12 A predictable σ-algebra is generated by the family of all continuous adapted processes or by the family of all left-continuous adapted processes. ∎

REMARK 1.13 If $\{X_t\}$ is predictable, then $X_T I_{[T<\infty]}$ is $\mathcal{F}_T$- measurable for every stopping time T. ∎

DEFINITION 1.16 *A process $\{X_t\}$ is said to be optional if the map $(t, \omega) \to X_t(\omega)$ is measurable with respect to Σ_0, the σ-algebra generated* *by the evanescent sets and all the stochastic intervals of the form* $[T, \infty)$ *for arbitrary stopping time* T.

REMARK 1.14 Suppose $\{X_t\}$ is an optional process. Then there is a predictable process $\{Y_t\}$ such that $[X \neq Y]$ is contained in the union of the graphs of a sequence of stopping times. ▮

DEFINITION 1.17 *Suppose* $\{X_t\}$, $t \in [0, \infty)$ *is a real valued* $\mathcal{B} \otimes \mathcal{F}$-*measurable stochastic process. The process* $\{X_t\}$ *is said to be bounded if* $X^*_\infty(\omega) = \sup_t |X_t(\omega)| \in L^\infty(\Omega, \mathcal{F}, P)$. *It is said to be locally bounded with respect to the filtration* $\{\mathcal{F}_t\}$ *if there is an increasing sequence* T_n *of stopping times* T_n *such that* $T_n \to \infty$ *a.s. and* $X^*_{T_n}(\omega) \leq n$ *a.s. for each* n *where* $X^*_t(\omega) = \sup_{s \leq t} |X_s(\omega)|$ *(here* $\mathcal{B}$ *is the Borel* σ-*algebra on* $[0, \infty)$*).*

Notation: Let $B(\mathcal{B} \otimes \mathcal{F})$ be the class of bounded measurable processes and $B(\sum_p)$ the space of bounded $\sum_p$-measurable processes.

THEOREM 1.1 (Projection theorem)
There exists a unique projection operator π_p *of* $B(\mathcal{B} \otimes \mathcal{F})$ *onto* $B(\sum_p)$ *such that for every* $X \in \mathcal{B}(\mathcal{B} \otimes \mathcal{F})$ *and* T *predictable stopping time,*

$$E\left(X_T I_{[T<\infty]}\right) = E\left(\left(\pi_p X\right)_T I_{[T<\infty]}\right) .$$

DEFINITION 1.18 $\pi_p X$ *is called the predictable projection of* X.

REMARK 1.15 If $X \in B(\sum_p)$, then, by the uniqueness the property, $\pi_p X = X$. If X is bounded or positive $\mathcal{B} \otimes \mathcal{F}$-measurable process, then for any T predictable stopping time,

$$E\left[X_T \; I_{[T<\infty]} | \mathcal{F}_{T-}\right] = \left(\pi_p X\right)_T \; I_{[T<\infty]} .$$

If $\{X_t\}$ and $\{Y_t\}$ are bounded measurable processes and $Y_t \in B(\Sigma_p)$, then

$$\pi_p(XY) = \left(\pi_p X\right) Y .$$

▮

Let $\{\mathcal{F}_t\}$ be a right-continuous complete filtration on a complete probability space $(\Omega, \mathcal{F}, P)$.

DEFINITION 1.19 *A $\mathcal{B} \otimes \mathcal{F}$-measurable process $\{A_t\}$, $t \in [0, \infty)$ with values in $[0, \infty)$ is called an increasing process if $A_t(\omega)$ is right-continuous and increasing for almost every $\omega \in \Omega$.*

Convention: $A_{0-} = 0$ a.s.

REMARK 1.16 Since an increasing process always has left limits, every increasing process is RCLL. Let $\mathcal{W}^+$ be the class of such processes. Let $\mathcal{W}$ be the class of processes of bounded variation. Every member of $\mathcal{W}$ is the difference of two members of $\mathcal{W}^+$. Let $\mathcal{V}^+$ denote the family of equivalence classes of processes (under P-indistinguishability) that admit a representative $\{A_t\}$ which is an increasing process adapted to $\{\mathcal{F}_t\}$ and such that $A_t < \infty$ a.s. Let $\mathcal{V}_0^+$ denote those processes $\{A_t\} \in \mathcal{V}^+$ such that $A_0 = 0$.

Let $\mathcal{V} = \mathcal{V}^+ - \mathcal{V}^+$. A process is in $\mathcal{V}$ if it is RCLL and adapted and if almost every sample path is of bounded variation on each compact subset of $[0, \infty)$. Let $\mathcal{V}_0$ denote the processes $\{A_t\} \in \mathcal{V}$ such that $A_0 = 0$. ▮

REMARK 1.17 If $\{A_t\} \in \mathcal{W}$, then $\{D_t\} \in \mathcal{W}^+$ where $D_t(\omega) = \int_0^t |dA_s(\omega)|$ is the total variation of $\{A_t\}$. ▮

REMARK 1.18 Suppose $\alpha : [0, \infty) \to [0, \infty]$ is an increasing right-continuous function. Define $r(t) = \inf\{s : \alpha(s) > t\}$ for $0 \le t < \infty$. Then, for every positive Borel measurable function f on $[0, \infty)$,

$$\int_0^\infty f(t)d\alpha(t) = \int_{\alpha(0)}^{\alpha(\infty)} f(r(t))dt = \int_{\alpha(0)}^{\infty} I_{[r(t)<\infty]} f(r(t))dt\,.$$

Here $\alpha(\infty)$ denotes the limit of $\alpha(s)$ as $s \to \infty$. This can be seen by establishing the result for $f(t) = I_{[0,s]}(t)$ where $0 \le s < \infty$ and extending. ▮

If $\alpha : [0, \infty) \to [0, \infty]$ is an increasing continuous function, then

$$\int_0^\infty f(\alpha(t))d\alpha(t) = \int_{\alpha(0)}^{\alpha(\infty)} f(t)dt \; .$$

Suppose $\{A_t\}$ is either increasing or a process of bounded variation and $\{X_t\}$ is a real-valued $\mathcal{B} \otimes \mathcal{F}$-measurable process. Consider the Stieltjes integral

$$(X \circ A)_t(\omega) \equiv \int_{[0,t]} X_s(\omega)dA_s(\omega)$$

whenever it exists. Note that $(X \circ A)_0 = X_0 A_0$.

DEFINITION 1.20 *Suppose $\{A_t\} \in \mathcal{V}^+$. For $t \in [0, \infty)$ and $\omega \in \Omega$, define*

$$C_t(\omega) = \inf\{s : A_s(\omega) > t\} \; .$$

Then $\{C_t\}$ is a stopping time and $\{C_t\}$ is called the stochastic time change associated with the process $\{A_t\}$.

Suppose $\{X_t\}$ is a bounded $\mathcal{B} \otimes \mathcal{F}$-measurable process and $\{A_t\}$ a $\mathcal{B} \otimes \mathcal{F}$-measurable process of integrable variation, that is, $E \int_0^\infty |dA_s| < \infty$. Define

$$(X, A) \equiv E\left[\int_0^\infty X_s(\omega)dA_s(\omega)\right] .$$

REMARK 1.19 Suppose $\{A_t\}$ is an increasing $\mathcal{B} \otimes \mathcal{F}$-measurable process such that $E[A_\infty] < \infty$. Then there exists a unique predictable increasing process $\pi_p^* A_t$ such that

$$E \int_0^\infty \left|d\pi_p^* A_t\right| < \infty$$

and for every bounded $\mathcal{B} \otimes \mathcal{F}$-measurable process $\{X_t\}$,

$$\left(\pi_p X, A\right) = \left(X, \pi_p^* A\right) = \left(\pi_p X, \pi_p^* A\right) .$$

▮

DEFINITION 1.21 $\{\pi_p^* A_t\}$ *is called the dual predictable projection (compensator) of* $\{A_t\}$.

REMARK 1.20 If $\{A_t\}$ is predictable increasing process such that $E(A_\infty) < \infty$, then $\pi_p^* A_t = A_t$. ▮

Martingales

Let $(\Omega, \mathcal{F}, P)$ be a probability space and $\{\mathcal{F}_t,\ t \in \tau\}$ be a filtration on $(\Omega, \mathcal{F})$ where τ is either $[0, \infty)$ or $\tau = \{1, 2, 3, \ldots\}$.

DEFINITION 1.22 *A process* $\{X_t\},\ t \in \tau$ *defined on* $(\Omega, \mathcal{F}, P)$ *is said to be a supermartingale with respect to the filtration* $\{\mathcal{F}_t\}$ *if*

(i) X_t *is* $\mathcal{F}_t$*-measurable, that is,* $\{X_t\}$ *is adapted to* $\{\mathcal{F}_t\}$,

(ii) $E|X_t| < \infty,\ t \in \tau$, *and*

(iii) $E(X_t|\mathcal{F}_s) \leq X_s$ *a.s. if* $s \leq t,\ s, t \in \tau$.

The process is said to be a submartingale if (i), (ii) hold and $E(X_t|\mathcal{F}_s) \geq X_s$ *a.s. if* $s \leq t$ *and a martingale if (i), (ii) hold and* $E(X_t|\mathcal{F}_s) = X_s$ *a.s. if* $s \leq t$.

LEMMA 1.1

(i) *If* $\{X_t\}$ *in an* $\{\mathcal{F}_t\}$*-martingale and* $\phi(\cdot)$ *is a convex function such that* $E(|\phi(X_t)|) < \infty$, *then* $\{\phi(X_t)\}$ *is an* $\{\mathcal{F}_t\}$*-submartingale.*

(ii) *If* $\{X_t\}$ *is an* $\{\mathcal{F}_t\}$*-submartingale and* $\phi(\cdot)$ *is a convex function and nondecreasing such that* $E(|\phi(X_t)|) < \infty$, *then* $\{\phi(X_t)\}$ *is an* $\mathcal{F}_t$*-submartingale.*

Martingales (Discrete Time)

We now consider the case when $\tau = N = \{1, 2, 3, \ldots\}$.

THEOREM 1.2 (Optional stopping)
Suppose $\{X_n,\ n \in N\}$ is an $\{\mathcal{F}_n\}$-supermartingale. If S and T are bounded $\{\mathcal{F}_n\}$-stopping times, then $E(X_T|\mathcal{F}_S) \le X_S$ a.s. if $S \le T$ a.s.

THEOREM 1.3
If $X = \{X_n,\ n \in N\}$ is an $\{\mathcal{F}_n\}$-supermartingale and S is a bounded $\mathcal{F}_n$-stopping time, then $X^S = \{X_{S\wedge n}\}$ is an $\{\mathcal{F}_n\}$-supermartingale.

THEOREM 1.4
Suppose $\{X_n, n \in N\}$ is an $\{\mathcal{F}_n\}$-supermartingale such that $\sup_n E(X_n^-) < \infty$. *Then $X_n \to X_\infty$ a.s. and $E|X_\infty| < \infty$.*

COROLLARY 1.1
Suppose $\{X_n,\ n \in N\}$ is a positive $\{\mathcal{F}_n\}$-supermartingale. Then $X_n \to X_\infty$ a.s.

COROLLARY 1.2
Suppose $\{X_n,\ n \in N\}$ is a uniformly integrable $\mathcal{F}_n$-supermartingale. Let $\mathcal{F}_\infty = \vee_n \mathcal{F}_n$. Then $\{X_n\}$, $n \in N \cup \{\infty\}$ is a supermartingale and $\lim_n \|X_n - X_\infty\|_1 = 0$.

THEOREM 1.5
If $\{X_n,\ n \in N\}$ is a uniformly integrable martingale, then $X_n \to X_\infty$ a.s. and $\|X_n - X_\infty\|_1 \to 0$. Furthermore,

$$X_n = E\left(X_\infty|\mathcal{F}_n\right),\ n \in N\ .$$

THEOREM 1.6
Suppose $\{\mathcal{F}_n,\ n \in N\}$ is a filtration and $\mathcal{F}_\infty = \vee_n \mathcal{F}_n$. If $Y \in L^1(\Omega, \mathcal{F}_\infty, P)$ and $X_n = E(Y|\mathcal{F}_n)$, then $\{X_n\}$ is uniformly integrable and $X_n \to Y$ a.s.

Inequalities for Supermartingales

THEOREM 1.7
Suppose $\{X_n,\ n \in N\}$ is a supermartingale. Then, for every $\alpha \geq 0$,

$$(i) \qquad \alpha\, P\left(\sup_n X_n \geq \alpha\right) \leq E\left(X_0\right) + \sup_n E\left(X_n^-\right)$$
$$\leq 2 \sup_n E\left|X_n\right| ,$$

and

$$(ii) \qquad \alpha\, P\left(\inf_n X_n \leq -\alpha\right) \leq \sup_n E\left(X_n^-\right) .$$

COROLLARY 1.3
Let $\{X_n,\ n \in N\}$ be a supermartingale. Then, for every $\alpha \geq 0$,

$$\alpha\, P\left\{\sup_n \left|X_n\right| \geq \alpha\right\} \leq 3 \sup_n E\left|X_n\right| .$$

Doob's Maximal Inequality

THEOREM 1.8
Let $\{X_n,\ n \in N\}$ be a martingale. Then, for every $\alpha \geq 0$,

$$\alpha\, P\left\{\sup_n \left|X_n\right| \geq \alpha\right\} \leq \sup_n E\left|X_n\right| .$$

Doob's L_p-Inequality

THEOREM 1.9
Suppose $\{X_n\}$ is a positive submartingale. Define $X^* = \sup_n X_n$. Then

$$(i) \qquad \left\|X^*\right\|_p < \infty \Leftrightarrow \sup_n \left\|X_n\right\|_p < \infty, \ 1 < p \leq \infty$$

and

$$(ii) \quad \|X^*\|_p \leq q \sup_n \|X_n\|_p \quad \text{with } \frac{1}{p} + \frac{1}{q} = 1, \ 1 < p < \infty .$$

COROLLARY 1.4
Let $\{X_n\}$ be a martingale. Then

$$\left\{ E\left[\sup_n |X_n|\right]^2 \right\}^{1/2} \leq 2 \sup_n \left(E\,|X_n|^2\right)^{1/2} .$$

DEFINITION 1.23 *Let $(\Omega, \mathcal{F}, P)$ be a probability space and $\{\mathcal{F}_n\},\ n \in N$ be a decreasing sequence of sub-σ-algebras of $\mathcal{F}$, that is, $\mathcal{F}_n \subset \mathcal{F}_m$ if $n \geq m$. The process $\{X_n,\ n \in N\}$ is said to be a reverse supermartingale if*

(i) *X_n is $\mathcal{F}_n$-measurable, $n \in N$,*

(ii) *$E|X_n| < \infty$, $n \in N$, and*

(iii) *$E[X_m|\mathcal{F}_n] \leq X_n$ a.s. if $n \geq m$.*

REMARK 1.21 Suppose $\{X_n\}$ is a reverse supermartingale with respect to a decreasing sequence σ-algebras $\{\mathcal{F}_n\},\ n \in N$. (i) If $\lim_n E(X_n) < \infty$, then $\{X_n\}$ is uniformly integrable. (ii) If $\lim_n E(X_n) < \infty$, then $X_n \to X_\infty$ a.s. and in L^1-mean. ∎

DEFINITION 1.24 *A random process $\{A_n,\ n \geq 0\}$ adapted to the filtration $\{\mathcal{F}_n,\ n \geq 0\}$ is called an increasing process if*

(i) *$0 = A_0 \leq A_1 \leq \cdots \leq A_n \leq \cdots$ a.s.*
and a natural increasing process if in addition

(ii) *A_{n+1} is $\mathcal{F}_n$-measurable, $n \geq 0$.*

(Remark: Note that A_n is $\mathcal{F}_n$-measurable by definition.)

THEOREM 1.10 **(Doob decomposition)**
Any supermartingale $\{X_n\}$ with respect to $\{\mathcal{F}_n,\ n \geq 0\}$ permits the unique decomposition

$$X_n = M_n - A_n, \quad n \geq 0 \quad a.s.$$

where $M = \{M_n, \mathcal{F}_n, n \geq 0\}$ is a martingale and $\{A_n, \mathcal{F}_n, n \geq 0\}$ is a natural increasing process.

PROOF Let $M_0 = X_0$, $M_{n+1} - M_n = X_{n+1} - E(X_{n+1}|\mathcal{F}_n)$ and $A_0 = 0$, $A_{n+1} - A_n = X_n - E(X_{n+1}|\mathcal{F}_n)$. This gives a decomposition. Suppose there is another such decomposition. Then $X_n = M'_n - A'_n$, $n \geq 0$. Hence, $A'_{n+1} - A'_n = (M'_{n+1} - M'_n) - X_{n+1} - X_n$. Taking conditional expectations with respect to $\mathcal{F}_n$ on both sides and using the fact that A'_n and A'_{n+1} are both $\mathcal{F}_n$-measurable, it follows that

$$A'_{n+1} - A'_n = X_n - E(X_{n+1}|\mathcal{F}_n)$$

$$= A_{n+1} - A_n ,$$

since $E(M'_{n+1} - M'_n|\mathcal{F}_n) = 0$ by the martingale property. But $A'_0 = A_0 = 0$. Hence, $A'_n = A_n$ and $M'_n = M_n$ a.s. This shows the uniqueness of the decomposition. ∎

COROLLARY 1.5

If a supermartingale $X = \{X_n, \mathcal{F}_n, n \geq 0\}$ dominates a submartingale $Y = \{Y_n, \mathcal{F}_n, n \geq 0\}$, then there exists a natural increasing process $\{A_n, n \geq 0\}$ and a martingale $\{M_n, \mathcal{F}_n, n \geq 0\}$ such that

$$X_n = M_n + E(A_\infty|\mathcal{F}_n) - A_n .$$

THEOREM 1.11

Let $0 = A_0 \leq A_1 \leq \cdots$ where the random variables A_n are $\mathcal{F}_n$-measurable and $EA_\infty < \infty$. In order that A_n be $\mathcal{F}_{n-1}$-measurable, $n \geq 1$, it is necessary and sufficient that for each bounded martingale $Y = \{Y_n, \mathcal{F}_n, n \geq 0\}$,

$$E\left\{\sum_{n=1}^{\infty} Y_{n-1}(A_n - A_{n-1})\right\} = EY_\infty A_\infty \tag{1.2.1}$$

where $Y_\infty = \lim Y_n$ a.s.

PROOF **Necessity:** Let A_n be $\mathcal{F}_{n-1}$ measurable and $EA_\infty < \infty$. Then

$$E[A_n Y_n] = E\{E(A_n Y_n|\mathcal{F}_{n-1})\}$$

$$= E\{A_n Y_{n-1}\},\ n \geq 1 \tag{1.2.2}$$

and hence,

$$E\left\{\sum_{n=1}^{\infty} Y_{n-1}(A_n - A_{n-1})\right\} = \lim_{N\to\infty} E\left\{\sum_{n=1}^{N} Y_{n-1}(A_n - A_{n-1})\right\}$$

(by the dominated convergence theorem)

$$= \lim_{N\to\infty} \sum_{n=1}^{N} \left[EY_{n-1}A_n - EY_{n-1}A_{n-1}\right]$$

$$= \lim_{N\to\infty} E[A_N Y_N] = E[Y_\infty A_\infty]$$

This proves (1.2.1).

Sufficiency: Suppose (1.2.1) holds. Then

$$E\left\{\sum_{n=1}^{\infty} A_n \left[Y_{n-1} - Y_n\right]\right\} = 0 \tag{1.2.3}$$

for any bounded martingale $Y = \{Y_n, \mathcal{F}_n, n \geq 0\}$. For any stopping time T, the stopped process Y^T is also a martingale. Take $T \equiv 1$ and apply (1.2.3). Then we have

$$E[A_1(Y_0 - Y_1)] = 0\,. \tag{1.2.4}$$

Applying (1.2.3) in turn for $T \equiv 2$, $T \equiv 3$ etc., we have (1.2.2) for any bounded martingale Y. Hence, from (1.2.2),

$$E\left\{\left[Y_n - Y_{n-1}\right]\left[A_n - E(A_n|\mathcal{F}_{n-1})\right]\right\} = 0\,. \tag{1.2.5}$$

Let $Y_{n+m} = Y_n$, $m \geq 0$ where $Y_n = sgn[A_n - E(A_n|\mathcal{F}_{n-1})]$ and $Y_k = E(Y_n|\mathcal{F}_k)$ for $k < n$. Then, from (1.2.5), we have

$$0 = E\left|A_n - E(A_n|\mathcal{F}_{n-1})\right|$$

and hence, $A_n = E(A_n|\mathcal{F}_{n-1})$, which implies that A_n is $\mathcal{F}_{n-1}$ -measurable. ∎

Martingales (Continuous Time)

We now discuss the theory of martingales when the index set $\tau = [0, \infty)$. Before we discuss the properties of continuous-time parameter martingales, we present some examples.

Example 1.2 **Processes with independent increments**
Let $\{X_t, t \geq 0\}$ be a real-valued process with independent increments defined on a probability space $(\Omega, \mathcal{F}, P)$. Let $\mathcal{F}_s^X = \sigma\{X_t, 0 \leq t \leq s\}$. Then $X_t - X_s$ is independent of $\mathcal{F}_s^X$ whenever $0 \leq s \leq t$. Suppose X_t is integrable and $E(X_t) = 0$. Then $\{X_t, t \geq 0\}$ is a martingale with respect to the filtration $\{\mathcal{F}_t^X, t \geq 0\}$. If, in addition, X_t^2 is integrable for all t, then $\{X_t^2 - E(X_t^2)\}$ is also a martingale with respect to $\{\mathcal{F}_t^X\}$. $\square$

Example 1.3 **Markov Chains**
Let $N_+ = \{1, 2, \ldots\}$. Suppose $\{X_t\} \geq 0$ is a right-continuous N_+-valued $\{\mathcal{F}_t\}$-Markov chain with left limits defined on a probability space $(\Omega, \mathcal{F}, P)$. Further suppose that $\{X_t\}$ admits a stationary distribution $Q = ((q_{ij}))_{N_+ \otimes N_+}$. Then, for $0 \leq s \leq t$, if f is a nonnegative function from $N_+ \otimes N_+$ into R_+, then

$$E\left[\sum_{s<u\leq t} f(X_{u-}, X_u) \,|X_s\right] = E\left[\sum_{j \neq X_u} \int_s^t q_{X_u j} f(X_u, j)\, du | X_s\right].$$

Hence, by the Markov property of $\{X_t\}$, that is, the conditional independence of $\mathcal{F}_s$ and $(X_u, u \geq s)$ given X_s, it follows that

$$E\left[\sum_{s<u\leq t} f(X_{u-}, X_u) \,|\mathcal{F}_s\right] = E\left[\int_s^t \sum_{j \neq X_u} q_{X_u j} f(X_u, j)\, du | \mathcal{F}_s\right].$$

If $f : N_+ \otimes N_+ \to R$ is such that, for all $t \geq 0$,

$$E\left[\int_0^t \sum_{j \neq X_s} q_{X_s j} \,|f(X_s, j)|\, ds\right] < \infty,$$

then

$$Y_t \equiv \sum_{0<s\leq t} f(X_{s-}, X_s) - \int_0^t \left\{ \sum_{j\neq X_s} q_{X_s j} f(X_s, j) \right\} ds$$

is an $\{\mathcal{F}_t\}$-martingale.

(a) In particular, let

$$f(i, j) = \begin{cases} 1 & \text{if } i = k, j = l \\ 0 & \text{otherwise.} \end{cases}$$

Then $\sum_{0<s\leq t} f(X_{s-}, X_s) = N_t(k, l)$ is the number of transitions from k to l in $[0, t]$. Hence,

$$N_t(k, l) - q_{kl} \int_0^t I[X_s = k] ds$$

is an $\{\mathcal{F}_t\}$-martingale. Note that

$$E\left[q_{kl} \int_0^t I[X_s = k] ds \right] = q_{kl} \int_0^t P[X_s = k] ds \leq q_{kl} t < \infty .$$

(b) Another choice of f is $f(i, j) = I[j = l]$. Then $\sum_{0<s\leq t} f(X_{s-}, X_s) = N_t(l)$ counts the numbers of entrances into the state l during $[0, t]$. Hence,

$$N_t(l) - \int_0^t \sum_{i\neq l} q_{il} I[X_s = i] ds$$

is an $\{\mathcal{F}_t\}$-martingale provided

$$\sum_{i\neq l} \int_0^t q_{il} P(X_s = i) ds < \infty, \ t \geq 0 .$$

(c) Suppose $\{Q_t\}$ is a birth and death process with parameters λ_n and μ_n such that $E[Q_0] < \infty$. If

$$\sum_{n=0}^{\infty} E\left[\int_0^t \lambda_n P[Q_s = n]\, ds\right] < \infty, \; t \geq 0,$$

then

$$A_t - \int_0^t \lambda_{Q_s} ds \text{ is an } \mathcal{F}_t^Q - \text{ martingale},$$

and

$$D_t - \int_0^t \mu_{Q_s} I(Q_s > 0) ds \text{ is an } \mathcal{F}_t^Q - \text{ martingale}.$$

where A_t is the number of upward jumps in $(0, t]$ and D_t is the number of downward jumps in $(0, t]$. □

Example 1.4 **Likelihood Ratios**
Let P and $\tilde{P}$ be two probability measures on $(\Omega, \mathcal{F})$ and $\{\mathcal{F}_t, t \geq 0\}$ be a filtration of $\mathcal{F}$. Let P_t and $\tilde{P}_t$ be the restrictions of P and $\tilde{P}$ to $\mathcal{F}_t$. Suppose $\tilde{P}_t << P_t$ for all $t \geq 0$ (in such a case $\tilde{P}$ is said to be *locally absolutely continuous* with respect to P). Let

$$L_t = \frac{d\tilde{P}_t}{dP_t}.$$

Then $\{L_t, \mathcal{F}_t, \; t \geq 0\}$ is a martingale. □

We now discuss some properties of continuous-time martingales. Hereafter let $\tau = [0, \infty)$ or $[0, \infty]$ and $(\Omega, \mathcal{F}, P)$ be a probability space with a filtration $\{\mathcal{F}_t\}$. We assume that $\mathcal{F}$ is complete under P and each $\mathcal{F}_t$ contains all the null sets of $\mathcal{F}$. Assume further that $\{\mathcal{F}_t\}$ is right-continuous.

THEOREM 1.12
(i) Suppose $\{X_t, \; t \in \tau\}$ is a right-continuous supermartingale and $J =$

$[u, v] \subset \tau$. Then, for any $\lambda > 0$.

$$(a) \quad \lambda\, P\left\{\sup_{t\in J} X_t \geq \lambda\right\} \leq E\left[X_u\right] + E\left[X_v^-\right],$$

and

$$(b) \quad \lambda\, P\left\{\inf_{t\in J} X_t \leq -\lambda\right\} \leq E\left\{|X_v|\right\}.$$

(ii) Suppose $\{X_t,\ t \in \tau\}$ is a right-continuous submartingale. If $1 < p < \infty$ and $X_v \in L^p$, then

$$\left\{E\left[\left|\sup_{t\in J} X_t\right|^p\right]\right\}^{1/p} \leq q \sup_{t\in J}\left\{E\,|X_t|^p\right\}^{1/p}$$

$$= q\left\{E\,|X_v|^p\right\}^{1/p}$$

where $1/p + 1/q = 1$.

THEOREM 1.13

Suppose $\{X_t,\ t \in \tau\}$ is a right-continuous supermartingale. Then $\{X_t\}$ has left limits a.s., that is, $\{X_t\}$ is RCLL. Furthermore, almost every path is bounded on every compact interval.

THEOREM 1.14

Suppose $\{X_t,\ t \in \tau\}$ is a supermartingale. Then the supermartingale $\{X_t,\ t \in \tau\}$ has a right-continuous modification if and only if the function $E(X_t)$ is right-continuous in t.

REMARK 1.22 Every martingale admits a right-continuous modification. ∎

THEOREM 1.15

Suppose $\{X_t,\ t \in \tau = [0, \infty)\}$ is a right-continuous martingale. If $\sup_{t\in\tau} E[X_t^-] < \infty$, then X_t converges a.s. to an integrable random variable as $t \to \infty$.

COROLLARY 1.6
Suppose $\{X_t,\ t \in [0,\infty)\}$ is a nonnegative right-continuous supermartingale. Then $X_t^- = 0$ a.s. and $\lim_t X_t = X_\infty$ exists a.s. Furthermore, $\{X_t,\ t \in [0,\infty]\}$ is a supermartingale.

COROLLARY 1.7
Suppose $\{X_t,\ t \in [0,\infty)\}$ is a right-continuous uniformly integrable supermartingale. Then

(i) $\sup_t E|X_t| < \infty$,

(ii) $X_t \to X_\infty$ *a.s.*,

(iii) $X_t \to X_\infty$ *in L^1-mean, and*

(iv) $\{X_t,\ t \in [0,\infty]\}$ *is a supermartingale.*

COROLLARY 1.8
Suppose $\{X_t,\ t \in [0,\infty)\}$ is a uniformly integrable martingale. Then $\{X_t,\ t \in [0,\infty]\}$ is a martingale.

THEOREM 1.16
Let $\mathcal{F}_\infty = \vee_t \mathcal{F}_t$ and Y be $\mathcal{F}_\infty$-measurable. Then $\{E[Y|\mathcal{F}_t],\ t \in [0,\infty]\}$ is a uniformly integrable martingale. There exists a right-continuous modification $\{Y_t\}$ of this martingale and $\lim_t Y_t = Y$ a.s. and in L^1-mean.

***THEOREM 1.17* (Optional stopping)**
Suppose $\{X_t,\ t \in [0,\infty]\}$ is a right-continuous supermartingale with respect to the filtration $\{\mathcal{F}_t,\ t \in [0,\infty]\}$. If S and T are stopping times such that $S \le T$ a.s., then X_S and X_T are integrable and $X_S \ge E(X_T|\mathcal{F}_S)$ a.s.

COROLLARY 1.9
If $\{X_t,\ t \in [0,\infty]\}$ is a right-continuous martingale and S and T are stopping times such that $S \le T$ a.s., then $X_S = E[X_T|\mathcal{F}_S]$ a.s.

***THEOREM 1.18* (Characterization of uniformly integrable martingales)**
Suppose $\{M_t,\ t \in [0,\infty]\}$ is an adapted right-continuous process such that for

every stopping time T, $E|M_T| < \infty$ and $E[M_T] = 0$. Then $M_t = E[M_\infty|\mathcal{F}_t]$ and $\{M_t,\ t \in [0, \infty]\}$ is an uniformly integrable $\mathcal{F}_t$-martingale.

1.3 Doob–Meyer Decomposition

DEFINITION 1.25 *A supermartingale $X = \{X_t, \mathcal{F}_t, t \geq 0\}$ with right-continuous paths is said to belong to the class D if $\{X_T,\ T \in \zeta\}$, where ζ is the set of stopping times with $P(T < \infty) = 1$, is uniformly integrable.*

THEOREM 1.19
Any uniformly integrable martingale $X = \{X_t, \mathcal{F}_t, t \geq 0\}$ with right continuous trajectories belongs to D.

DEFINITION 1.26 *A process $\{A_t, \mathcal{F}_t, t \geq 0\}$ is said to be increasing if A_t is $\mathcal{F}_t$-measurable, $A_0 = 0$ and $A_s \leq A_t$ a.s. if $s \leq t$. It is said to be a natural increasing process if for any bounded positive right-continuous martingale $Y = \{Y_t, \mathcal{F}_t, t \geq 0\}$ having left limits*

$$E\int_0^\infty Y_{s-} dA_s = EY_\infty A_\infty .$$

The process is said to be integrable if $EA_\infty < \infty$.

THEOREM 1.20
Let $A \equiv \{A_t, \mathcal{F}_t, t \geq 0\}$ be an integrable increasing process. Then A is natural if and only if for any bounded martingale, right-continuous and having limits from the left, $Y \equiv \{Y_t, \mathcal{F}_t, t \geq 0\}$,

$$E\left\{\int_0^T Y_s dA_s\right\} = E\left\{\int_0^T Y_{s-} dA_s\right\} \tag{1.3.1}$$

for any $T > 0$.

THEOREM 1.21 **(Doob–Meyer decomposition)**
Let $X = \{X_t, \mathcal{F}_t, t \geq 0\}$ be a right-continuous supermartingale belonging to the class D. Then there exists a right-continuous uniformly integrable martingale $M = \{M_t, \mathcal{F}_t, t \geq 0\}$ and an integrable natural increasing process $A = \{A_t, \mathcal{F}_t, t \geq 0\}$ such that

$$X_t = M_t - A_t \text{ a.s., } t \geq 0 ,$$

and this decomposition (with the natural increasing process) is unique a.s.

DEFINITION 1.27 *A process $M = \{M_t, \mathcal{F}_t, t \geq 0\}$ is a local martingale if there exists an increasing sequence of stopping times T_n with respect to $\{\mathcal{F}_t, t \geq 0\}$ such that $P(T_n \to \infty) = 1$, $T_n \leq n$ a.s. and the sequence*

$$\left\{M_t^{T_n} = M_{t \wedge T_n}, \ \mathcal{F}_t, t \geq 0\right\}, n \geq 1$$

is a sequence of uniformly integrable martingales.

THEOREM 1.22 **(Doob–Meyer decomposition)**
Let $X = \{X_t, \mathcal{F}_t, t \geq 0\}$ be a right-continuous nonnegative supermartingale. Then there exists a right-continuous local martingale M and a natural increasing process $A = (A_t, \mathcal{F}_t, t \geq 0\}$ such that

$$X_t = M_t - A_t \quad \text{a.s.}$$

and the decomposition is unique.

REMARK 1.23 Recall that in the case of discrete time, the increasing process $A = \{A_n, \mathcal{F}_n, n \geq 0\}$ is called natural if A_{n+1} is $\mathcal{F}_n$-measurable. It can be shown that if $A = \{A_t, \mathcal{F}_t, t \geq 0\}$ is a natural increasing process according to definition given above, then A_t is $\mathcal{F}_{t-}$-measurable. In fact, if T is a stopping time with respect to $\{\mathcal{F}_t\}$, then $A_T = A_{T(\omega)}(\omega)$ is $\mathcal{F}_{T-}$-measurable where $\mathcal{F}_{T-}$ is the σ-algebra generated by sets of the form $[T > t] \cap \Lambda_t$ where $\Lambda_t \in \mathcal{F}_t, t \geq 0$. ∎

Doob–Meyer Decomposition for the Square of a Square Integrable Martingale

Let m_T^2 be the family of square integrable martingales on $[0, T]$, that is, right-continuous martingales $X = \{X_t, \mathcal{F}_t, \ 0 \le t \le T\}$ with $\sup_{0 \le t \le T} E(X_t^2) < \infty$. Then the process $Z = \{X_t^2, \mathcal{F}_t, \ 0 \le t \le T\}$ is a nonnegative submartingale in the class D. By the Doob–Meyer decomposition applied to the submartingale Z, we have the following result.

THEOREM 1.23
Let $X \in m_T^2$. Then there exists a unique natural increasing process $A_t \equiv \langle X \rangle_t, \ 0 \le t \le T$ such that

$$X_t^2 = M_t + \langle X \rangle_t \ \ a.s. \ [P]$$

where $M \equiv \{M_t, \mathcal{F}_t, \ 0 \le t \le T\}$ is a martingale. Furthermore,

$$E\left[(X_t - X_s)^2 \,|\mathcal{F}_s\right] = E\left[\langle X \rangle_t - \langle X \rangle_s | \mathcal{F}_s\right], \ s \le t \,. \tag{1.3.2}$$

PROOF We have already seen that the decomposition holds. Since M and X are martingales, $E(M_t - M_s|\mathcal{F}_s) = 0$ and $E(X_t^2 - X_s^2|\mathcal{F}_s) = E\{(X_t - X_s)^2|\mathcal{F}_s\}$ for $0 \le s \le t \le T$. Hence,

$$E\left[(X_t - X_s)^2 \,|\mathcal{F}_s\right] = E\left[\langle X \rangle_t - \langle X \rangle_s | \mathcal{F}_s\right] .$$

∎

Example 1.5
If $X = \{W_t, \mathcal{F}_t, \ 0 \le t\}$, where W is a Wiener process (see Section 1.4), then $\langle W \rangle_t = t$ a.s. $[P]$ since $\{W_t^2 - t, \ 0 \le t \le T\}$ is a martingale. □

Example 1.6
Let $a(t, \omega) \in m_T^2$ and $X = (X_t, \mathcal{F}_t, \ 0 \le t \le T)$ where

$$X_t = \int_0^t a(s, \omega) dW_s$$

(for stochastic integrals, see Section 1.4 (cf. [5, 16]). Then X is a continuous martingale, and it can be shown that

$$X_t^2 - \int_0^t a^2(s,\omega)ds$$

is a martingale and $\int_0^t a^2(s,\omega)ds$ is the natural increasing process. Hence,

$$\langle X\rangle_t = \int_0^t a^2(s,\omega)ds\ .$$

□

The following result is an analogue of (1.3.2) in Theorem 1.23 for products of square integrable martingales.

THEOREM 1.24
Let $X \in m_T^2$ and $Y \in m_T^2$. Then there exists a unique process $\langle X, Y\rangle_t$ which is the difference of two natural increasing processes and a martingale $\{M_t, \mathcal{F}_t\}$ such that

$$X_t Y_t = M_t + \langle X, Y\rangle_t \ \ a.s.\ [P]$$

and

$$E\left[(X_t - X_s)(Y_t - Y_s)\,|\mathcal{F}_s\right] = E\left[\langle X, Y\rangle_t - \langle X, Y\rangle_s|\mathcal{F}_s\right],\ t \geq s\ .$$

PROOF Since $X - Y$ and $X + Y$ are square integrable martingales,

$$(X_t - Y_t)^2 = Z_t + \langle X - Y\rangle_t;\ \ (X_t + Y_t)^2 = W_t + \langle X + Y\rangle_t$$

where Z_t and W_t are martingales. Define

$$\langle X, Y\rangle_t = \frac{1}{4}\left[\langle X + Y\rangle_t - \langle X - Y\rangle_t\right]$$

and

$$M_t = X_t Y_t - \langle X, Y \rangle_t .$$

Note that $\langle X, Y \rangle_t$ is a difference of two natural increasing processes. We shall show that $\{M_t\}$ is a martingale. Using the relation

$$ab = \frac{1}{4}\left[(a+b)^2 - (a-b)^2\right],$$

we have

$$\begin{aligned}
E\left[X_t Y_t - X_s Y_s | \mathcal{F}_s\right] &= E\left[(X_t - X_s)(Y_t - Y_s) | \mathcal{F}_s\right] \\
&= \frac{1}{4} E\left[\{(X_t + Y_t) - (X_s + Y_s)\}^2 \right. \\
&\quad \left. - \{(X_t - Y_t) - (X_s - Y_s)\}^2 | \mathcal{F}_s\right] \\
&= \frac{1}{4}\left(E\left\{\left[\langle X+Y \rangle_t - \langle X+Y \rangle_s\right] | \mathcal{F}_s\right\}\right. \\
&\quad \left. - E\left\{\left[\langle X-Y \rangle_t - \langle X-Y \rangle_s\right] | \mathcal{F}_s\right\}\right) \\
&= E\left[\langle X, Y \rangle_t - \langle X, Y \rangle_s | \mathcal{F}_s\right] \quad \text{(by Theorem 1.23)} .
\end{aligned}$$

Hence, $\{M_t, \mathcal{F}_t\}$ is a martingale. Uniqueness can be established after some arguments (for details, see [44]). ▮

REMARK 1.24 In general, $\langle X+Y \rangle_t \neq \langle X \rangle_t + \langle Y \rangle_t$. If $X \perp Y$, that is, $\langle X, Y \rangle_t = 0,\ t \leq T$, then $\langle X+Y \rangle_t = \langle X \rangle_t + \langle Y \rangle_t$. Furthermore, $\langle X, Y \rangle_t = 0$ if and only if XY is a martingale. This follows from the uniqueness of the decomposition in Theorem 1.24. ▮

1.4 Stochastic Integration

Stochastic Integrals with Respect to a Wiener Process

DEFINITION 1.28 Let $(\Omega, \mathcal{F}, P)$ be a probability space and $\{\mathcal{F}_t, t \geq 0\}$ be a filtration. A stochastic process $W = \{W_t, \mathcal{F}_t, t \geq 0\}$ is called a standard Wiener process relative to the filtration $\{\mathcal{F}_t, t \geq 0\}$ if
(i) the trajectories $W_t(\omega)$, $t \geq 0$ are continuous (P-a.s.) in t,
(ii) $W = \{W_t, \mathcal{F}_t, t \geq 0\}$ is a square integrable martingale with $W_0 = 0$, and
(iii) $E(W_t - W_s)^2|\mathcal{F}_s] = t - s,\ t \geq s$.

DEFINITION 1.29 A stochastic process $B = \{B_t, t \geq 0\}$ on a probability space $(\Omega, \mathcal{F}, P)$ is called the standard Brownian motion if
(i) $B_0 = 0$ a.s. $[P]$,
(ii) B is a process with stationary independent increments, and
(iii) $B_t - B_s \simeq N(0, |t-s|)$, where $N(0, \sigma^2)$ denotes the Gaussian distribution with mean 0 and variance σ^2.

THEOREM 1.25
Any Wiener process is a Brownian motion.

THEOREM 1.26
Let $0 \equiv t_0^{(n)} < t_1^{(n)} < \cdots < t_n^{(n)} \equiv T$ be a subdivision of $[0, T]$ such that $\max_i |t_{i+1}^{(n)} - t_i^{(n)}| \to 0$ as $n \to \infty$. Then

$$\lim_{n\to\infty} \sum_{i=0}^{n-1} \left[W_{t_{i+1}^{(n)}} - W_{t_i^{(n)}} \right]^2 = T \quad a.s.$$

and

$$l.i.m._{n\to\infty} \sum_{i=0}^{n-1} \left[W_{t_{i+1}^{(n)}} - W_{t_i^{(n)}} \right]^2 = T\ .$$

(Here l.i.m. denotes convergence in the quadratic mean).

PROOF It is clear that

$$E\left[\sum_{i=0}^{n-1}\left\{W_{t_{i+1}^{(n)}} - W_{t_i^{(n)}}\right\}^2\right] = T$$

and hence, it is sufficient to prove that

$$\operatorname{var}\left[\sum_{i=0}^{n-1}\left\{W_{t_{i+1}^{(n)}} - W_{t_i^{(n)}}\right\}^2\right] \to 0\,.$$

This follows from the fact that the process W has independent increments with $\operatorname{var}[W_t - W_s] = |t-s|$. Almost sure convergence property can be proved by the Borel–Cantelli lemma in the special case $t_i^{(n)} = \frac{i}{n}T$. ∎

DEFINITION 1.30 *A stochastic process $\{f(t,\omega), t \geq 0, \omega \in \Omega\}$ is called nonanticipative with respect to the filtration $\{\mathcal{F}_t\}$ if for each t, $f(t,.)$ is $\mathcal{F}_t$-measurable.*

DEFINITION 1.31 *A stochastic process $f(t,\omega)$ is called simple nonanticipative if there exists a finite subdivision $0 = t_0 < t_1 < \cdots < t_n = T$ of the interval $[0,T]$ where*

$$f(t,\omega) = \alpha_0(\omega)\ \ (\mathcal{F}_0\text{-measurable})\ \ \textit{for } t = t_0 = 0$$

$$= \alpha_i(\omega)\ \ \left(\mathcal{F}_{t_i}\text{-measurable}\right)\ \ \textit{for } t_i < t \leq t_{i+1}, 0 \leq i \leq n-1$$

and $E[\int_0^T f^2(t,\omega)dt] < \infty$.

For f simple, define

$$I_T(f) \equiv \int_0^T f dW \equiv \int_0^T f(t,\omega)dW(t,\omega) = \Sigma_{i=0}^{n-1}\alpha_i\left[W(t_{i+1}) - W(t_i)\right]\,.$$

LEMMA 1.2
Let $\mathcal{M}_T$ denote the class of random processes f which are nonanticipative and for which $E[\int_0^T f^2(t,\omega)dt] < \infty$. Then there exists sequence f_n of simple

functions such that

$$E\left\{\int_0^T [f - f_n]^2\, dt\right\} \to 0 \quad as\ n \to \infty\,.$$

REMARK 1.25 For $f \in \mathcal{M}_T$, we define

$$I_T(f) = \text{l.i.m.}_{n\to\infty}\, I_T(f_n)\ .$$

Let f be nonanticipative such that

$$P\left\{\int_0^T f^2(t,\omega)dt < \infty\right\} = 1\,.$$

Denote this class by $\mathcal{P}_T$. ∎

LEMMA 1.3
Let $f \in \mathcal{P}_T$. Then there exist $f_n \in \mathcal{M}_T$ such that $\int_0^T [f_n - f]^2 dt \to 0$ as $n \to \infty$ in probability.

REMARK 1.26 For $f \in \mathcal{P}_T$, define

$$I_T(f) = \int_0^T f dW = \lim_n -p \int_0^T f_n dW$$

where $\{f_n\}$ is as given in Lemma 1.3 and $\lim_n - p$ denotes the limit in probability. It can be shown that the limit $I_T(f)$ does not depend on the choice of the sequence $\{f_n\}$ in Lemma 1.3. This integral is called the *Ito stochastic integral* of f with respect to the Wiener process. ∎

Properties of an Ito Stochastic Integral

(i) If $f \in \mathcal{M}_T$, then

$$E\left[\int_0^T f dW\right] = 0\,,$$

and

$$E\left[\int_0^T f dW\right]^2 = E\left[\int_0^T f^2(t)dt\right].$$

(ii) If f_1 and $f_2 \in \mathcal{M}_T$, then

$$E\left[\int_0^T f_1 dW \int_0^T f_2 dW\right] = E\left[\int_0^T f_1 f_2 dt\right].$$

(iii) If $f \in \mathcal{P}_T$, then, for any $\varepsilon > 0$ and $\delta > 0$,

$$P\left\{\left|\int_0^T f(t)dW(t)\right| > e\right\} \le P\left\{\int_0^T f^2(t)dt > \delta\right\} + \frac{\delta}{\varepsilon^2}.$$

All the above properties are proved for simple functions at first and then extended by a limiting argument.

REMARK 1.27 Note that

$$\int_0^T W dW = \frac{W^2(T) - T}{2}.$$

The calculus of an Ito integral is not the same as ordinary calculus. Stratanovich defined a symmetric version of a stochastic integral that adapts to the ordinary calculus. ∎

Further properties of an Ito Stochastic integral

Let

$$I_t(f) = \int_0^t f dW, f \varepsilon \mathcal{P}_T, 0 \le t \le T.$$

Then $\{I_t(f), \mathcal{F}_t, 0 \le t \le T\}$ is a martingale with continuous sample paths almost surely. Further, for any $\varepsilon > 0$ and $\delta > 0$,

$$P\left\{\sup_{0 \le t \le T} |\int_0^t f(s)dW(s)| > \varepsilon\right\} \le P\left\{\int_0^T f^2(t)dt > \delta\right\} + \frac{\delta}{\varepsilon^2}.$$

In particular, if $f \in \mathcal{M}_T$, then

$$P\left\{\sup_{0\le t\le T}\left|\int_o^t f(s)dW(s)\right| > \varepsilon\right\} \le \frac{1}{\varepsilon^2}\int_o^T Ef^2(t)dt\,.$$

Suppose τ is a stopping time with respect to $\{\mathcal{F}_t\}$. Then

$$\int_o^\tau f dW$$

stands for the process $I_t(f)$ stopped at time τ. If $\tau_1 \le \tau_2$ are stopping times, then

$$\int_{\tau 1}^{\tau_2} f dW = \int_0^{\tau_2} f dW - \int_0^{\tau_1} f dW\,.$$

If $f \in \mathcal{M}_T$, then

i) $E\left[\int_{\tau_1}^{\tau_2} f dW\right] = 0\,,$

ii) $E\left[\int_{\tau_1}^{\tau_2} f dW\right]^2 = E\left[\int_{\tau_1}^{\tau_2} f^2 dt\right]\,,$

iii) $E\left[\int_{\tau_1}^{\tau_2} f dW|\mathcal{F}_{\tau_1}\right] = 0$ a.s., and

iv) $E\left[\left\{\int_{\tau_1}^{\tau_2} f dW\right\}^2|\mathcal{F}_{\tau_1}\right] = E\left[\int_{\tau_1}^{\tau_2} f^2 dt|\mathcal{F}_{\tau_1}\right]$ a.s.

LEMMA 1.4
For any $f \in \mathcal{M}_1$ and $\alpha > 0, \beta > 0$,

$$P\left[\sup_{0\le t\le 1}\left\{\int_0^t f(s)dW(s) - \frac{\alpha}{2}\int_0^t f^2(s)ds\right\} > \beta\right] \le e^{-\alpha\beta}\,.$$

This lemma follows from the fact that

$$\mathcal{Z}(t) = \exp\{\int_0^t f(s)dW(s) - \frac{1}{2}\int_0^t f^2(s)ds\}, 0 \le t \le 1$$

is a supermartingale.

REMARK 1.28 As a consequence of the above properties, it follows that if $W = \{W_t, \mathcal{F}_t\}$ is a Wiener process, then for any stopping time τ adapted to $\{\mathcal{F}_t\}$ with $E\tau < \infty$,

$$EW_\tau = 0, \quad \text{and} \quad EW_\tau^2 = E\tau .$$

∎

DEFINITION 1.32 *A random process $X = \{X_t, \mathcal{F}_t, 0 \le t \le T\}$ is called an Ito process (relative to the Wiener process $W = \{W_t, \mathcal{F}_t, 0 \le t\}$) if there exist nonanticipative processes $a = \{a_t, \mathcal{F}_t, 0 \le t \le T\}$ and $b = \{b_t, \mathcal{F}_t, 0 \le t \le T\}$ such that*

$$P\left\{\int_0^T |a_t|\,dt < \infty\right\} = 1 ,$$

$$P\left\{\int_0^T b_t^2 dt < \infty\right\} = 1 ,$$

and

$$X_t = X_0 + \int_0^t a(s,\omega)ds + \int_0^t b(s,\omega)dW(s) \quad a.s.$$

for all $0 \le t \le T$. Then X is said to have the stochastic differential

$$dX_t = a(t,\omega)dt + b(t,\omega)dW(t), 0 \le t \le T .$$

REMARK 1.29 If $a(s,.)$ and $b(s,.)$ are measurable with respect to the σ-algebra generated by $X(t), 0 \le t \le s$, then the process X is said to be of the *diffusion type.* If $a(s,\omega)$ and $b(s,\omega)$ are of the form $A(s, X(s,\omega))$ and $B(s, X(s,\omega))$, then X is said to be a *diffusion.* ∎

***THEOREM 1.27* (Ito's lemma)**
Let $F(t,x)$ be a continuous function with continuous derivatives $F_t(t,x)$, $F_x(t,x)$, $F_{xx}(t,x)$, and $X(.)$ be an Ito process with stochastic differential

$$dX_t = a(t,\omega)dt + b(t,\omega)dW(t), \quad X_0 = X_0, \quad 0 \le t \le T .$$

Then the random process $Z(t) = F(t, X(t))$ is an Ito process with stochastic differential

$$dZ(t) = \left[F_t(t, X(t)) + F_x(t, X(t))a(t) + \frac{1}{2} F_{xx}(t, X(t))b^2(t) \right] dt$$

$$+ F_x(t, X(t))b(t)dW(t) ,$$

where $Z(0) = F(0, X_0), 0 \leq t \leq T$.

REMARK 1.30 Applying the Ito's lemma, one can prove the following:

(i) If $\{B_t\}$ is a continuous martingale such that $B_t^2 - t$ is a martingale, then $\{B_t\}$ is a Brownian motion.

(ii) If $\{Q_t\}$ is a purely discontinuous martingale with jumps equal to 1 and $Q_t^2 - t$ is a martingale, then $P_t = Q_t + t$ is a Poisson process.

∎

The following theorem gives a representation for square integrable martingales in terms of stochastic integrals.

THEOREM 1.28 **[44] (Representation for square integrable martingales)** *Suppose $X = \{X_t, \mathcal{F}_t\}$, $0 \leq t \leq T$ is a square integrable martingale and $W = \{W_t, \mathcal{F}_t\}$, $0 \leq t \leq T$ is a Wiener process. (Assume as usual that $\mathcal{F}_t$ is right-continuous and complete with $\mathcal{F}_{T+} \equiv \mathcal{F}_T$.) Then there exists a random process $\{a(t, \omega), \mathcal{F}_t\}$ with*

$$E\left[\int_0^T a^2(t, \omega)dt \right] < \infty$$

and

$$\langle X, W\rangle_T = \int_0^t a(s, \omega)ds. \quad a.s. \;\; [P] .$$

In fact,

$$X_t = \int_0^t a(s,\omega)dW(s) + Z_t \quad a.s.\ [P],\ 0 \le t \le T$$

where $Z = \{Z_t, \mathcal{F}_t\} \in m_T^2$. *Furthermore,* $Z \perp Y$, *that is,* $\langle Z, Y\rangle_t = 0,\ 0 \le t \le T$, *where*

$$Y_t = \int_0^t a(s,\omega)dW(s)\,.$$

REMARK 1.31 Note that any random process $X = \{X_t, \mathcal{F}_t\}, 0 \le t \le T$ of the form

$$X_t = \int_0^t a(s,\omega)dW(s),\ E\left[\int_0^T a^2(s,\omega)ds\right] < \infty$$

is a square integrable martingale. ∎

We now obtain a maximal inequality.

THEOREM 1.29

Let a random process $\{f(u,t), \mathcal{F}_t, 0 \le t \le T\}$ *be continuously differentiable with respect to u with probability one and* $\left|\frac{\partial f(u,t)}{\partial u}\right| < c_1 < \infty$. *Let*

$$\zeta(u) = \int_0^T f(u,t)dW(t)\,.$$

Then there exists a constant $C > 0$ *such that for any* $N > 0$,

$$P\left(\sup_{A<u<B} |\zeta(u) - \zeta(A)| > N\right) \le 8K_2 \exp\left\{-\frac{N}{2K_1(B-A)}\right\} \qquad (1.4.1)$$

where $K_1 = C\sqrt{T}$ *and* $K_2 = 1 + \exp\{-\frac{1}{2}K_1^2\}$.

PROOF Note that

$$P\{\zeta(u+h)-\zeta(u)>c\} = P\left\{\int_0^T [f(u+h,t)-f(u,t)]dW(t) > c\right\}$$

$$\leq P\left\{\int_0^T \frac{\Delta f}{h} dW - \frac{1}{2}\int_0^T \left(\frac{\Delta f}{h}\right)^2 dt > \frac{c}{2h}\right\}$$

$$+ P\left\{\frac{1}{2}\int_0^T \left(\frac{\Delta f}{h}\right)^2 dt > \frac{c}{2h}\right\}$$

$$< e^{-\frac{c}{2h}}\left(1 + E\exp\left\{\frac{1}{2}\int_0^T \left(\frac{\Delta f}{h}\right)^2 dt\right\}\right)$$

$$\leq K_2 e^{-\frac{c}{2h}}$$

by the Chebyshev's inequality and the inequality

$$E\left[\exp\left\{\int_0^T g(t)dW(t) - \frac{1}{2}\int_0^T g^2(t)dt\right\}\right] \leq 1$$

(cf. [44] or [5]) where $\Delta f = f(u+h,t) - f(u,t)$. See Lemma 1.4. Let a_n be a sequence such that

$$\sum_{n=1}^{\infty} a_n 2^{-n} = N' = \frac{N}{K_1(B-A)}. \tag{1.4.2}$$

Then, a lemma due to Burnashev (cf. [17, p. 240]) shows that

$$P\left\{\sup_{A<u<B} [\zeta(u) - \zeta(A)] > N\right\} \leq \frac{K_2}{2}\sum_{n=1}^{\infty} 2^n e^{-a_n}. \tag{1.4.3}$$

Minimizing, the quantity on the right side of the inequality (1.4.3) with respect to $\{a_n\}$ under the condition (1.4.2), by the method of Lagrange multipliers, we obtain that $a_n = N' - 4\log 2 - 2n\log 2$, and substituting this value for a_n into (1.4.3), we arrive at the inequality (1.4.1). ∎

THEOREM 1.30

Let $f(.)$ be a nonanticipative process with $E\left[\exp\left\{\frac{1}{2}\|f\|^2\right\}\right] < \infty$. Then

$$E\left[\exp\left\{i\int_0^T f(t)dW(t) + \frac{1}{2}\|f\|^2\right\}\right] = 1\,.$$

Here $\|f\|^2 = \int_0^T f^2(t)dt$.

PROOF Let

$$V(T) = \exp\left\{i\int_0^T f(t)dW(t) + \frac{1}{2}\int_0^T f^2(t)dt\right\}\,.$$

Apply the Ito's lemma. Then

$$V(t) = 1 + i\int_0^T V(t)f(t)dW(t)\,.$$

Note that $E[V(T)] = 1$ provided

$$E\left[\int_0^T V(t)f(t)dW(t)\right] = 0\,.$$

It can be checked that

$$E\left[\int_0^T |V(t)|^2|f(t)|^2 dt\right] < \infty$$

under the condition

$$E\left[\exp\left(\frac{1}{2}\|f\|^2\right)\right] < \infty\,.$$

Hence, it follows that

$$E\left[\int_0^T V(t)f(t)dW(t)\right] = 0 \text{ and } E[V(t)] = 1\,.$$

∎

COROLLARY 1.10
Let $\tilde{f}(.) \in \mathcal{P}_T$ and suppose that $P\{\int_0^T \tilde{f}^2(t)dt \geq \sigma^2\} = 1$. Define

$$\tau = \inf\left\{t : \int_0^t \tilde{f}^2(s)ds = \sigma^2\right\}, \quad \textit{and}$$

$$\zeta = \int_0^\tau \tilde{f}(t)dW(t). \quad \textit{Then } \zeta \textit{ is } N\left(0, \sigma^2\right).$$

This result follows from the above lemma: choose

$$f(t) = \lambda \tilde{f}(t) I[t \leq \tau]\,.$$

Then

$$E\left[\exp\left\{i\lambda \int_0^\tau \tilde{f}(t)\; I[t \leq \tau]dW(t) + \frac{1}{2}\lambda^2\sigma^2\right\}\right] = 1\,,$$

that is,

$$E\left[e^{i\lambda\zeta}\right] = e^{-\frac{1}{2}\lambda^2\sigma^2}\,.$$

Hence, ζ is $N(0, \sigma^2)$.

THEOREM 1.31 **[38] (Central Limit Theorem)**

If $f_\varepsilon(.) \in \mathcal{P}_{T_\varepsilon}$, and

$$\int_0^{T_\varepsilon} f_\varepsilon^2(t)dt \xrightarrow{P} \sigma^2 \quad \textit{as } \varepsilon \to 0\,,$$

then

$$\int_0^{T_\varepsilon} f_\varepsilon(t)dW_\varepsilon(t) \xrightarrow{\mathcal{L}} N\left(0, \sigma^2\right) \quad \textit{as } \varepsilon \to 0$$

where $\{W_\varepsilon(t), 0 \le t \le T_\varepsilon\}$ *is a standard Wiener process.*

PROOF Define

$$g_\varepsilon(t) = f_\varepsilon(t) \text{ for } 0 \le t \le T_\varepsilon$$

$$= \sigma \quad \text{for} \quad T_\varepsilon < t \le T_\varepsilon + 1 ,$$

$$\tau_\varepsilon = \inf\left\{t : \int_0^t g_\varepsilon^2(s)ds = \sigma^2\right\} ,$$

and

$$\zeta_\varepsilon = \int_0^{\tau_\varepsilon} g_\varepsilon(t)dW_\varepsilon(t) .$$

Let

$$F_\varepsilon = \int_0^{T_\varepsilon} f_\varepsilon(t)dW_\varepsilon(t) .$$

Note that $\pounds(\zeta_\varepsilon)$ is $N(0, \sigma^2)$ by Corollary 1.10. It is sufficient to prove that

$$\zeta_\varepsilon - F_\varepsilon \xrightarrow{p} 0 \text{ as } \varepsilon \to 0 .$$

For $a > 0$ and $b > 0$,

$$P(|\zeta_\varepsilon - F_\varepsilon| > a) = P\left\{\left|\int_0^{\tau_\varepsilon} f_\varepsilon(t)dW_\varepsilon(t) - \int_0^{T_\varepsilon} g_\varepsilon(t)dW_\varepsilon(t)\right| > a\right\}$$

$$= P\left\{\left|\int_0^{T_\varepsilon+1} g_\varepsilon(t)\left[I_{\{t\le T_\varepsilon\}} - I_{\{t\le\tau_\varepsilon\}}\right]dW_\varepsilon(t)\right| > a\right\}$$

$$\le \frac{b}{a^2} + P\left\{\int_0^{T_\varepsilon+1} g_\varepsilon^2(t)\left|I_{\{t\le T_\varepsilon\}} - I_{\{t\le\tau_\varepsilon\}}\right|dt > b\right\} .$$

Let $A = \{\omega : \int_0^{T_\varepsilon+1} g_\varepsilon^2(t)|I_{\{t\le T_\varepsilon\}} - I_{\{t\le\tau_\varepsilon\}}|dt > b\}$.

Note that $P\{A \cap [\tau_\varepsilon = T_\varepsilon]\} = 0$, and

$$P(A \cap [\tau_\varepsilon < T_\varepsilon]) = P\left\{ \int_0^{T_\varepsilon + 1} g_\varepsilon^2(t) \left| I_{\{t \le T_\varepsilon\}} - I_{\{t \le \tau_\varepsilon\}} \right| dt > b, \tau_\varepsilon < T_\varepsilon \right\}$$

$$= P\left\{ ||f_\varepsilon||^2 - \sigma^2 > b, \tau_\varepsilon < T_\varepsilon \right\}.$$

Similarly,

$$P(A \cap [\tau_\varepsilon > T_\varepsilon]) = P\left\{ \sigma^2 - ||f_\varepsilon||^2 > b, \tau_\varepsilon > T_\varepsilon \right\}.$$

Hence,

$$P(A) \le P\left\{ \left| \|f_\varepsilon\|^2 - \sigma^2 \right| > b \right\} \to 0 \; as \; \varepsilon \to 0 .$$

Hence, the theorem is proved. ∎

Stochastic Integration with Respect to a Square Integrable Martingale

Let $(\Omega, \mathcal{F}, P)$ be a complete probability space and $\{\mathcal{F}_t, t \ge 0\}$ be a right-continuous complete filtration.

Let a function $f = f(t, \omega) \in \Phi_1$, if it is *nonanticipative,* that is, $f(t, \omega)$ is $\mathcal{F}_t$-measurable for each $t \ge 0$ and the function $f = f(t, \omega) \in \Phi_2$, if it is *strongly nonanticipative,* that is, $f(\tau, \omega)$ is $\mathcal{F}_\tau$-measurable for each stopping time τ with respect to $\{\mathcal{F}_t, t \ge 0\}$.

Suppose the function $f = f(t, \omega) \in \Phi_3$, if it is nonanticipative and measurable with respect to the σ algebra on $R^+ \times \Omega$ generated by the nonanticipative processes with left-continuous paths. In other words $f \in \Phi_3$ if f is *nonanticipative and predictable.*

REMARK 1.32 $\Phi_1 \supset \Phi_2 \supset \Phi_3$. ∎

Notation: Let $\mathcal{M}$ denote the class of square integrable martingales on $[0, \infty]$, which is right-continuous martingales X with $\sup\limits_{0 \le t \le \infty} E[X_t^2] < \infty$. Let $\mathcal{M}^c$ denote the class of elements of $\mathcal{M}$ that have continuous paths a.s. Let $X \in \mathcal{M}$

and $A_t =< X >_t$ be the associated natural increasing process. Let $L_A^2(\Phi_i)$ denote the functions $f \in \Phi_i$ such that

$$E \int_0^\infty f^2(s, \omega) dA_s < \infty, 1 \le i \le 3 .$$

DEFINITION 1.33 *A function $f \in L_A^2(\Phi_1)$ is called simple if there exists $0 = t_0 < t_1 < \cdots < t_n < \infty$ such that*

$$f(t, \omega) = \sum_{k=0}^{n-1} f(t_k, \omega) \, I_{(t_k, t_{k+1}]}(t) .$$

Let $\mathcal{E}$ be the class of simple functions in $L_A^2(\Phi_1)$.

DEFINITION 1.34 *A function $f \in L_A^2(\Phi_2)$ is called simple stochastic if there exists an increasing sequence of stopping times*

$$0 = \tau_0 < \tau_1 < \cdots < \tau_n < \infty$$

with respect to $\{\mathcal{F}_t, t \ge 0\}$ such that $f(t, \omega) = \sum_{k=0}^{n-1} f(\tau_k, \omega) I_{(\tau_k, \tau_{k+1}]}(t)$.

Let $\mathcal{E}_s$ be the class of simple stochastic functions in $L_A^2(\Phi_2)$.

DEFINITION 1.35 *For $f \in \mathcal{E}_s$, define*

$$I(f) = \sum_{k=0}^{n-1} f(\tau_k, \omega) \left[X_{\tau_{k+1}} - X_{\tau_k} \right] .$$

and denote $I(f)$ by $\int_0^\infty f(s) dX_s$.

LEMMA 1.5
Let $X \in \mathcal{M}$ and $A_t =< X >_t, t \ge 0$ be the natural increasing process corresponding to X. Then the space $\mathcal{E}_s$ is dense in $L_A^2(\Phi_3)$.

LEMMA 1.6
Let $X \in \mathcal{M}$ and $A_t = \langle X \rangle_t, t \geq 0$ be the natural increasing process corresponding to X. Suppose A_t is continuous with probability one. Then the space $\mathcal{E}$ is dense in $L_A^2(\Phi_2)$.

REMARK 1.33 If $X \in \mathcal{M}$ is left quasi-continuous, that is, $X_{\tau_n} \to X_\tau$ if $\tau_n \uparrow \tau$ with $P(\tau < \infty) = 1$ and τ_n and τ are stopping times, then A is continuous. ∎

LEMMA 1.7
Let $X \in \mathcal{M}$ and $A_t = \langle X \rangle_t, t \geq 0$ be the natural increasing process corresponding to X. Suppose A_t is absolutely continuous with probability one. Then the space $\mathcal{E}$ is dense in $L_A^2(\Phi_1)$.

REMARK 1.34 Using Lemmas 1.5 through 1.7, one can define the integral

$$I(f) = \int_0^\infty f(t)dX(t)$$

for $X \in \mathcal{M}$ and $f \in L_A^2(\Phi_i), 1 \leq i \leq 3$ depending on the properties of the associated natural increasing process $A_t = \langle X \rangle_t$. Furthermore,

$$E\left[\int_0^\infty f(t)dX(t)\right] = 0$$

and

$$E\left[\int_0^\infty f(t)dX(t)\right]^2 = E\left[\int_0^\infty f^2(t)dA_t\right]$$

whenever the stochastic integrals are defined. ∎

REMARK 1.35 If the martingale $X \in \mathcal{M}^c$, that is, $X \in \mathcal{M}$ and has continuous paths, then for $f \in L_A^2(\Phi_2)$, the stochastic integral $I_t(f) = \int_0^t f(s)dX_s$ has a continuous modification. ∎

REMARK 1.36 It $X \in \mathcal{M}$ and $f \in L_A^2(\Phi_3)$ then the process $\{I_t(f), \mathcal{F}_t, t \geq$

0} will be a square integrable martingale and has a right-continuous modification. ∎

REMARK 1.37 As in the case of Ito integrals, if $X \in \mathcal{M}$ and A is continuous, then one can define

$$I_t(f) = \int_0^t f(s) dX_s$$

for $f \in \Phi_2$ satisfying

$$P\left(\int_0^\infty f^2(t, \omega) dA_t < \infty\right) = 1 .$$

∎

THEOREM 1.32

If $X \in \mathcal{M}$ has continuous paths and

$$A_t = < X >_t = \int_0^t a^2(s, \omega) ds$$

where $a(s, \omega)$ is nonanticipative and $a^2(s, \omega) > 0$ a.s. on $[0, T] \times \Omega$, then there exists a Wiener process W with respect to $\{\mathcal{F}_t\}$ on $[0, T]$ such that

$$X_t = X_o + \int_0^t a(s, \omega)\, dW_s, 0 \le t \le T .$$

Quadratic Characteristic and Quadratic Variation Processes

Suppose $M \in m^{(2)}$, the class of square integrable martingales on $[0, \infty]$. Then

$$M_\infty^* = \sup_{0 \le t \le \infty} |M_t| \in L^2(\Omega)$$

and $\{M_t^2\}$ is a submartingale. Let $X_t = E[M_\infty^2 | F_t] - M_t^2$. Then $X_t \ge 0$ and $X_\infty = 0$ and X_t is a supermartingale with a right-continuous version. Hence, X is a potential of class D. Let $< M, M >$ be the unique predictable increasing process given by the Doob–Meyer decomposition of the potential $X_t = E[M_\infty^2 | F_t] - M_t^2$. The process $\langle M, M \rangle$ is called the *(predictable quadratic variation) quadratic characteristic* of M. Note that $M_t^2 - < M, M >_t$ is a martingale and $< M, M >_0 = M_0^2$. We denote $< M, M >_t$ by $< M >_t$ for convenience at times.

REMARK 1.38 Suppose $M \in m^{(2)}$. Then, $M = M^c + M^d$ where $M^c \perp M^d$. Here, $M^c \in m_0^{2.c}$ and $M^d \in m^{2.d}$ where $m_0^{2.c}$ is the class of continuous square integrable martingales with $M_0 = 0 = M_{0-}$ and $m^{2.d}$ is the class of purely discontinuous martingales. Note that M^d is the sum of compensated jump martingales. Define

$$[M, M]_t = < M^c, M^c >_t + \sum_{s \leq t} \Delta M_s^2 .$$

The process $[M, M]_t$ is called the *(optional quadratic variation) quadratic variation* of M. We denote $[M, M]_t$ by $[M]_t$ for convenience at times. ∎

REMARK 1.39 Observe that if $M \in m^{2.c}$, then $[M, M]_t = \langle M, M \rangle_t$. Since $M = M^c + M^d \in m^{(2)}$, it follows that

$$M_t^2 = M_t^{c^2} + 2M_t^c M_t^d + M_t^{d^2} .$$

Since M^c and M^d are orthogonal martingales, $M^c M^d$ is a martingale. It can be shown that

$$M_t^{d^2} - \sum_{s \leq t} \Delta M_s^2 \text{ is a martingale in } m_0^1$$

and

$$M_t^{c^2} - \langle M^c, M^c \rangle_t \text{ is a martingale in } m_0^1 .$$

Therefore, $M_t^2 - [M, M]_t$ is a martingale in m_0^1.

Note that m_0^1 is the space of martingales $M \in \mathcal{M}$ such that $M_0 = 0$ and $\| M \|_{m^1} = \| M_\infty^* \|_1 < \infty$ where $M_\infty^* = \sup_{0 \leq t \leq \infty} |M_t|$. Here $\mathcal{M}$ is the class of uniformally integrable martingales. ∎

Properties of Quadratic Characteristic and Quadratic Variation Processes

- If $M \in m^{(2)}$, then $< M, M > - [M, M]$ is a martingale.

- $< M^c, M^c >$ is continuous.
- For, $M, N \in m^{(2)}$, define

$$\langle M, N\rangle = \frac{1}{2}(< M+N, M+N > - < M, M > - < N, N >)$$

and

$$[M, N] = \frac{1}{2}([M+N, M+N] - [M, M] - [N, N]) .$$

Note that

$$[M, N]_t =< M^c, N^c >_t + \sum_{s\leq t} \Delta M_s \Delta N_s .$$

- M, N are orthogonal if and only if $\langle M, N\rangle = 0$.
- $E[\langle M, M\rangle_\infty - \langle M, M\rangle_T | \mathcal{F}_T] = E[M_\infty^2 | F_T] - M_T^2$ for any stopping time T since $M^2 - < M, M >$ is a martingale.
- **Cauchy–Schwarz-type inequality:**
if $M, N \in m^{(2)}$ and $H, K \in B(\mathcal{B} \otimes \mathcal{F})$, then

$$\int_0^\infty |H_s||K_s| \quad |d\langle M, N\rangle_s|$$

$$\leq \left(\int_0^\infty H_s^2 d\langle M, M\rangle_s\right)^{\frac{1}{2}} \left(\int_0^\infty K_s^2 d\langle N, N\rangle_s\right)^{\frac{1}{2}}$$

and

$$\int_0^\infty |H_s||K_s| \quad |d[M, N]_s|$$

$$\leq \left(\int_0^\infty H_s^2 d[M, M]_s\right)^{\frac{1}{2}} \left(\int_0^\infty K_s^2 d[N, N]_s\right)^{\frac{1}{2}}$$

(note that the integrals are well defined as the processes $[M, M]$ and $\langle M, M\rangle$ are increasing processes).

- If $1 < p < \infty$ and $1/p + 1/q = 1$, then

$$E\left[\int_0^\infty |H_s||K_s| \ |d < M, N >_s|\right]$$

$$\leq \left\| \left(\int_0^\infty H_s^2 d\langle M, M\rangle_s\right)^{\frac{1}{2}} \right\|_p \left\| \left(\int_0^\infty K_s^2 d < N, N >_s\right)^{\frac{1}{2}} \right\|_q$$

and

$$E\left[\int_0^\infty |H_s||K_s| \ |d[M, N]_s|\right]$$

$$\leq \left\| \left(\int_0^\infty |H_s^2 \ d[M, M]_s^{\frac{1}{2}}\right) \right\|_p \left\| \left(\int_0^\infty K_s^2 d[N, N]_s\right)^{\frac{1}{2}} \right\|_q .$$

For proofs of the above properties, see [12].

We now obtain an inequality linking a martingale $\{M_t\}$ and its predictable quadratic variation $\langle M\rangle_t$ as an application of the following result known as Lenglart's inequality.

THEOREM 1.33 **[39]**
Let X, Y be adapted right-continuous nonnegative processes such that Y is predictable and nondecreasing with $Y_0 = 0$. Suppose that Y dominates X in the sense that for every finite stopping time T, $E(X_T) \leq E(Y_T)$. Then, for each $\varepsilon > 0$ and $\eta > 0$ and every finite stopping time T,

$$P\left[\sup_{t \leq T} X_t \geq \varepsilon\right] \leq \frac{\eta}{\varepsilon} + P\left(Y_T \geq \eta\right) .$$

Applications:

(i) Suppose M is a square integrable martingale, that is, M is a martingale such that $\sup_t E[M_t^2] < \infty$. Note that the predictable quadratic variation process $\langle M\rangle$ is the unique increasing predictable process such that $M^2 -$

$\langle M\rangle$ is a mean zero martingale. The process $\langle M\rangle$ dominates M^2 in the sense of Theorem 1.33. Hence,

$$P\left[\sup_{t\le T} M_t^2 \ge \varepsilon\right] \le \frac{\eta}{\varepsilon} + P\left[< M >_T \ge \eta\right]$$

for every $\varepsilon > 0$ and $\eta > 0$ and for every finite stopping time T.

(ii) Suppose N is a point process with a stochastic intensity λ. Then

$$M_t = N_t - \int_0^t \lambda(s)ds$$

is a martingale and M_t^2 is dominated by $< M >_t = \int_0^t \lambda(s)ds$. Hence,

$$P\left[\sup_{t\le T}\left[N_t - \int_0^t \lambda(s)ds\right]^2 \ge \varepsilon\right] \le \frac{\eta}{\varepsilon} + P\left[\int_0^t \lambda(s)ds \ge \eta\right]$$

for every $\varepsilon > 0$ and $\eta > 0$ and for every finite stopping time T.

DEFINITION 1.36 *Let X and Y be two optional processes (the optional $\sigma-$ algebra is the $\sigma-$ algebra on $\Omega \times R_+$ generated by all RCLL adapted processes as mappings on $\Omega \times R_+$). The process X is said to be L-dominated by Y if $E(|X_T|) \le E(|Y_T|)$ for all finite stopping times T.*

The following result is a slight variation of Theorem 1.33.

THEOREM 1.34 [39]
Let X be an RCLL process which is L-dominated by an increasing process A. Then, for all finite stopping time T, $\varepsilon > 0$ and $\eta > 0$, if A is adapted, then

$$P\left(\sup_{s\le T}|X_s| \ge \varepsilon\right) \le \frac{1}{\varepsilon}\left[\eta + E\left(\sup_{s\le T}\Delta A_s\right)\right] + P\left[(A_T \ge \eta)\right]$$

where $\Delta A_s = A_s - A_{s-}$.

Central Limit Theorem

The following result is a central limit theorem for sequences of martingales.

THEOREM 1.35 **[57, 58, 59]**
Let $\{M^n(t), \mathcal{F}_t^n, 0 \le t \le 1\}, n \ge 1$ be a sequence of square integrable martingales not necessarily defined over the same probability space. Let V be a continuous nondecreasing function on [0, 1] such that $V_0 = 0$ and suppose that

(i) for each t, $< M^n >_t \to V_t$ in probability as $n \to \infty$, and

(ii) there exists constants $b_n \downarrow 0$ such that

$$P\left(\sup_{t\le 1} |\Delta M_t^n| \le b_n\right) \to 1 \text{ as } n \to \infty .$$

Then there is a continuous nonzero Gaussian martingale M with $\langle M\rangle_t = V_t$ such that $M^n \xrightarrow{w} M$ on $D[0, 1]$ as $n \to \infty$. In fact the martingale M is a mean zero Gaussian process with independent increments and $E[M_t M_s] = V_{t\wedge s}$.

For a sketch of the proof, see [35, p. 179] The condition (ii) in the above theorem can be replaced by the weaker Lindeberg-type condition: for every $\varepsilon > 0$ (ii)$'$ $\lim E\left[\sum_{t\le 1}(\Delta M_t^{(n)})^2 I(|\Delta M_t^{(n)}| \ge \varepsilon)\right] = 0$.

We now state another *central limit theorem for martingales* due to Hutton and Nelson [21, 22, 23].

THEOREM 1.36
Let $\boldsymbol{M}(u)$ be a k-dimensional square integrable martingale on a probability space $(\Omega, \mathcal{F}, P)$, that is, each component of $\boldsymbol{M}(u)$ is a real-valued square integrable martingale and $b(u)$ be a scalar function such that

(i) $b(u) \to \infty$,

(ii) $\{\sup_{t\le u}|\Delta \boldsymbol{M}(t)|\}/b^{\frac{1}{2}}(u) \xrightarrow{p} 0$,

(iii) $\frac{[\boldsymbol{M}](u)}{b(u)} \xrightarrow{p} \Psi$, *and*

(iv) $E([\boldsymbol{M}](u))/b(u) \to \boldsymbol{D}$

as $u \to \infty$ where D is a positive definite matrix and Ψ is an $\mathcal{F}$-measurable random matrix. Then

$$\frac{M(u)}{b^{\frac{1}{2}}(u)} \text{ converges stably to } Z^*$$

where the characteristic function of Z^* equals $E\{\exp(-s\Psi s^T/2)\}$. Furthermore, $M(u)[M]^{-\frac{1}{2}}(u)$ is asymptotically $N(0, I)$-distributed conditionally on $F \cap [\Psi > 0]$ if $F \in \mathcal{F}$. Here $\Delta M(t)$ denotes the jump size of the process M at the time t.

(For the definition of stable convergence, see Section 1.8).

Burkholder's Inequality

Let M be a square integrable martingale. Let $[M]$ be the quadratic variation process, that is,

$$[M]_t = \langle M^{(c)} \rangle t + \sum_{s \le t} \Delta M_s^2$$

and $\langle M \rangle$ be its quadratic characteristic. Here $M^{(c)}$ is the continuous martingale component of M and $\Delta M_t = M_t - M_{t-}$. Note that $[M] - \langle M \rangle$ is a zero-mean martingale.

The following result is due to Burkholder [7]. This gives an upperbound linking the expectation of $\sup_{0 \le t \le 1} M_t^2$, $[M]_1$ and $\langle M \rangle_1$ for a martingale M.

THEOREM 1.37

For each $p > 1$, there exist constants c_p, C_p such that for every martingale M,

$$c_p E\left\{[M]_1^{p/2}\right\} \le E\left\{\sup_{t \le 1} |M_t|^p\right\} \le C_p E\left\{[M]_1^{\frac{p}{2}}|\right\} .$$

In particular, for $p = 2$,

$$E\left\{\sup_{t \le 1} M_t^2\right\} \le c_2 E\{[M]_1\} = c_2 E\{\langle M \rangle_1\} .$$

REMARK 1.40 For another version dealing with continuous martingales, see Theorem A.2 of Appendix A. ∎

1.5 Local Martingales

We have discussed earlier the definition and some properties of local martingales. We recall some notation for clarity and discuss stochastic integration with respect to a local martingale in this section.

Let $(\Omega, \mathcal{F})$ be a measurable space and $\mathcal{F} = \{\mathcal{F}_t, t \geq 0\}$ be a nondecreasing family of σ-algebras contained in $\mathcal{F}$. Suppose $\mathcal{F}_t$ is right continuous and $\mathcal{F} = \vee_{t \geq 0} \mathcal{F}_t$.

Let $\mathcal{W}$ be the σ-algebra generated by the mappings $Y_t(\omega)$ defined on $[0, \infty] \times \Omega$ that are $\mathcal{F}_t$-measurable right continuous with left limits ($RCLL$) and $\mathcal{P}$ be the σ-algebra generated by the mappings $Y_t(\omega)$ defined on $[0, \infty] \times \Omega$ that are $\mathcal{F}_t$-measurable and continuous. It is known that $\mathcal{P} \subset \mathcal{W}$. The σ-algebra $\mathcal{P}$ is the *predictable* σ-algebra and $\mathcal{W}$ is called the well-measurable σ-algebra (cf. [10]).

Let P be a probability measure on $(\Omega, \mathcal{F})$ and $\mathcal{F}(P)$ be the completion of $\mathcal{F}$ with respect to P. Let $\mathcal{N}(P)$ be the σ-algebra generated by the sets in $\mathcal{F}(P)$ of P-measure zero. Define $\mathcal{F}_t(P) = \mathcal{F}_t \vee \mathcal{N}(P)$ and $F(P) = \{\mathcal{F}_t(P),\ t \geq 0\}$. We introduce some more notation. By a martingale with a continuous time parameter, we always assume that the martingale is $RCLL$ P-a.s.

Denote by

(i) $\zeta(P)$ = the set of all stopping times relative to $F(P) = \{\mathcal{F}_t(P),\ t \geq 0\}$,

(ii) $m(P)$ = the set of all uniformly integrable martingales $X = (X_t, \mathcal{F}_t(P),\ t \geq 0)$, and

(iii) $m^{(c)}(P)$ = the set of all uniformly integrable continuous martingales in $m(P)$.

Note that a martingale $X = (X_t, \mathcal{F}_t(P),\ t \geq 0)$ is said to be *continuous* if it has continuous paths P-a.s.

Further let us denote

(iv) $m^{(d)}(P)$ = the set of all uniformly integrable purely discontinuous martingales, and

(v) $m^{(2)}(P) = \{X \in m(P) : \|X\|^2 = \sup_{t \geq 0} E|X_t|^2 < \infty\}$.

Localization: For any class K of random processes $X = (X_t, \mathcal{F}_t(P),\ t \geq 0)$, let K_{loc} be the class of processes $Y = (Y_t, \mathcal{F}_t(P),\ t \geq 0)$ with the property that for each $Y \in K_{loc}$, there is a nondecreasing sequence of stopping times $\tau_n \in \bar{\tau}(P)$, $n \geq 0$, $\tau_n \uparrow \infty$ a.s. $[P]$ such that the "stopped" process

$$Y^{\tau_n} = \left(Y_{t \wedge \tau_n}, \mathcal{F}_t(P),\ t \geq 0\right)$$

belongs to the class K.

DEFINITION 1.37 *The processes in the class $m_{loc}(P)$ are called local martingales and the processes in the class $m_{loc}^{(2)}(P)$ are called local square-integrable martingales.*

REMARK 1.41 Every martingale is a local martingale. For if $M = \{M_t\}$ is a $\mathcal{F}_t$-adapted martingale, define $M_s^n = E\{M_n | \mathcal{F}_s\}$. Then $\{M_s^n\}$ is a martingale and $\{M_t\}$ is a local martingale. ∎

Let $\mathcal{V}^+(P) = \{A = (A_t, \mathcal{F}_t(P),\ t \geq 0) : A_0 = 0, A_t < \infty$ a.s. $[P]$ and A has nondecreasing right continuous paths a.s. $[P]\}$.

For any $A \in \mathcal{V}^+(P)$, let $A_\infty = \lim_{t \to \infty} A_t$. Let

$$\mathcal{V}(P) = \mathcal{V}^+(P) - \mathcal{V}^+(P)$$

$$= \left\{A - B : A \in \mathcal{V}^+(P),\ B \in \mathcal{V}^+(P)\right\},$$

$$\alpha^+(P) = \left\{A \in \mathcal{V}^+(P) : EA_\infty < \infty\right\},$$

and

$$\alpha(P) = \alpha^+(P) - \alpha^+(P).$$

Recall that two processes X and Y are said to be *P-indistinguishcable* if

$$P\,[\omega : \text{ there exists } t,\ X_t(\omega) \neq Y_t(\omega)] = 0.$$

The processes are said to be *P-equivalent* if

$$P\,[X_t \neq Y_t] = 0 \text{ for all } t \geq 0.$$

THEOREM 1.38
Let $A \in \alpha_{loc}(P)$. Then there exists one and only one predictable process $\tilde{A} \in \alpha_{loc}(P)$ (to within P-indistinguishability) called the dual predictable projection of A or the compensator of A such that

$$A - \tilde{A} \in m_{loc}(P) ,$$

that is, $A - \tilde{A}$ is a local martingale.

For proof, see [10].

For any process $X = \{X_t, \ t \geq 0\}$ and $A \in \mathcal{V}(P)$, define the process $X \circ A$ by the relation

$$(X \circ A)_t = \int_0^t X_s(\omega) dA_s(\omega) \text{if} \int_0^t |X_s(w)| \ |dA_s(\omega)| < \infty$$

$$= \infty \text{ otherwise} .$$

Note that $(X \circ A)_t$ is a Lebesgue–Stieltjes integral defined for each $\omega \in \Omega$.

THEOREM 1.39
If τ is a stopping time, X is a nonnegative predictable process and $A \in \alpha_{loc}^{+}(P)$, then

$$E\left[(X \circ A)_\tau\right] = \left[(X \circ \tilde{A})_\tau\right]$$

For proof, see [10].

Quadratic Characteristic

Let $X \in m_{loc}^{(2)}(P)$, the class of locally square integrable martingales. By the Doob–Meyer decomposition (cf. [49]) there exists one and only one predictable process (to within P-indistinguishability) (denoted by $\langle X, X \rangle$ or $\langle X \rangle$) in the class $\alpha_{loc}^{+}(P)$ for which

$$X^2 - \langle X, X \rangle \ \in \ m_{loc}(P) .$$

The process $\langle X, X \rangle$ is called the *quadratic characteristic* of X.

If X and Y belong to $m_{loc}^{(2)}(P)$, then the process

$$\langle X, Y\rangle = \frac{1}{4}(\langle X+Y,\ X+Y\rangle - \langle X-Y,\ X-Y\rangle)$$

is the unique predictable process in $\alpha_{loc}(P)$ for which

$$XY - \langle X, Y\rangle \in m_{loc}(P)\ .$$

Decomposition of a Local Martingale

Every process $X \in m_{loc}^{(2)}(P)$ can be *uniquely* decomposed into a sum

$$X = X^{(c)} + X^{(d)}$$

where $X^{(c)} \in m_{loc}^{(c)}(P)$ and $X^{(d)} \in m_{loc}^{(d)}(P) \cap m_{loc}^{(2)}(P)$. The process $X^{(c)}$ is called the *continuous* component of X and $X^{(d)}$ the *purely discontinuous* component, which is the sum of the jumps (cf. [48]).

Each local martingale X admits a decomposition [48]

$$X = U + V \tag{1.5.1}$$

where $U \in m^{(2)}(P)$ and $V \in \alpha_{loc}(P)$. This decomposition is not unique in general.

Applying the above two representations, it can be shown that if $X \in m_{loc}(P)$, then it admits a unique decomposition

$$X = X^{(c)} + X^{(d)} \tag{1.5.2}$$

where $X^{(c)} \in m_{loc}^{(c)}(P)$ and $X^{(d)} \in m_{loc}^{(d)}(P)$.

Quadratic Variation

Given $X \in m_{loc}(P)$, let us consider the process $[X, X] \in \mathcal{V}^{+}(P)$ defined by

$$[X, X]_t = \langle X^c, X^c\rangle_t + \sum_{s\leq t}(\Delta X_s)^2$$

where $\Delta X_s = X_s - X_{s-}$, $X_{0-} = 0$. This is called the *quadratic variation* of the process X.

For $X, Y \in m_{loc}(P)$, define

$$[X, Y] = \frac{1}{4}([X+Y,\ X+Y] - [X-Y,\ X-Y]) .$$

Then, it can be shown that

$$[X, Y]_t = \langle X^{(c)}, Y^{(c)} \rangle_t + \sum_{s \le t} \Delta X_s \Delta Y_s .$$

Furthermore,

$$XY - [X, Y] \in m_{loc}(P), \quad [X, X]^{1/2} \in \alpha_{loc}^{+}(P) .$$

THEOREM 1.40 **(Davis inequality)**
For any $X \in m_{loc}(P)$, there exist constants C_1 and C_2 such that

$$C_1 E[X, X]_T^{1/2} \le E \sup_{0 \le t \le T} |X_t| \le C_2 E[X, X]_T^{1/2} .$$

For proof, see [48].

Stochastic Integral with Respect to a Local Martingale

Suppose that $X \in m^{(2)}(P)$ and let $L^2(X)$ be the class of predictable processes $H = (H_t, \mathcal{F}_t(P), t \ge 0)$ for which

$$E\left[\int_0^\infty H_s^2 d\langle X, X \rangle_s\right] < \infty .$$

Then the stochastic integral $H \bullet X$ defined by

$$(H \bullet X)_t \equiv \int_0^t H_s dX_s, \ t > 0$$

is that element in $m^{(2)}(P)$ for which

$$\langle H \bullet X, Y \rangle = H \circ \langle X, Y \rangle$$

for any $Y \in m^{(2)}(P)$.

It can be shown that this stochastic integral coincides with the Ito stochastic integral in the following way (cf. [44]).

Suppose H is a simple process in $L^2(X)$ in the sense that H takes only a finite number of values. Then $H \bullet X$ is defined to be the corresponding integral sum. In general, if $H \in L^2(X)$, then there exists a sequence of simple processes $(H^{(n)},\ n \geq 0)$ in $L^2(X)$ such that

$$E\left[\int_0^\infty \left|H_s^{(n)} - H_s\right|^2 d\langle X\rangle_s\right] \to 0 \text{ as } n \to \infty .$$

Here $H \bullet X$ is defined to be the l.i.m. $H^{(n)} \bullet X$ where l.i.m. denotes the convergence in quadratic mean.

There are two ways for defining a stochastic integral with respect to a local martingale. The first is based on the decomposition (1.5.1) and the second on (1.5.2). Both approaches lead to the same integral.

Suppose H is a bounded predictable process and $X \in m_{loc}(P)$. By (1.5.1), there exists $U \in m_{loc}^{(2)}(P)$ and $V \in \alpha_{loc}(P)$ such that

$$X = U + V .$$

Assume without loss of generality that $\{\tau_n\}$ is a common localizing sequence for X, U and V. Define the integral $H \bullet X^{\tau_n}$ by

$$H \bullet X^{\tau_n} = H \bullet U^{\tau_n} + H \circ V^{\tau_n} \tag{1.5.3}$$

where $H \bullet U^{\tau_n}$ is the stochastic integral with respect to the square integrable martingale U^{τ_n} and $H \circ V^{\tau_n}$ is the Lebesgue–Stieltjes integral defined earlier. Relation (1.5.3) implies that there exist a process $H \bullet X \in m_{loc}(P)$ with the properties

$$(H \bullet X)^{\tau_n} = H \bullet X^{\tau_n} , \tag{1.5.4}$$

$$(H \bullet X)^{(c)} = H \bullet X^{(c)} , \tag{1.5.5}$$

$$\Delta(H \bullet X)_t = H_t \Delta X_t , \tag{1.5.6}$$

and

$$[H \bullet X, Y]_t = H \circ [X, Y] \tag{1.5.7}$$

for all $Y \in m_{loc}(P)$. If H is not necessarily bounded but predictable, the integral $H \bullet X$ is defined using (1.5.7). In fact, define

$$L^{(1)}_{loc}(X) = \left\{ H \in \mathcal{P} : \left(H^2 \circ [X, X] \right)^{1/2} \in \alpha^+_{loc}(P) \right\} ,$$

where $\mathcal{P}$ is the class of predictable processes with respect to $F(P) = \{\mathcal{F}_t(P),\ t \geq 0\}$. If $X \in m_{loc}(P)$ and $H \in L^{(1)}_{loc}(X)$, then there exists an element $J \in m_{loc}(P)$ such that for every local martingale Y

$$[J, Y] = H \circ [X, Y] . \tag{1.5.8}$$

The element J can be obtained by considering bounded martingales Y only. Furthermore H is unique up to P-indistinguishability (cf. [49]). The element J here after denoted by $H \bullet X$ is called the *stochastic integral of H with respect to the local martingale X*. The properties (1.5.3) to (1.5.7) continue to hold for $H \bullet X$.

Another approach is due to Jacod [26] following the decomposition (1.5.2). Suppose $H \in L^{(1)}_{loc}(X)$. Then

$$H^2 \circ [X, X] = H^2 \circ \langle X^{(c)}, X^{(c)} \rangle + \sum_{s \leq \cdot} (H_s \Delta X_s)^2 , \tag{1.5.9}$$

$$\left(H^2 \circ \langle X^{(c)}, X^{(c)} \rangle \right)^{1/2} \in \alpha^+_{loc}(P) , \tag{1.5.10}$$

and

$$\sum_{s \leq \circ} (H_s \Delta X_s)^{1/2} \in \alpha^+_{loc}(P) . \tag{1.5.11}$$

The condition (1.5.11) implies that

$$H^2 \circ \langle X^{(c)}, X^{(c)} \rangle \in \alpha^+_{loc}(P) ,$$

and one can define the integral $H \bullet X^{(c)}$ as given by the first approach. Since

$$X = X^{(c)} + X^{(d)} ,$$

it is natural to have

$$H \bullet X = H \bullet X^{(c)} + H \bullet X^{(d)}$$

and hence,

$$\Delta(H \bullet X) = \Delta\left(H \bullet X^{(c)}\right) + \Delta\left(H \bullet X^{(d)}\right)$$

assuming that $\Delta(H \bullet X^{(c)}) = 0$ which in turn shows that

$$\Delta\left(H \bullet X^{(d)}\right) = H \Delta X .$$

It can be shown ([8, 40]) that there exists a process $X' \in m_{loc}^{(d)}(P)$ such that

$$\Delta X'_t = H_t \Delta X_t , \; t \geq 0 .$$

Define

$$H \bullet X = H \bullet X^c + X' .$$

It is known that $H \bullet X$ so defined satisfies the property

$$[H \bullet X, Y] = H \circ [X, Y]$$

for all bounded $Y \in m_{loc}(P)$ and the integral $H \bullet X$ is the same as the one derived from the first approach [26].

Some Inequalities for Local Martingales

REMARK 1.42 **(Application of the Lenglart domination property)** Suppose X is a locally integrable increasing process. Then there exists a predictable locally integrable increasing process $\tilde{X}$ such that

(i) $X - \tilde{X}$ is a local martingale,

(ii) $E(X_T) = E(\tilde{X}_T)$ for all stopping times T, and

(iii) $E[(H \bullet \tilde{X})_\infty] = E[(H \bullet X)_\infty]$ for all nonnegative predictable processes H.

∎

Here $\tilde{X}$ is the compensator for X. Applying Lenglart's inequality, it follows that

$$P\left[\sup_{s\leq T}|X_s|\geq\varepsilon\right]\leq\frac{\eta}{\varepsilon}+P\left[\tilde{X}_T\geq\eta\right]$$

for all stopping times T, $\varepsilon>0$ and $\eta>0$.

Let $(\Omega, \mathcal{F}, P)$ be a probability space and $\{M_t, t\geq 0\}$ be a locally square integrable martingale with respect to the filtration $\{\mathcal{F}_t, t\geq 0\}$. Suppose that $M_0=0$. We assume that $\{\mathcal{F}_t\}$ satisfies the "usual" conditions. Let $V_t=\langle M\rangle_t$ be the predictable quadratic characteristic and $\Delta M_t=M_t-M_{t-}$ for $t>0$.

The following result is given in Shorrock and Wellner [67].

THEOREM 1.41
Suppose that $|\Delta M_t|\leq K$ for all $t>0$ and some $0\leq K<\infty$. Then, for each $a>0$ and $b>0$.

$$P\left(M_t\geq a \ \ and \ V_t\leq b^2 \ \ for\ some \ \ t\right)\leq\exp\left(-\frac{a^2}{2\left(aK+b^2\right)}\right).$$

Consider the mth order process $\sum_{s\leq t}|\Delta M_s|^m$ with the compensator $V_{m,t}$ for $m=3,4,\ldots$. Then $V_{2,t}\equiv V_t$ for $t\geq 0$. An extended result over Theorem 1.41 is given below.

***THEOREM 1.42* [69]**
Suppose that for all $t\geq 0$ and some $0<K<\infty$,

$$V_{m,t}\leq\frac{m!}{2}K^{m-2}R_t, \ m\geq 2$$

where $\{R_t\}$ is a predictable process. Then, for each $a>0$ and $b>0$,

$$P\left(M_t\geq a \ \ and \ R_t\leq b^2 \ for\ some\ t\right)<\exp\left(\frac{-a^2}{2\left(aK+b^2\right)}\right).$$

For proof, see [69].

Let $\{g_t\}$ be a predictable process bounded from below and $\{M_t\}$ be a martingale with process $\{V_{m,t}\}$ as defined above. We now give an extension of Theorem 1.42 to martingales of the form $\int_0^t g\,dM$. If $g_t \geq -L$, then

$$|g_t|^m \leq \frac{m!}{2} C_L^2 \frac{1}{2} \left(e^{g_t} - 1\right)^2, \ m \geq 2$$

where

$$C_L^2 = \frac{4\left(e^L - 1 - L\right)}{\left(e^{-L} - 1\right)^2}$$

(cf. [70]). This inequality is a consequence of the fact that $\frac{e^t - 1 - t}{(1-e^{-t})^2}$ is increasing in t and $\frac{e^t-1-t}{(1-e^t)^2}$ is decreasing in t. Furthermore if $|\Delta M_t| \leq K$ for all $t > 0$, then

$$\int_0^t |g|^m \, dV_m \leq K^{m-2} \int_0^t |g_t|^m \, dV$$

$$\leq \frac{m!}{2} K^{m-2} C_L^2 \frac{1}{2} \int_0^t \left(e^g - 1\right)^2 dV, \ m \geq 2 \, .$$

Let

$$d_t^2(g, 0) = \frac{1}{2} \int_0^t \left(e^g - 1\right)^2 dV, \ t > 0 \, .$$

Then the following corollary holds.

COROLLARY 1.11

Suppose that $|\Delta M_t| \leq K$ for all $t > 0$ and some $0 < K < \infty$. Let $\{g_t\}$ be a predictable process satisfying $g_t \geq -L$ for all $t \geq 0$. Then, for each $a > 0$ and $b > 0$,

$$P\left\{\left|\int_0^t g\,dM\right| \geq a \ \text{and} \ d_t^2(g, 0) \leq b^2 \ \text{for some} \ t\right\}$$

$$\leq 2\exp\left[-\frac{a^2}{2\left(aK + C_L^2 b^2\right)}\right].$$

This gives an inequality for stochastic integrals with respect to a locally square integrable martingale.

Application of Corollary 1.11 to Counting Processes

Suppose $\{N_t,\ t \geq 0\}$ is a counting process with compensator $\{A_t,\ t \geq 0\}$. We assume $\{A_t\}$ to be continuous. Let $0 < T \leq \infty$ and $\{g_t\}$ be a predictable process with $g_t \geq -L$ for all t. Then, by the Corollary 1.11, it follows that

$$P\left[\int_0^T g d(N - A) \geq a \text{ and } d_T^2(g, 0) \leq b^2\right]$$

$$\leq 2\exp\left[\frac{-a^2}{2\left(a + C_L^2 b^2\right)}\right].$$

For any predictable processes g, $\tilde{g}$, define

$$d_T(g, \tilde{g}) = \left(\frac{1}{2}\int_0^T \left(e^g - e^{\tilde{g}}\right)^2 dA\right)^{1/2}.$$

Let $\zeta \subseteq \Lambda$ where Λ is the class of all predictable processes with $g_t \geq -L$ for all t.

DEFINITION 1.38 **(Entropy)** *Given $b > 0$, $\delta > 0$ and a measurable set $B \subset \Omega$, let $\{[g_j^L, g_j^U],\ i \leq j \leq m\} \subset \Lambda \times \Lambda$ be a collection of pairs of predictable functions such that for each $g \in \zeta$ there exists a $J_g \in \{1, \ldots, m\}$ such that*

(i) the map $g \to J_g$ is nonrandom,

(ii) $g_{J_g}^L \leq g \leq g_{J_g}^U$ for all t and $\omega \in B$, and

(iii) $d_T(g_{J_g}^L,\ g_{J_g}^U) \leq \delta$ on $\{d_T(g, 0) \leq b\} \cap B$.

Let $N(\delta, b, B)$ be the smallest value of m for which such a bracketing set $\{[g_j^L, g_j^U]\}_{j=1}^m$ exists. Let $H(\delta, b, B)$ be greater than $\max(\log N(\delta, b, B), 1)$ and continuous in $\delta > 0$. Then $H(\delta, b, B)$ is an upper bound for the δ-entropy with bracketing of ζ locally at a sphere with radius b around the origin on the set B. The quantity $H(\delta, b, B)$ is also called as the entropy of ζ. If no finite bracketing exists, then we take $H(\delta, b, B) = \infty$.

The following theorem for counting processes N with a compensator A is due to Van de Geer [69].

THEOREM 1.43

Let ζ be a class of predictable functions with $g_t \geq -L$ for all $t \geq 0$, $g \in \zeta$ and let $H(\delta, b, B)$ be the entropy of ζ where $B \subseteq [A_T \leq \sigma_T^2]$. Then there exist constants C_i, $1 \leq i \leq 4$ depending on L such that for $0 \leq \varepsilon \leq 1$ and

$$\frac{\varepsilon b^2}{C_1} \geq \int_{\varepsilon b^2/(C_2\sigma_T)\wedge b/8}^{b} \sqrt{H(x, b, B)}dx \vee b\,,$$

the following inequality holds:

$$P\left(\left\{\left|\int_0^T gd(N-A)\right| \geq \varepsilon b^2 \text{ and } d_T^2(g,0) \leq b^2 \text{ for some } g \in \zeta\right\} \cap B\right)$$

$$\leq C_3 \exp\left(\frac{-\varepsilon^2 b^2}{C_4}\right)$$

(here $a \vee b = \max(a, b)$ and $a \wedge b = \min(a, b)$).

Strong Law of Large Numbers

Let $(\Omega, \mathcal{F}, P)$ be a probability space and $\{\mathcal{F}_t,\ t \geq 0\}$ be a filtration associated with it. Suppose further that the "usual" conditions hold for the probability space and the filtration. Let $\{M_t,\ t \geq 0\}$ and $\{A_t, t \geq 0\}$, RCLL processes with $A_0 = M_0 = 0$, be a local martingale and a nondecreasing predictable process, respectively.

THEOREM 1.44 [41]
If $\{M_t\}$ is a local square integrable martingale, then

$$\frac{M_t}{A_t} \to 0 \ a.s. \ as\ t \to \infty \tag{1.5.12}$$

for

$$A_t = f(\langle M\rangle_t), \lim_{t\to\infty} A_t = \infty \ and \int_0^\infty \frac{dt}{f^2(t)} < \infty . \tag{1.5.13}$$

REMARK 1.43 [10] For a local martingale M, the relation (1.5.12) also holds when $M = B - \tilde{B}$ with $A = \tilde{B}$, $\lim_{t\to\infty} A_t = \infty$ where $B = \{B_t, t \geq 0\}$ is a non-negative nondecreasing local integrable $RCLL$ process with $B_0 = 0$ and $\tilde{B} = \{\tilde{B}_t, t \geq 0\}$ is the dual predictable projection (compensator) for B assuming that

$$E\left\{\sup_t \left(B_t - \lim_{s\downarrow t} B_s\right)\right\} < \infty .$$

∎

Let $m_t = (1+A_t)^{-1}M_t$. Then $m_t \in m_{loc}$. If $M \in m_{loc}^{(2)}$, then $m \in m_{loc}^2$ and

$$\langle m\rangle_t = (1+A_t)^{-2}\, O(\langle M\rangle_t) .$$

Let $M \in m_{loc}$ and $m = m^{(c)} + m^{(d)}$ be the decomposition of m into a continuous component $m^{(c)}$ and a purely discontinuous compact $m^{(d)}$.

Define

$$G_t = \langle m^{(c)}\rangle_t + \sum_{s\leq t} \frac{(\Delta m_s)^2}{1+|\Delta m_s|}$$

and

$$Q_t = \langle m^{(c)}\rangle_t + \sum_{s\leq t} (\Delta m_s)^2 \wedge 1 .$$

THEOREM 1.45 [43]
Let $P[A_\infty = \infty] = 1$. Then

$$P\left[\frac{M_t}{A_t} \to 0 \ as\ t \to \infty\right] = 1 \tag{1.5.14}$$

under any of the following conditions:

(i) *Suppose $M \in m_{loc}^{(2)}$. Then (1.5.14) holds provided $P[\langle m\rangle_\infty < \infty] = 1$.*

(ii) *Suppose $M \in m_{loc}$. Then (1.5.14) holds provided $P[\tilde{G}_\infty < \infty] = 1$ where $\{\tilde{G}_t\}$ denotes the compensator of the process $\{G_t\}$.*

For additional alternate conditions for validity of (1.5.14), see [43].

REMARK 1.44 The following analogue of Kronecker's lemma for local martingales is useful in proving Theorem 1.45. ▮

THEOREM 1.46 **[43]**
Let $M \in m_{loc}$ and m be as defined above. Then

$$[A_\infty = \infty] \cap \left[\lim_{t\to\infty} m(t) < \infty\right] \subset \left[\frac{M_t}{A_t} \to 0 \ as\ t \to \infty\right]. \tag{1.5.15}$$

REMARK 1.45 For additional results of this nature, see [62]. ▮

REMARK 1.46 Let $M(u)$ be a scalar local square integrable martingale and let $A(u)$ be any process such that $A(u) \geq \langle M\rangle_u$ for all u. Then

$$\frac{M(u)}{A(u)} \to 0 \text{ a.s. on the set } [\lim A(u) = \infty]\,.$$

This can be seen as follows. Theorem 1.44 implies that $M(u)$ has a finite limit if $\langle M\rangle_\infty < \infty$ and that

$$\frac{M(u)}{\langle M\rangle_u} \to 0 \text{ a.s. on the set } [\lim\langle M\rangle_u = \infty]\,. \tag{1.5.16}$$

This implies that

$$\frac{M(u)}{A(u)} \to 0 \text{ a.s. on the set } [\lim A(u) = \infty]\,. \tag{1.5.17}$$

▮

A Martingale Conditional Law

Let $(\Omega, \mathcal{F}, P)$ be a complete probability space and $\{M(t), \mathcal{F}_t, t \geq 0\}$ be a real-valued continuous local martingale. Let $\langle M \rangle_t$ be its quadratic characteristic. Suppose h is a nondecreasing, positive continuous function on $[0, \infty)$ such that

$$\lim_{t\to\infty} \frac{\sqrt{t \log\log t}}{h(t)} = 0\,. \tag{1.5.18}$$

If

$$\int_{\varepsilon}^{\infty} \frac{1}{h^2(t)} dt < \infty \tag{1.5.19}$$

for all $\varepsilon > 0$, then

$$\lim_{t\to\infty} \frac{M(t)}{h\left(\langle M \rangle_t\right)} = 0 \text{ a.s. on } [\langle M \rangle_t \uparrow \infty] \tag{1.5.20}$$

by Revuz and Yor [62]. Define the process.

$$Y^t(\cdot) = \frac{M(t+\cdot)}{h(\langle M \rangle_{t+\cdot})}\,, \tag{1.5.21}$$

for any fixed t, which is $C[0, \infty)$-valued. The following theorem is due to Levanony [42].

THEOREM 1.47
If the condition (1.5.18) holds, then

$$\lim_{t\to\infty} Q_t \log P\left(\|Y^t\|_\infty > \lambda \,|\, \mathcal{F}_t\right) = -\frac{1}{2}\lambda^2 \;\; a.s.\ [P] \tag{1.5.22}$$

for all $\lambda > 0$ where

$$Q_t = Q_t(\langle M \rangle) = \sup_{s\geq 0} \frac{s}{h^2\left(\langle M \rangle_t + s\right)}\,. \tag{1.5.23}$$

Proof of this result is based on the following theorem for the Wiener process due to Levanony [42]. We omit the proof.

THEOREM 1.48

Let $\{W(t), \zeta_t\}$ *be a standard Wiener process and* $T_t \uparrow \infty$ *be a continuous* ζ_t*-stopping time process. Define*

$$X^t(\cdot) = \frac{W(T_t + \cdot)}{h(T_t + \cdot)} \tag{1.5.24}$$

where $0 < h \in C[0,\infty)$ *is nondecreasing and satisfies the condition (1.5.18). Then*

$$\lim_{t\to\infty} Q_t \log P\left(\|X^t\|_\infty > \lambda \,|\zeta_{T_t}\right) = -\frac{1}{2}\lambda^2 \ \ a.s.\ [P] \tag{1.5.25}$$

for all $\lambda > 0$ *where* $Q_t = Q_t(T) = \sup_{s\geq 0} \frac{s}{h^2(T_t+s)}$.

PROOF (for Theorem 1.47) Consider the martingale $\{M(t), \mathcal{F}_t, t \geq 0\}$. Let

$$\tau(t) = \inf\{s > 0 | \langle M\rangle_s > t\} \ ,$$

$$\zeta_t = \mathcal{F}_{\tau(t)}$$

and

$$W(t) = M(\tau(t)) \ .$$

Then $\{W(t), \zeta_t, t \geq 0\}$ is a standard Wiener process (cf. [34, 62]) and $\langle M\rangle_s$ is a $\{\zeta_t\}$-stopping time. Let $T_t = \langle M\rangle_t$. Then $P(\cdot|\mathcal{F}_t) = P(\cdot|\zeta_{\langle M\rangle_t})$ for all $t \geq 0$. This together with a random time change argument, that is, $M(t+s) = W(\langle M\rangle_{t+s})$ for all $s, t \geq 0$ a.s. $[P]$ imply that, outside an W-null set independent of λ and t,

$$P\left(\|Y^t\|_\infty > \lambda\,|\mathcal{F}_t\right) = P\left(\sup_{s\geq 0} \frac{|M(t+s)|}{h(\langle M\rangle_{t+s})} > \lambda\,|\mathcal{F}_t\right)$$

$$= P\left(\sup_{s\geq 0} \frac{|W(\langle M\rangle_{t+s})|}{h(\langle M\rangle_{t+s})} > \lambda\,|\zeta_{\langle M\rangle_t}\right)$$

$$= P\left(\sup_{r\geq 0}\frac{|W(\langle M\rangle_t + r|}{h(\langle M_t\rangle + r)} > \lambda \,|\zeta_{\langle M\rangle_t}\right)$$

$$= P\left(\|X^t\|_\infty > \lambda|\zeta_{T_t}\right).$$

Note that we have made use of the fact that h, W and $\langle M\rangle$ are continuous and that $\langle M\rangle_t \uparrow \infty$ in the next to last equality concerning the supremum. An application of Theorem 1.48 completes the proof. ∎

Limit Theorems for Continuous Local Martingales

Recall that $m_{loc}^{(c)}$ and $m_{loc}^{(2)}$ denote the classes of local continuous and locally square integrable martingales $M = \{M_t, t \geq 0\}$, respectively, with respect to a filtration $\{\mathcal{F}_t, t \geq 0\}$ on a complete probability space $(\Omega, \mathcal{F}, P)$. We assume that the filtration satisfies the "usual" conditions as earlier (cf. [10]). Suppose $M_0 = 0$. It is clear that $m_{loc}^{(c)} \subset m_{loc}^{(2)}$. For $M \in m_{loc}^{(2)}$ denote by $\langle M\rangle = \{\langle M\rangle_t, t \geq 0\}$ the quadratic characteristic with respect to $\{\mathcal{F}_t\}$ and P.

THEOREM 1.49
Let $M = \{M_t,\ t \geq 0\} \in m_{loc}^{(c)}$ and suppose that

$$\mathcal{L}\left[Q_t^2\langle M\rangle_t\,|P\right] \xrightarrow{w} \mathcal{L}\left(K^2\,|P\right) \tag{1.5.26}$$

as $t \to \infty$ where $Q_t \to 0$ as $t \to \infty$, K is a random variable such that $P(K > 0) = 1$, $\mathcal{L}(\cdot|P)$ denotes the distribution under P and $\xrightarrow{w}$ denotes the weak convergence. Then

$$\mathcal{L}\left[Q_tM_t\,|P\right] \xrightarrow{w} \mathcal{L}[K\eta|P] \tag{1.5.27}$$

as $t \to \infty$ where η is a Gaussian random variable with mean 0 and variance 1 and K is independent of η under P.

PROOF Condition (1.5.26) and the fact that $P(K > 0) = 1$ imply that $\lim_{t\to\infty}\langle M\rangle_t = \infty$ a.s. $[P]$. From the results on the representation of locally square integrable continuous martingales (cf. [24, Ch. II, Theorem 7.2]), if follows that $M_t = \tilde{W}(\langle M\rangle_t)$, $t \geq 0$ where $\{\tilde{W}_t, t \geq 0\}$ is a standard Wiener process with respect to the filtration $\{\mathcal{F}_{\tau_t}, t \geq 0\}$ under P where $\tau_t = \inf\{s : \langle M\rangle_s > t\}$. Hence, we have the representation

$$Q_t M_t = Q_t \tilde{W}(\langle M\rangle_t), \; t \geq 0 . \tag{1.5.28}$$

Suppose that, instead of (1.5.26), we have the more stringent condition

$$Q_t^2 \langle M\rangle_t \xrightarrow{P} K^2 \text{ as } t \to \infty \tag{1.5.29}$$

under P. Then (1.5.27) follows from the results in [20]. Let us now consider the general condition (1.5.26).

By a theorem of Skorokhod (cf. [24, Ch. I, Theorems 2.7 and 4.2]), there exist a probability space $(\hat{\Omega}, \hat{\mathcal{F}}, \hat{P})$ and random variables ν_t and $\hat{K}$ and a standard Wiener process $\hat{W} = (\hat{W}_t, t \geq 0)$ defined on it such that

$$\mathcal{L}\left(\left(\tilde{W}, \langle M\rangle_t \,|P\right) = \mathcal{L}\left(\left(\hat{W}, \nu_t\right)\middle|\hat{P}\right), \; t \geq 0$$

and

$$\mathcal{L}(K|P) = \mathcal{L}\left(\hat{K}|\hat{P}\right), \; \hat{P}\left\{\lim_{t\to\infty} Q_t^2 \nu_t = \hat{K}^2\right\} = 1 .$$

Hence, in view of the first part of the proof, it follows that

$$\mathcal{L}\left(Q_t \hat{W}(\nu_t)\,|\hat{P}\right) \xrightarrow{w} \mathcal{L}\left(\hat{K}\hat{\eta}|\hat{P}\right) \text{ as } t \to \infty$$

under the condition (1.5.29) where $\hat{\eta}$ is a Gaussian random variable, with mean 0 and variance 1 with respect to $\hat{P}$, that is independent of $\hat{K}$. Since

$$\mathcal{L}\left(Q_t \tilde{W}(\langle M\rangle_t)\,|P\right) = \mathcal{L}\left(Q_t \hat{W}(\nu_t)\,|\hat{P}\right) ,$$

we obtain (1.5.26). ∎

A multidimensional version of Theorem 1.49 is as follows.

Let $M = (M^{(1)}, \cdots, M^{(k)})'$ where $M^{(i)} = \{M_t^{(i)}, \; t \geq 0\} \in m_{loc}^{(2)}$. Denote by $\langle M, M\rangle = \{\langle M, M\rangle_t, \; t \geq 0\}$ the matrix process with elements $\langle M^{(i)}, M^{(j)}\rangle = (\langle M^{(i)}, M^{(j)}\rangle_t, \; t \geq 0)$ that are the quadratic characteristics between $M^{(i)}$ and $M^{(j)}$ with respect to $\{\mathcal{F}_t, \; t \geq 0\}$ under P.

THEOREM 1.50
Suppose that $M \in m_{loc}^{(c)}$ and

$$\mathcal{L}\left(Q_t^2\langle M, M\rangle_t | P\right) \xrightarrow{w} \mathcal{L}\left(K^2 | P\right) \tag{1.5.30}$$

as $t \to \infty$ where $Q_t \to 0$ as $t \to \infty$ and K is a symmetric positive definite random matrix such that $P(\lambda' K \lambda > 0) = 1$ for all $\lambda \in R^k$, $|\lambda| \neq 0$. Then

$$\mathcal{L}(Q_t M_t | P) \xrightarrow{w} \mathcal{L}(K\eta | P) \tag{1.5.31}$$

as $t \to \infty$ where $\eta = (\eta_1 \cdots, \eta_k)'$ is a random vector with independent $N(0, 1)$ components under P and are independent of the matrix K.

PROOF It is sufficient to prove that as $t \to \infty$

$$\mathcal{L}(Q_t (M_t, \lambda) | P) \xrightarrow{w} \mathcal{L}((K\eta, \lambda) | P) \tag{1.5.32}$$

for $\lambda \in R^k$, $|\lambda| \neq 0$ where (a, b) denotes the inner product of vectors $a, b \in R^k$. Note that $(M, \lambda) = \{(M_t, \lambda),\ t \geq 0\} \in m_{loc}^{(c)}$ for all $\lambda \in R^k$ with quadratic characteristic $\lambda' \langle M, M\rangle \lambda$. Condition (1.5.30) implies that

$$\mathcal{L}\left(Q_t^2 \lambda' \langle M, M\rangle_t \lambda | P\right) \xrightarrow{w} \mathcal{L}\left(\lambda' K^2 \lambda | P\right) \tag{1.5.33}$$

for all $\lambda \in R^k$. Hence, by Theorem 1.49, it follows that

$$\mathcal{L}(Q_t (M_t, \lambda) | P) \xrightarrow{w} \mathcal{L}\left(\left(\lambda' K^2 \lambda\right)^{1/2} \eta^{(\lambda)} | P\right) \tag{1.5.34}$$

as $t \to \infty$ for all $\lambda \in R^k$, $|\lambda| \neq 0$ where $\eta^{(\lambda)}$ is $N(0, 1)$ under P and is independent of $\lambda' K^2 \lambda$. It is easy to see that

$$\mathcal{L}((K\eta, \lambda) | P) = \mathcal{L}\left(\left(\lambda' K^2 \lambda\right)^{1/2} \eta^{(\lambda)} | P\right)$$

for all $\lambda \in R^k$, $|\lambda| \neq 0$. Hence, the relation (1.5.31) follows from (1.5.33).
∎

REMARK 1.47 It is easy to see by analogous arguments that

$$\mathcal{L}\left(Q_t M_t,\, Q_t^2 \langle M_t \rangle | P\right) \xrightarrow{w} \mathcal{L}\left(K\eta,\, K^2 | P\right) \tag{1.5.35}$$

as $t \to \infty$ under the conditions of Theorem 1.49 in the one-dimensional case and

$$\mathcal{L}\left(Q_t M_t,\, Q_t^2 \langle M, M \rangle_t | P\right) \xrightarrow{\omega} \mathcal{L}\left(K\eta,\, K^2 | P\right) \tag{1.5.36}$$

as $t \to \infty$ under the conditions of Theorem 1.50 in the multidimensional case. ∎

Some Additional Results on Stochastic Integrals with Respect to Square Integrable Local Martingales

LEMMA 1.8

Every $\{X_t\} \in m_{loc}^{(2)}$, the space of regular right continuous L^2-martingales, converges a.s. to a finite limit on the set $[\langle X \rangle_\infty < \infty]$.

The proof of this lemma follows as in the discrete case (cf. [49, Proof VII-2.3]) by noting that $\{\langle X \rangle_t\}$ is locally bounded since it is regular right continuous and predictable and that for every stopping time τ, $\langle X_t^\tau \rangle = \langle X_{t \wedge \tau} \rangle$ is the natural increasing process associated with the stopped local martingale $\{X_t^\tau\} = \{X_{t\wedge\tau}\}$.

THEOREM 1.51

Let $\{\phi_t\}$ be a predictable and locally bounded process and suppose that $\{M_t\} \in m_{loc}^{(2)}$. Then, for every $\alpha > \frac{1}{2}$,

$$\int_0^t \phi_s dM_s = o\left(\left[\int_0^t \phi_s^2 d\langle M \rangle_s\right]^\alpha\right) + O(1) \ a.s.$$

PROOF Let $X_t = \int_0^t \phi_s dM_s$. Then $\{X_t\} \in m_{loc}^{(2)}$ and

$$\langle X \rangle_t = \int_0^t \phi_s^2 d\langle M \rangle_s \,.$$

Hence, by Lemma 1.8, it follows that X_t converges to a finite limit a.s. on $[\int_0^t \phi_s^2 d\langle M\rangle_s < \infty]$. Let $A_t = \langle X\rangle_t$ and define

$$Y_t = \int_0^t \frac{\phi_s}{1+A_s^\alpha} dM_s .$$

Then $\{Y_t\}$ is local L^2-martingale and

$$\langle Y\rangle_t = \int_0^t \frac{\phi_s^2}{\left(1+A_s^\alpha\right)^2} dA_s < \infty \text{ a.s.}$$

Hence, Y_t converges to a finite limit a.s. Applying a stochastic variant of Kronecker's lemma (cf. [43]), it follows that

$$\frac{X_t}{A_t^\alpha} \to 0 \text{ a.s. on } [A_\infty = \infty] .$$

∎

REMARK 1.48 The above result is valid of A_t^α if replaced by $f(A_t)$ where f is any nonnegative increasing function such that

$$\int_0^\infty \frac{1}{[1+f(t)]^2} dt < \infty .$$

∎

THEOREM 1.52
Let $\{X_t\} \in m_{loc}^{(2)}$. Then, for every $\delta > 0$.

$$[X]_t = o\left(\langle X\rangle_t^{1+\delta}\right) + O(1) \quad a.s.$$

PROOF Assume that $X_0 = 0$. Let $A_t = \langle X \rangle_t$ and $V_t = [X]_t$. Since $\{A_t\}$ is regular right-continuous and predictable, there exist stopping times τ_N such that $A_t^N = A_{t \wedge \tau_N} \le N$ for all t and $[\tau_N = \infty] \supset \{A_\infty < N\}$. Then, for the stopped martingale $\{X_t^N\} = \{X_t^{\tau_N}\}$, $\langle X^N \rangle_t = A_t^N$, and $[X^N]_t = V_t^N \equiv V_{t \wedge \tau_N}$. Hence, it follows, from the identity $EV_t^N = E(X_t^N)^2 = EA_t^N$ that $\int_{[\tau_N = \infty]} V_\infty^N dP \le N$ which implies that $V_\infty = V_\infty^N < \infty$ a.s. on $[\tau_N = \infty]$. Let $N \uparrow \infty$. Then $V_\infty < \infty$ a.s. on $[A_\infty < \infty]$.

Consider the nonegative submartingale

$$Y_t = \int_0^t \frac{dV_s}{1 + A_s^{1+\delta}} .$$

Let λ be the Doleans measure of $\{X_t^2\}$ (cf. Appendix A, [47]). Then

$$E(Y_T) = \int I_{[0,T]} \left(1 + A^{1+\delta}\right)^{-1} d\lambda$$

$$= E \int_0^T \frac{dA_s}{1 + A_s^{1+\delta}} \le C < \infty$$

for all T. Hence, by the Doob's inequality,

$$P\left\{ \sup_{0 \le t \le T} |Y_t| \ge N \right\} \le \frac{C}{N} ,$$

which proves that Y_T converges a.s. to a finite limit. But

$$Y_t = \int_0^t \frac{d(V_s - A_s)}{1 + A_s^{1+\delta}} + \int_0^t \frac{dA_s}{1 + A_s^{1+\delta}} .$$

Since Y_t and the second term on the right-hand side of the above equation converge to finite limits, so does the first term on the right-hand side. Since $V_s - A_s$ is a local L^2-martingale, we obtain by Kronecker's lemma (cf. [43]), that

$$\left(1 + A_t^{1+\delta}\right)^{-1} (V_t - A_t) \to 0 \text{ a.s. on } [A_\infty = \infty)$$

and hence,

$$\left(1 + A_t^{1+\delta}\right)^{-1} V_t \to 0 \text{ a.s on } [A_\infty = \infty] .$$

∎

THEOREM 1.53
Let $\{\theta_t\}$ be a predictable right-continuous nondecreasing process. Let $\{\phi_t\}$ be predictable and locally bounded. Suppose that $\{N_t\} \in m^{2,d}$, the space of purely discontinuous L^2-martingales, that is, regular right-continuous L^2-martingales $\{M_t\}$ which are strongly orthogonal to every $\{J_t\} \in m_c^2$, that is, $\{M_t J_t\}$ is a martingale where m_c^2 is the space of continuous L^2-martingales. Then

$$\int_0^t \phi_s \Delta\theta_s dN_s = \sum_{s \le t} \phi_s \Delta\theta_s \Delta N_s$$

for all $t \geq 0$.

For proof, see [9, p. 265].

1.6 Semimartingales

DEFINITION 1.39 *A stochastic process $X = (X_t, \mathcal{F}_t(P), t \geq 0)$ is called a semimartingale if it admits a representation*

$$X_t = X_0 + M_t + A_t \quad a.s.\ [P] \tag{1.6.1}$$

with $M \in m_{loc}(P)$, $A \in \mathcal{V}(P)$, and $M_0 = A_0 = 0$.

The above representation is not unique in general. If there is another representation

$$X_t = X_0 + \bar{M}_t + \bar{A}_t \quad \text{a.s.}[P]$$

with $\bar{M} \in m_{loc}(P)$and $\bar{A} \in \mathcal{V}(P)$, then it can be shown that the continuous components of $M^{(c)}$ and $\bar{M}^{(c)}$ are the same (cf. [48]). The component $M^{(c)}$

is called the *continuous martingale component* of the semimartingale X and is denoted b $X^{(c)}$. As before, we define the *quadratic variation* of X over $[0, t]$ to be the process

$$[X, X]_t = \langle X^{(c)}, X^{(c)} \rangle_t + \sum_{s \leq t} (\Delta X_s)^2 \tag{1.6.2}$$

where $\Delta X_s = X_s - X_{s-}$.

Another representation for a semimartingale: Let X be a semimartingale. Then

$$X = X^{(1)} + X^{(2)} \tag{1.6.3}$$

where

$$X_t^{(1)} = X_0 + \sum_{s \leq t} \Delta X_s \ I\,(|\Delta X_s| > 1)\,, \tag{1.6.4}$$

and

$$X_t^{(2)} = X_t - X_t^{(1)}\,. \tag{1.6.5}$$

Here $I(A)$ denotes the indicator function of the set A. Since $X^{(1)} \in \mathcal{V}(P)$, the process $X^{(2)}$ is also a semimartingale of the form

$$X^{(2)} = M^{(2)} + A^{(2)}\,.$$

Note that the semimartingale $X^{(2)}$ has bounded jumps, that is, $|\Delta X^{(2)}| \leq 1$. Hence,

$$\left[\sum_{s \leq \cdot} (\Delta X_s)^2 \right]^{1/2} \in \alpha_{loc}^{+}(P)$$

and the process $A^{(2)} \in \alpha_{loc}(P)$ (cf. [48]). Theorem 1.38 implies that there exists a predictable process $\{\alpha_t, \mathcal{F}_t(P),\ t \geq 0\}$ such that

$$A^{(2)} - \alpha \in m_{loc}(P)\,.$$

Hence,

$$X^{(2)} = M + \alpha$$

where $M \in m_{loc}(P)$ and $\alpha \in \alpha_{loc}(P) \cap \mathcal{P}$, which implies in turn that

$$X_t = X_0 + \sum_{s \leq t} \Delta X_s \ I\,(|\Delta X_s| > 1) + M_t + \alpha_t\,. \tag{1.6.6}$$

If we decompose M into $M^{(c)} + M^{(d)}$, then $X^{(c)} = m^{(c)}$ and hence,

$$X_t = X_0 + \alpha_t + X_t^{(c)} + \sum_{s \le t} \Delta X_s \, I\left(|\Delta X_s| > 1\right) + M_t^{(d)} . \tag{1.6.7}$$

The representations (1.6.6) and (1.6.7) are unique, unlike (1.6.1).

REMARK 1.49 The following processes are examples of semimartingales:

(i) local martingales,

(ii) processes of bounded variation,

(iii) supermartingales (by Doob–Meyer decomposition),

(iv) an adapted RCLL process X_t with independent increments satisfying the property that

$$Q(u,t) = E\left[e^{iuX_t}\right], -\infty < u < \infty$$

as a function of t is a function of bounded variation on any compact set (cf. [27]), and

(v) point (counting) processes.

∎

Stochastic Integral with Respect to a Semimartingale

If $X = X_0 + M + A$ is a semimartingale and H is a predictable locally bounded process, then the stochastic integral $H \bullet X$ is defined to be the process

$$H \bullet X \equiv H \bullet M + H \circ A \tag{1.6.8}$$

and it can be shown that $H \bullet X$ does not depend on the choice of the representation of X.

DEFINITION 1.40 *The quadratic covariation of two semimartingales X and Y is the process*

$$[X,Y] = XY - X_0Y_0 - X_- \bullet Y - Y_- \bullet X \tag{1.6.9}$$

(here X_- denotes the process X_{t-} given the process $X = \{X_t\}$). If $Y = X$, then the process reduces to the quadratic variation of the process X. Note that

$$[X, Y]_0 = 0 , \tag{1.6.10}$$

$$[X, Y] = [X - X_0, Y - Y_0] , \tag{1.6.11}$$

and

$$[X, Y]_t = \frac{1}{4}\left([X + Y, X + Y]_t - [X - Y, X - Y]_t\right) . \tag{1.6.12}$$

Product Formulae for Semimartingales

THEOREM 1.54
If X and Y are semimartingales, then the product XY is a semimartingale and

$$X_t Y_t = \int_{[0,t]} X_{s-} dY_s + \int_{[0,t]} Y_{s-} dX_s + [X, Y]_t ; \tag{1.6.13}$$

that is,

$$d(XY)_t = X_{t-} dY_t + Y_{t-} dX_t + d[X, Y]_t . \tag{1.6.14}$$

REMARK 1.50 For a brief discussion on product formulae in the Lebesgue–Stieltjes calculus, see Appendix D. ▮

THEOREM 1.55
If X is a process of bounded variation and Y is a semimartingale, then

$$d(XY)_t = X_{t-} dY_t + Y_t dX_t . \tag{1.6.15}$$

THEOREM 1.56
If M and N are local martingales such that $MN - [M, N]$ is a square integrable local martingale with $M_0 N_0 - [M, N]_0 = 0$, then

$$M_t N_t - [M, N]_t = \int_0^t M_{s-} dN_s + \int_0^t N_{s-} dM_s . \tag{1.6.16}$$

THEOREM 1.57
If M is a local martingale, then the process

$$M_t^2 - [M, M]_t = 2 \int_{[0,t]} M_{s-} dM_s \tag{1.6.17}$$

is a local martingale that is 0 at $t = 0$.

All the above results are consequences of the following Ito's lemma for semimartingales.

THEOREM 1.58
Suppose X is a process with values in R^d, each of whose components X^i is a semimartingale. Suppose that F is a real-valued function twice continuously differentiable on R^d. Then $F(X_t)$ is a semimartingale and

$$\begin{aligned} F(X_t) &= F(X_0) + \sum_{i=1}^{d} \int_{(0,t]} \frac{\partial}{\partial x^i} F(X_{s-})\, dX_s^i \\ &+ \frac{1}{2} \sum_{i,j=1}^{d} \int_{(0,t]} \frac{\partial^2}{\partial x^i \partial x^j} F(X_{s-})\, d\langle X^{ic}, X^{jc}\rangle_s \\ &+ \sum_{0<s\leq t} \left(F(X_s) - F(X_{s-}) - \sum_{i=1}^{d} \frac{\partial}{\partial x^i} F(X_{s-})\, \Delta X_s^i \right) \end{aligned} \tag{1.6.18}$$

(here X^{ic} denotes the continuous component of X^i and $\Delta X_s^i = X_s^i - X_{s-}^i$).

For proofs, see [12] or [47].

Generalized Ito–Ventzell Formula

Let $X = \{X_t, t \geq 0\}$ be a d-dimensional $\mathcal{F}_t$ -adapted process defined on a probability space $(\Omega, \mathcal{F}, P)$. Suppose X is a continuous local semimartingale, that is, X_t is continuous and has the representation

$$X_t = X_0 + M_t + B_t \tag{1.6.19}$$

for all t a.s. where $E|X_0| < \infty$, $\{M_t\}$ is a continuous local martingale, and $B_t \in \mathcal{V}_{loc}$. Here $\mathcal{V}_{loc}$ is the class of d-dimensional processes B_t, with $B_t = (B_t', \ldots, B_t^d)$ satisfying the conditions (i) $B_0^i = 0$ a.s., $1 \leq i \leq d$, and (ii) $t \to B_t^i(\omega)$ is for almost all ω continuous and of bounded variation in every finite interval.

THEOREM 1.59

Let $X_t = X_0 + M_t + B_t$ be a continuous local semimartingale. Let $G(t, x)$, $(t, x) \in R_+ \times R^d$ be a real-valued function continuous together with its partial derivatives $\frac{\partial G}{\partial t}, \frac{\partial G}{\partial x}, \frac{\partial^2 G}{\partial x^i \partial x^j}$. Then the process $G(t, X_t)$ is a continuous local semimartingale and

$$\begin{aligned} G(t, X_t) - G(0, X_0) = &\int_0^t \frac{\partial G}{\partial s}(s, X_s)\, ds \\ &+ \sum_i \int_0^t \frac{\partial G}{\partial x^i}(s, X_s)\, dM_s^i \\ &+ \sum_i \int_0^t \frac{\partial G}{\partial x^i}(s, X_s)\, dB_s^i \\ &+ \frac{1}{2} \sum_i \int_0^t \frac{\partial^2 G}{\partial x^i \partial x^j}(s, X_s)\, d\langle M^i, M^j \rangle_s . \end{aligned} \tag{1.6.20}$$

For a proof, see [33, p. 86] or [37, p. 92].

Convergence of Quadratic Variation of Semimartingales

Let $X^{(n)}$ be a d-dimensional semimartingale defined on a probability space $(\Omega^n, \mathcal{F}^n, P^n)$ with the filtration $\{\mathcal{F}_t^n\}$. We call $(\Omega^n, \{\mathcal{F}_t^n\}, \mathcal{F}^n, P^n)$ a *stochastic basis*. Let $[X^{(n)}, X^{(n)}]$ denote the $R^d \otimes R^d$-valued process whose components are the co-quadratic variations $[X^{n,j}, X^{n,k}]$. Let X be another d-dimensional semimartingale on a stochastic basis $(\Omega, \{\mathcal{F}_t\}, \mathcal{F}, P)$ with $[X, X]$ the associated process as above.

Let h be a truncation function (cf. [30]). Denote by $(B^{(n)}(h), C^n, \nu^n)$ and $(B(h), C, \nu)$ the local characteristics of X_n and X, respectively. For the definition of local characteristics for a semimartingale, see the discussion later in this section (cf. [30]).

THEOREM 1.60 **[30, p. 341]**
Consider the following conditions:

$$\begin{aligned} &(i)\quad X_n \xrightarrow{\mathcal{L}} X\,, \\ &(ii)_h\ \lim_{b\uparrow\infty} \sup_n P^n\left[\,Var\left(B^{n,j}(h)\right)_t > b\right] = 0 \\ &\qquad\qquad \text{for all } t > 0,\ 1 \le j \le d\,. \end{aligned} \tag{1.6.21}$$

(a) *If (i) holds, then for all the truncations functions h, the conditions $(ii)_h$ are equivalent.*

(b) *If (i) and $(ii)_h$ hold for some truncation function h, then*

$$\left(X^{(n)}, \left[X^{(n)}, X^{(n)}\right]\right) \xrightarrow{\mathcal{D}} (X, [X, X]) \quad \text{as } n \to \infty \tag{1.6.22}$$

(in the space $D(R^d \times (R^d \otimes R^d))$ with the Skorokhod topology).

In particular,

$$\left[X^{(n)}, X^{(n)}\right] \xrightarrow{\mathcal{L}} [X, X] \quad \text{as } n \to \infty\,. \tag{1.6.23}$$

(Here $Var(B(h))_t$ denotes the total variation of the process $B(h)$ over $[0, t]$).

REMARK 1.51 The assumption that the limiting process X is a semimartingale is not needed in the above result. The conditions (i) and $(ii)_h$ imply that the limit X is a semimartingale. ∎

DEFINITION 1.41 *A semimartingale* $\{X_t, t \geq 0\}$ *is called a special semimartingale if there is a decomposition*

$$X_t = X_0 + M_t + A_t, \quad M \in m_{loc}, A \in \mathcal{V}(P)$$

where A *is locally of integrable variation with* $M_0 = A_0 = 0$.

THEOREM 1.61

The following conditions are equivalent:

(a) *X is a special semimartingale,*

(b) *the increasing process* $\sum_{0<s\leq t} \Delta X_s^2$ *is locally integrable,*

(c) *for every decomposition of X as in the definition, the process A is locally of integrable variation, and*

(d) *there is a decomposition of the above form for which A is predictable.*

Further, if X is a semimartingale, the decomposition as given in (d) with A predictable is unique.

PROOF See [12, p. 148]. ∎

Yoerup's Theorem for Local Martingales

Suppose $(\Omega, \mathcal{F}, P)$ is a probability space with a complete right-continuous filtration $\{\mathcal{F}_t, t \geq 0\}$. Suppose Q is a probability measure on $(\Omega, \mathcal{F})$ and Q is equivalent to P. Then the evanescent sets and predictable σ-algebra are both the same under Q and P. Let $M_\infty = \frac{dQ}{dP}$ and $M_t = E_P[M_\infty | \mathcal{F}_t]$. We can assume that $\{M_t\}$ is RCLL and uniformly integrable with respect to P. Then $\{M_t\}$ is a positive martingale and M_T is the density of the restriction of Q to $\mathcal{F}_T$ with respect to P for any stopping time T.

LEMMA 1.9

A process $\{X_t, t \geq 0\}$ *is a local martingale under the measure Q if and only if the process* $\{X_t M_t, t \geq 0\}$ *is a local martingale under P.*

REMARK 1.52 The process $\{M_t^{-1}, t \geq 0\}$ is a martingale under the measure Q. ∎

THEOREM 1.62
A process $\{X_t, t \geq 0\}$ is a semimartingale under the measure Q if and only if it is a semimartingale under P.

COROLLARY 1.12
If $\{X_t, t \geq 0\}$ is a local martingale under P, then $\{X_t, t \geq 0\}$ is a semimartingale under Q.

THEOREM 1.63
(Yoeurp) (i) Suppose $\{X_t\}$ is a P-local martingale with $X_0 = 0$ a.s. Then the process

$$X_t - \int_0^t M_s^{-1} d[X, M]_s. \quad t \geq 0$$

is a local martingale under Q. (ii) The process X is a special semimartingale under Q if $\langle X, M\rangle_t$ exists, and then the canonical decomposition of X under Q is

$$X_t = \left(X_t - \int_0^t M_s^{-1} d\langle X, M\rangle_s \right) + \int_0^t M_s^{-1} d\langle X, M\rangle_s . \tag{1.6.24}$$

Hence, the first term is a local martingale under Q and the second term is a predictable process of finite variation.

REMARK 1.53 Suppose P and Q are equivalent probability measures and $\{X_t\}$ is a semimartingale under P. Then $\{X_t\}$ is also a semimartingale under Q. Suppose $\{H_t\}$ is a predictable locally bounded process. Then the stochastic integrals $(H \bullet X)_t$ under P and $(H \bullet X)_t$ under Q are indistinguishable as processes. ∎

THEOREM 1.64 (Exponential formula for semimartingales) (Doléans–Dade)
Suppose $\{X_t, t \geq 0\}$ is a semimartingale with $X_0 = 0$ a.s. Then there exists a

unique semimartingale $\{Z_t, t \geq 0\}$ *such that*

$$Z_t = Z_{0-} + \int_{[0,t]} Z_{s-} dX_s \,. \tag{1.6.25}$$

Furthermore,

$$Z_t = Z_{0-} \exp\left(X_t - \frac{1}{2}\langle X^{(c)}, X^{(c)}\rangle_t\right) \Pi_{0 \leq s \leq t} (1 + \Delta X_s)\, e^{-\Delta X_s} \tag{1.6.26}$$

for $t \geq 0$*, and the infinite product is convergent a.s. (here* $X^{(c)}$ *is the continuous component of* X *and* $\Delta X_s = X_s - X_{s-}$*).*

REMARK 1.54 (i) If $X_0 = 0$ and $Z_{0-} = Z_0$, then

$$Z_t = Z_0 + \int_0^t Z_{s-} dX_s \,. \tag{1.6.27}$$

(ii) If

$$V_t = \Pi_{0 \leq s \leq t} (1 + \Delta X_s)\, e^{-\Delta X_s} \,, \tag{1.6.28}$$

then V_t is a purely discontinuous process of finite variation.

(iii) We write $Z_t = Z_{0-}\ \mathcal{E}(X)_t$ (exponential formula).

(iv) It can be shown that if X and Y are semimartingales, then

$$\mathcal{E}(X)\mathcal{E}(Y) = \mathcal{E}(X + Y + [X, Y]) \,. \tag{1.6.29}$$

∎

REMARK 1.55 Suppose $\{X_t\}$ is a semimartingale with $X_0 = 0$. Let $\{Z_t\}$ be the unique solution of the equation

$$Z_t = 1 + \int_0^t Z_{s-} dX_s \tag{1.6.30}$$

with $Z_{0-} = 1$. In view of Theorem 1.64,

$$Z_t = \exp\left(X_t - \frac{1}{2}\langle X^{(c)}, X^{(c)}\rangle_t\right) \Pi_{0\le s\le t} (1 + \Delta X_s) e^{-\Delta X_s}, \qquad (1.6.31)$$

and in our notation $Z_t = \mathcal{E}(X)_t$. Furthermore, if $\{X_t, t \ge 0\}$ is a local martingale, then $\{Z_t, t \ge 0\}$ is also a local martingale and $Z > 0$ a.s. if and only if $\Delta X_s > -1$ a.s. for all s. ∎

Suppose Z is a uniformly integrable positive martingale. Then $Z_\infty = \lim_{t\to\infty} Z_t$ exists a.s. and

$$E[Z_\infty] = E[Z_0] = 1 .$$

Define a new probability measure Q on $(\Omega, \mathcal{F})$ through the relation

$$\frac{dQ}{dP} = Z_\infty .$$

Clearly, Q is equivalent to P if and only if $Z_\infty > 0$ a.s.

***THEOREM 1.65* (Vanschuppen and Wong)**
Define $Z_t = \mathcal{E}(X_t)$ as above. Suppose $\{Z_t\}$ is a uniformly integrable positive martingale. Define Q by

$$\frac{dQ}{dP} = Z_\infty .$$

If $\{N_t\}$ is a local martingale under P and $\langle N, X\rangle_t$ exists under P, then

$$\tilde{N}_t = N_t - \langle N, X\rangle_t$$

is a local martingale under Q. In other words, the Radon–Nikodym derivative of the measures Q' and P' generated by $\tilde{N}_t$ and N_t, respectively, is $\frac{dQ'}{dP'} = Z_\infty$. Furthermore, if N is continuous, then $\tilde{N}$ is continuous and the quadratic variation of $\tilde{N}$ under Q is the same as that of N under P.

REMARK 1.56 This is an extension of Girsanov's theorem for Ito integrals. ∎

Sketch of proof of Theorem 1.65: Note that $\{Z_t\}$ is a uniformly integrable positive martingale and $E[Z_\infty] = 1$. Further, $\frac{dQ}{dP} = Z_\infty$ and

$$Z_t = 1 + \int_0^t Z_{s-} dX_s \,.$$

Hence,

$$\langle N, Z \rangle_t = \int_0^t Z_{s-} d\langle N, X \rangle_s$$

and

$$\langle N, X \rangle_t = \int_0^t Z_{s-}^{-1} d\langle N, Z \rangle_s \,.$$

Hence,

$$\tilde{N}_t = N_t - \langle N, X \rangle_t = N_t - \int_0^t Z_{s-}^{-1} d\langle N, Z \rangle_s$$

is a local martingale under Q by Theorem 1.65.

Example 1.7 [6]
Suppose that $\{N_t\}$ and $\{A_t\}$ are $\mathcal{F}_t$-adapted processes such that for every $\theta \in R$,

$$Y_t^{(\theta)} = \exp\left[\theta N_t - \frac{1}{2}\theta^2 A_t\right] \tag{1.6.32}$$

is a martingale and there exists a neighborhood I of $\theta = 0$ such that

$$\begin{aligned} &\text{(i)} \quad \left|Y_t^{(\theta)}\right| \le a, \\ &\text{(ii)} \quad \left|\tfrac{d}{d\theta} Y_t^{(\theta)}\right| \le b, \quad \text{and} \\ &\text{(iii)} \quad \left|\tfrac{d^2}{d\theta^2} Y_t^{(\theta)}\right| \le c \end{aligned} \tag{1.6.33}$$

where a, b, c are independent of t. Then $\{N_t\}$ and $\{N_t^2 - A_t\}$ are martingales. This can be checked in the following way.

Note that

$$\frac{d}{d\theta} Y_t^\theta \mid_{\theta=0} = N_t$$

and

$$\frac{d^2}{d\theta^2} Y_t^\theta \Big|_{\theta=0} = N_t^2 - A_t \ .$$

Let $s \leq t$ and $A \in \mathcal{F}_s$. Then

$$\int_A E\left[\left(\frac{d}{d\theta} Y_t^\theta\right)_{\theta=0} |\mathcal{F}_s\right] dP = \int_A \left(\frac{d}{d\theta} Y_t^\theta\right)_{\theta=0} dP$$

$$= \left(\frac{d}{d\theta} \int_A Y_t^\theta \ dP\right)_{\theta=0}$$

$$= \left(\frac{d}{d\theta} \int_A Y_s^\theta \ dP\right)_{\theta=0}$$

$$= \int_A \left(\frac{d}{d\theta} Y_s^\theta\right)_{\theta=0} dP \ .$$

Hence,

$$E\left[\left(\frac{d}{d\theta} Y_t^\theta\right)_{\theta=0} |\mathcal{F}_s\right] = \left(\frac{d}{d\theta} Y_s^\theta\right)_{\theta=0} \quad \text{a.s.} \ [P] \ .$$

which shows that $\{N_t\}$ is a martingale. Similarly, it can be shown that $\{N_T^2 - A_t\}$ is a martingale by using $\left(\frac{d^2}{d\theta^2} Y_t^\theta\right)_{\theta=0}$. □

Example 1.8 **[6]**

Suppose $\{N_t\}$ and $\{A_t\}, t \geq 0$ are continuous stochastic processes where $N_0 = 0$ a.s. Then

(i) $\{N_t, t \geq 0\}$ is a continuous local martingale with $\langle N, N\rangle_t = A_t$ if and only if

(ii) $Y_t^\theta = \exp[\theta N_t - \frac{1}{2}\theta^2 A_t]$ is a continuous local martingale for every $\theta \in R$.

This can be seen as follows.

Suppose (i) holds. Then, by the differentiation rule, it follows that

$$dY_t^\theta = Y_t^\theta d\left(\theta N_t - \frac{1}{2}\theta^2 A_t\right) + \frac{1}{2}Y_t^\theta \theta^2 d\langle N, N\rangle_t \tag{1.6.34}$$

and hence,

$$Y_t^\theta = 1 + \int_0^t \theta Y_s^\theta dN_t \,. \tag{1.6.35}$$

Therefore, Y_t^θ is a continuous local martingale of the form given in (ii).

Conversely, suppose that (ii) holds. Define the stopping time

$$T_n = \inf\{t : |N_t| \geq n \text{ or } |A_t| \geq n\} \,.$$

Then

$$Y_{t\wedge T_n}^\theta = \exp\left[\theta N_{t\wedge T_n} - \frac{1}{2}\theta^2 A_{t\wedge T_n}\right]$$

and for θ in some neighborhood of zero and $Y_{t\wedge T_n}^\theta$ satisfies the conditions in Example 1.7. Hence, $\{N_{t\wedge T_n}\}$ and $\{N_{t\wedge T_n}^2 - A_{t\wedge T_n}\}$ are continuous martingales. Since $T_n \uparrow \infty$ as $n \to \infty$, the result follows. □

REMARK 1.57 Let $N_t = (N_t^1, \ldots, N_t^m), t \geq 0$ be continuous stochastic processes with values in R^m and $N_0 = 0$ a.s. Let $A_t = (A_t^{ij}), t \geq 0,\ 1 \leq i,\ j \leq m$ be a continuous matrix-valued process. Then $\{N_t\}$ is a continuous R^m-valued local martingale with $\langle N^i, N^j\rangle_t = A_t^{ij}$ if and only if for all $\theta \in R^m$,

$$Y_t^\theta = \exp\left\{(\theta, N_t) - \frac{1}{2}(\theta, A_t\theta)\right\} \tag{1.6.36}$$

is a continuous local martingale. ∎

THEOREM 1.66 **(Vector version of Theorem 1.65)**
Suppose that $\{X_t\}$ is a continuous local martingale under P such that $Z_t = \mathcal{E}(X_t)$ is a uniformly integrable positive martingale under P. Let Q be defined by $dQ/dP = Z_\infty$ on $(\Omega, \mathcal{F})$. Let $\{N_t\} = \{(N_t^1, \ldots, N_t^m)\}$ be an m-dimensional continuous local martingale with values in R^m under P. Let $\tilde{N}_t^i = N_t^i - \langle N^i, X\rangle_t$. Then $\{\tilde{N}_t\} = \{(\tilde{N}_t^1, \ldots, \tilde{N}_t^m)\}$ is a continuous local martingale under Q. Furthermore, the quadratic variation of $\tilde{N}$ under Q is equal to the quadratic variation of N under P, that is,

$$\langle \tilde{N}^i, \tilde{N}^j\rangle_t^Q = \langle N^i, N^j\rangle_t^P .$$

COROLLARY 1.13
(Girsanov's theorem) *Suppose that $\{B_t\} = \{(B_t^1, \ldots, B_t^m)\}$, $0 \leq t \leq T$ is an m-dimensional Brownian motion on a probability space $(\Omega, \mathcal{F}, P)$ with filtration $\{\mathcal{F}_t\}$. Let $f : \Omega \times [0, T] \to R^m$ be a predictable process such that*

$$\int_0^T \|f_t\|^2 \, dt < \infty \quad a.s. \tag{1.6.37}$$

Let

$$\xi_0^t(f) = \exp\left[\sum_{i=1}^m \int_0^t f_s^i B_s^i - \frac{1}{2}\int_0^t \|f_s\|^2 \, ds\right], \quad 0 \leq t \leq T \tag{1.6.38}$$

and suppose that

$$E\left[\xi_0^T(f)\right] = 1 . \tag{1.6.39}$$

Define Q by $dQ/dP = \xi_0^T(f)$. Then $\{\tilde{B}_t\} = \{(\tilde{B}_t^1, \ldots, \tilde{B}_t^m)\}$ is a Brownian motion on $(\Omega, \mathcal{F}, Q)$ with filtration $\{\mathcal{F}_t\}$ where

$$\tilde{B}_t^i = B_t^i - \int_0^t f_s^i \, ds . \tag{1.6.40}$$

REMARK 1.58 It can be shown that $E[\xi_0^T(f)] = 1$ if $E[\exp \frac{1}{2}(\int_0^T \|f_s\|^2 ds)] < \infty$. This condition cannot be weakened in general. ∎

Proof of Girsanov's Theorem (Sketch for $m \mathcal{D}$ 1)

Let $N = B$ in Theorem 1.65 where B is the Wiener process on $[0, T]$. Clearly, B is a continuous local martingale (in fact a continuous martingale) under P. Let $X = \int_0^t f dB$. Then X is a continuous martingale. Furthermore,

$$\tilde{N}_t = N_t - \langle N, X\rangle_t = B_t - \langle B, X\rangle_t = B_t - \int_0^t f ds \tag{1.6.41}$$

is a continuous local martingale under Q since $\frac{dQ}{dP} = Z_T$, $Z_t = \mathcal{E}(X)_t$ is a uniformly integrable positive martingale on $[0, T]$ and $E[Z_T] = 1$. Note that

$$Z_t = \exp\left[\int_0^t f dB - \frac{1}{2}\int_0^t f^2 dt\right] . \tag{1.6.42}$$

Furthermore, $\langle \tilde{N}, \tilde{N}\rangle_t$ under $Q = \langle N, N\rangle_t$ under $P = t$ since $N = B$. Hence, $\tilde{N}$ is a continuous martingale with $\langle \tilde{N}, \tilde{N}\rangle$ under $Q = t$. Therefore $\tilde{N}$ is a Wiener process on $(\Omega, \mathcal{F}, Q)$ and

$$\tilde{N}_t = B_t - \int_0^t f\, ds . \tag{1.6.43}$$

Stochastic Differential Equations

DEFINITION 1.42 *Let $(\Omega, \mathcal{F}, P)$ be a complete probability space and $\{\mathcal{F}_t\}$ be a right-continuous complete filtration. Let $\mathcal{F}_\infty = \vee_t \mathcal{F}_t$. Let Z be a R^n-valued semimartingale with $Z_0 = 0$. Let $\mathcal{D}$ be the space of RCLL functions on R_+ with values in R^d. Let $f : R_+ \times \Omega \times \mathcal{D} \to L(R^m, R^d)$ (the set of $n \times d$ matrices) such that*

$$(foX)(t, \omega)x = f\,(t, \omega, X_.(\omega))$$

is locally bounded and predictable if X is RCLL adapted. Let H be a given RCLL adopted process. A process $\{X_t\}$ is said to be a strong solution of the *stochastic differential equation*

$$X_t = H_t + \int_0^t (foX)_s dZ_s \text{ a.s. } t \geq 0 \tag{1.6.44}$$

if X is RCLL adapted satisfying (1.6.44).

Symbolically,

$$dX_t = dH_t + (foX)_t dZ_t,\ t \geq 0\,. \tag{1.6.45}$$

THEOREM 1.67

($n = d = 1$). *Suppose Z is a semimartingale with $Z_0 = 0$ a.s. and H is an adapted RCLL process. Then the stochastic differential equation*

$$X_t(\omega) = H_t(\omega) + \int_0^t f(s, \omega, X_.(\omega))\, dZ_s(\omega) \tag{1.6.46}$$

has a unique strong solution X if f satisfies the definition given above and, for any two RCLL adapted processes X and Y,

$$|f(s, \omega, X_.(\omega)) - f(s, \omega, Y_.(\omega))| \leq K\left(|X(\omega) - Y(\omega)|_s^*\right) \tag{1.6.47}$$

where

$$|X(\omega)|_s^* = \sup_{0 \leq u \leq s} |X_u(\omega)|\ . \tag{1.6.48}$$

Random Measures

Let $(\Omega, \mathcal{F})$ be a measurable space and E be a Borel set in a compact metric space. Let $\mathcal{E}$ be a σ-algebra of Borel sets in E. Let $\mathcal{W}$ and $\mathcal{P}$ be the well measurable and predictable σ-algebras.

Let $\tilde{\Omega} = \Omega \times R_+ \times E$, $\tilde{E} = R_+ \times E$, $\tilde{\mathcal{W}} = \mathcal{W} \otimes \mathcal{E}$, $\tilde{\mathcal{P}} = \mathcal{P} \otimes \mathcal{E}$, and $\tilde{\mathcal{E}} = \mathcal{B} \otimes \mathcal{E}$ where $R_+ = (0, \infty)$ and $\mathcal{B}$ is the σ-algebra of Borel subsets of R_+. Here $\otimes$ denotes the product σ-algebra.

DEFINITION 1.43 *A nonnegative function $\mu = \mu(\omega, \tilde{A})$, $\omega \in \Omega$, $\tilde{A} \in \tilde{\mathcal{E}}$ is called a random measure on $\tilde{E}$ if* *(i)* $\mu(\cdot, \tilde{A})$ *is* $\mathcal{F}$*-measurable for each* $\tilde{A} \in \tilde{\mathcal{E}}$*, and*

(ii) $\mu(\omega, \cdot)$ *is a* σ*-finite measure on* $(\tilde{E}, \tilde{\mathcal{E}})$ *for each* $\omega \in \Omega$*.*

REMARK 1.59 A random measure μ is said to be *integer-valued* if

(i) $\mu(\omega, \tilde{A}) \in \{0, 1, 2, \ldots; +\infty\}, \tilde{A} \in \tilde{\mathcal{E}}$, and

(ii) $\mu(\omega, \{t\}, E) \leq 1, \ t > 0.$

∎

Let $Z = Z(\omega, t, x)$ be a nonnegative $(\mathcal{F} \otimes \mathcal{B} \otimes \tilde{\mathcal{E}})$-measurable function. Define a new random measure

$$(Zo\mu)(\omega, dt, dx) = Z(\omega, t, x)\mu(\omega, dt, dx)$$

and a nondecreasing process $Z * \mu$ by

$$\begin{aligned}(Z * \mu)_t &= (Zo\mu)(\omega, (0, t], E) \\ &= \int_{(0,t]\times E} Z(\omega, s, y)\mu(\omega, ds, dy). \qquad (1.6.49)\end{aligned}$$

DEFINITION 1.44 *A random measure* μ *is said to be well measurable (predictable) if the random process* $Z * \mu$ *is well measurable (predictable) for each* $\tilde{\mathcal{W}}$*-measurable* ($\tilde{\mathcal{P}}$*-measurable) nonnegative function* Z *on* $\tilde{\Omega}$*.*

Given a probability measure P on $(\Omega, \mathcal{F})$ and a random measure μ, define a measure $M_\mu^P = M_\mu^P(d\omega, dt, dx)$ on $(\tilde{\Omega}, \tilde{\mathcal{W}})$ by the relation

$$M_\mu^P(Z) = E[Z * \mu]_\infty \qquad (1.6.50)$$

where Z runs through the set of nonnegative $\tilde{\mathcal{W}}$-measurable functions on $\tilde{\Omega}$. Then

$$M_\mu^P(Z) = E\left[\int_{(0,\infty)\times E} Z(\omega, s, y)\mu(\omega, ds, dy)\right]$$

$$= \int_{\Omega\times(0,\infty)\times E} Z(\omega, s, y)\mu(\omega, ds, dy)P(d\omega) .$$

THEOREM 1.68 [25]
Suppose the random measure μ is such that M_μ^P is $\tilde{\mathcal{P}} - \sigma$-finite, that is, the restriction of the measure M_μ^P to $(\tilde{\Omega}, \tilde{\mathcal{P}})$ is σ-finite. Then there exists a unique (to within a set of P-measure zero) predictable random measure $\nu = \nu(\omega, dt, dx)$ such that

$$M_\mu^P(Z) = M_\nu^P(Z) \tag{1.6.51}$$

for every nonnegative $\tilde{\mathcal{P}}$-measurable function Z. If, in addition μ is well measurable, then for each nonnegative $\mathcal{P}$-measurable function Z such that $Z * \mu \in \alpha_{loc}(P)$,

$$Z * \mu - Z * \nu \in m_{loc}(P) . \tag{1.6.52}$$

If μ is an integer-valued random measure μ with $\tilde{\mathcal{P}} - \sigma$-finite measure M_μ^P, then there exists a modification of the measure ν such that $\nu(\omega, \{t\}, E) \le 1$.

REMARK 1.60 The random measure ν described in the above theorem is called the *dual predictable projection* or the *compensator* for μ (cf. Section 1.2). ∎

From the properties of the random measure ν, it follows that

$$\mu(\omega, (0, t], B) - \nu(\omega, (0, t], B)$$

is a local martingale for each $B \in \mathcal{E}$ such that

$$(\nu(\omega, (0, t], B))_{t\ge 0} \in \alpha_{loc}^+(P) .$$

DEFINITION 1.45 *Suppose X is a real-valued process with paths that are right-continuous with left limits. Define*

$$\mu(\omega, dt, dx) = \sum_s I(\Delta X_s \ne 0)\varepsilon_{(s,\Delta X_s)}(dt, dx) \tag{1.6.53}$$

where ε_α is the Dirac measure concentrated at a. The random measure μ is called the saltus measure of the process X.

DEFINITION 1.46 *If X is a Markov process, then the compensator of the saltus measure is called the Lévy system.*

Stochastic Integral with Respect to the Measure $\mu - \nu$

Suppose μ is a well-measurable integer-valued random measure and that M_μ^P is σ-finite. Then there exists the compensator ν of the random measure μ by Theorem 1.68. For simplicity, we suppress the variable ω in the following discussion. We would like to define the stochastic integral

$$\int_0^t \int_E U(s,x)(\mu - \nu)(ds, dx), \quad t \geq 0, \tag{1.6.54}$$

hereafter denoted by $U * (\mu - \nu)$ for a collection of functions U which are $\tilde{\mathcal{P}}$-measurable in such a way that $U * (\mu - \nu) \in m_{loc}(P)$. If each of the processes

$$|U| * \mu \text{ and } |U| * \nu$$

belong to $\alpha^+(P)$, then we define

$$U * (\mu - \nu) = U * \mu - U * \nu . \tag{1.6.55}$$

We further require that

$$\Delta(U * (\mu - \nu))_t = \int_E U(s,x)\mu(\{t\}, dx) - \int_E U(s,x)\nu(\{t\}, dx) . \tag{1.6.56}$$

Let

$$a_t = \nu(\{t\}, E) , \tag{1.6.57}$$

$$\hat{U}_t = \int_E U(t,x)\nu(\{t\}, dx) , \tag{1.6.58}$$

$$G(t,U) = \int_0^t \int \frac{\left(U(s,x) - \hat{U}_s\right)^2}{1 + \left|U(s,x) - \hat{U}_s\right|} d\nu$$

$$+\sum_{s\leq t}\frac{\hat{U}_s^2}{1+\left|\hat{U}_s\right|}(1-a_s)\ , \tag{1.6.59}$$

and

$$G^{(i)}(t,U)=\int_0^t\int_E\left|U(s,x)-\hat{U}_s\right|^i d\nu$$

$$+\sum_{s\leq t}\left|\hat{U}_s\right|^i(1-a_s),\quad i=1,2\,. \tag{1.6.60}$$

Let $\zeta_{loc}(P)$ denote the class of $\tilde{\mathcal{P}}$-measurable functions $U(\cdot)$ such that, for each $t>0$, the quantities $\hat{U}_t$ are defined and $G(t,U)\in\alpha^+_{loc}(P)$. Define $\zeta^{(i)}_{loc}(P)$ analogously for $G^{(i)}(t,U)$. Then it is known that

$$\zeta^{(i)}_{loc}(P)\subset\zeta_{loc}(P)\,.$$

THEOREM 1.69 **[25]**
Let $U\in\zeta_{loc}(P)$. Then there exists a unique (up to sets of P-measure zero) random process denoted by $U*(\mu-\nu)$ that is a purely discontinuous local martingale for which (1.6.56) holds. Furthermore,

(i) $U\in\zeta^{(1)}_{loc}(P)\Leftrightarrow Var[U*(\mu-\nu)]\in\alpha^+_{loc}(P)$,

(ii) $U\in\zeta^{(2)}_{loc}(P)\iff U*(\mu-\nu)\in m^2_{loc}(P)$ *and* $\langle U*\mu-\nu\rangle_t=G^{(2)}(t,U)$, *and*

(iii) if $U\in\zeta_{loc}(P)$, *then there are* $U_i\in\zeta^{(i)}_{loc}(P)$ *such that* $U=U_1+U_2$.

Note that

$$G^{(2)}(t,U)=\int_0^t\int_E U^2(s,x)d\nu-\sum_{s\leq t}\hat{U}_s^2\,. \tag{1.6.61}$$

REMARK 1.61 In part (iii) of Theorem 1.69, the functions U_1 and U_2 can be chosen to be

$$U_1=(U-\hat{U})\ I(|U-\hat{U}|>1/2)+\hat{U}I(|\hat{U}|>1/2)$$

and

$$U_2 = (U - \hat{U})\ I(|U - \hat{U}| \leq 1/2) + \hat{U} I(|\hat{U}| \leq 1/2)\,.$$

(ii) Starting from the decomposition $U = U_1 + U_2$, the stochastic integral $U * (\mu - \nu)$ can also be defined as the sum of $U_1 * (\mu - \nu)$ and $U_2 * (\mu - \nu)$ where the first one is a Lebesgue–Stieltjes integral and the second is the limit in probability of the integrals $U_2^{(n)} * (\mu - \nu)$ of step functions $U_2^{(n)} \in \zeta_{loc}^{(2)}(P)$ such that

$$G^{(2)}(t, \left|U - \hat{U}_n\right|) \to 0 \text{ as } n \to \infty$$

for each $t > 0$. ∎

REMARK 1.62 For a proof of Theorem 1.69, see [25]. An alternate proof was given by Kabanov et al. [31]. ∎

Decomposition of Local Martingales Using Stochastic Integrals

Let X be a local martingale. Then $X = X^{(c)} + X^{(d)}$ by (1.5.2). Let μ be the saltus measure of X or equivalently that of $X^{(d)}$ and ν be its compensator. It can be shown that

$$X_t^{(d)} = \int_0^t \int_{R-\{0\}} x d(\mu - \nu) \text{ a.s. } [P] \tag{1.6.62}$$

following Theorem 1.69, and hence,

$$X_t = X_t^{(c)} + \int_0^t \int_{R-\{0\}} x d(\mu - \nu), t \geq 0 \text{ a.s. } [P]\,. \tag{1.6.63}$$

(For details see [25, 31]).

Predictable Local Characteristics of Semimartingales

It was mentioned earlier (see (1.6.7)) that every semimartingale X can be uniquely represented in the form

$$X_t = X_0 + \alpha_t + X_t^{(c)} + \sum_{s \leq t} \Delta X_s\ I(|X_s| > 1) + M_t^{(d)} \tag{1.6.64}$$

where $M^{(d)} \in m_{loc}^{(d)}(P)$ and $\alpha \in \alpha_{loc}(P) \cap \mathcal{P}$. Let μ be the saltus measure of X, ν its compensator, and $\beta = \langle X^{(c)}, X^{(c)} \rangle$.

The triple (α, β, ν) is called the triplet of P-*local characteristics* of the semimartingale X and the decomposition given by (1.6.64) is called the *canonical decomposition* [19, 25]. The triplet (α, β, ν) is uniquely determined by the process X. It can be shown that

$$M_t^{(d)} = \int_0^t \int_{|x| \leq 1} x d(\mu - \nu) \tag{1.6.65}$$

and

$$X_t = X_0 + \alpha_t + X_t^{(c)} + \int_0^t \int_{|x|>1} x d\mu + \int_0^t \int_{|x| \leq 1} x d(\mu - \nu) \tag{1.6.66}$$

where μ is the saltus measure of the semimartingale X and ν its compensator. For proof, see [31].

REMARK 1.63 It is known that a semimartingale X is a process with independent increments if and only if α and β and the random measure ν do not depend on ω (cf. [19]). In such a case,

$$E\left\{e^{i\lambda(X_t - X_s)} | \mathcal{F}_s(P)\right\} = \exp\left\{ i\lambda(\alpha_t^{(c)} - \alpha_s^{(c)}) - \frac{\lambda^2}{2}(\beta_t - \beta_s) \right.$$

$$+ \int_s^t \int_{|x| \leq 1} \left(e^{i\lambda x} - 1 - i\lambda x\right) d\nu^{(c)}$$

$$\left. + \int_s^t \int_{|x|>1} \left(e^{i\lambda x} - 1\right) d\nu^{(c)} \right\}$$

$$\Pi_{s<u\leq t}\left\{1+\int\limits_{R-\{0\}}\left(e^{i\lambda x}-1\right)\nu(\{u\},dx)\right\}$$

(1.6.67)

where

$$\alpha_t^{(c)}=\alpha_t-\sum_{s\leq t}\int\limits_{|x|\leq 1}x\nu(\{s\},dx)\,, \tag{1.6.68}$$

and

$$\nu^{(c)}([0,t],\Gamma)=\nu([0,t],\Gamma)-\sum_{s\leq t}\nu(\{s\},\Gamma)\,. \tag{1.6.69}$$

(cf. [19]). ∎

In particular, if X is a semimartingale with $X_0 = 0$, then it is a standard Wiener process if and only if $\alpha \equiv 0,\ \beta_t \equiv t,\ \nu \equiv 0$ and it is a Poisson process with parameter λ if and only if $\alpha_t = \lambda t,\ \beta \equiv 0$ and ν is the measure assigning the mass λ to $\{1\}$.

Suppose $\{X_t\}$ is a real-valued process with independent increments with $X_0 = 0$ and without fixed points of discontinuity. It is well known that X_t has an infinitely divisible distribution and its characteristic function is of the form

$$E\left[e^{iuX_t}\right]=e^{\psi_t(u)} \tag{1.6.70}$$

with

$$\psi_t(u)=iu\alpha_t-\frac{u^2}{2}b_t+\int(e^{iux}-1-iuh(x))F_t(dx) \tag{1.6.71}$$

(Lévy–Khintchin representation) where $\alpha_t \in R, b_t \in R_+$, F_t is a positive measure that integrates $\min(1, x^2)$ and h is a bounded Borel-measurable function with compact support "behaving like x" near the origin. Furthermore,

$$\left\{\frac{e^{iuX_t}}{e^{\psi_t(u)}},t\geq 0\right\} \tag{1.6.72}$$

is a martingale relative to the obvious filtration.

Suppose $\{X_t\}$ is a semimartingale. The problem is to find two processes α_t and β_t and a random measure ν such that if we define the process $\psi_t(u)$ by (1.6.71) with a_t, b_t, F_t replaced by α_t, β_t and $\nu([0,t] \times dx)$, then (1.6.72) holds. It is known that such a triplet (α, β, ν) exists which is unique and predictable. This triplet is called the triplet of *local characteristics* of X. If X is a semimartingale with independent increments, then the triplet is deterministic [30, p. 75].

Let X be a semimartingale and $h : R \to R$ be a bounded function with compact support such that $h(x) = x$ in a neighborhood of zero. Such a function $h(\cdot)$ is called a truncation function. Note that $\Delta X_s - h(\Delta X_s) \neq 0$ only if $|\Delta X_s| > b$ for some $b \neq 0$.

Let

$$\tilde{X}(h)_t = \sum_{s \leq t} [\Delta X_s - h(\Delta X_s)], \quad t \geq 0 \tag{1.6.73}$$

and

$$X(h) = X - \tilde{X}(h). \tag{1.6.74}$$

Then the process $\tilde{X}(h) \in \mathcal{V}(P)$ and the process $X(h)$ is a semimartingale. Note that $X(h)$ is a special semimartingale [12, p. 148] since $\Delta X(h) = h(\Delta X)$, which is bounded [30, p. 44]. Hence

$$X(h) = X_0 + M(h) + \alpha(h) \tag{1.6.75}$$

where $\alpha(h)$ is predictable and belongs to $\mathcal{V}(P)$, and $M(h)$ is a local martingale with $M(h)_0 = 0$.

DEFINITION 1.47 *Let h be fixed. The triplet (α, β, ν) consisting of*

(i)

$$\alpha = \alpha(h) \quad \textit{as given above}, \tag{1.6.76}$$

(ii) $\beta = \langle X^{(c)}, X^{(c)} \rangle$ *where* $X^{(c)}$ *is the continuous martingale component of* X,

(iii) ν *is the predictable random measure on* $R_+ \times R$, *the compensator of the random measure* μ *associated with jumps of* X, *namely*

$$\mu(\omega; dt, dx) = \sum_s I(\Delta X_s \neq 0)\, \varepsilon_{(s, \Delta X_s)}(dt, dx) \tag{1.6.77}$$

is called the local characteristics of X.

REMARK 1.64 It is clear from the above definition that β and ν do not depend on h while α does. It can be shown that there exists a "good" version of the characteristics (α, β, ν) where
(i) α is a predictable process,
(ii) β is continuous and belong to $\mathcal{V}^+(P)$, and
(iii) ν is a predictable random measure. ▮

THEOREM 1.70 [30, p. 77]
Let $X = (X^1, \ldots, X^d)$ be a d-dimensional semimartingale. There exists a version of the local characteristics (α, β, ν) of X which is of the form

$$\begin{aligned} &(i)\ \ \alpha^{(i)} = b^{(i)} \circ A\,, \\ &(ii)\ \ \beta^{(i,j)} = c^{(i,j)} \circ A\,, \\ &(iii)\ \nu(\omega; dt, dx) = dA_t(\omega) K_{\omega,t}(dx) \end{aligned} \tag{1.6.78}$$

where

(i) *A is a predictable process in $\alpha^+_{loc}(P)$ that may be chosen continuous if and only if X is quasi-left continuous (that is, $\Delta X_T = 0$ a.s. on the set $[T < \infty]$ for every predictable stopping time T),*

(ii) *$b = (b^{(i)})_{i \le d}$ is a d-dimensional predictable process,*

(iii) *$c = ((c^{ij}))_{i,j \le d}$ is a predictable process with values in the set of all symmetric nonnegative definite $d \times d$ matrices, and*

(iv) *$K_{\omega,t}(dx)$ is a transition kernel from $(\Omega \times R_+ \times \mathcal{P}) \to (R^d, \mathcal{B}_d)$ that satisfies*

$$\begin{aligned} &(i)\ \ K_{\omega,t}(\{0\}) = 0, \qquad \int_{\min(\|x\|^2,1)} K_{\omega,t}(dx) \le 1\,, \\ &(ii)\ \ \Delta A_t(\omega) > 0 \Rightarrow b_t(\omega) = \int h(x) K_{\omega,t}(dx), \text{ and} \\ &(iii)\ \Delta A_t(\omega) K_{\omega,t}(R^d) \le 1\,. \end{aligned} \tag{1.6.79}$$

REMARK 1.65 As a consequence of (iii) and (1.6.79), it follows that, for $s \le t$, $((\beta_t^{(i,j)} - \beta_s^{(i,j)}))_{i,j \le d}$ is a symmetric nonnegative definite matrix, $\min(|x|^2, 1) * \nu \in \alpha_{loc}(P)$ and $\Delta\beta_t = \int h(x)\nu(\{t\} \times dx)$. (Here, for any function f and measure η, $f * \eta \equiv \int f(x)\eta(dx)$.) ▮

1.7 Girsanov's Theorem

We now study the transformation of the predictable local characteristics under a locally absolutely continuous change of measures.

Suppose P and $\tilde{P}$ are probability measures on $(\Omega, \mathcal{F})$. Let $Q = \frac{P+\tilde{P}}{2}$. As before, let $\mathcal{F}(P)$ denote the completion of $\mathcal{F}$ with respect to P. It is easy to see that each of P and $\tilde{P}$ can be extended to $\mathcal{F}(Q)$. We denote the extensions by the same letters.

For any stopping time τ relative to $\{\mathcal{F}_t(Q), t \geq 0\}$, denote by P_τ, the restriction of P to $\mathcal{F}_\tau(Q)$.

DEFINITION 1.48 *Suppose $\{\tau_n\}$ is a sequence of stopping times with respect to $\{\mathcal{F}_t(Q), t \geq 0\}$ such that $\tau_n \uparrow \infty$ a.s. $[Q]$ and $\tilde{P}_{\tau_n} << P_{\tau_n}, n \geq 0$. Then $\tilde{P}$ is said to be locally absolutely continuous with respect to P (we write $\tilde{P} \overset{loc}{<<} P$).*

Suppose $\tilde{P} \overset{loc}{<<} P$. It is known that there exists a unique (to within $\tilde{P}$- and P- indistinguishability) process $Z = (Z_t, \ t \geq 0)$ with paths that are right-continuous and have left limits such that

$$\begin{aligned} &\text{(i) the process } \left\{Z^{\tau_n} = Z_{t\wedge\tau_n}, \ t \geq 0\right\} \text{ is uniformly} \\ &\quad\ \text{integrable for every } n \geq 1\,, \\ &\text{(ii) } Z^{\tau_n} = \frac{d\tilde{P}_{\tau_n}}{dP_{\tau_n}} \ \text{a.s.} \ [P]\,. \end{aligned} \tag{1.7.1}$$

The process Z is called the *local density* of the measure $\tilde{P}$ with respect to P. Let

$$M_t = \int_0^t Z_{s-}^{\oplus} dZ_s \tag{1.7.2}$$

where $a^{\oplus} = a^{-1}$ if $a \neq 0$ and $a^{\oplus} = 0$ if $a = 0$. Then the process $M = (M_t, \mathcal{F}_t(Q), t \geq 0)$ is a local martingale and

$$Z_t = Z_0 + \int_0^t Z_{s-} dM_s\,. \tag{1.7.3}$$

Let $M = M^{(c)} + M^{(d)}$ be the decomposition of M into the continuous and purely discontinuous components, $\langle M^{(c)}\rangle$ the predictable process in the Doob–Meyer decomposition of the process $(M^{(c)})^2$, $\mu^{(M)}$ the saltus measure of M with $\nu^{(M)}$ as its compensator. Define

$$B_t(M) = \langle M^c\rangle_t + \int_0^t \int_{R-\{0\}} \frac{x^2}{1+x} d\nu^{(M)}, \quad t \geq 0 . \tag{1.7.4}$$

The following result gives conditions for the absolute continuity or the singularity of $\tilde{P}$ with respect to P.

THEOREM 1.71 [31]
Suppose $\tilde{P} \overset{loc}{<<} P$. *Then*

$$\tilde{P} << P \Leftrightarrow \tilde{P}\,(B_\infty(M) < \infty) = 1 , \tag{1.7.5}$$

and

$$\tilde{P} \perp P \Leftrightarrow \tilde{P}\,(B_\infty(M) = \infty) = 1 , \tag{1.7.6}$$

where

$$B_\infty(M) = \lim_{t\to\infty} B_t(M) . \tag{1.7.7}$$

The following theorem gives the relation between the predictable local characteristics of semimartingales when the semimartingales are transforms of one another under a locally absolutely continuous change of measures.

THEOREM 1.72 [28]
Let $X = (X_t, \mathcal{F}_t(Q), P), t \geq 0$ *be a semimartingale with the canonical representation*

$$X_t = X_0 + \alpha_t + X_t^{(c)} + \int_0^t \int_{|x|>1} x d\mu + \int_0^t \int_{|x|\leq 1} x d(\mu - \nu) \tag{1.7.8}$$

(note that $\beta = \langle X^{(c)}, X^{(c)}\rangle$ *and* (α, β, ν) *are the predictable local characteristics of* X*). If* $\tilde{P} \overset{loc}{<<} P$ *and* Z *is the local density, then the process*

$X = (X_t, \mathcal{F}_t(Q), \tilde{P}), t \geq 0$ is also a semimartingale with the triplet of $\tilde{P}$-local predictable characteristics $(\tilde{\alpha}, \tilde{\beta}, \tilde{\nu})$ connected with (α, β, ν) by the formulae

$$\tilde{\alpha}_t = \alpha_t + \int_0^t Z_{s-}^{-1} \frac{d < Z^{(c)}, X^{(c)} >_s}{d\beta_s} d\beta_s$$

$$+ \int_0^t \int_{|x|\leq 1} x(Y(s,x) - 1) d\nu_s \tag{1.7.9}$$

and

$$\tilde{\beta} = \beta \; ; \; d\tilde{\nu} = Y d\nu \, , \tag{1.7.10}$$

where

$$Y(s,x) = Z_{s-}^{\oplus}(Z_{s-} + x) \, . \tag{1.7.11}$$

For proof, see [31, p. 663].

We can write the above relations in the form

$$\tilde{\nu} = Y * \nu, \; \tilde{\alpha} = \alpha + Z_0 \circ \beta + [U(Y-1)] * \nu \, , \; \tilde{\beta} = \beta \tag{1.7.12}$$

where Y is a nonnegative $\tilde{\mathcal{P}}$-measurable function and Z_0 is a predictable process such that

$$\tilde{P}(|U(Y-1)| * \nu_t < \infty) = \tilde{P}\,(Z_0 \circ \beta_t < \infty)$$

$$= \tilde{P}\left(Z_0^2 \circ \beta_t < \infty\right) = 1 \, . \tag{1.7.13}$$

Here

$$U(\omega; t, x) = x \, I(|x| \leq 1), \omega \in \Omega, t > 0, \; x \in R \, . \tag{1.7.14}$$

Let

$$\zeta_{loc}^{(1)}(\mu, P) = \{U : U \text{ is } \mathcal{P}\text{-measurable and } |U| * \nu \in \upsilon_{loc}(P)\} \, , \tag{1.7.15}$$

$$\zeta_{loc}^{(2)}(\mu, P) = \left\{U : U \text{ is } \mathcal{P}\text{-measurable and } U^2 * \nu \in \upsilon_{loc}(P)\right\} , \tag{1.7.16}$$

and

$$\zeta_{loc}(\mu, P) = \left\{U : U_1 + U_2,\ U_1 \in \zeta_{loc}^{(1)}(\mu, P), U_2 \in \zeta_{loc}^{(2)}(\mu, P)\right) . \tag{1.7.17}$$

The class $\upsilon_{loc}(P)$ is similar to $\nu_{loc}(P)$ defined in Section 1.5. If $U \in \zeta_{loc}(\mu, P)$, then the stochastic integral $U * (\mu - \nu)$ is well defined and forms a local martingale with respect to $\{\mathcal{F}_t(P),\ t \geq 0\}$ and P. Moreover, if $U \in \zeta_{loc}^{(1)}(\mu, P)$, then

$$U * (\mu - \nu) \in m_{loc}(\mathcal{P}) \cap \upsilon_{loc}(P) ;$$

if $U \in \zeta_{loc}^{(2)}(\mu, P)$, then

$$U * (\mu - \nu) \in m_{loc}^{(2)}(P),\ [U * |\mu - \nu|]^{(c)} = 0 ,$$

$$\Delta(U * (\mu - \nu))_s = U\,(s, \Delta X_s)\, I(\Delta X_s \neq 0) ,$$

and if $U * \mu$ and $U * \nu$ belong to $\alpha(P)$, then

$$U * (\mu - \nu) = U * \mu - U * \nu .$$

REMARK 1.66 We have seen above that if $\tilde{P} \overset{loc}{<<} P$, then $\tilde{\beta} = \beta$ and $\tilde{\nu} << \nu$. In particular, if X is continuous under P and $\tilde{P} \overset{loc}{<<} P$, then X is also continuous under $\tilde{P}$, $\tilde{\nu} = \nu = 0$, $\tilde{\beta} = \beta$ and $\tilde{\alpha} - \alpha$ is absolutely continuous with respect to β. ∎

The following theorem allows the construction of a probability law $\tilde{P}$ with $\tilde{P} \overset{loc}{<<} P$ if Y and Z_0 are prespecified. This theorem is a generalization of the Girsanov's theorem for Wiener process and it follows from [28, p. 15].

Girsanov's Theorem for Semimartingales

THEOREM 1.73

Suppose X is a semimartingale under P with local characteristics (α, β, ν). Let Y be a $\tilde{P}$-measurable nonnegative function such that $Y - 1 \in \zeta_{loc}(\mu, P)$ and Z_0 be a predictable process for which $Z_0^2 \circ \beta \in \upsilon_{loc}(P)$. Define

$$N = Z_0 \bullet X^{(c)} + (Y - 1) * (\mu - \nu) , \tag{1.7.18}$$

$$W_t = \exp\left(N_t - \frac{1}{2} Z_0^2 \bullet \beta_t \ \Pi_{s \le t}\left[(1 + \Delta N_s)\exp(-\Delta N_s)\right]\right),$$

$$t \ge 0 . \tag{1.7.19}$$

Assume that $EW_t = 1$ for all $t \ge 0$. Define P^* on $\mathcal{F}_t$ by

$$P^*(A) = \int_A W_t dP, \ A \in \mathcal{F}_t . \tag{1.7.20}$$

Then there exists a probability measure $\tilde{P}$ on $\mathcal{F}$, the σ-algebra generated by $\{\mathcal{F}_s; s \ge 0\}$, with $\tilde{P}|\mathcal{F}_t = P^*$ for all $t > 0$. Moreover, X is a semimartingale under $\tilde{P}$ having the local characteristics $(\tilde{\alpha}, \tilde{\beta}, \tilde{\nu})$ with

$$\tilde{\nu} = Y \circ \nu, \ \tilde{\alpha} = \alpha + Z_0 \circ \beta + [U(Y - 1)] * \nu, \ \tilde{\beta} = \beta . \tag{1.7.21}$$

REMARK 1.67 The existence of a probability measure $\tilde{P}$, given $\tilde{P}_t$, $t \ge 0$, can be ensured if $(\Omega, \mathcal{F})$ is a standard measurable space. If $\tilde{P}$ is a probability measure on $(\Omega, \mathcal{F})$ such that under $\tilde{P}$ the process X is a semimartingale with the local characteristics $(\tilde{\alpha}, \tilde{\beta}, \tilde{\nu})$ given by (1.7.21), then $\tilde{P}$ is not necessarily absolute continuous with respect to P. This is because the local characteristics of a semimartingale do not in general determine the measure P on $(\Omega, \mathcal{F})$ uniquely. This can be seen from the following example. ∎

Example 1.9

Let $(\Omega, \mathcal{F}, P)$ be a probability space with a filtration $\{\mathcal{F}_t, t \ge 0\}$. Let $W =$

$\{W_t, t \geq 0\}$ be $\mathcal{F}_t$-Wiener process. Define

$$a(x) = \begin{cases} a_1 > 0 \text{ if } x > 0 \\ \frac{a_1+a_2}{2} \text{ if } x = 0 \\ a_2 > 0 \text{ if } x < 0 \end{cases}$$

where $a_1 \neq a_2$. Consider the stochastic differential equation

$$dX_t = a\,(X_t)\,dW_t\ .$$

The solution of this equation is a strong Markov process with

$$\langle X \rangle_t = \int_0^t a^2\,(X_s)\,ds\ .$$

Let $T_u = \inf\{t : \langle X \rangle_t \geq u\}$. Then T_u is an $\{\mathcal{F}_t\}$-stopping time. Let $\bar{W} = X_t = \{X_{T_u}, u \geq 0\}$. Then $\bar{W}$ is an $\{\mathcal{F}_{T_u}\}$-adapted Wiener process (cf. [36]). Extending $(\Omega, \mathcal{F}, \{\mathcal{F}_{T_u}\}, P)$ if necessary, we can assume that it carries a second Wiener process $\bar{W}_{(2)}$ independent of $\bar{W}$ that is adapted to $\{\mathcal{F}_{T_u}\}$. Since $\langle X \rangle_t$ is an $\{\mathcal{F}_{T_u}\}$-stopping time for every $t > 0$, the process $X = \bar{W}_{\langle X \rangle}$ and $\bar{W}_{(2)\langle X \rangle}$ are continuous local $\mathcal{F}_t$-martingales with the same local characteristics $(0, \langle X \rangle, 0)$. But these processes are different in distribution. The first process is strongly Markov, whereas the second process is not (cf. [13]). □

Girsanov's Theorem for Semimartingales (Multidimensional Version)

Let $\boldsymbol{X} = (X^1, \ldots, X^d)$ be a d-dimensional semimartingale with the local characteristics (B, C, ν) relative to a given truncation function h as given by Theorem 1.70. Let $\boldsymbol{X}^{(c)}$ be the continuous martingale component of $\boldsymbol{X}$ and A be an increasing predictable process such that

$$C^{(i,j)} = c^{(i,j)} \circ A\ . \tag{1.7.22}$$

THEOREM 1.74 [30, p. 159]

Suppose $\tilde{P} \overset{loc}{<<} P$ and $\boldsymbol{X}$ is as described above. Then there exists a $\tilde{P}$-measurable nonnegative function Y and a predictable process $\beta = (\beta^{(i)})_{i \leq d}$

satisfying

$$\begin{aligned}&(i)\ \ |h(x)(Y-1)*\nu_t| < \infty \ \ a.s.\ [\tilde{P}],\ t \in R_+\,,\\ &(ii)\ \ \left|\textstyle\sum_{j\le d} c^{(i,j)}\beta^{(j)}\right| \circ A_t < \infty \ \ a.s.\ [\tilde{P}],\ t\in R_+,\ and\\ &(iii)\ (\textstyle\sum_{j,k\le d}\beta^{(k)}c^{(j,k)}\beta^{(k)}| \circ A_t < \infty\ a.s\ [\tilde{P}],\ t \in R_+\end{aligned} \tag{1.7.23}$$

and such that a version of the local characteristics of X relative to $\tilde{P}$ are

$$\begin{aligned}&(i)\ \ B'^{(i)} = B^{(i)} + \left(\textstyle\sum_{j\le d} c^{(i,j)}\beta^{(j)}\right)\circ A + h(x)(Y-1)*\nu\,,\\ &(ii)\ \ C' = C,\ and\\ &(iii)\ \nu' = Y*\nu\,.\end{aligned} \tag{1.7.24}$$

Furthermore, Y and β satisfy the above conditions only if

$$YZ_- = M^P_{\mu^X}\left(Z|\tilde{P}\right), \tag{1.7.25}$$

and

$$\langle Z^{(c)}, X^{i(c)}\rangle = \left(\sum_{j\le d} c^{(i,j)}\beta^{(j)}Z_-\right)\circ A \tag{1.7.26}$$

up to a P-null set where Z is the density process, $Z^{(c)}$ is its continuous martingale part relative to P, and $\langle Z^c, X^{i(c)}\rangle$ is the predictable quadratic covariation relative to P (which is also equal to $[Z, X^{i(c)}]$). Furthermore, it is possible to choose Y so that

$$\begin{aligned}\nu\left(\omega:\{t\}\times R^d\right) = 1 \Rightarrow \nu'&\left(\omega:\{t\}\times R^d\right)\\ &= \int Y(\omega;t,x)\nu(\omega:\{t\}\times dx) = 1\,.\end{aligned} \tag{1.7.27}$$

Let X be a d-dimensional semimartingale defined on a probability space $(\Omega, \mathcal{F}, P)$ with the filtration $\{\mathcal{F}_t, t \ge 0\}$, which is increasing and right-continuous. Suppose (α, β, ν) are the local characterization of X and $X^{(c)}$ the continuous martingale part. Let μ^X be the measure associated with the jumps of X and ν be its compensator. Can we say that every $\mathcal{F}_t$ local martingale is the sum of a stochastic integral with respect to $X^{(c)}$ and a stochastic integral with respect to $\mu - \nu$?

DEFINITION 1.49 *A local martingale M is said to have the representation property relative to X if it has the form*

$$M = M_0 + H \bullet X^{(c)} + W * (\mu - \nu) \tag{1.7.28}$$

where $H = (H^i)_{i \leq d} \in L^2_{loc}(X^{(c)})$ and $W \in \zeta_{loc}(P)$. Here $L^2_{loc}(X^{(c)})$ is the set of all predictable processes H such that $H^2 \circ \langle X^c, X^c \rangle$ is locally integrable.

We refer the reader to [30] for a discussion on this topic.

Gaussian Martingales

DEFINITION 1.50 *A Gaussian martingale is an R^d-valued martingale X such that $X_0 = 0$ and the distribution of any finite family $(X_{t_1}, \ldots, X_{t_n})$ is Gaussian.*

Let $c_{ij}(t) = E(X_t^{(i)} X_t^{(j)})$ and $c(t) = ((c_{ij}^{(t)}))_{i,j \leq d}$. It is easy to see that, for all $s \leq t$, $c(t) - c(s)$ is a symmetric nonnegative definite matrix and for all $\boldsymbol{u} \in R^d$,

$$E\left[e^{i\boldsymbol{u}\cdot(X_t - X_s)}\right] = \exp\left\{-\frac{1}{2}\boldsymbol{u}^T(c(t) - c(s))\boldsymbol{u}\right\},$$

where $\boldsymbol{u}^T$ denotes the transpose of $\boldsymbol{u}$.

Furthermore, $E[X_s^{(j)} X_t^{(k)}] = c_{jk}(s)$ from the martingale property of X. In fact, $X_t - X_s$ is uncorrelated with $\{X_r, r \leq s\}$. Since the process is Gaussian, it follows that $X_t - X_s$ is independent of $\{X_r, r \leq s\}$ for $t \geq s$. Therefore,

$$E\left[\exp\left\{\sum_{j \leq n} i\boldsymbol{u}_j \cdot \left(X_{t_j} - X_{t_{j-1}}\right)\right\}\right]$$

$$= \exp\left[-\frac{1}{2}\sum_{j \leq n} \boldsymbol{u}_j^T \left(c\left(t_j\right) - c\left(t_{j-1}\right)\right) \boldsymbol{u}_j\right]$$

whenever $0 = t_0 \leq t_1 \leq \cdots \leq t_n$ and $\boldsymbol{u}_j \in R^d$. Hence, the process X is completely determined by the function c.

Suppose

$$X = X^{(c)} + X^{(d)} \text{ where } X^d = \sum_{s \leq t,\ s \in J} \Delta X_s$$

where J is the set of fixed times of discontinuity of X. Then $X^{(c)}$ and $X^{(d)}$ are independent Gaussian martingales and $X^{(c)}$ is a Wiener process.

1.8 Limit Theorems for Semimartingales

Stable Convergence and Semimartingales

DEFINITION 1.51 *Suppose $\{X_n\}$ is a sequence of random variable on a probability space $(\Omega, \mathcal{F}, P)$ converging in distribution to a random variable X. We say that the convergence is stable if for all the continuity points x of the distribution function of X and all $A \in \mathcal{F}$,*

$$\lim_{n \to \infty} P\left[\{X_n \leq x\} \cap A\right] = Q_x(A)$$

exists and if $Q_x(A) \to P(A)$ as $x \to \infty$. We denote this convergence by

$$X_n \xrightarrow{\mathcal{L}} X \ \textit{(stably)}\,.$$

The concept of stability was introduced by Renyi [61].

DEFINITION 1.52 *A sequence of random variables $\{Y_n\}$ on $(\Omega, \mathcal{F}, P)$ with $E|Y_n| < \infty$ is said to converge weakly in $L_1(\Omega, \mathcal{F}, P)$ to a random variable Y if for all $A \in \mathcal{F}$,*

$$E\left[Y_n I(A)\right] \to E[Y I(A)] \ \text{ as } n \to \infty\,.$$

We denote this type of convergence by

$$Y_n \to Y \ \left(\textit{ weakly in } L_1\right)\,.$$

Note that if $\exp(itY_n) \to \exp(itY)$ weakly in L_1 for every real t, then $Y_n \xrightarrow{\mathcal{L}} Y$.

THEOREM 1.75
Suppose $Y_n \xrightarrow{\mathcal{L}} Y$ where $\{Y_n\}$ are defined on a probability space $(\Omega, \mathcal{F}, P)$. Then $Y_n \to Y$ (stably) if and only if there exists a random variable Y' defined on an extension of $(\Omega, \mathcal{F}, P)$ with the same distribution as that of Y such that for all real t

$$\exp(itY_n) \to Z(t) = \exp\left(itY'\right) \ (\textit{weakly in } L_1) \quad \textit{as } n \to \infty$$

and $E(Z(t)I(A))$ is a continuous function of t for all $A \in \mathcal{F}$.

REMARK 1.68 This theorem follows from the continuity theorem for characteristic functions. This allow us to identify the stably convergent sequences in terms of the characteristic functions, which is usually simpler to check. ∎

DEFINITION 1.53 *If the random variable Y' in Theorem 1.75 can be taken to be independent of $A \in \mathcal{F}$, then the limit is said to be mixing (in the sense of [60]). In this case*

$$P\left(\{Y_n \le y\} \cap A\right) \to P(Y \le y)P(A) \quad \textit{as } n \to \infty$$

for all $A \in \mathcal{F}$ and for all the points of continuity y of the distribution function of Y. We write, in such an event,

$$Y_n \xrightarrow{\mathcal{L}} Y \ (\textit{mixing})\,.$$

REMARK 1.69 Aldous and Eagleson [1] and Eagleson [11] discussed mixing and stability of limit theorems. Rootzen [63, 64, 65, 66] studied the relationship between the mixing condition and fluctuation of sequences which converge in distribution. Unlike convergence in distribution, stable convergence is not a property of the sequence of distribution functions but a property of the sequence of random variables. For example, let X and X' be independent and identically distributed random variables. Let

$$Z_n = X \text{ if } n \text{ is odd}$$

$$= X' \text{ if } n \text{ is even}.$$

Then $Z_n \xrightarrow{\mathcal{L}} X$ but Z_n does not converge to X stably. The following results are due to Aldous and Eagleson [1]. We omit the proofs. ∎

THEOREM 1.76

Suppose $Y_n \xrightarrow{\mathcal{L}} Y$. Then the following conditions are equivalent:

(i) *$Y_n \Rightarrow Y$ stably,*

(ii) *for all fixed $\mathcal{F}$-measurable random variable Z, the sequence (Y_n, Z) has a limiting distribution,*

(iii) *for each $t \in R$, the sequence of random variables e^{itY_n} converges weakly in L_1,*

(iv) *for all fixed k and $B \in \sigma(Y_1, \ldots, Y_k)$, $P(B) > 0$, $\lim_{n\to\infty} P(Y_n \leq x|B)$ exists for a countable dense set of points x.*

As a corollary to Theorem 1.76, the following result can be proved.

COROLLARY 1.14

Suppose that $Y_n \Rightarrow Y$ (stably). Let $g(x, y)$ be a continuous function of (x, y) and let Z be any $\mathcal{F}$-measurable random variable. Then $g(Y_n, Z)$ converges stably.

Stable Convergence of Semimartingales

Let $(\Omega, \mathcal{F}, P)$ be a probability space and $X^{(n)} = \{(X_t^{(n)}, \mathcal{F}_t^{(n)}, 0 \leq t \leq 1)\}$ be a sequence of RCLL semimartingales defined on $(\Omega, \mathcal{F}, P)$ with $X_0^{(n)} = 0$. We assume that $\mathcal{F}^{(n)} = \{\mathcal{F}_t^{(n)}, 0 \leq t \leq 1\}$ satisfies the usual conditions for each n and $\mathcal{F}_t^{(n)} \subset \mathcal{F}$ for all n. We assume further that the following nesting condition is satisfied by $\mathcal{F}^{(n)}, n \geq 1$.
(N) There exists a sequence $t_n \downarrow 0$ such that

$$\text{(i) } \mathcal{F}_{t_n}^n \subseteq \mathcal{F}_{t_{n+1}}^{n+1} \text{ and (iv) } \vee_n \mathcal{F}_{t_n}^n = \vee_n \mathcal{F}_1^n .$$

We shall call such a sequence $\{t_n\}$ an (N)-sequence.

Further suppose that $X = \{X_t, \mathcal{F}_t, 0 \le t \le 1\}$ is another semimartingale with $X_0 = 0$ and with ζ-conditionally independent increments where $\zeta \subset \mathcal{F}_0$. In other words, if $(B, \langle X^{(c)} \rangle, \nu)$ forms the triplet of local (predictable) characteristics of X, then it is ζ-measurable and if

$$h(x) = x\, I(|x| \le 1)\,, \tag{1.8.1}$$

$$G_t^X(\lambda) = i\lambda B_t - \frac{1}{2}\lambda^2 \langle X^c \rangle_t$$

$$+ \int_0^t \int_{R-\{0\}} \left(e^{i\lambda x} - 1 - i\lambda h(x)\right) \nu(ds, dx)\,, \tag{1.8.2}$$

then

$$E\left(e^{i\lambda X_t} | \zeta\right) = \mathcal{E}_t(G(\lambda)),\ 0 \le t \le 1 \tag{1.8.3}$$

where $\mathcal{E}_t$ is the Doléans–Dade exponential given by

$$\mathcal{E}_t(G) = e^{G_t} \Pi_{0<s\le t} \left\{(1 + \Delta G_s)\, e^{-\Delta G_s}\right\}\,. \tag{1.8.4}$$

Let D be a subset of $[0, 1]$ with $\{1\} \subset D$ and let Φ_D be the class of all left-continuous piecewise constant nonrandom functions H with a finite number of discontinuites all of which belong to D.

For $H \in \Phi_D$ and f RCLL, let us define $H \circ f$ as the process

$$(H \circ f)_t = \int_0^t H df,\ 0 \le t \le 1$$

where the integral is the appropriate finite sum. For any $H \in \Phi_D$, define

$$M^{(n)} = H \circ X^{(n)} \text{ and } M = H \circ X\,. \tag{1.8.5}$$

Denote by G^n, $\tilde{G}^{(n)}$ and $\tilde{G}$ the predictable processes

$$G^{(n)}(\lambda) = G^{X^{(n)}}(\lambda),\ \tilde{G}^{(n)} = G^{M^{(n)}}(1),\ \tilde{G} = G^M(1) \tag{1.8.6}$$

following (1.8.2). Observe that

$$E\left(e^{iM_t}|\zeta\right) = \mathcal{E}_t(\tilde{G}) \tag{1.8.7}$$

from (1.8.4). Define

$$Z_t^{M_n} = e^{iM_t^n}\mathcal{E}_t^{-1}\left(\tilde{G}^{(n)}\right). \tag{1.8.8}$$

Then $\{Z_t^{M_n}\}$ is a local martingale on $[0, T]$ where $T = \inf\{t : |\mathcal{E}_t(\tilde{G}^{(n)}))| = 0\}$ with $Z_0^{M_n} = 1$ (cf. [29]).

Let us denote the convergence of the finite dimensional distributions with time points in D by

$$\mathcal{L}_f\left(X^{(n)}; D\right) \to \mathcal{L}_f(X; D). \tag{1.8.9}$$

If this convergence is stable, that is,

$$E\left[\xi\left(e^{iM_t^{(n)}} - \mathcal{E}_t(\tilde{G})\right)\right] \to 0 \tag{1.8.10}$$

for all bounded $\mathcal{F}$-measurable functions ξ and for all $H \in \Phi_D$ (following Theorem 1.76), we write

$$\mathcal{L}_f\left(X^{(n)}; D\right) \to \mathcal{L}_f(X; D) \text{ (stably)}. \tag{1.8.11}$$

Let

$$Z^{X^{(n)}} = \left(Z_t^{X^{(n)}}(\lambda) \equiv e^{i\lambda X_t^{(n)}}\mathcal{E}_t^{-1}\left(G^{(n)}(\lambda)\right), \mathcal{F}_t^{(n)}\right). \tag{1.8.12}$$

Then $Z^{X^{(n)}}$ is a local martingale based on the $X^{(n)}$ process following (1.8.8). The following theorem is due to Feigin [14]. We omit the proof.

THEOREM 1.77
Suppose $\mathcal{I} = \vee_n \mathcal{F}_1^n$ and, for all $\lambda \in R$,

$$\begin{aligned}
&(i)\ \ \mathcal{E}_t\left(G^{(n)}(\lambda)\right) \xrightarrow{p} \mathcal{E}_t\left(G^X(\lambda)\right),\ t \in D,\\
&(ii)\ \left|\mathcal{E}_t\left(G^X(\lambda)\right)\right| > 0 \text{ a.s., and}\\
&(iii)\ \left|Z_{t_n}^{X^{(n)}}(\lambda) - 1\right| \xrightarrow{p} 0 \text{ for an } (N)\text{-sequence } \{t_n\}.
\end{aligned} \tag{1.8.13}$$

Then, if $\zeta \subset \mathcal{I}$, *then*

$$\mathcal{L}_f\left(X^{(n)}; D\right) \to \mathcal{L}_f(X; D) \text{ stably}. \tag{1.8.14}$$

An application of this theorem is the following stable central limit theorem for one-dimensional martingales.

THEOREM 1.78

Suppose $M = \{M_t, m_t, 0 < t < \infty\}$ *is a scalar square integrable martingale and let* $I_T = E(M_T^2)$. *If*

$$\begin{aligned}&(i)\ I_T \to \infty \text{ as } T \to \infty,\\ &(ii)\ E\left\{I_T^{-1/2} \sup_{t\le T} |\Delta M_t|\right\} \to 0 \text{ as } T \to \infty, \text{ and}\\ &(iii)\ I_T^{-1}[M]_T \xrightarrow{p} \eta^2 \text{ as } T \to \infty,\end{aligned} \tag{1.8.15}$$

then

$$\mathcal{L}\left(I_T^{-1/2} M_T\right) \to \mathcal{L}(X) \text{ (stably)} \tag{1.8.16}$$

where

$$E\left[e^{i\lambda X}\right] = E\left[e^{-\frac{1}{2}\lambda^2\eta^2}\right], \ \lambda \in R. \tag{1.8.17}$$

PROOF Let

$$X^T = \left(X_t^T, \mathcal{F}_t^T\right) = \left(I_T^{-1/2} M_{tT}, m_{tT}, \ 0 \le t \le 1\right) \tag{1.8.18}$$

with the local characteristics $(B^T, X^{T^{(c)}}, \nu^T)$. Then

$$N_t(\varepsilon) = \int_0^1 \int_{|x|>\varepsilon} |x| \nu^T(ds, dx) \xrightarrow{p} 0 \text{ for all } \varepsilon > 0. \tag{1.8.19}$$

This can be seen from the following argument.

By the Lenglart's inequality (cf. Theorem 1.34), if μ^T denotes the jump measure of X^T, it follows that for all $\beta > 0$,

$$P\left(N_t(\varepsilon) > \delta\right) \le \frac{1}{\delta}\left[\beta + E\left(\sup_{0\le t\le 1} \Delta X^T\right)\right]$$

$$+ P\left(\int_0^1 \int_{|x|>\varepsilon} |x| \mu^T(ds, dx) \geq \beta\right), \tag{1.8.20}$$

and (1.8.19) follows from condition (ii), which implies that

$$\int_0^1 \int_{|x|>\varepsilon} |x|^k \mu^T(ds, dx) \xrightarrow{P} 0 \text{ for all } \varepsilon > 0 \text{ and } k \geq 0\,. \tag{1.8.21}$$

Furthermore, relation (1.8.19) implies that

$$\sum_{0 \leq s \leq 1} \left| \int_{|x| \leq \varepsilon} x \nu^T(\{\lambda\}, dx) \right| \xrightarrow{P} 0 \text{ for all } \varepsilon > 0\,, \tag{1.8.22}$$

since $\int x \nu^T(\{s\}, dx) = 0$. Note that

$$H_t^T(\varepsilon) = \int_0^t \int_{|x| \leq \varepsilon} x^2 \left(\mu^T - \nu^T\right)(ds, dx) \tag{1.8.23}$$

is a purely discontinuous martingale for which

$$E\left(H_t^T(\varepsilon)\right)^2 = E\left(\left[H^T(\varepsilon)\right]_t\right)$$

$$\leq 2\varepsilon^2 \cdot 2E\left(|X_1^T|\right)^2 = 4\varepsilon^2\,. \tag{1.8.24}$$

As consequence of (1.8.21) and condition (iii) of the hypothesis, it follows that

$$\langle X^{T^{(c)}} \rangle_1 + \int_0^1 \int_{|x| \leq \varepsilon} x^2 \mu^T(ds, dx) \xrightarrow{P} \eta^2 \text{ for all } \varepsilon > 0\,. \tag{1.8.25}$$

Choose a family $\varepsilon_T \downarrow 0$ and $\varepsilon_T < 1/2$ so that (1.8.21), (1.8.22), and (1.8.25) hold for $\varepsilon = \varepsilon_T$. Let

$$X^T = Y^T + R^T \tag{1.8.26}$$

where

$$Y_t^T = Y_t^{T(c)} + \int_0^t \int_{|x|\le\varepsilon_T} x\left(\mu^T - \nu^T\right)(ds, dx), \tag{1.8.27}$$

and

$$R_t^T = B_t^{X^T} + \int_0^t \int_{\varepsilon_T<|x|\le 1} x\left(\mu^T - \nu^T\right)(ds, dx)$$

$$+\int_0^t \int_{|x|>1} x\mu^T(ds, dx). \tag{1.8.28}$$

Then $R_t^T \xrightarrow{p} 0$ from (1.8.19) and (1.8.21) and the fact that, for X^T a martingale,

$$B_t^{X^T} = -\int_0^1 \int_{|x|>1} x\nu^T(ds, dx). \tag{1.8.29}$$

Using the relations (1.8.24) and (1.8.25), following the techniques of Liptser and Shiryayev [46], it can be proved that for the martingale Y^T

$$\langle Y^T\rangle_t \xrightarrow{p} \eta^2,\ G_1^{Y^T}(\lambda) \xrightarrow{p} -\frac{1}{2}\lambda^2\eta^2 \tag{1.8.30}$$

and

$$\sum_{s\le 1}\left(\Delta G_s^{Y^T}(\lambda)\right)^2 \xrightarrow{p} 0$$

so that

$$\mathcal{E}_t\left(G^{Y^T}(\lambda)\right) \xrightarrow{p} \exp\left(-\frac{1}{2}\lambda^2\eta^2\right). \tag{1.8.31}$$

Note that

$$\nu^{Y^T}(ds, dx) \leq I\left(|x| \leq 2\varepsilon_T\right) \nu^T(ds, dx)\,, \tag{1.8.32}$$

and the relation (1.8.22) implies that

$$\sum_{t \leq 1} \left|\Delta B_t^{X^T}\right|^2 \xrightarrow{p} 0\,. \tag{1.8.33}$$

Furthermore, since $\varepsilon_T \leq \varepsilon_{t_T T}$ whenever $t_T \downarrow 0$, it follows that

$$\langle Y^T \rangle_{t_T} \leq I_{t_T T} \{I_T\}^{-1} \langle Y^{t_T T} \rangle_1 \xrightarrow{p} 0 \tag{1.8.34}$$

as long as $\frac{I_{t_T T}}{I_T} \to 0$ for the (N)-sequence $\{t_T : T \geq 0\}$. This can be arranged by choosing t_T appropriately. It can be shown that these arguments imply that the conditions (i)–(iii) of Theorem 1.77 hold and we obtain Theorem 1.78. ∎

THEOREM 1.79
If the conditions of Theorem 1.78 hold and if in addition

$$\frac{I_{tT}}{I_T} \to f(t) \quad \text{as } T \to \infty \text{ for all } t \in [0, 1] \tag{1.8.35}$$

where f is continuous, then

$$\mathcal{L}\left(I_T^{(-1/2)} M_{tT};\ 0 \leq t \leq 1\right) \to \mathcal{L}(X) \ \text{(stably)} \tag{1.8.36}$$

where X, conditionally on η^2, is a continuous Gaussian martingale with $\langle X \rangle_t = f(t)\eta^2$.

PROOF Condition (1.8.35) implies that

$$\mathcal{E}_t\left(G^{Y^T}(\lambda)\right) \xrightarrow{p} \exp\left(-\frac{1}{2}\lambda^2\eta^2 f(t)\right) \tag{1.8.37}$$

and

$$\sup_{0 \leq t \leq 1} \left|\langle Y^T \rangle_t - \eta^2 f(t)\right| \xrightarrow{p} 0\,. \tag{1.8.38}$$

The result follow from Theorem 1.78 and the tightness following arguments in in [46]. ∎

We now consider a multidimensional version of Theorem 1.78 due to Sorensen [68].

THEOREM 1.80

Let $M = (M, \ldots, M_n)^T$ be an n-dimensional square integrable $\{\mathcal{F}_t\}$-martingale with the quadratic variation matrix $[M]$ and set $H_t = E(M_t M_t^T)$. Suppose there exists a nonrandom vector-valued function $k_t = (k_{1t}, \ldots, k_{nt})^T$ with $k_{it} > 0$ and $k_{it} \to \infty$ as $t \to \infty$ for $1 \le i \le n$ such that the following conditions hold as $t \to \infty$:

$$\text{(i)}\ \ k_{it}^{-1} E\left(\sup_{s \le t} |\Delta M_{is}|\right) \to 0,\ \ 1 \le i \le n\,, \tag{1.8.39}$$

$$\text{(ii)}\ \ K_t^{-1}[M]_t K_t^{-1} \xrightarrow{p} \eta^2 \ \text{ where } \ K_t = \ \text{diag}\ (k_{1t}, \ldots, k_{nt})$$

and η^2 is a random nonnegative definite matrix, and

$$\text{(iii)}\ \ K_t^{-1} H_t K_t^{-1} \to \Sigma$$

where $\sum$ is a positive definite matrix.

Then

$$K_t^{-1} M_t \to Z \ \ (\text{stably}) \tag{1.8.40}$$

where the distribution of Z is the normal variance mixture with characteristic function

$$Q(u) = E\left(\exp\left(-\tfrac{1}{2}u^T \eta^2 u\right)\right),\ \ u = (u_1, \ldots, u_n)^T\,, \tag{1.8.41}$$

$$\mathcal{L}\left(\left(K_t[M]_t^{-1}K_t\right)^{1/2} K_t^{-1} M_t \,|\, \left(\det\,(\eta^2) > 0\right)\right)$$

$$\to N\,(0, I_n)\ \ (\textit{mixing})\,, \tag{1.8.42}$$

and

$$\mathcal{L}\left(M_t^T [M]_T^{-1} M_t \middle| \left(\det\left(\eta^2\right) > 0\right)\right) \to \chi_n^2 \text{ (mixing)}. \tag{1.8.43}$$

Here $N(0, I_n)$ denotes the multivariate normal distribution with mean zero and the covariance matrix I_n and χ_n^2 denotes the chi-square distribution with n degree of freedom. Here I_n is the identity matrix of order n.

PROOF The first part follows from Theorem 1.78 by the Cramér–Wold device. The stability in (1.8.40) implies that, conditionally on $[\det(\eta^2) > 0)$,

$$\eta^{-1} K_t^{-1} M_t \to N(0, I_n) \text{ (stably)} \tag{1.8.44}$$

and the results (1.8.42) and (1.8.43) follow from this under condition (ii) of (1.8.39). ∎

1.9 Diffusion-Type Processes

Diffusion Processes

Let X_t denote the coordinates of a sufficiently small particle suspended in a liquid at an instant t. Suppose the velocity of the motion of the liquid at the point x and the instant t is equal to $g(t, x)$. Further suppose that the fluctuational component of the displacement is a random variable whose distribution depends on the position x of the particle, the instant t at which the displacement is observed and the quantity Δt which is the length of the interval of the time during which its displacement is observed. Suppose the average value of this displacement is zero independent of t, X_t, and Δt. Thus, the displacement of the particle can be written in the form

$$X_{t+\Delta t} - X_t = g(t, X_t)\, dt + \varepsilon_{t, X_t, \Delta t} \tag{1.9.1}$$

where $E\varepsilon_{t,X_t,\Delta t} = 0$. If $g(t, x) = 0$ and the distribution of $\varepsilon_{t,x_t,\Delta t}$ is independent of x and t as in the case of the Wiener process, then $E\varepsilon_{t,X_t,\Delta t}^2 = \ell \Delta t$ for some constant $\ell > 0$. It is natural to assume that the properties of the medium change slightly for small changes in t and x. This leads to a homogeneous process, and it may be assumed that

$$\varepsilon_{t,X_t,\Delta t} = \sigma(t, X_t)\varepsilon_{t,\Delta t} \tag{1.9.2}$$

where $\sigma(t, x)$ characterizes the properties of the medium at the point x at the instant t and $\varepsilon_{t,\Delta t}$ is the value of the increment that is obtained in the homogeneous case under the condition $\sigma(t, x) = 1$. In other words, $\varepsilon_{t,\Delta t}$ must be distributed like the increment of a Wiener process $W(t)$, namely $W(t + \Delta t) - W(t)$. Hence,

$$X_{t+\Delta t} - X_t \simeq g(t, X_t)\,\Delta t + \sigma(t, X_t)[W(t + \Delta t) - W(t)]\,. \tag{1.9.3}$$

This leads to the stochastic differential equation

$$dX_t = g(t, X_t)\,dt + \sigma(t, X_t)\,dW(t), \quad t \geq 0\,, \tag{1.9.4}$$

which can be taken as the starting point of a diffusion process. The stochastic differential equation is interpreted in the form

$$X_t - X_0 = \int_0^t g(s, X_s)\,ds + \int_0^t \sigma(s, X_s)\,dW(s), \quad t \geq 0\,, \tag{1.9.5}$$

and the precise meaning in which the stochastic integral

$$\int_0^t \sigma(s, X_s)\,dW(s)$$

is defined and the properties of such integrals are discussed earlier in this chapter (cf. [5]).

Ito Stochastic Differential Equation

Consider the stochastic differential equation

$$X_t = X_s + \int_s^t \sigma(u, X_u)\,dW_u + \int_s^t g(u, X_u)\,du, \; 0 \leq s \leq t, \; t \geq 0 \tag{1.9.6}$$

where X_s is an $\mathcal{F}_s$-measurable random variable independent of $\{W_u - W_v,\ u \geq v \geq s\}$. Let $\mathcal{F}_{s,t}$ be the σ-algebra generated by X_s and $\mathcal{F}^W_{s,t}$ where $\mathcal{F}^W_{s,t}$ is the completion of the σ-algebra generated by $\{W_u - W_v,\ t \geq u \geq v \geq s\}$. Let $\mathcal{F}_t \equiv \mathcal{F}_{0,t}$. Suppose that $g : R_+ \times R^d \to R^d$ is measurable and σ is a $d \times m$ ordered measurable matrix with $\sigma_{ij} : R_+ \times R^d \to R$ such that there exists a constant $K > 0$ satisfying

$$|g(u,x) - g(u,y)| \leq K\,|x - y|\,,$$

$$|\sigma(u,x) - \sigma(u,y)| \leq K|x - y|, \text{ and} \tag{1.9.7}$$

$$|g(u,x)|^2 + |\sigma(u,x)|^2 \leq K\left(1 + |x|^2\right)\,.$$

Under the conditions stated above, the equation

$$dX_t = g\,(t, X_t)\,dt + \sigma\,(x, X_t)\,dW_t,\ \ X_s = x \in R^d,\ t \geq s \tag{1.9.8}$$

has a unique, continuous, strong solution $X_t = X_{s,x}(t,\omega)$ and for each $t \geq s,\ X_t$ is $\mathcal{B}(R^d) \times \mathcal{F}^W_{s,t}$-measurable.

Let $s = 0$ and $X_0 = x_0 \in R^d$ where x_0 is independent of $\mathcal{F}^W_{0,T}$. Then the equation

$$dX_t = g\,(t, X_t)\,dt + \sigma(t, X_t)\,dW_t,\ \ 0 \leq t \leq T,\ \ X_0 = x_0 \in R^d \tag{1.9.9}$$

has a unique solution $\{X_t\}$ that is a continuous Markov process relative to $\{\mathcal{F}_t\}$. If g and σ depend only on X_t but not t, then the process is a homogeneous Markov process.

Stochastic Taylor's formula: The following result leading to a stochastic Taylor expansion of a diffusion process $X^{(\varepsilon)}$ satisfying the stochastic differential equation

$$dX_t^{(\varepsilon)} = \mu\left(\varepsilon, X_t^{(\varepsilon)}\right)dt + \varepsilon\ \sigma\left(X_t^{(\varepsilon)}\right)\,dW_t, X_0^{(\varepsilon)} = x \in (l, r) \tag{1.9.10}$$

is due to Azencott [3, p. 239–285]. Suppose $\mu(0,u) > 0$ for all $u \in (l,r)$. Here $\{W_t\}$ is the standard Wiener process.

THEOREM 1.81

If $\mu(\varepsilon, u) \in C^{N+1}([0,\infty) \times (l, r))$ and $\sigma(u)$ is $C^{N+1}((l, r))$, then there exist real-valued semimartingales $g_j = (g_j(t))$, $1 \le j \le N$, with $g_j(0) = 0$, continuous on $[0, \zeta^0)$ and such that

$$X_t^{(\varepsilon)} = x(t) + \sum_{j=1}^{N} \varepsilon^j g_j(t) + \varepsilon^{(N+1)} R_{N+1}^{(\varepsilon)}(t)$$

$$\text{if } t < \zeta^{(0)} \wedge \zeta^{(\varepsilon)}, \tag{1.9.11}$$

$$R_{N+1}^{(\varepsilon)}(t) = \infty \quad \text{if } t \ge \zeta^{(0)} \wedge \zeta^{(\varepsilon)}, \tag{1.9.12}$$

and for all $T \in (0, \zeta^0)$,

$$\lim_{\varepsilon \to 0,\, r \to +\infty} P\left(\sup_{0 \le s \le T} \left| R_{N+1}^{(\varepsilon)}(s) \right| \ge r \right) = 0 . \tag{1.9.13}$$

Moreover,

$$\lim_{\varepsilon \to 0} P\left(\zeta^{(\varepsilon)} < \zeta^{(0)} \right) = 0 . \tag{1.9.14}$$

Here $x(t)$ is the solution of the ordinary differential equation

$$dx(t) = \mu(0, x(t))dt, \quad x(0) = x, \tag{1.9.15}$$

$$\zeta^{(\varepsilon)} = \inf\left\{ t \ge 0 \,:\, X_t^{(\varepsilon)} \notin (l, r) \right\}, \tag{1.9.16}$$

and $\zeta^{(0)}$ is the deterministic explosion time of $x(t)$; that is, $\zeta^{(0)} = t(r-)$ where

$$t(a) = \int_x^a \frac{du}{\mu(u, 0)} = x^{-1}(a) . \tag{1.9.17}$$

REMARK 1.70 It is known that the semimartingales $g_j, 1 \le j \le N$ in (1.9.11) are uniquely determined as the solutions of a system of stochastic differential equations. The process g_1 is defined for $t < \zeta^0$ by

$$dg_1(t) = \left[\mu'_\varepsilon(0, X(t)) + \mu'_u(0, X(t))g_1(t)\right]dt$$

$$+\sigma(X(t))dW_t, \ g_1(0) = 0 \tag{1.9.18}$$

where μ'_ε and μ'_u are the partial derivatives of $\mu(\varepsilon, u)$ with respect to ε and u, respectively, and $\{W_t\}$ is the standard Wiener process. In fact,

$$g_1(t) = \mu(0, X(t))\left[\int_0^t \frac{\mu'_\varepsilon(0, X(s))}{\mu(0, X(s))}ds + \int_0^t \frac{\sigma(X(s))}{\mu(0, X(s))}dW_s\right], \tag{1.9.19}$$

and hence, $\{g_1(t), \ t < \zeta^0\}$ is a Gaussian process. ∎

In our discussion later, we will come across instances where we need to build martingales out of diffusion processes. A method that will lead to appropriate function $Y_t = f(X_t)$ to be martingales, using eigen functions, is discussed next.

Eigen Functions and Martingales

The generator L of a Markov process $\{X_t\}$ is defined by

$$Lf(x) = \lim_{\Delta\to 0} \frac{1}{\Delta}\left[(\pi_\Delta f)(x) - f(x)\right] \tag{1.9.20}$$

whenever it exists where $(\pi_\Delta f)(x) = E[f(X_\Delta)|X_0 = x]$. It is known that the domain $\mathcal{D}$ of the generator L of a diffusion contains the class of bounded, twice continuously differentiable functions with bounded derivatives. For an eigenfunction $\phi \in \mathcal{D}$, with the eigenvalue λ, one can prove, using the Markov property, that

$$\frac{\partial}{\partial\Delta}\pi_\Delta(\phi)(x) = L\pi_\Delta(\phi)(x) = -\lambda\pi_\Delta(\phi)(x)\,, \tag{1.9.21}$$

and hence

$$(\pi_\Delta\phi)(x) = e^{-\lambda\Delta}\phi(x)\,. \tag{1.9.22}$$

The domain $\mathcal{D}$ can be extended to $\mathcal{D}^*$ consisting of the class of twice continuously differentiable functions f for which the process

$$N_t = f(X_t) - f(X_0) - \int_0^t Lf(X_s)\,ds \tag{1.9.23}$$

is a martingale. If

$$dX_t = a(X_t)\,dt + \sigma(X_t)\,dW_t\,, \tag{1.9.24}$$

then, by the Ito's lemma,

$$N_t = \int_0^t f'(X_s)\,\sigma(X_s)\,dW_s\,, \tag{1.9.25}$$

and hence a sufficient condition for $f \in \mathcal{D}^*$ is that

$$\int_0^t E\left[f'(X_s)^2\,\sigma^2(X_s)\right]ds < \infty\,. \tag{1.9.26}$$

Let $\mathcal{D}^{**}$ be the class of twice continuously differentiable functions satisfying (1.9.26). Let

$$Y_t = e^{\lambda t}\phi(X_t) \tag{1.9.27}$$

where ϕ is an eigen function corresponding to the eigen value λ. By the Ito's formula,

$$Y_t = Y_0 + \int_0^t e^{\lambda s}\left[L\phi(X_s) + \lambda\phi(X_s)\right]ds + \int_0^t e^{\lambda s}\phi'(X_s)\,\sigma(X_s)\,dW_s$$

$$= Y_0 + \int_0^t e^{\lambda s}\phi'(X_s)\,\sigma(X_s)\,dW_s\,. \tag{1.9.28}$$

Hence, if $\phi \in \mathcal{D}^{**}$, then Y is a martingale.

REMARK 1.71 There are many physical, biological, economic, and social phenomena that can be reasonably modeled by diffusion processes. For instance (cf. [32]),

(i) molecular motions of particles subject to interaction,

(ii) security price fluctuations in a perfect market,

(iii) communications systems with noise,

(iv) neurophysiological activity with disturbances,

(v) community relationships,

(vi) gene substitution in evolutionary development.

∎

It is convenient to consider a diffusion process from a different viewpoint for stochastic modeling purposes. A diffusion process can also be defined as a continuous time parameter process with continuous paths almost surely and with a strong Markov property. Every diffusion process has the property that for every $\varepsilon > 0$

$$\lim_{h \downarrow 0} \frac{1}{h} P(|X(t+h) - X(t)| > \varepsilon | X(t) = x) = 0$$

for all x in the state space. Suppose

$$\mu(x,t) = \lim_{h \downarrow 0} \frac{1}{h} E[(X(t+h) - X(t)) | X(t) = x]$$

and

$$\sigma^{(2)}(x,t) = \lim_{h \downarrow 0} \frac{1}{h} E\left[(X(t+h) - X(t))^2 | X(t) = x\right]$$

exist for all x and t. The coefficient $\mu(x,t)$ is called the *drift* and $\sigma(x,t)$ is the *diffusion.*

Let $X(t)$ be a time homogeneous diffusion with drift $\mu(x)$ and diffusion $\sigma(x)$. Consider

$$Y^{\lambda}(t) = \exp\left[\lambda X(t) - \lambda \int_0^t \mu(X(s))ds - \frac{1}{2}\lambda^2 \int_0^t \sigma^2(X(s))ds\right] \quad (1.9.29)$$

for $t > 0$. Then, under some regularity conditions, $\{Y^\lambda(t), \mathcal{F}_t, t > 0\}$ is a martingale where $\mathcal{F}_t = \sigma\{X(s) : s \le t\}$. If $X(t)$ is a standard Wiener process, then $\mu(x) \equiv 0$ and $\sigma^2(x) \equiv 1$ and

$$Y^\lambda(t) = \exp\left[\lambda X(t) - \lambda^2 t/2\right]. \tag{1.9.30}$$

Conversely, if Y^λ is a martingale for every λ real, then X is a diffusion. In fact $\{X(t), t \ge 0\}$ is a diffusion process (under some regularity conditions) with drift $\mu(x)$ and diffusion $\sigma(x)$ if and only if for every bounded and twice continuously differentiable f, the process

$$Z_f(t) = f(X(t)) - f(X(0))$$

$$-\int_0^t \left[\frac{1}{2}\sigma^2(X(s))f''(X(s)) + \mu(X(s))f'(X(s))\right] ds \tag{1.9.31}$$

is a martingale. Similar result also holds for nonhomogeneous diffusion processes (see [32, p. 377]).

Stochastic Modeling (cf. [32]

Stochastic models evolving in discrete time have the structure

$$Z_{k+1} = f(Z_k, s_k) + \varepsilon_k$$

or in general

$$Z_{k+1} = f(Z_k, s_k, \varepsilon_k)$$

where Z_{k+1} is the characteristic at the $(k+1)$th generation, with Z_{k+1} depending on (i) Z_k, the state at the kth generation, (ii) s_k, other randomly varying or fixed parameter at kth generation, and (iii) ε_k, "noise." Here, the variables s_k and ε_k can be interpreted as, for instance, $\{s_k\}$, the stochastic or deterministic effect and $\{\varepsilon_k\}$, the demographic or sampling effect in the biological context.

A continuous time version of such a process (as an extension of the above discrete set up) can be modeled by a diffusion process.

Examples of Diffusion Processes (cf. [32])

Example 1.10 **(Wiener process (Brownian motion) W with drift)**
Here, $W(0) = 0$ and $W(t) - W(s)$ is normal with independent increments with mean $E[W(t) - W(s)] = \mu|s - t|$ and $var[W(t) - W(s)] = \sigma^2|t - s|$.
□

Example 1.11 **(Ornstein–Uhlenbeck process V)**
The process V is a solution of the stochastic differential equation

$$dV(t) = -\alpha V(t)dt + \sigma dW(t), \quad t \geq 0, \quad \alpha > 0\,. \tag{1.9.32}$$

A Wiener process models the position of a particle in a liquid, whereas the Ornstein–Uhlenbeck process models the velocity of the particle. Two factors affect the velocity in a short time: (i) the functional resistance of the surrounding medium, which is assumed to reduce the magnitude of velocity by a proportional amount, and (ii) the change in velocity due to random collisions with neighboring particles. □

One can build from one diffusion process several other diffusions by using Ito's lemma.

(i) **Geometric Wiener process Y:** Suppose $\{W(t), t \geq 0\}$ is a Wiener process with drift μ and diffusion σ. Then the process $Y = e^W$ is called the *geometric Wiener process (geometric Brownian motion)*. Applying the Ito's formula, we obtain

$$dY(t) = \left(\mu + \frac{1}{2}\sigma^2\right) Y(t)dt + \sigma\, Y(t)\, dW^*(t) \tag{1.9.33}$$

where W^* is a standard Wiener process. This is used to model prices of shares in a perfect market. Note that prices are nonnegative and exhibit long-run exponential growth (or decay). This process is also used for modeling population growth. Note that

$$dW = \mu dt + \sigma dW^*(t)\,.$$

Since $Y = e^W \equiv F(W)$, it follows that

$$dY = F_w(W)dW + \frac{1}{2}F_{ww}(W)\sigma^2 dt$$

$$= e^W dW + \frac{1}{2}e^W \sigma^2 dt$$

$$= Y\left[\mu dt + \sigma dW^*(t)\right] + \frac{1}{2}Y\sigma^2 dt$$

$$= \left(\mu Y + \frac{1}{2}\sigma^2 Y\right) dt + \sigma Y dW^*(t).$$

(ii) **Bessel process Y:** Let $Y(t) = \sqrt{Z(t)}, t \geq 0$ where $Z(t) = W_1^2(t) + \cdots + W_n^2(t)$ and W_i, $1 \leq i \leq n$ are independent, standard Wiener processes. Then $\{Y(t), t \geq 0\}$ is a diffusion process with the drift coefficient $\frac{n-1}{2Y(t)}$ and the diffusion coefficient unity.

Example 1.12 **(Modeling sunspot activity) [2]**

After subtracting the mean, the number of sun spots $\zeta(t)$ can be modeled by the stochastic differential equation

$$d\zeta'(t) + \left(a_1\zeta'(t) + a_2\zeta(t)\right) dt = dW(t) \tag{1.9.34}$$

where $\zeta'(t)$ denotes the derivative of $\zeta(t)$ in an appropriate sense. □

For other examples of stochastic modeling by diffusion processes with applications to biology, demography, economics, etc., see [32].

REMARK 1.72 In the models described above, one of the basic problems of interest is the estimation of the parameters involved in the stochastic differential equation on the basis of a realization of the process satisfying the stochastic differential equation. ∎

Diffusion-Type Processes

Let $(\Omega, \mathcal{F}, P)$ be a probability space and $\{\mathcal{F}_t, 0 \leq t \leq T\}$ be a right-continuous complete filtration. Let $W = \{W_t, \mathcal{F}_t\}$ be a Wiener process defined on $(\Omega, \mathcal{F}, P)$. Let $Z = \{Z_t, \mathcal{F}_t, 0 \leq t \leq T\}$ be a supermartingale of the form

$$Z_t = \exp\left(\int_0^t \beta_s dW_s - \frac{1}{2}\int_0^t \beta_s^2 ds\right) \tag{1.9.35}$$

where $\{\beta_t, \mathcal{F}_t, t \geq 0\}$ is nonanticipative and

$$P\left(\int_0^T \beta_s^2 ds < \infty\right) = 1 . \tag{1.9.36}$$

Suppose that

$$P\left(\int_0^t \gamma^2 ds < \infty\right) = 1,\ V_s = Z_s \rho_s .$$

Then, by Ito's lemma,

$$Z_t = 1 + \int_0^t \gamma_s dW_s \text{ with } \gamma_s = Z_s \beta_s . \tag{1.9.37}$$

If $EZ_T = 1$, then the process

$$\tilde{W}_t = W_t - \int_0^t \beta_s ds,\ 0 \leq t \leq T \tag{1.9.38}$$

is a Wiener process on $(\Omega, \mathcal{F}, \tilde{P})$ with respect to the filtration $\{\mathcal{F}_t, 0 \leq t \leq T\}$, and the probability measure $\tilde{P}$ is given by

$$\left(d\tilde{P}/dP\right)(\omega) = Z_T(\omega) . \tag{1.9.39}$$

by Girsanov's theorem. See Corollary 1.13. In fact, the following result is a rephrasing of the result given in Corollary 1.13.

Let $(\Omega, \mathcal{F}, P)$ be a probability space and $\{\mathcal{F}_t, 0 \leq t \leq T\}$ be a right-continuous complete filtration. Let $W = \{W_t, \mathcal{F}_t, \ 0 \leq t \leq T\}$ be a Wiener process defined on $(\Omega, \mathcal{F}, P)$. Let $Z = \{Z_t, \mathcal{F}_t, \ 0 \leq t \leq T\}$ be a nonnegative continuous supermartingale with

$$Z_t = 1 + \int_0^t \gamma_s dW_s \tag{1.9.40}$$

where $\gamma = \{\gamma_t, \mathcal{F}_t, \ 0 \leq t \leq T\}$ is such that

$$P\left[\int_0^T \gamma_s^2 ds < \infty\right] = 1 \,. \tag{1.9.41}$$

Suppose $EZ_T = 1$. Then $Z = \{Z_t, \mathcal{F}_t, \ 0 \leq t \leq T\}$ is a martingale and define $d\tilde{P} = Z_T(\omega)dP$. Then, on $(\Omega, \mathcal{F}, \tilde{P})$, the process $\tilde{W} = \{\tilde{W}_t, \mathcal{F}_t, \ 0 \leq t \leq T\}$ defined by

$$\tilde{W}_t = W_t - \int_0^t Z_s^* \gamma_s ds \tag{1.9.42}$$

is a Wiener process (here $Z_s^* = Z_s^{-1}$ if $Z_s > 0$ and $Z_s^* = 0$ if $Z_s = 0$).

REMARK 1.73 Let $\gamma_i = \{\gamma_i(t), \mathcal{F}_t, \ 0 \leq t \leq T\}, \ 1 \leq i \leq n$ be random processes with $P[\int_0^T \gamma_i^2(t)dt < \infty] = 1, \ 1 \leq i \leq n$ and let $W = \{(W_1(t), \ldots, W_n(t)), \mathcal{F}_t, \ 0 \leq t \leq T\}$ be an n-dimensional Wiener process. Define

$$Z_t = 1 + \int_0^t \sum_{i=1}^{n-1} \gamma_i(s)dW_i(s), \ 0 \leq t \leq T \,. \tag{1.9.43}$$

If $E[Z_T] = 1$, then the process

$$\tilde{W}_t = W_t - \int_0^t Z_s^* \gamma_s \, ds \tag{1.9.44}$$

is a Wiener process with respect to the measure $d\tilde{P} = Z_T dP$ and the filtration $\{\mathcal{F}_t\}$. ∎

Let $(\Omega, \mathcal{F}, P)$ be a probability space and $\{\mathcal{F}_t,\ 0 \le t \le T\}$ be a filtration satisfying the usual conditions. Let $W = \{W_t, \mathcal{F}_t\},\ 0 \le t \le T$ be a Wiener process. Recall that a continuous random process $\zeta = \{\zeta_t, \mathcal{F}_t,\ 0 \le t \le T\}$ is called an *Ito process* if there exists nonanticipative processes $\alpha = \{\alpha_t, \mathcal{F}_t\}$ and $\beta = \{\beta_t, \mathcal{F}_t\},\ 0 \le t \le T$ such that $P\{\int_0^T |\alpha_t| dt < \infty\} = 1,\ P\{\int_0^T \beta_t^2 dt < \infty = 1\}$ and

$$\zeta_t = \zeta_0 + \int_0^t \alpha(s, \omega) ds + \int_0^t \beta(s, \omega) dW_s \text{ a.s.}, \tag{1.9.45}$$

and the process $\{\zeta_t\}$ is said to have the stochastic differential

$$d\zeta_t = \alpha(t, \omega)dt + \beta(t, \omega)dW_t\ . \tag{1.9.46}$$

Furthermore, an Ito process $\zeta = \{\zeta_t, \mathcal{F}_t,\ 0 \le t \le T\}$ is a *process of diffusion type* if $\alpha(t, \omega)$ and $\beta(t, \omega)$ are $\mathcal{F}_t^\zeta$-measurable for almost all $t,\ 0 \le t \le T$ where $\mathcal{F}_t^\zeta = \sigma\{\zeta(s),\ 0 \le s \le t\}$.

THEOREM 1.82 **[44]**
Let $\zeta = (\zeta_t, \mathcal{F}_t\}$ be an Ito process with the stochastic differential

$$d\zeta_t = A_t(\omega)dt + b_t(\zeta)dW_t \tag{1.9.47}$$

and $\eta = (\eta_t, \mathcal{F}_t)$ be a process of diffusion type with

$$d\eta_t = a_t(\eta)dt + b_t(\eta)dW_t,\ \eta_0 = \zeta_0\ , \tag{1.9.48}$$

where ζ_0 is an $\mathcal{F}_0$-measurable random variable with $P(|\zeta_0| < \infty) = 1$. Suppose the following conditions hold:
(i) $a_t(x)$ and $b_t(x)$ satisfy the following conditions for $t \in [0, T]$, $x \in C_T = C[0, T]$:

$$(\alpha)\ |a_t(x)) - a_t(y)|^2 + |b_t(x) - b_t(y)|^2$$

$$\leq L_1 \int_0^t |x_s - y_s|^2 \, dK(s) + L_2 \, |x_t - y_t|^2 \,,$$

$$(\beta) a_t^2(x) + b_t^2(x) \leq L_1 \int_0^t \left(1 + x_s^2\right) dK(s) + L_2 \left(1 + x_t^2\right)$$

(1.9.49)

where $L_1 > 0, L_2 > 0$ and $0 \leq K(s) \leq 1$, with $K(\cdot)$ nondecreasing and right-continuous,
(ii) for any $0 \leq t \leq T$, the equation

$$b_t(\zeta)\alpha_t(\omega) = A_t(\omega) - a_t(\zeta) \tag{1.9.50}$$

has (P-a.s.) bounded solution,
(iii) $P\left(\int_0^T \alpha_t^2(\omega)dt < \infty\right) = 1$, and
(iv) $E \exp\left(-\int_0^T \alpha_t(\omega)dW_t - \frac{1}{2}\int_0^T \alpha_t^2(\omega)dt\right) = 1$.
Then μ_ζ is equivalent to μ_η and (P-a.s.)

$$\frac{d\mu_\eta}{d\mu_\zeta}(\zeta) = E\left\{\exp\left(-\int_0^T \alpha_t(\omega)dW_t - \frac{1}{2}\int_0^T \alpha_t^2(\omega)dt\right) |\mathcal{F}_T^\zeta\right\} \tag{1.9.51}$$

(here μ_ζ and μ_η are the probability measures generated by the processes ζ and η, respectively, on $C[0, T]$ endowed with the sup norm).

Sketch of proof: Note that

$$\alpha_t(\omega) = b_t^*(\zeta)\,[A_t(\omega) - a_t(\zeta)] \tag{1.9.52}$$

where $b_t^* = b_t^{-1}$ if $b_t \neq 0$ and $b_t^* = 0$ if $b_t = 0$.

Let

$$Z_t = \exp\left(-\int_0^t \alpha_s(\omega)dW_s - \frac{1}{2}\int_0^t \alpha_s^2(\omega)ds\right), \tag{1.9.53}$$

and

$$d\tilde{P} = Z_t dP\,. \tag{1.9.54}$$

Then, by the Girsanov theorem,

$$\tilde{W}_t = W_t + \int_0^t \alpha_s(\omega)ds \tag{1.9.55}$$

is a Wiener process with respect to $\{\mathcal{F}_t,\ 0 \le t \le T\}$ under $\tilde{P}$. But

$$\eta_0 + \int_0^t a_s(\zeta)ds + \int_0^t b_s(\zeta)d\tilde{W}_s$$

$$= \eta_0 + \int_0^t a_s(\zeta)ds + \int_0^t b_s(\zeta)\alpha_s(\omega)ds + \int_0^t b_s(\zeta)dW_s$$

$$= \eta_0 + \int_0^t a_s(\zeta)ds + \int_0^t b_s(\zeta)b_s^*(\zeta)\,[A_s(\omega) - a_s(\zeta)]\,ds + \int_0^t b_s(\zeta)dW_s$$

$$= \eta_0 + \int_0^t A_s(\omega)ds + \int_0^t b_s(\zeta)dW_s = \zeta_t\,. \tag{1.9.56}$$

Hence, $\zeta = \{(\zeta_t, \mathcal{F}_t)\}$ on $(\Omega, \mathcal{F}, \tilde{P})$ satisfies the same stochastic differential equation that $\eta = \{(\eta_t, \mathcal{F}_t)\}$ on $(\Omega, \mathcal{F}, P)$ does. Therefore,

$$\tilde{P}(\zeta \in A) = P(\eta \in A) \tag{1.9.57}$$

and

$$\mu_\eta(A) = P(\eta \in A) = \tilde{P}(\zeta \in A) = \int_{[\zeta \in A]} Z_T(\omega) dP(\omega)$$

$$= \int_A E\left[Z_T | \mathcal{F}_T^\zeta\right]_{\zeta = x} d\mu_\zeta(x) . \tag{1.9.58}$$

Hence, $\mu_\eta << \mu_\zeta$ and the relation (1.9.58) holds. One can check that $\mu_\zeta <<$ μ_η using the fact that $\frac{d\tilde{P}}{dP} = Z_T$ and $P(Z_T = 0) = 0$.

1.10 Point Processes

Univariate Point Process (Simple)

A realization of a point process over $[0, \infty)$ can be described by a sequence τ_n with values in $[0, \infty)$ such that $\tau_0 = 0$, $\tau_n < \tau_{n+1}$ if $\tau_n < \infty$. Let $\tau_\infty = \lim_{n\to\infty} \tau_n$. Define

$$N_t = \begin{cases} n & \text{if } t \in [\tau_n, \tau_{n+1}) , \ n \geq 0 \\ +\infty & \text{if } \quad t \geq \tau_\infty . \end{cases}$$

Observe that $\{N_t\}$ is a right-continuous step function such that $N_0 = 0$, and its jumps are upward jumps of size 1. If $\{\tau_n\}$ are random variables defined on a probability space $(\Omega, \mathcal{F}, P)$, then $\{\tau_n, \ n \geq 1\}$ is called a *point process,* and the process $\{N_t, \ t \geq 0\}$ is called a *counting process* (or *point process*). If $E[N_t] < \infty$, $t \geq 0$, then the point process is said to be integrable. Assume that $P[N_t < \infty, \ t \geq 0] = 1$ (equivalently $\tau_\infty = \infty$ a.s.).

More formally, we have the following definition.

DEFINITION 1.54 *Let $(\Omega, \mathcal{F}, P)$ be a complete probability space and $F = \{\mathcal{F}_t, t \geq 0\}$ be a right-continuous complete filtration. Let $T = \{\tau_n, n \geq 1\}$ be a sequence of stopping times with respect to $F = \{\mathcal{F}_t, t \geq 0\}$ such that*
(i) $\tau_1 > 0$ a.s.-P,
(ii) $\tau_n < \tau_{n+1}$ on $[\tau_n < \infty]$, a.s.-P, and
(iii) $\tau_n = \tau_{n+1}$ on $[\tau_n = \infty]$, a.s.-P.

Let $N_t = \sum_{n \geq 1} I_{[\tau_n \leq t]}$ with $\tau_\infty = \lim_{n\to\infty} \tau_n$, $\tau_0 \equiv 0$. Then N_t is called a point process corresponding to $T = \{\tau_n, n \geq 1\}$.

Properties

(i) The process N_t has right-continuous trajectories with left limits, and they are piecewise constant with unit jumps.

(ii) If $N = \{N_t, \mathcal{F}_t, t \geq 0\}$ is a point process and σ is a stopping time with respect to $\{\mathcal{F}_t, t \geq 0\}$, then the process $\{N_{t\wedge\sigma}, \mathcal{F}_t, t \geq 0\}$ is a point process.

(iii) Note that $N_{\tau_n} \leq n$ a.s. $[P]$. Define $N^{(n)} = \{N_{t\wedge\tau_n}, \mathcal{F}_t, t \geq 0\}$. Then $P(0 \leq N_{t\wedge\tau_n} \leq n) = 1$ for all $t \geq 0$, and hence, $N^{(n)}$ is a bounded nondecreasing submartingale of class D. Hence, it has a Doob–Meyer decomposition

$$N_{t\wedge\tau_n} = m_t^{(n)} + A_t^{(n)}$$

where $\{m_t^{(n)}, \mathcal{F}_t, t \geq 0\}$ is a uniformly integrable martingale and $\{A_t^{(n)}, \mathcal{F}_t, t \geq 0\}$ is a natural increasing process. It can be shown that $N = \{N_t, \mathcal{F}_t, t \geq 0\}$ admits, for all $t < \tau_\infty$, the unique decomposition $N_t = m_t + A_t$ where $m = (m_t, \mathcal{F}_t, t < \tau_\infty)$ is a local martingale and $A = (A_t, \mathcal{F}_t, t \geq 0)$ is a natural increasing (and predictable) process (cf. Theorem 1.88). A is called the *compensator* of the point process N.

(iv) Let $N = \{N_t, \mathcal{F}_t, t \geq 0\}$ be a point process and $N_t = m_t + A_t$ be its Doob–Meyer decomposition for $t < \tau_\infty$. Let $\mathcal{F}_t^N$ be the σ-algebra generated by $\{N_s, s \leq t\}$. Let $\bar{N} = \{N_t, \mathcal{F}_t^N, t \geq 0\}$. Suppose $N_t = \bar{m}_t + \bar{A}_t$ is the Doob–Meyer decomposition for $\bar{N}$. This representation of N_t is called the *minimal representation* of the point process N. From now on, we assume that $\{\mathcal{F}_t^N, t \geq 0\}$ is a right continuous family (see Lemma 18.4 of [45, p. 244]).

(v) Let $N_t = \sum_{n\geq 1} I_{\{\tau_n \leq t\}}, t \geq 0$ be a point process and let $\theta = \theta(\omega)$

be a stopping time with respect to the filtration $\{\mathcal{F}_t^N, t \geq 0\}$ such that $P(\theta < \tau_\infty) = 1$. Then there exist Borel measurable functions $Q_n = Q_n(t_1, \ldots, t_n), n \geq 1$ and a constant Q_0 such that

$$\theta(\omega) = \sum_{n \geq 1} I_{[\tau_{n-1} \leq \theta < \tau_n]} Q_{n-1}(\tau_1, \ldots, \tau_{n-1}) ,$$

that is, $\theta(\omega)$ is constant on $[\theta < \tau_1]$ and

$$\theta(\omega) = Q_{n-1}(\tau_1(\omega), \ldots, \tau_{n-1}(\omega)) \text{ on } \tau_{n-1} \leq \theta < \tau_n,\ n > 1 .$$

In other words, there exist a Borel measurable function $\theta_n(t, \ldots, t_n)$ and a constant θ_0 such that

$$\theta = \theta_{n-1}(\tau_1, \ldots \tau_{n-1}) \text{ on the set } [\theta < \tau_n] .$$

In particular

$$\theta \wedge \tau_n = \theta_{n-1} \wedge \tau_n \text{ and } \theta \wedge \tau_k = \theta_{n-1} \wedge \tau_k \text{ if } k < n .$$

LEMMA 1.10

Let $X = \{X_t, \mathcal{F}_t, t \geq 0\}$ be an integrable process. Then X_t is a martingale if for every two-valued stopping time τ_n with respect to $\{\mathcal{F}_t\}$,

$$EX_\tau = EX_0 .$$

PROOF Let s and t be such that $s < t$ and define $A = \{\omega : X_s < E(X_t|\mathcal{F}_s)\}$. Suppose $P(A) > 0$. Define $\tau = t\, I_A + s I_{A^c}$. Observe that $[\tau = t] = A$ and $[\tau = s] = A^c$. Furthermore, $A \in \mathcal{F}_s$ by definition. Note that τ is a stopping time with respect to $\{\mathcal{F}_t\}$ and

$$\begin{aligned} E[X_\tau] &= E[I_A X_t + I_{A^c} X_s] \\ &= E[I_A X_t] + E[I_{A^c} X_s] \\ &= E[I_A E(X_t|\mathcal{F}_s)] + E[I_{A^c} X_s] \\ &< E[I_A X_s + I_{A^c} X_s] = EX_s \text{ (since } P(A) > 0) , \end{aligned}$$

which contradicts the assumption. Hence, $P(A) = 0$. Similarly, we show that $P(B) = 0$ where $B = \{\omega : X_s > E(X_t|\mathcal{F}_s)\}$. Hence, $X_s = E(X_t|\mathcal{F}_s),\ s < t$. ∎

As an application of the above lemma, the following theorem can be proved.

THEOREM 1.83 **[45]**
Let $F_1(t) = P(\tau_1 \le t)$ and suppose that

$$F_i(t) = P\left(\tau_i \le t|\tau_1, \ldots, \tau_{i-1}\right),\ i \ge 2$$

exists as a well-defined regular conditional distribution. Define

$$\bar{A}_t = \sum_{i \ge 1} \bar{A}_t(i)$$

where

$$\bar{A}_t^{(i)} = \int_0^{t \wedge \tau_i} \frac{dF_i(u)}{1 - F_i(u-)},\ i \ge 1\ .$$

Then $\bar{A} = (\bar{A}_t, \mathcal{F}_t^N),\ 0 \le t < \tau_\infty$ is the compensator of the point process $\bar{N} = (N_t, \mathcal{F}_t^N, t \ge 0)$.

REMARK 1.74 Let $N = \{N_t, \mathcal{F}_t, t \ge 0\}$ be a point process with the compensator $A = \{A_t, \mathcal{F}_t, t \ge 0\}$. Suppose

$$A_t = \int_0^t \lambda_s db_s$$

where $\lambda = \{\lambda_t, \mathcal{F}_t, t \ge 0\}$ is predictable nonnegative process and b_t is nonrandom, nonnegative, right-continuous, nondecreasing function. Then the point process N is said to be of the *Poisson type.* ∎

REMARK 1.75 Suppose $N = \{N_t, \mathcal{F}_t, t \ge 0\}$ is a Poisson-type point process. Let $A = \{A_t, \mathcal{F}_t, t \geq 0\}$ be its compensator with

$$A_t = \int_0^t \lambda_s db_s .$$

Suppose $\bar{A} = (\bar{A}_t, \mathcal{F}_t^N, t \geq 0)$ is the compensator in the minimal representation. Then

$$\bar{A}_t = \int_0^t \bar{\lambda}_s db_s$$

where $\bar{\lambda}_s = E(\lambda_s | \mathcal{F}_{s-}^N)$. ∎

THEOREM 1.84 **(Structure of local martingales)**
Let $\bar{N} = (N_t, \mathcal{F}_t^N, t \geq 0)$ be a point process with the compensator $\bar{A} = (\bar{A}_t, \mathcal{F}_t^N, t \geq 0)$ and let $Y = (Y_t, \mathcal{F}_t^N)$ be a $\tau_{\infty}-$-local martingale with right-continuous sample paths. Then Y has the representation

$$Y_t = Y_0 + \int_0^t f_s d\left[N_s - \bar{A}_s\right] \tag{1.10.1}$$

where $f = \{f_t, \mathcal{F}_t^N, t \geq 0\}$ is predictable with

$$P\left(\int_0^t |f_s|\, d\bar{A}_s = \infty,\ t < \tau_{\infty}\right) = 0 .$$

For proof, see [45, p. 282].

REMARK 1.76 Let $\mathcal{X}$ be the space of piecewise constant functions $x = (x_t, t \geq 0)$ such that $x_0 = 0$, $x_t = x_{t-} + (0 \text{ or } 1)$. Let $\mathcal{B}$ be the σ-algebra generated on $\{x : x_s, s \geq 0\}$ and $\mathcal{B}_t = \sigma\{x : x_s, s \leq t\}$. Let $\tau_i(x) = \inf\{s \geq 0 : x_s = i\}$. Set $\tau_i(x) = \infty$ if $\lim_{t\to\infty} x_t < i$. Let $\tau_{\infty}(x) = \lim_{i\to\infty} \tau_i(x)$. Observe that for each $x = (x_t, t \geq 0)$, $x_t = \sum_{i\geq 1} I_{[\tau_i(x)\leq t]}$ and the family of σ-algebras $\mathcal{B}_t$ is right-continuous. ∎

Let μ and $\tilde{\mu}$ be probability measures on $(\mathcal{X}, \mathcal{B})$. Then $X = (x_t, \mathcal{B}_t, \mu)$ and $\tilde{X} = (x_t, \mathcal{B}_t, \tilde{\mu})$ are point processes on $(\mathcal{X}, \mathcal{B})$. Let $A = (A_t(x), \mathcal{B}_t, \mu)$ and $\tilde{A} = (\tilde{A}_t(x), \mathcal{B}_t, \tilde{\mu})$ be the compensators of X and $\tilde{X}$, respectively. Note that

$$A_t(x) = \sum_{i \geq 1} A_t^{(i)}(x) , \tag{1.10.2}$$

where

$$A_t^{(i)}(x) = \int_0^{t \wedge \tau_i(x)} \frac{dF_i(u)}{1 - F_i(u-)} , \tag{1.10.3}$$

$$F_1(t) = \mu \{x : \tau_1(x) \leq t\} , \tag{1.10.4}$$

and

$$F_i(t) = \mu \{x : \tau_i(x) \leq t | \tau_{i-1}(x), \ldots, \tau_1(x)\} . \tag{1.10.5}$$

Hence, μ completely determines the compensator A. Furthermore, the compensator A satisfies the following properties for almost every $x \in \mathcal{X}$ and $t \geq 0$.

$$\begin{aligned}
&\text{(i)} \ A_0(x) = 0 , \\
&\text{(ii)} \ A_s(x) \leq A_t(x) \text{ if } s \leq t , \\
&\text{(iii)} \ A_t(x) = A_{t \wedge \tau_\infty(x)}(x) , \\
&\text{(iv)} \ A_{\tau_\infty(x)-}(x) = A_{\tau_\infty(x)}(x) , \\
&\text{(v)} \ \text{almost all samples paths of } A \text{ are right continuous, and} \\
&\text{(vi)} \ \Delta A_t(x) = A_t(x) - A_{t-}(x) \leq 1 .
\end{aligned} \tag{1.10.6}$$

In the following, we consider processes where A and $\tilde{A}$ are related by the equation

$$\tilde{A}_t(x) = \int_0^t \lambda_s(x) dA_s(x) \text{ on } [t < \tau_\infty] , \text{ a.s. } [\tilde{\mu}] , \tag{1.10.7}$$

where $\lambda = (\lambda_t(x), \mathcal{B}_t)$ is nonnegative predictable and

$$\Delta A_t(x) = 1 \text{ implies } \lambda_t(x) = I(t < \tau_\infty(x)) \text{ a.s. } [\tilde{\mu}] .$$

Note that $\lambda_t(x)$ can be chosen so that

$$\lambda_t(x) \le (\Delta A_t(x))^{-1} . \tag{1.10.8}$$

***THEOREM 1.85* (Absolute continuity of measures)**
A necessary and sufficient condition for the measure $\tilde{\mu}$ to be absolutely continuous with respect to the measure μ is that $\tilde{\mu}$-a.s.,

$$\begin{aligned}
&(i)\ \ \tilde{A}_t(x) = \int_0^t \lambda_s(x) dA_s(x), \quad t < \tau_\infty ,\\
&(ii)\ \ \Delta A_t(x) = 1 \ \textit{implies}\ \Delta\tilde{A}_t(x) = 1, \quad t < \tau_\infty,\ \textit{and}\\
&(iii)\ \int_0^{\tau_\infty} \left(1 - \sqrt{\lambda_t(x)}\,\right)^2 dA_t(x)\\
&\qquad + \sum_{\substack{t<\tau_\infty\\ 0<\Delta A_t(x)<1}} \left(1 - \sqrt{\frac{1-\Delta\tilde{A}_t(x)}{1-\Delta A_t(x)}}\right)^2 (1 - \Delta A_t(x)) < \infty .
\end{aligned} \tag{1.10.9}$$

For proof, see [45, p. 310].

As a special case of Theorem 1.85, we have the following result.

THEOREM 1.86
Suppose the point process $X = (x_t, \mathcal{B}_t, \mu)$ has a continuous compensator. Then, a necessary and sufficient condition for a measure $\tilde{\mu}$ to be absolutely continuous with respect to μ is that $\tilde{\mu}$-a.s.,

$$\tilde{A}_t(x) = \int_0^t \lambda_s(x) dA_s(x),\ t < \tau_\infty , \tag{1.10.10}$$

and

$$\int_0^{\tau_\infty} \left(1 - \sqrt{\lambda_s(x)}\right)^2 dA_s(x) < \infty \tag{1.10.11}$$

where $(\lambda_s(x), \mathcal{B}_s)$ is a nonnegative predictable process. In this case,

$$\frac{d\tilde{\mu}}{d\mu}(t,x) = \exp\left\{ \int_0^t \log \frac{d\tilde{A}_s(x)}{dA_s(x)} dx_s - \left[\tilde{A}_t(x) - A_t(x)\right] \right\} . \tag{1.10.12}$$

COROLLARY 1.15

Let $X = (x_t, \mathcal{B}_t, \mu_\pi)$ be a Poisson process with mean one and $\tilde{X} = (x_t, \mathcal{B}_t, \tilde{\mu})$ be a point process with measure $\tilde{\mu} << \mu_\pi$. Then the process $\tilde{X}$ has the compensator

$$\tilde{A}_t(x) = \int_0^t \lambda_s(x) ds \tag{1.10.13}$$

where $(\lambda_t(x), \mathcal{B}_t)$ is a nonnegative predictable process such that

$$\int_0^\infty \left(1 - \sqrt{\lambda_s(x)}\right)^2 ds < \infty, \ \tilde{\mu} - \ a.s. \tag{1.10.14}$$

In this case

$$\frac{d\tilde{\mu}}{d\mu_\pi}(t, x) = \exp\left\{\int_0^\infty \log \lambda_s(x) dx_s + \int_0^t (1 - \lambda_s(x))\, ds\right\}. \tag{1.10.15}$$

Multivariate Point Process

Let $\{\tau_n\}$ be a point process defined on $(\Omega, \mathcal{F}, P)$ and let $\{Z_n, n \geq 1\}$ be a sequence of random variables taking values in $\{1, 2, \ldots, k\}$ defined on $(\Omega, \mathcal{F}, P)$. Let

$$N_t(i) = \sum_{n \geq 1} I\,[\tau_n \leq t]\, I\,[Z_n = i]\ . \tag{1.10.16}$$

Then $N_t = (N_t(1), \ldots, N_t(k))$ and the sequence $\{(T_n, Z_n), n \geq 1\}$ is called a *k-variate point process.*

Doubly Stochastic Poisson Process

Let $\{N_t\}$ be a point process adapted to $\{\mathcal{F}_t\}$ and λ_t be a nonnegative measurable process defined on $(\Omega, \mathcal{F}, P)$. Suppose that λ_t is $\mathcal{F}_0$-measurable for all $t \geq 0$ and $\int_0^t \lambda_s ds < \infty$ a.s. P. If, for all $0 \leq s \leq t,\ u \in R$,

$$E\left[e^{iu(N_t - N_s)} | \mathcal{F}_s\right] = \exp\left\{\left(e^{iu} - 1\right) \int_s^t \lambda_v dv\right\}, \tag{1.10.17}$$

then the process $\{N_t\}$ is called a *doubly stochastic Poisson process* and λ_t is called its *stochastic intensity*.

Observe that

$$P[N_t - N_s = k|\mathcal{F}_s] = e^{-\int_s^t \lambda_u du} \frac{\left(\int_s^t \lambda_u du\right)^k}{k!}, k \geq 1 . \tag{1.10.18}$$

If $\lambda_t = f(t, Y_t)$ for some measurable function f and for some measurable process $\{Y_t\}$ and if $\mathcal{F}_0$ contains $\sigma\{Y_s : s \geq 0\}$, the $\{N_t\}$ is called a doubly stochastic Poisson process driven by $\{Y_t\}$. This means that a trajectory of a "driving" process Y_t is observed and then the Poisson process is generated with intensity $f(t, Y_t)$.

THEOREM 1.87 **(Characterization)**
Let $\{N_t\}$ be a point process adapted to $\mathcal{F}_t$ and let λ_t be a nonnegative measurable process such that

$$\begin{aligned} &(a)\ \lambda_t \text{ is } \mathcal{F}_0\text{-measurable, and} \\ &(b)\ \int_0^t \lambda_s ds < \infty,\ P - \text{ a.s.} \end{aligned} \tag{1.10.19}$$

Then

$$E\left[\int_0^\infty C_s dN_s\right] = E\left[\int_0^\infty C_s \lambda_s ds\right] \tag{1.10.20}$$

for all $\mathcal{F}_t$-predictable processes C_t if and only if N_t is a doubly stochastic Poisson process with $\mathcal{F}_t$-intensity λ_t (here $\int_0^\infty C_s dN_s = \sum_{n\geq 1} C_{T_n} I_{(T_n<\infty)}$).

For proof, see [6].

DEFINITION 1.55 *Let $\{N_t\}$ be a point process adapted to $\{\mathcal{F}_t\}$ and $\{\lambda_t\}$ be a nonnegative $\mathcal{F}_t$-progressively measurable process, that is, for all $t \geq 0$, $(s, \omega) \to \lambda_s(\omega)$ from $[0, t] \times \Omega \to R_+$ is $\mathcal{B}([0, t]) \times \mathcal{F}_t$-measurable. Suppose that*

$$\int_0^t \lambda_s ds < \infty,\ P - \text{ a.s.}$$

and

$$E\left[\int_0^\infty C_s dN_s\right] = E\left[\int_0^\infty C_s \lambda_s ds\right]$$

for all $\mathcal{F}_t$- predictable process C_t. Then the process $\{N_t\}$ is said to admit the $(P, \mathcal{F}_t)$-intensity λ_t.

REMARK 1.77 Suppose the $\{N_t\}$ admits the $\mathcal{F}_t$-intensity $\{\lambda_t\}$. Then

(i) $M_t = N_t - \int_0^t \lambda_s ds$ is an $\mathcal{F}_t$-local martingale,

(ii) if X_t is $\mathcal{F}_t$-predictable such that $E\left[\int_0^t |X_s| \lambda_s ds\right] < \infty$, then $\int_0^t X_s dM_s$ is an $\mathcal{F}_t$-martingale, and

(iii) if X_t is $\mathcal{F}_t$-predictable such that $\int_0^t |X_s|\lambda_s ds < \infty$ a.s., then

$$\int_0^t X_s dM_s \text{ is an } \mathcal{F}_t\text{-local martingale .}$$

∎

THEOREM 1.88
Let N_t be a point process with $\mathcal{F}_t$-intensity λ_t. Then there exists an $\mathcal{F}_t$-intensity $\tilde{\lambda}_t$ that is $\mathcal{F}_t$-predictable.

PROOF Let $\tilde{\lambda}_t(\omega)$ be the Radon–Nikodym derivative of $P(d\omega)N_t(\omega)dt$ on $\mathcal{P}(\mathcal{F}_t)$ with respect to $P(d\omega)dt$ on $\mathcal{P}(\mathcal{F}_t)$ (here $\mathcal{P}(\mathcal{F}_t)$ is the $\mathcal{F}_t$-predictable σ-algebra generated by sets of the form $(s, t) \times A,\ 0 \le s \le t,\ A \in \mathcal{F}_s$). Then, by definition,

$$E\left[\int_0^\infty C_s dN_s\right] = E\left[\int_0^\infty C_s \lambda_s ds\right] = E\left[\int_0^\infty C_s \tilde{\lambda}_s ds\right] \qquad (1.10.21)$$

for all nonnegative $\mathcal{F}_t$-predictable processes. Furthermore, $(t, \omega) \to \tilde{\lambda}_t(\omega)$ is $\mathcal{P}(\mathcal{F}_t)$-measurable and hence, predictable. Note that

$$\tilde{\lambda}_t(\omega) = \left(\frac{dPdN_u}{dPdu}\right)(t, \omega) \text{ on } \mathcal{P}(\mathcal{F}_t) \ . \tag{1.10.22}$$

Here, $\tilde{\lambda}_t$ is the predictable $(P, \mathcal{F}_t)$ -intensity of N_t. It can be seen that $\tilde{\lambda}_t$ is unique a.s. $P(d\omega)\, dN_t(\omega)$, that is, $\lambda_t^*(\omega) = \tilde{\lambda}_t(\omega)\ P(d\omega)dN_t(\omega)$ a.s. if λ_t^* and $\tilde{\lambda}_t$ are both $\mathcal{F}_t$-predictable $(P, \mathcal{F}_t)$-intensities of N_t. In other words,

$$\lambda_t^*(\omega) = \tilde{\lambda}_t(\omega) \quad \lambda_t^*(\omega)dt - \text{a.s. and}$$

$$\lambda_t^*(\omega) = \tilde{\lambda}_t(\omega) \quad \tilde{\lambda}_t(\omega)dt - \text{a.s.}$$

For proof, apply (1.10.21) for $C_s = I(\lambda_s^* > \tilde{\lambda}_s)I(s \leq a)$ for every $a \in R$. ∎

Stochastic Time Change

THEOREM 1.89 **(Characterization)**
Let N_t be a point process with $\mathcal{F}_t$-intensity λ_t and ζ_t-intensity $\tilde{\lambda}_t$ where $\mathcal{F}_t$ and ζ_t are filtrations of N_t such that

$$\mathcal{F}_t^N \subset \zeta_t \subset \mathcal{F}_t \ . \tag{1.10.23}$$

Suppose that $N_\infty = \infty \quad P-$ a.s. Define the ζ_t stopping time $\tau(t)$ for each t by

$$\int_0^{\tau(t)} \tilde{\lambda}_s ds = t \ . \tag{1.10.24}$$

Then the point process $\tilde{N}_t$ defined by

$$\tilde{N}_t = N_{\tau(t)} \tag{1.10.25}$$

is a standard Poisson process with intensity unity.

PROOF See [6]. ∎

THEOREM 1.90 **(An exponential supermartingale)**
Let $(N_t(1), \ldots, N_t(k))$ be a k-variate point process adapted to $\mathcal{F}_t$ and let $\lambda_t(i), 1 \le i \le k$ be the predictable $(P, \mathcal{F}_t)$-intensities of $N_t(i), 1 \le i \le k$. Let $T_n(i)$ denote n-th the jump time for the i-th component $N_t(i)$ of the point process. Let $\mu_t(i), 1 \le i \le k$ be $\mathcal{F}_t$-predictable nonnegative processes such that for all $t \ge 0$ and $1 \le i \le k$.

$$\int_0^t \mu_s(i)\lambda_s(i)ds < \infty \ a.s.\ [P]\,. \tag{1.10.26}$$

Define

$$L_t = \Pi_{i=1}^k L_t(i) \tag{1.10.27}$$

where

$$L_t(i) = \begin{cases} \exp\left(\int_0^t (1 - \mu_s(i)\lambda_s(i))\, ds\right) & \text{if } t < T_1(i) \\ \left\{\Pi_{n\ge1}\mu_{T_n(i)}(i) I\,(T_n(i) \le t)\right\} \exp\left(\int_0^t (1 - \mu_s(i)\lambda_s(i))\, ds\right) & \\ \qquad \text{if } t \ge T_1(i) & \end{cases} \tag{1.10.28}$$

Hereafter, we write in short,

$$L_t(i) = \left\{\Pi_{n\ge1}\mu_{T_n(i)}(i) I\,(T_n(i) \le t)\right\} \exp\left(\int_0^t (1 - \mu_s(i)\lambda_s(i))\, ds\right). \tag{1.10.29}$$

Then L_t is a $(P, \mathcal{F}_t)$-nonnegative local martingale and a $(P, \mathcal{F}_t)$-super martingale.

PROOF By the exponential formula

$$L_t = 1 + \sum_{i=1}^{k} \int_0^t L_s\,(\mu_s(i) - 1)\,dM_s(i) \tag{1.10.30}$$

where

$$M_t(i) = N_t(i) - \int_0^t \lambda_s(i)ds\ . \tag{1.10.31}$$

Let

$$S_n = \inf\left\{ t | L_{t-} \geq n \text{ or } \sum_{i=1}^{k} \int_0^t (\mu_s(i) + 1)\,\lambda_s(i)ds \geq n \right\}$$

$$\text{if } \{\cdots\} \neq \phi$$

$$= +\infty \quad \text{otherwise}\ . \tag{1.10.32}$$

Then $L_{t\wedge S_n}$ is a $(P, \mathcal{F}_t)$-martingale by the remarks made earlier. It can be checked that $S_n \uparrow \infty$ a.s. and hence, L_t is a $(P, \mathcal{F}_t)$-local martingale. But L_t is nonnegative. Hence, L_t is a $(P, \mathcal{F}_t)$-supermartingale (from Fatou's lemma). ▮

THEOREM 1.91
With the same notation as above, suppose that $E(L_1) = 1$. *Define* $\tilde{P}$ *by* $d\tilde{P}/dP = L_1$. *Then* $\{N_t(i)\}$ *has the* $(\tilde{P}, \mathcal{F}_t)$*-intensity*

$$\tilde{\lambda}_t(i) = \mu_t(i)\lambda_t(i) \;\; over \;\; [0, 1] \; for\ 1 \leq i \leq k\ .$$

PROOF See [6]. ▮

REMARK 1.78 If $\mu_{T_n(i)}(i) > 0$ on $[T_n(i) < \infty]$ $P-$ a.s. for all $n \geq 0$ and all $1 \leq i \leq k$, then $L_t > 0$ P a.s. for all $t \geq 0$. In particular $L_1 > 0$ a.s. $[P]$ and $P << \tilde{P}$ with $\frac{dP}{d\tilde{P}} = \frac{1}{L_1}$. ▮

REMARK 1.79 With the same notation as above, suppose that $\lambda_t(i) = 1$ for $1 \leq i \leq k$ and $\mu_t(i)$ is bounded. Then $E(L_1) = 1$. This can be seen as follows for $k = 1$. Let $\mu_t, \tilde{\lambda}_t, N_t$ be $\mu_t(1), \tilde{\lambda}_t(1)$ and $N_t(1)$, respectively. Note that

$$L_t \leq K^{N_t} e^{Kt}$$

where K is an upper bound of μ_t. Hence,

$$E\left[\int_0^1 |L_{s-}(\mu_s - 1)|\, ds\right] \leq E\left[(K+1)K^{N_t}e^K\right] < \infty\,, \qquad (1.10.33)$$

since N_t is a Poisson random variable. Hence L_t is a martingale by the remarks made earlier. Therefore $E(L_1) = 1$ (note that (1.10.33) implies that $\int_0^t L_{s-}(\mu_s - 1)dM_s$ is an $\mathcal{F}_t$-martingale). ∎

REMARK 1.80 *(Heuristics for the computation of the Radon-Nikodym derivative of two doubly stochastic Poisson processes with intensities $\lambda(\cdot)$ and $\eta(\cdot)$).* Suppose $N = \{N_t, t \geq 0\}$ is a doubly stochastic Poisson process with intensity $\lambda(\cdot)$ and suppose that the jumps occur at times

$$0 = T_0 < T_1 < \cdots < T_{n-1} < T_{n+1} = t\,.$$

Then the likelihood under $\lambda(\cdot)$ is proportional to

$$\Pi_{j=0}^{n}\left\{\left(\int_{T_j}^{T_{j+1}} \lambda(s)ds\right)\exp\left(-\int_{T_j}^{T_{j+1}} \lambda(s)ds\right)\right\}.$$

$$\simeq \left\{\Pi_{j=0}^{n}\left(T_{j+1} - T_j\right)\lambda(T_j)\right\}\exp\left\{-\int_0^t \lambda(s)ds\right\}\,. \qquad (1.10.34)$$

Hence, the likelihood ratio corresponding to intensities $\lambda(\cdot)$ and $\eta(\cdot)$ is approximately

$$\frac{\left(\Pi_{j=0}^{n}\left(T_{j+1}-T_{j}\right)\lambda\left(T_{j}\right)\right)\exp\left\{-\int\limits_{0}^{t}\lambda(s)ds\right\}}{\left(\Pi_{j=0}^{n}\left(T_{j+1}-T_{j}\right)\eta\left(T_{j}\right)\right)\exp\left\{-\int\limits_{0}^{t}\eta(s)ds\right\}}$$

$$=\left(\Pi_{j=0}^{n}\frac{\lambda\left(T_{j}\right)}{\eta\left(T_{j}\right)}\right)\exp\left\{-\int\limits_{0}^{t}(\lambda(s)-\eta(s))ds\right\}$$

$$=\left(\Pi_{j=0}^{n}\mu(T_{j})\right)\exp\left\{\int\limits_{0}^{t}(1-\mu(s))\eta(s)ds\right\}$$

where

$$\mu(\cdot)=\lambda(\cdot)/\eta(\cdot)\,. \tag{1.10.35}$$

∎

References

[1] Aldous, D. and Eagleson, G. (1978) On mixing and stability of limit theorems, *Ann. Probab.*, **6**, 325–331.

[2] Arato, M. (1982) *Linear Stochastic Systems with Constant Coefficients: A Statistical Approach*, Lecture Notes in Control and Information Sciences, **45**, Springer, Berlin.

[3] Azencott, R. (1982) Formulae de Taylor stochastique et development as asymptotique d'integrales de Feynman, *Lecture Notes in Mathematics*, **921**, 239–285.

[4] Basawa, I.V. and Prabhu, N.U. (1994) *Statistical Inference in Stochastic Processes,* Special issue of *J. Stat. Plan. Inf.*, **39**.

[5] Basawa, I.V. and Prakasa Rao, B.L.S. (1980) *Statistical Inference for Stochastic Processes,* Academic Press, London.

[6] Bremaud, P. (1981) *Point Processes and Queues: Martingale Dynamics,* Springer, Berlin.

[7] Burkholder, D. (1973) Distribution function inequalities for martingales, *Ann. Probab.*, **1**, 19–42.

[8] Chou, C.S. (1977) Le processus des sauts d'une martingale locale, *Seminaire de Probabilites XI,* Lecture Notes in Math., **581**, Springer, Berlin, 356–361.

[9] Christopeit, N. (1986) Quasi-least-squares estimation in semimartingale regression models, *Stochastics,* **16**, 255–278.

[10] Dellacherie, C. (1972) *Capacites et Processus Stochastiques,* Springer, Berlin.

[11] Eagleson, G. (1976) Some simple conditions for limit theorems to be mixing, *Theory Probab. Appl.*, **21**, 637–643.

[12] Elliott, R.J. (1982) *Stochastic Calculus and Applications,* Springer, New York.

[13] Engelbert, H. and Schmidt, W. (1985) On solutions of one-dimensional stochastic differential equation without drift, *Z. Warsch. verw Gebiete,* **68**, 287–314.

[14] Feigin, P.D. (1985) Stable convergence of semimartingales, *Stoch. Proc. Appl.*, **19**, 125–134.

[15] Fleming, T.R. and Harrington, D.P. (1991) *Counting Processes and Survival Analysis,* Wiley, New York.

[16] Gikhman, I.I. and Skorokhod, A.V. (1972) *Stochastic Differential Equations,* Springer, Berlin.

[17] Gikhman, I.I. and Skorokhod, A.V. (1977) *The Theory of Stochastic Process,* Springer, Berlin.

[18] Grenander, U. (1981) *Abstract Inference,* Wiley, New York.

[19] Grigelionis, B. (1977) Martingale characterization of random processes by independent increments, *Lithuanian Math. J.*, **17**, 52–60.

[20] Hall, P. and Heyde, C. (1980) *Martingale Limit Theory and Its Applications,* Academic Press, New York.

[21] Hutton, J.E. and Nelson, P.I. (1984) Interchanging the order of differentiation and stochastic integration, *Stoch. Proc. Appl.*, **18**, 371–377.

[22] Hutton, J.E. and Nelson, P.I. (1984) A mixing and stable central limit theorem for continuous time martingales, *Tech. Report,* **42**, Kansas State University.

[23] Hutton, J.E. and Nelson, P.I. (1986) Quasi-likelihood estimation for semimartingales, *Stoch. Proc. Appl.*, **22**, 245–257.

[24] Ikeda, N. and Watanabe, S. (1981) *Stochastic Differential Equations and Diffusion Processes,* North-Holland, Amsterdam.

[25] Jacod, J. (1976) Un theoreme de representation pour les martingales discontinues, *Z. Warsch. verw Gebiete,* **34**, 225–244.

[26] Jacod, J. (1977) Sur la construction des integrales stochastiques et les sous-espaces stables de martingales, *Seminaire de Probabilites XI,* Lecture Notes in Math., **581**, Springer, Berlin, 390–410.

[27] Jacod, J. (1979) *Calcul Stochastique et Problemes de Martingales,* Lecture Notes in Math., **714**, Springer, Berlin.

[28] Jacod, J. and Memin, J. (1976) Caracteristiques locales et conditions de cotinuite absolue pour les martingales, *Z. Warsch. verw Gebiete,* **35**, 1–37.

[29] Jacod, J., Kloptowoski, A., and Memin, J. (1982) Theoreme de la limite centrale et convergence fonctionelle vers un processus a accroissemenets independants: la methode des martingales, *Ann. Inst. Henri Poincare,* **18**, 1–45.

[30] Jacod, J. and Shiryayev, A.N. (1987) *Limit Theorems for Stochastic Processes,* Springer, Heidelberg.

[31] Kabanov, Yu.M., Liptser, R.S., and Shiryayev, A.N. (1979) Absolute continuity and singularity of locally absolutely continuous distributions. I, *Math. USSR Sbornik,* **35**, 631–680.

[32] Karlin, S. and Taylor, H. (1981) *A Second Course in Stochastic Processes,* Academic Press, New York.

[33] Kallianpur, G. (1980) *Stochastic Filtering Theory,* Springer, New York.

[34] Karatzas,I. and Shreve, S. (1988) *Brownian Motion and Stochastic Calculus,* Springer, Berlin.

[35] Karr, A.F. (1991) *Point Processes and Their Statistical Inference,* Marcel Dekker, New York.

[36] Kazamaki, N. (1972) Changes of time, stochastic integrals, weak martingales, *Z. Warsch. verw Gebiete,* **22**, 25–32.

[37] Kunita, H. (1990) *Stochastic Flows and Stochastic Differential Equations,* Cambridge University Press, Cambridge.

[38] Kutoyants, Y.A. (1984) *Parameter Estimation for Stochastic Processes,* (Trans. and Ed. B.L.S. Prakasa Rao), Heldermann, Berlin.

[39] Lenglart, E. (1977) Relation de domination entre deux processus, *Ann. Inst. Henri Poincaré,* **13**, 171–179.

[40] Lepingle, D. (1977) Sur la representation des sauts des martingales, *Seminaire de Probabilites XI,* Lecture Notes in Math., **581**, Springer, Berlin, 418–434.

[41] Lepingle, D. (1978) Sur les comportement asymptotique des martingales locales, *Seminaire de Probabilites XII,* Lecture Notes in Math., **649**, Springer, Berlin, 148–161.

[42] Levanony, D. (1994) Conditional tail probabilities in continuous-time martingale LLN with application to parameter estimation in diffusions, *Stoch. Proc. Appl.,* **51**, 117–134.

[43] Liptser, R.S. (1980) A strong law of large numbers for local martingales, *Stochastics,* **3**, 217–228.

[44] Liptser, R.S. and Shiryayev, A.N. (1977) *Statistics of Random Processes: General Theory,* Springer, New York.

[45] Liptser, R.S. and Shiryayev, A.N. (1978) *Statistics of Random Processes: Applications,* Springer, New York.

[46] Liptser, R.S. and Shiryayev, A.N. (1980) A functional central limit theorem for semimartingales, *Theory Probab. Appl.*, **25**, 667–688.

[47] Metivier, M. (1982) *Semimartingales,* de Gryuter, Berlin.

[48] Meyer, P. (1976) Uncours sur les integrales stochastiques, *Seminaire de Probabilites X,* Lecture Notes in Math., **511**, Springer, Berlin.

[49] Neveu, J. (1975) *Discrete-Parameter Martingales,* North-Holland, Amsterdam.

[50] Prabhu, N.U. (1988) *Statistical Inference from Stochastic Processes,* In *Contemporary Mathematics,* **80**, Amer. Math. Soc., Providence, RI.

[51] Prabhu, N.U. and Basawa, I.V. (1991) *Statistical Inference in Stochastic Processes,* Marcel Dekker, New York.

[52] Prakasa Rao, B.L.S. (1983) *Nonparametric Functional Estimation,* Academic Press, Orlando, FL.

[53] Prakasa Rao, B.L.S. (1987) *Asymptotic Theory of Statistical Inference,* Wiley, New York.

[54] Prakasa Rao, B.L.S. (1995) *Statistics and Its Applications,* Indian National Science Academy, New Delhi.

[55] Prakasa Rao, B.L.S. (1999) *Statistical Inference for Diffusion Type Processes,* Arnold, London.

[56] Prakasa Rao, B.L.S. and Bhat, B.R. (1996) *Stochastic Processes and Statistical Inference,* New Age International, New Delhi.

[57] Rebolledo, R. (1977) Remarks on the convergence in distribution of martingales to a continuous martingale, *C. R. Acad. Sci. Paris A,* **285**, 465–468.

[58] Rebolledo, R. (1977) Remarks on the convergence in distribution of martingales towards continuous martingale limit, *C. R. Acad. Sci. Paris A,* **285**, 517–520.

[59] Rebolledo, R. (1980) Central limit theorems for local martinggales, *Z. Warsch. verw Gebiete,* **51**, 269–286.

[60] Renyi, A. (1958) On mixing sequences of sets, *Acta Math. Acad. Sci. Hung.,* **9**, 215–228.

[61] Renyi, A. (1963) On stable sequences of events, *Sankhya Ser. A*, **25**, 293–302.

[62] Revuz, D. and Yor, M. (1991) *Continuous Martingales and Brownian Motion*, Springer, Berlin.

[63] Rootzen, H. (1976) A note on the central limit theorem for doubly stochastic Poisson processes, *J. Appl. Prob.*, **13**, 809–813.

[64] Rootzen, H. (1976) Fluctuations of sequences which converge in distribution, *Ann. Probab.*, **4**, 456–463.

[65] Rootzen, H. (1977) On the functional central limit theorem for martingales, *Z. Warsch. verw Gebiete*, **38**, 199–210.

[66] Rootzen, H. (1977) A note on convergence to mixtures of normal distributions, *Z. Warsch. verw Gebiete*, **38**, 211–216.

[67] Shorrock, G.R. and Wellner, J. (1986) *Empirical Processes with Applications to Statistics*, Wiley, New York.

[68] Sorensen, M. (1991) Likelihood methods for diffusions with jumps, *Statistical Inference in Stochastic Processes*, (Ed. N.U. Prabhu and I.V.Basawa), 67–105.

[69] Van de Geer, S. (1995) Exponential ineqalities for martingales, with application to maximum likelihood estimation for counting processes, *Ann. Statist.*, **23**, 1779–1801.

[70] Wong, W.H. and Shen, X. (1995) Probability inequalities for likelihood ratios and covergence rates of sieve MLE's, *Ann. Statist.*, **23**, 339–362.

Chapter 2

Exponential Families of Stochastic Processes

2.1 Introduction

A parametric family of models for a stochastic process is called an exponential family if for every $t > 0$, the likelihood function corresponding to the process observed continuously over the time interval $[0, t]$ has the structure of an exponential family. We now discuss a semimartingale approach for the study of such families following Kuchler and Sorensen [6]. We have seen earlier that a semimartingale is the sum of a local martingale and a process with bounded variation. The local martingales generalize the class of martingales, whereas the class of semimartingales covers almost all stochastic processes. Let us first discuss some examples of stochastic processes where the likelihood function belongs to an exponential family.

Example 2.1

Let $\{X_t\}$ be a Wiener process with drift parameter θ and diffusion coefficient unity observed over $[0, t]$. Then the likelihood function is given by

$$L_t(\theta) = \exp\left(\theta X_t - \frac{1}{2}\theta^2 t\right), \theta \in R,$$

which belongs to an exponential family. Note that X_t is a sufficient statistic for θ. $\square$

Example 2.2

Let $\{X_t\}$ be a counting process with intensity $\{\theta Z_t\}$ with $\theta > 0$ where the process $\{Z_t\}$ is predictable with respect to the filtration generated by $\{X_t\}$. Then the likelihood function for the process $\{X_t\}$ observed over $[0, t]$ is

$$L_t(\theta) = \exp\left(\log\theta\ (X_t - X_0) - \theta \int_0^t Z_s ds\right).$$

If $Z_t = 1$ for all t, then we have a Poisson process. If $Z_t = X_t$, then the process is a pure birth process. □

Example 2.3

Let $\{X_t\}$ be a stochastic process with independent stationary increments. It is possible to characterize such processes through the Levy characteristic $(\nu_\theta, \sigma^2, \gamma_\theta)$ (cf. [5]). Suppose

$$\nu_\theta(dx) = \exp(\theta x)\nu(dx),$$

$$\gamma_\theta = \gamma + \theta\sigma^2 - \int_{R-\{0\}} y\left(1+y^2\right)^{-1}(1-\exp(\theta y))\nu(dy)$$

where (ν, σ^2, γ) are the Levy characteristics of $\{X_t\}$ under a fixed reference measure and

$$\theta \in \left\{u \in R : \int_{R-\{0\}} x^2\left(1+x^2\right)^{-1} \exp(ux)\nu(dx) < \infty\right\}.$$

In this case, the likelihood function is given by

$$L_t(\theta) = \exp\left(\theta X_t - v(\theta)\, t\right)$$

with

$$v(\theta) = \theta\gamma + \frac{1}{2}\theta^2\sigma^2 + \int_{R-\{0\}} \left(e^{\theta x} - 1 - \frac{\theta x}{1+x^2}\right)\nu(dx).$$

Note that X_t is a sufficient statistic for θ. For $\nu = 0$, the process X reduces to the Wiener process with drift. If $\sigma^2 = 0, \gamma = \frac{1}{2}$ and $\nu_A = I_A\,(1)$, then we have the Poisson process, and if $\sigma^2 = 0, \gamma = \frac{1}{2}\pi$ and $\nu(dz) = z^{-1} I_{R+}(z)dz$, then we obtain the gamma process. Here I_A denotes the indicator function of a set A. □

Example 2.4

The Wiener process with drift is a particular example of a family of solutions of stochastic differential equations of the form

$$dX_t = \theta\, a\,(t, X_t)\, dt + b\,(t, X_t)\; dW_t, X_o = x, t \geq 0\,.$$

Here $\{W_t\}$ is a standard Wiener process, $a(.,.)$ and $b(.,.) > 0$ are Borel measurable functions, and $\theta \in \Theta \subset R$. We suppose that this equation has a strong solution for every $\theta \in \Theta$. Under suitable conditions on $a(.,.)$ and $b(.,.)$, the likelihood function is given by

$$L_t(\theta) = \exp\left\{\theta \int_0^t \left[a\,(s, X_s)\,/b^2\,(s, X_s)\right] dX_s \right.$$

$$\left. -\frac{1}{2}\theta^2 \int_0^t \left[a^2\,(s, X_s)\,/b^2\,(s, X_s)\right] ds \right\}$$

(cf. [3]). A special case is the family of Ornstein–Uhlenbeck processes that are solutions of the stochastic differential equation

$$dX_t = \theta X_t dt + dW_t, \theta \in R, t \geq 0\,.$$

In this case, the likelihood function is given by

$$L_t(\theta) = \exp\left\{\frac{1}{2}\theta X_t^2 - \frac{1}{2}\theta^2 \int_0^t X_s^2 ds - \frac{1}{2}\theta t\right\}\,.$$

□

In a number of statistical models for stochastic processes, the likelihood function corresponding to an observation of the process over $[0, t]$ has the form

$$L_t(\theta) = a_t \exp\{\theta A_t - f(\theta)B_t\}, \theta \in \Theta \subseteq R$$

where a_t, A_t, and B_t depend on the observed process and f is a deterministic function.

The exponential families of processes discussed in the above examples so far were specified in terms of the diffusion coefficients, jump intensities, and drift coefficients. We now describe a unified approach to the study of exponential families through semimartingales via their local characteristics. However, the local characteristics do not in general determine the semimartingale uniquely, as we have seen earlier.

2.2 Exponential Families of Semimartingales

Let $(\Omega, \mathcal{F})$ be a measurable space with the filtration $\{\mathcal{F}_t, t \geq 0\}$ that is right-continuous and suppose that $\mathcal{F}$ is the σ-algebra generated by $\{\mathcal{F}_s, s \geq 0\}$. Let $\mathcal{P} = \{P_\theta, \theta \in \Theta\}$, $\Theta \subset R^n$ be a family of probability measures on $(\Omega, \mathcal{F})$. Let P_θ^t denote the restriction of P_θ to $\mathcal{F}_t$.

DEFINITION 2.1 *A family $\mathcal{P}$ of probability measures on $(\Omega, \mathcal{F})$ is said to be an exponential family if there exists a probability measure P on $(\Omega, \mathcal{F})$ such that*

$$(i) P_\theta <<^{loc} P \text{ for all } \theta \in \Theta, \text{ and}$$

$$(ii) L_t(\theta) = \frac{dP_\theta^t}{dp^t} = a_t(\theta)\, q_t \exp\left[\sum_{i=1}^{m} \gamma_t^{(i)}(\theta) B_t^{(i)}\right], t > 0$$

$$= a_t(\theta)\, q_t \exp\left(\gamma_t^T(\theta) B_t\right) \tag{2.2.1}$$

where a and $\gamma^{(i)}$, $1 \leq i \leq m$ are nonrandom functions of θ and t, while $q_t \geq 0$ and $B_t^{(i)}$, $i = 1, \ldots, m$ are processes adapted to $\{\mathcal{F}_t\}$. Here,

$\gamma_t(\theta) = (\gamma_t^{(1)}(\theta), \ldots, \gamma_t^{(m)}(\theta))^T$, $B_t = (B_t^{(1)}, \ldots, B_t^{(m)})^T$, *and* α^T *denotes the transpose of a vector* α.

Note that for any fixed $t \geq 0$, the family $\{P_\theta^t, \theta \in \Theta\}$ is an exponential family in the classical sense (cf. [1]). If the functions $\gamma^{(i)}$, $1 \leq i \leq m$ do not depend on t, then we will call the exponential family *time homogeneous*.

Let P be a fixed probability measure on $(\Omega, \mathcal{F})$ and let $X = \{X_t, t \geq 0\}$ be a stochastic process defined on $(\Omega, \mathcal{F})$ with the RCLL (right-continuous with left limits) property. Suppose that the filtration $\{\mathcal{F}_t, t \geq 0\}$ is generated by X and that the process X is a semimartingale under P. Then X is also a semimartingale under $P_\theta <<^{loc} P$ for every $\theta \in \Theta$ and we call $\mathcal{P} = \{P_\theta\}$ an *exponential family of semimartingales.* For simplicity, we assume that $X_0 = x_0$ a.s $[P]$ and a.s $[P_\theta]$ for all $\theta \in \Theta$ where x_0 is nonrandom. Let

$$X_t^{(1)} = x_0 + \sum_{s \leq t} \Delta X_s I\left(|\Delta X_s| > 1\right)$$

where $\Delta X_s = X_s - X_{s-}$ as before. Consider the random measure μ on $R_+ \times E$ associated with the jumps of X defined by

$$\mu(\omega; dt, dx) = \sum_{s \geq 0} I\left(\Delta X_s(\omega) \neq 0\right) \varepsilon_{(s, \Delta X_s(\omega))}(dt, dx)$$

where $E = R - \{0\}$ and ε_a is the Dirac measure at a. We have seen earlier that there exists another random measure ν on $R_+ \times E$ such that $U * (\mu - \nu)$ is a local martingale under P where $U(\omega; t, x) = x\ I(|x| \leq 1)$. Furthermore, there exists a continuous local martingale $m^{(c)}$ and a predictable process α such that

$$X = X^{(1)} + m^{(c)} + U * (\mu - \nu) + \alpha\ .$$

Here $U * (\mu - \nu)$ represent the jumps that are not too big, while α can be thought of as the integrated drift. Let $\beta =< m^{(c)} >$ be the quadratic characteristic of $m^{(c)}$. The triple (α, β, ν) is the set of local characteristics of X with respect to P. Note that β generalizes the integrated diffusion coefficient and ν can be thought of as a generalized jump intensity.

For the Wiener process with drift in Example 2.1, the local characteristics are $(\theta t, t, 0)$. For the counting process in Example 2.2, the local characteristics are $(\theta \int_0^t Z_s ds, 0, \theta Z_t dt \epsilon_1(dx))$. The process with independent stationary increments in Example 2.3 has the local characteristics $(t\alpha_\theta, t\sigma^2, e^{\theta x}\nu(dx))$ where

$$\alpha_\theta = \gamma + \theta\sigma^2 - \int_{R-\{0\}} y\left(1+y^2\right)^{-1}\nu(dy) + \int_{[-1,1]\cap R-\{0\}} y\left(e^{\theta y}-1\right)\nu(dy)$$

and the local characteristics of the exponential family of diffusion type processes in the Example 2.4 are

$$\left(\theta\int_0^t a(s,X_s)\,ds, \int_0^t b^2(s,X_s)\,ds, 0\right).$$

The triplet of local characteristics is uniquely determined by P and X. However, P is not uniquely determined by the triplet. For our present discussion, we assume that necessary conditions hold so that P is uniquely determined by the triplet of local characteristics. We consider the structure of the exponential family of semimartingales in terms of their local characteristics later in this Chapter.

We call a representation of the form (2.2.1) an *exponential representation* of $\mathcal{P}$. This representation need not be unique. The process $\boldsymbol{B}_t$ in the representation (2.2.1) is called the *canonical process*. Note that for all $A \in \mathcal{F}$,

$$P_\theta^t(A) = 0 \Longleftrightarrow P^t\left(A \cap [q_t > 0]\right) = 0\,.$$

Hence, for a fixed t, the measures $\{P_\theta^t\}$ are equivalent. Let us assume without loss of generality that $0 \in \Theta$ and that $\gamma_t(0) = 0$. Then

$$L_t^{(\theta)} = \frac{dP_\theta^t}{dP_0^t} = \exp\left(\gamma_t(\theta)^T\boldsymbol{B}_t - \phi_t(\theta)\right) \tag{2.2.2}$$

where $\phi_t(\theta)$ is nonrandom and

$$\phi_t(\theta) = \log\left(a_t(\theta)/a_t(0)\right)\,. \tag{2.2.3}$$

Note that for fixed $t \geq 0$, $\boldsymbol{B}_t$ is a sufficient statistic for $\gamma_t(\theta)$ with respect to $\{\mathcal{F}_t\}$.

REMARK 2.1 The existence of the exponential representation (2.2.1) depends on the filtration $\{\mathcal{F}_t\}$. The measures $\{P_\theta \mid \zeta_t\}$ need not have the exponential representation under a different filtration $\{\zeta_t\}$. The measures $\{P_\theta \mid \zeta_t, t \geq 0\}$ need not be equivalent even if $\{P_\theta \mid \mathcal{F}_t, t \geq 0\}$ are. This can be seen from the following example. ▮

Example 2.5

Consider the following family of stochastic differential equations:

$$dX_t = \theta\, a\,(t, X_t)\, dt + dW_t,\, t \geq 0,\, X_0 = x_0 \tag{2.2.4}$$

where W is the standard Wiener process and $\theta \in \Theta$. Suppose that (2.2.4) has a strong solution for all $\theta \in \Theta$. The class of solutions of (2.2.4) induces a family of probability measures $\{P_\theta, \theta \in \Theta\}$ on the space $C[0, \infty)$ of continuous functions from $[0, \infty)$ into R equipped with the σ-algebra generated by the cylinder sets. Let $\mathcal{F}_t$ be the σ-algebra generated by the cylinder sets in $C[0, t]$. Suppose

$$P_\theta \left(\int_0^t a^2\,(s, X_s)\, ds = \infty \right) > 0$$

for some θ and t. This is the case if $a(t, x) = \frac{1}{x}$ (Bessel process) and $\theta < \frac{1}{2}$. Then the measures $\{P_\theta^t, \theta \in \Theta\}$ are not equivalent, and hence $\{P_\theta^t, \theta \in \theta\}$ is not an exponential family with respect to the filtration $\{\mathcal{F}_t\}$. Let

$$\tau_u = u \wedge \inf \left\{ t : \int_0^t a^2\,(s, X_s)\, ds \geq u \right\} .$$

Then τ is an $\{\mathcal{F}_t\}$-stopping time. The class of likelihood functions,

$$\frac{dP_\theta^{\tau_u}}{dP_0^{\tau_u}} = \exp \left(\theta \int_0^{\tau_u} a\,(s, X_s)\, dX_s - \frac{1}{2}\theta^2 \int_0^{\tau_u} a^2\,(s, X_s)\, ds \right)$$

where $P_\theta^{\tau_u}$ denotes the restriction of P_θ to $\mathcal{F}_{\tau_u}$ forms an exponential family relative to the filtration $\{\mathcal{F}_{\tau_u}, u > 0\}$. □

Consider a time-homogeneous exponential family. In other words, $\gamma(.)$ is independent of t in the representation (2.2.2). Consider the canonical representation

$$\frac{dP^t_\Phi}{dP^t_0} = \exp\left[\Phi^T B_t - \phi_t(\gamma^{-1}(\Phi))\right], t \geq 0, \Phi \in \Gamma \tag{2.2.5}$$

where $\Gamma = \{\gamma(\theta) : \theta \in \Theta\}$. The term Φ is called a *canonical parameter.* In general Γ is a submanifold in R^m of dimension lower than m. If the interior of Γ is nonempty, then the family is said to have a *nonempty kernel.*

THEOREM 2.1

Consider a time-homogeneous exponential family with the exponential representation (2.2.5). If int Γ is nonempty, then the canonical process $\{B_t\}$ has independent increments under P_θ for all $\theta \in \Theta$. Furthermore, for all $\lambda \in \Gamma - \gamma(\theta)$,

$$E_{\Phi_0}\left[e^{\lambda^T(B_t - B_s)}\right] = \exp\left\{\left[\phi_t\left(\gamma^{-1}(\lambda + \gamma(\theta))\right) - \phi_t(\theta)\right]\right.$$
$$\left. - \left[\phi_s\left(\gamma^{-1}(\lambda + \gamma(\theta))\right) - \phi_s(\theta)\right]\right\}. \tag{2.2.6}$$

In particular, the process $\{B_t\}$ has stationary increments if and only if $\phi_t(\theta)$ is of the form $\phi_t(\theta) = t\ f(\theta)$.

PROOF Let $\Phi_0 \in \Gamma$. Then

$$\frac{dP^t_\Phi}{dp^t_{\Phi_0}} = \exp\left[(\Phi - \Phi_0)^T B_t - \phi_t\left(\gamma^{-1}(\Phi)\right) + \phi_t\left(\gamma^{-1}(\Phi_0)\right)\right], \tag{2.2.7}$$

and it is a P_{Φ_0}- martingale for every $\Phi \in \Gamma$. Hence,

$$E_{\Phi_0}\left(\exp\left[\lambda^T\left(\boldsymbol{B}_t-\boldsymbol{B}_s\right)\right] \mid \mathcal{F}_s\right)$$

$$=\exp\left\{\left[\phi_t\left(\gamma^{-1}\left(\lambda+\Phi_0\right)\right)-\phi_s\left(\gamma^{-1}\left(\lambda+\Phi_0\right)\right)\right]\right.$$

$$\left.-\left[\phi_t\left(\gamma^{-1}\left(\Phi_0\right)\right)-\phi_s\left(\gamma^{-1}\left(\Phi_0\right)\right)\right]\right\} \tag{2.2.8}$$

for all $\lambda \in \Gamma - \Phi_0$. This shows that the conditional Laplace transform of $\boldsymbol{B}_t - \boldsymbol{B}_s$ under P_{Φ_0} is equal to a nonrandom function for λ in the open set int $(\Gamma - \Phi_0)$. By analytic continuation, the conditional Laplace transform is nonrandom in its entire domain. Hence, $\boldsymbol{B}_t - \boldsymbol{B}_s$ is independent of $\mathcal{F}_s$ under P_{Φ_0}. Relation (2.2.8) implies (2.2.6). ∎

Example 2.6

Let $\{W_t\}$ be the standard Wiener process and $X_t = \exp(W_t + \theta t), \theta \in R$. The process $X \equiv \{X_t, t \geq 0\}$ is the geometric Brownian motion. The likelihood function corresponding to an observation of the process X on $[0, t]$ is

$$L_t(\theta) = \exp\left[\left(\theta + \frac{1}{2}\right) \log X_t + \frac{1}{2}\left(\theta + \frac{1}{2}\right)^2 t\right], \theta \in R\,.$$

Here, $\Gamma = R \neq \phi$ and $B_t = \log X_t = W_t + \theta t$. Note that B_t has independent increments for all $\theta \in R$. The process X is a solution of the stochastic differential equation

$$dX_t = \left(\theta + \frac{1}{2}\right) X_t dt + X_t dW_t, t \geq 0, X_0 = 1$$

by Ito's formula. But $\{X_t\}$ does not have independent increments. □

Example 2.7

Consider the class of Ornstein–Uhlenbeck processes consisting of the solutions of (2.2.4) with $a(t, X_t) = X_t$ and $\Theta = R$. This is a time-homogeneous exponential family with the canonical process

$$B_t = \left(\frac{1}{2} \left(X_t^2 - x_0^2 \right), S_t \right)$$

where

$$S_t = \int_0^t X_s^2 ds$$

(cf. Example 2.4). However, B_t does not have independent increments, and hence, there does not exist a larger class of time-homogeneous exponential family on $(\Omega, \mathcal{F})$ with the nonempty kernel containing the class of Ornstein–Uhlenbeck processes. □

Example 2.8

Consider the class of counting processes with intensity $\lambda_t(\alpha, \theta) = h_t(\alpha, \theta) X_{t-}$ where X is the observed process and

$$h_t(\alpha, \theta) = \alpha \left\{ \left(e^{(\theta - 1)t} - 1 \right) \left(1 - \alpha(1-\theta)^{-1} \right) + 1 \right\}^{-1}$$

with $\theta < 1$ and $0 < \alpha \leq 1 - \theta$. For $\alpha = 1 - \theta$, the process is a pure birth process. The likelihood function corresponding to an observation of X on $[0, t]$ is

$$L_t(\alpha, \theta) = \exp \left\{ \theta \int_0^t X_s ds + \log \left(h_t(\alpha, \theta) \right) \, X_s - x_0 \, \log \alpha \right\}$$

where $X_0 = x_0$. This model is an exponential family with

$$\gamma_t(\alpha, \theta) = (\theta, \log h_t(\alpha, \theta)), \theta < 1, 0 < \alpha \leq 1 - \theta$$

$$= (\gamma_1, \gamma_2), \gamma_1 < 1, \gamma_2 \leq \log(1 - \gamma_1)$$

and the kernel Γ is nonempty. The exponential family is however not time-homogeneous and the increments of the process

$$B_t = \left(\int_0^t X_s ds, X_t \right)$$

are not independent. $\square$

Continuous Semimartingales

Suppose a process Y is a continuous semimartingale on $(\Omega, \mathcal{F}, P)$ with the filtration $\{\mathcal{F}_t\}$. Let its local characteristics be (α, β, ν). Then the random measure $\nu \equiv 0$ and α is continuous. Furthermore, $M = Y - \alpha$ is a continuous local martingale with $< M >= \beta$. Let $Z^{(i)}, 1 \leq i \leq n$ be predictable processes such that

$$Z^{(i)} Z^{(j)} \circ \beta \in \nu_{loc}(P), 1 \leq i, j \leq n \tag{2.2.9}$$

where $\nu_{loc}(P)$ denotes the set of all processes with locally integrable variation. Let $\nu_{loc}^{+}(P)$ denote the nondecreasing elements of $\nu_{loc}(P)$. Suppose that $\{P_\theta, \theta \in \Theta\}, \Theta \subset R^n$ is a family of probability measures on $(\Omega, \mathcal{F})$ such that, under P_θ, the process X is a semimartingale with local characteristics

$$\mathcal{L}_\theta = \left(\alpha + \sum_{i=1}^{n} \theta_i Z^{(i)}, \beta, 0 \right) . \tag{2.2.10}$$

Then it follows from the Girsanov's theorem that the condition

$$P_\theta \left(|Z^{(i)} Z^{(j)} \circ \beta_t| < \infty \right) = 1, t > 0, 1 \leq i, j \leq n, \theta \in \Theta \tag{2.2.11}$$

implies that $P_\theta \ll^{loc} P$ and $L_t(\theta) = H_t^\theta$ a.s. $[P]$ where

$$H_t^\theta = \exp \left[\sum_{i=1}^{n} \theta_i Z^{(i)} \bullet M_t - \frac{1}{2} \sum_{i,j=1}^{n} \theta_i \theta_j Z^{(i)} Z^{(j)} \circ \beta_t \right], t > 0 . \tag{2.2.12}$$

In view of the Girsanov's theorem for semimartingales, one can construct an exponential family of probability measures that is locally absolutely continuous with respect to P and such that the corresponding local characteristics of X are $\mathcal{L}_\theta, \theta \in \Theta$ as given by (2.2.10) provided

$$E\left(H_t^\theta\right) = 1, t > 0, \theta \in \Theta \tag{2.2.13}$$

and the corresponding likelihood function is given by (2.2.12). The exponential families of continuous semimartingales are curved and have the same structure as the special case of exponential families of diffusion-type processes discussed in Example 2.4. For $n = 1$, the family is a (2,1)-curved exponential family (cf. [2]). The process

$$\theta = X - \alpha - \sum_{i=1}^{n} \theta_i Z^{(i)} \circ \beta \tag{2.2.14}$$

is a continuous local martingale with $< M^\theta >= \beta$ under P_θ and the process X is a solution of the stochastic differential equation

$$dX_t = d\alpha_t + \left(\sum_{i=1}^{n} \theta_i Z^{(i)} \right) d\beta_t + dM_t^\theta , t \geq 0 . \tag{2.2.15}$$

Semimartingales with Jumps

Suppose the process X is quasi-left-continuous, that is, $X_T = X_{T-}$ almost surely for every predictable stopping time T. Recall that a stopping time is predictable if there exists a sequence $\{T_k\}$ of stopping times such that $T_k < T$ and $T_k \uparrow T$ almost surely. Let (α, β, ν) be the local characteristics of X under P. Then the processes α and β are continuous.

Let $Z^{(i)}, 1 \leq i \leq n$ be predictable processes satisfying (2.2.9). Further suppose that $\{Y_{\boldsymbol{k}}, \boldsymbol{k} \in K \subset R^l\}$ is a family of strictly positive predictable random functions from $\Omega \times R_+ \times E$ into R_+ such that

$$\left(Y_{\boldsymbol{k}} - 1\right) I \left(Y_{\boldsymbol{k}} > 2\right) * \nu \in \nu_{loc}^{+}(P), \boldsymbol{k} \in K \tag{2.2.16}$$

and

$$\left(Y_{\boldsymbol{k}} - 1\right)^2 I \left(Y_{\boldsymbol{k}} \leq 2\right) * \nu \in \nu_{loc}^{+}(P), \boldsymbol{k} \in K . \tag{2.2.17}$$

Further we assume that $Y_{\boldsymbol{k}}(\omega, s, 0) \equiv 1$ a.s. $[P]$. Here $E = R - \{0\}$ as before. Suppose $\{P_{\theta,\boldsymbol{k}} : (\theta, \boldsymbol{k}) \in \Theta \times K\}$ is a family of probability measures on $(\Omega, \mathcal{F})$ such that, under $P_{\theta,\boldsymbol{k}}$, X is a semimartingale with the local characteristics

$$\mathcal{L}_{\theta,K} = \left(\alpha + \sum_{i=1}^{n} \theta_i Z^{(i)} \circ \beta + \left[U\left(Y_{\boldsymbol{k}} - 1\right) * \nu\right], \beta, Y_{\boldsymbol{k}} * \nu \right) \tag{2.2.18}$$

where $U(x) = x \ I(|x| \leq 1)$. In addition we assume that the following conditions hold:

$$P_{\theta,\boldsymbol{k}} \left(| \left(Z^{(i)} Z^{(j)} \right) \circ \beta_t | < \infty \right) = 1;$$

$$(t > 0, 1 \le i, j \le n, (\theta, \boldsymbol{k}) \in \Theta \times K), \tag{2.2.19}$$

$$P_{\theta,\boldsymbol{k}}\left(\left((Y_{\boldsymbol{k}} - 1)\, I\,(Y_{\boldsymbol{k}} > 2) * \nu_t < \infty\right) = 1;\right.$$

$$(t > 0, (\theta, \boldsymbol{k}) \in \Theta \times K), \tag{2.2.20}$$

and

$$P_{\theta,\boldsymbol{k}}\left((Y_{\boldsymbol{k}} - 1)^2\, I\,(Y_{\boldsymbol{k}} \le 2) * \nu_t < \infty\right) = 1; (t > 0, (\theta, \boldsymbol{k}) \in \Theta \times K). \tag{2.2.21}$$

Then it can be shown that

$$P_{\theta,\boldsymbol{k}} <<^{loc} P\ ,\ (\theta, \boldsymbol{k}) \in \Theta \times K\ , \tag{2.2.22}$$

and

$$L_t(\theta, \boldsymbol{k}) = H_t^{\theta}\ \exp\left((Y_{\boldsymbol{k}} - 1) * (\mu - \nu)_t + \left(\log(Y_{\boldsymbol{k}}) - Y_{\boldsymbol{k}} + 1\right) * \mu_t\right)\ , \tag{2.2.23}$$

where

$$H_t^{\theta} = \exp\left[\sum_{i=1}^{n} \theta_i Z^{(i)} \bullet X_t^{(c)} - \frac{1}{2}\sum_{i,j=1}^{n} \theta_i\theta_j \left(Z^{(i)} Z^{(j)} \circ \beta_t\right)\right] \tag{2.2.24}$$

(cf. [6]).

Some Types of Exponential Families

Suppose we choose $Y_{\boldsymbol{k}}$ to be of the form

$$Y_{\boldsymbol{k}}(\omega, s, x) = \psi(\omega, s, x) \exp\left(\sum_{i=1}^{\ell} k_i\ \Phi_i(s, x)\right)$$

$$= \psi\ exp\left(\boldsymbol{k}^T \Phi\right) \tag{2.2.25}$$

where $\Phi_i, 1 \le i \le \ell$ are nonrandom functions and ψ is a strictly positive predictable process from $R_+ \times E$ into R_+. Suppose that $\Phi_i(s, 0) = 0, 1 \le i \le \ell$

and $\psi(\omega, s, 0) = 1$ a.s.$[P]$. Here $\boldsymbol{k}^T = (k_1, \ldots, k_\ell)$ and $\Phi^T = (\Phi_1, \ldots, \Phi_\ell)$. Let $\mathcal{E}$ denote the σ-algebra of Borel subsets of E.

Suppose the following conditions hold: (R) There exists a set $A \in \mathcal{B}_{[0,\infty]} \otimes \mathcal{E}$ such that, for all $t \geq 0$,

(i) $|\Phi_i| I_A * (\mu - \nu)_t < \infty$ a.s. $[P]$, $1 \leq i \leq \ell$,

(ii) $|\Phi_i| I_{A^c} * \mu_t < \infty$ a.s. $[P]$, $1 \leq i \leq \ell$,

(iii) $|\log \psi + \boldsymbol{k}^T \Phi - \psi \exp(\boldsymbol{k}^T \Phi) + 1| I_A * \nu_t < \infty$ a.s. $[P]$, $\boldsymbol{k} \in K$, and

(iv) $|\psi e^{\boldsymbol{k}^T \Phi} - 1| I_{A^c} * \nu_t < \infty$ a.s. $[P]$, $\boldsymbol{k} \in K$,

where the set A is chosen so that $A^c \subset \{(s, x) : |x| > 1\}$. If this is the case, then the conditions (ii)–(iv) hold since X has only finitely many jumps with $|\Delta X_0| > 1$ in a finite time interval and ν and $Y_{\boldsymbol{k}} * \nu$ are the local characteristics under P and $P_{\theta,\boldsymbol{k}}$, respectively. Hence,

$$\left(|x|^2 \wedge 1\right) * \nu \in \nu_{loc}^{+}(P), \left(\left(|x|^2 \wedge 1\right) Y_{\boldsymbol{k}}\right) * \nu \in \nu_{loc}^{+}(p) .$$

Under these regularity conditions (R) on ψ and Φ (cf. [6]), the likelihood function can be written in the form

$$L_t(\theta, \boldsymbol{k}) = a_t \, H_t^{\theta} \, \exp\left(\boldsymbol{k}^T B_t - g(\boldsymbol{k}, t)\right) , \tag{2.2.26}$$

where

$$a_t = \exp\left((\log \psi) I_A * (\mu - \nu)_t + (\log \psi) I_{A^c} * \mu_t\right) ,$$

$$g(\boldsymbol{k}, t) = \left(\psi e^{\boldsymbol{k}^T \Phi} - 1 - \log \psi - \boldsymbol{k}^T \Phi\right) I_A * \nu_t + \left(\psi e^{\boldsymbol{k}^T \Phi} - 1\right) I_{A^c} * \nu_t ,$$

$$\boldsymbol{B}_t = \left(B_t^{(1)}, \ldots, B_t^{(\ell)}\right)^T \text{ , and}$$

$$B_t^{(i)} = \Phi_i I_A * (\mu - \nu)_t + \Phi_i I_{A^c} * \mu_t, 1 \leq i \leq \ell .$$

Here the set A can be chosen to be $[0, \infty) \times [-1, 1]$ satisfying the conditions in [6]. They consider five types of exponential families of semimartingales with jumps. For instance, if $g(\boldsymbol{k}, t)$ is nonrandom, then we obtain an exponential family. Clearly, $g(\boldsymbol{k}, t)$ is nonrandom when ψ and ν are nonrandom. The families $\{P_{\theta,\boldsymbol{k}} : (\theta, \boldsymbol{k}) \in \Theta \times K\}$ in this case include the exponential families of processes with independent increments. For other special cases, see [6].

2.3 Stochastic Time Transformation

Let us consider an exponential family of processes with a likelihood function of the form

$$\frac{dP_\theta^t}{dP_0^t} = \exp\left(\theta^T \boldsymbol{A}_t - k(\theta)\, S_t\right) \tag{2.3.1}$$

where $\theta \in \Theta \subset R^k$ and intΘ is nonempty. Here $k(\theta)$ and S_t are one-dimensional and $\boldsymbol{A}_t$ is a k-dimensional RCLL process. Further suppose that S_t is a nondecreasing continuous random function such that $S_0 = 0$ and $S_t \to \infty$ as $t\infty$ a.s. There are several important special cases of stochastic processes covered by this model.

Example 2.9 **(Exponential families of counting process)**
This is a class of counting processes $\{X_t\}$ with intensity of the form

$$\lambda_t = kH_t, k > 0, t > 0 \tag{2.3.2}$$

where H_t is a predictable positive process with $E[H_t] < \infty$ for all $t < \infty$ and the likelihood function is given by (2.3.1) with $\theta = \log k, k(\theta) = \exp(\theta) - 1, A_t = X_t$ and $S_t = \int_0^t H_s ds$. $\square$

Let us now consider an exponential family with the likelihood function of the form (2.2.15). Consider a class of stopping times defined by the relation

$$\tau_u = \inf\{t : S_t > u\}, u \geq 0. \tag{2.3.3}$$

Since S_t is nondecreasing and continuous, the function $u \to \tau_u$ is strictly increasing and has an inverse function. In fact $S_{\tau_u} = u$. Furthermore, τ_u is right-continuous with limits from the left. Note that τ is continuous in the intervals where it is strictly increasing and $\tau_u < \infty$ with $\tau_u \to \infty$ as $u \to \infty$.

Consider the filtration $\{\mathcal{F}_{\tau_u}, u \geq 0\}$ where $\mathcal{F}_{\tau_u}$ is the σ-algebra of events corresponding to the stopping time τ_u. Recall that $D \in \mathcal{F}_{\tau_u}$ if and only if $D \cap [\tau_u \leq t] \in \mathcal{F}_t$ for all $t \geq 0$. By the fundamental identity of sequential analysis (see Appendix C),

$$\frac{dP_\theta^{\tau_u}}{dP_0^{\tau_u}} = \exp\left(\theta^T \boldsymbol{A}_{\tau_u} - k(\theta)\,\tau_u\right) \tag{2.3.4}$$

where $P_\theta^{\tau_u}$ denotes the restriction of P_θ to $\mathcal{F}_{\tau_u}$. Hence, $\{P_\theta, \theta \in \Theta\}$ is also an exponential family with respect to the filtration $\{\mathcal{F}_{\tau_u}\}$. The family has a nonempty kernel with respect to the filtration $\{\mathcal{F}_{\tau_u}\}$, and hence, by Theorem 2.1, the canonical process

$$\boldsymbol{B}_u = \boldsymbol{A}_{\tau_u} \tag{2.3.5}$$

is a Levy process under P_θ. If S is strictly increasing then $\tau_{S_t} = t$ and the process $\boldsymbol{A}_t$ is a stochastic time transformation of a Levy process, that is,

$$\boldsymbol{A}_t = \boldsymbol{B}_{S_t}\,. \tag{2.3.6}$$

Furthermore,

$$[S_t \leq u] = [\tau_u \geq t] \in \mathcal{F}_{\tau_u} \tag{2.3.7}$$

and hence S_t is a stopping time with respect to the filtration $\{\mathcal{F}_{\tau_u}\}$.

Construction of Exponential Families by a Stochastic Time Transformation

Suppose $k(\boldsymbol{s})$ is a k-dimensional infinitely divisible cumulant transform, that is, $t\,k(.)$ is a cumulant transform for all $t \geq 0$. Then there exists a probability space $(\Omega, \mathcal{F}, P)$ and a process $\boldsymbol{X} = (X^{(1)}, \ldots, X^{(k)})$ with independent stationary increments such that the cumulant transform of $\boldsymbol{X}_t - \boldsymbol{X}_u$ is $(t-u)\,k(.)$, for $t > u$. Suppose $\boldsymbol{X}$ is an RCLL process. The natural exponential family of processes generated by $\boldsymbol{X}$ is given by

$$\frac{dP_\theta^t}{dP^t} = \exp\left(\theta^T \boldsymbol{X}_t - t\,k(\theta)\right)\,. \tag{2.3.8}$$

Let $\{\mathcal{F}_t\}$ be the filtration generated by $\boldsymbol{X}$. This filtration is right-continuous by assumption. Let $u \to \tau_u, u \geq 0$ be a right-continuous increasing process such that for each u, τ_u is a stopping time with respect to $\{\mathcal{F}_t\}$. Further assume that τ_u is finite P_θ-a.s. for all $u \geq 0$ and all $\theta \in \Theta$ and that $\tau_u \to \infty$ as $u \to \infty$

a.s. $[P_\theta]$ for all $\theta \in \Theta$. The family of stopping times $\{\tau_u\}$ defines a stochastic time transformation. The family $\{P_\theta : \theta \in \Theta\}$, by the fundamental identity of sequential analysis, is also an exponential family with respect to the filtration $\{\mathcal{F}_{\tau_u}\}$ with the likelihood function

$$L_{\tau_u}(\theta) = \frac{dP_\theta^{\tau_u}}{dP_0^{\tau_u}} = \exp\left(\theta^T \boldsymbol{X}_{\tau_u} - \tau_u k(\theta)\right) . \tag{2.3.9}$$

Since $\boldsymbol{X}$ is infinitely divisible under P, its cumulant transform is of the form

$$k(\boldsymbol{s}) = \delta^T \boldsymbol{s} + \frac{1}{2}\boldsymbol{s}^T \Sigma \boldsymbol{s} + \int\limits_{E^k} \left[e^{\boldsymbol{s}^T \boldsymbol{x}} - 1 - \boldsymbol{s}^T \gamma(\boldsymbol{x})\right] \nu(d\boldsymbol{x}) \tag{2.3.10}$$

where $\gamma_i(\boldsymbol{x}) = \frac{x_i}{1+x_i^2}$, $1 \leq i \leq k$, $\gamma^T(\boldsymbol{x}) = (\gamma_1(\boldsymbol{x}), \ldots, \gamma_k(\boldsymbol{x}))$, and (δ, Σ, ν) are the Levy characteristics of $\boldsymbol{X}$ under P. In view of (2.3.9), the Levy characteristics under P_θ are $(\delta_\theta, \Sigma_\theta, \nu_\theta)$ given by

$$\nu_\theta(d\boldsymbol{x}) = e^{\theta^T \boldsymbol{x}} \nu(d\boldsymbol{x}) , \tag{2.3.11}$$

$$\Sigma_\theta = \Sigma , \tag{2.3.12}$$

$$\delta_\theta = \delta + \Sigma\, \theta + \int\limits_{E^k} \gamma(\boldsymbol{x}) \left[e^{\theta^T \boldsymbol{x}} - 1\right] \nu(d\boldsymbol{x}) . \tag{2.3.13}$$

Note that

$$\int\limits_{E^k} \frac{\boldsymbol{x}^T \boldsymbol{x}}{1 + \boldsymbol{x}^T \boldsymbol{x}} \nu(d\boldsymbol{x}) < \infty . \tag{2.3.14}$$

If τ is $\boldsymbol{X}$-continuous, that is, if $\boldsymbol{X}$ is constant on all intervals $[\tau_u, \tau_{u-}]$, $u > 0$, then the local characteristics $(\alpha_\theta, \beta_\theta, \nu_\theta)$ of the semimartingale Y defined by $Y_u = \boldsymbol{X}_{\tau_u}$ under P_θ are given by

$$\nu_\theta(du, d\boldsymbol{x}) = d\tau_{u-} e^{\theta^T \boldsymbol{x}} \nu(d\boldsymbol{x}) , \tag{2.3.15}$$

$$\beta_\theta(u) = \Sigma\, \tau_{u-} , \tag{2.3.16}$$

and

$$\alpha_\theta(u) = \left(\delta + \Sigma\, \theta - \int_{E^k} \gamma(\boldsymbol{x}) \left(1 - e^{\theta^T \boldsymbol{x}}\right) \nu(d\boldsymbol{x}) \right) \tau_{u-} \,, \tag{2.3.17}$$

where

$$\gamma_i(\boldsymbol{x}) = \frac{x_i}{1 + x_i^2},\ 1 \le i \le k,\ \gamma^T(\boldsymbol{x}) = (\gamma_1(\boldsymbol{x}), \ldots, \gamma_k(\boldsymbol{x}))\,. \tag{2.3.18}$$

This follows from [4, Chapter 10.1]. Here the drift α_θ, the quadratic characteristic β_θ, and the intensity with which the jumps occur have been changed, but the distribution of the jump size is preserved under the stochastic time transformation.

Example 2.10
Suppose X is a Poisson process. Then $k(s) = e^s - 1$. Suppose we want to generate a family of counting processes with intensity proportional to the state of the process after a stochastic time transformation. We assume that $X_0 = 1$. From the relations (2.3.15) to (2.3.17), we can obtain this if we can find a family of stopping times $\{\tau_u\}$ satisfying

$$d\tau_u = X_{\tau_u} du\,. \tag{2.3.19}$$

Define

$$Z_t = \int_0^t X_s^{-1} ds$$

$$= \sum_{i=1}^{X_t - 1} i^{-1} (T_i - T_{i-1}) + X_t^{-1} \left(t - T_{X_t - 1}\right) \tag{2.3.20}$$

where $T_i,\ i \ge 1$ are the jump times. Then $\{Z_t\}$ is a strictly increasing continuous $\mathcal{F}_t$-adapted process. Let τ_u be the inverse of Z_t. Then (2.3.19) holds and

$$[\tau_u \le t] = [u \le Z_t] \in \mathcal{F}_t\,. \tag{2.3.21}$$

Hence, τ_u is an $\{\mathcal{F}_t\}$-stopping time and the process $\{X_{\tau_u}\}$ is a pure birth process with intensity $e^{\theta} X_{\tau_u}$ under P_θ. □

For other examples, see [8].

REMARK 2.2 The theory of general exponential families of Levy processes and their relations to semimartingales is given in [6, 8]. Kuchler and Sorensen [7] discuss exponential families of Markov processes. They consider the case when the underlying observed process is Markovian under a dominating measure. It is shown that the observed process is Markovian under another probability measure in the exponential family if and only if the corresponding log-likelihood function forms an additive functional. ▮

References

[1] Barndorff-Nielsen, O.E. (1978) *Information and Exponential Families,* Wiley, Chichester.

[2] Barndorff-Nielsen, O.E. (1980) Conditionality resolutions, *Biometrika,* **67**, 293–310.

[3] Basawa, I.V. and Prakasa Rao, B.L.S. (1980) *Statistical Inference for Stochastic Processes,* Academic Press, London.

[4] Jacod, J. (1979) *Calcul Stochastique et problemes de martingales,* Lecture Notes in Mathematics, No. **714**, Springer, Berlin.

[5] Kuchler, I. and Kuchler, U. (1981) An analytical treatment of exponential families of processes with independent stationary increments, *Math. Nachr.,* **103**, 21–30.

[6] Kuchler, U. and Sorensen, M. (1989) Exponential families of stochastic proceses : a unifying semimartingale approach, *Internat. Statist. Rev.,* **57**, 123–144.

[7] Kuchler, U. and Sorensen, M. (1991) On exponential families of Markov processes, *Reseach Report,* No. 233, Aarhus University.

[8] Kuchler, U. and Sorensen, M. (1994) Exponential families of stochastic processes and Levy processes, *J. Stat. Plan. Inf.,* **39**, 211–237.

Chapter 3

Asymptotic Likelihood Theory

3.1 Introduction

Here we discuss the asymptotic likelihood theory for stochastic processes following Barndorff–Nielsen and Sorensen [3] and Kuchler and Sorensen [13]. It is well known that the Fisher information plays a major role in the asymptotic theory of statistical inference for stochastic processes. For instance, see [4]. The score function is standardized using the information, and the asymptotic normality of the maximum likelihood estimator is investigated. The information used for normalization could be either the observed information or the expected information depending on the nature of the problem. In addition to the observed and the expected information, we consider two additional information measures, which will be called incremental observed information and incremental expected information in the general context of stochastic processes.

Different Types of Information and Their Relationships

Let $(\Omega, \mathcal{F}, P_\theta)$ be a probability space with $\theta \in \Theta$ an open subset of R^k. Consider a family of experiments defined on $(\Omega, \mathcal{F}, P_\theta)$ indexed by $t \in R_+ = [0, \infty)$. Let $\mathcal{F}_t$ denote the σ-algebra generated by the experiment with index t. Note that $\mathcal{F}_t \subset \mathcal{F}$. We assume that the amount of data increases with t. This can be formulated mathematically as $\mathcal{F}_s \subset \mathcal{F}_t$ if $s \leq t$. We assume that $\cap_{s \geq t} \mathcal{F}_s = \mathcal{F}_t$. In other words the family $\{\mathcal{F}_t\}$ is assumed to be right-continuous. For simplicity, we assume that $\mathcal{F}_0 = \{\phi, \Omega\}$.

Suppose the family of probability measures $\{P_\theta\}$ is locally dominated by a probability measure P on $\mathcal{F}$, that is,

$$P_\theta^t << P^t, \; t \geq 0, \theta \in \Theta \tag{3.1.1}$$

where P_θ^t denotes the restriction of P_θ to $\mathcal{F}_t$. The likelihood function corresponding to the experiment with index t is

$$L_t(\theta) = \frac{dP_\theta^t}{dP^t}. \tag{3.1.2}$$

It is easy to see that

$$\int_A L_t(\theta)dP = \int_A dP_\theta^t = \int_A dP_\theta^s = \int_A L_s(\theta)dP$$

for all $s \le t$ and $A \in \mathcal{F}_s$. Hence,

$$E\,(L_t(\theta)|\,\mathcal{F}_s) = L_s(\theta) \text{ a.s.}\,; \tag{3.1.3}$$

where $E(\cdot)$ denotes expectation under P. This shows that $\{L_t(\theta), \mathcal{F}_t, t \ge 0\}$ is a martingale under P, and hence, $\{\log L_t(\theta), \mathcal{F}_t, t \ge 0\}$ is a supermartingale under P. These facts have been extensively used in [4] in their study of statistical inference for stochastic processes.

We assume that the following condition holds in the sequel for all $t \ge 0$:

(R1) (i) $L_t(\theta)$ is continuously differentiable with respect to θ.

(ii) The class of likelihood gradients $\{\dot{L}_t(\theta);\, \theta \in \Theta\}$ is locally dominated integrable under P; that is, for each $1 \le i \le k$ and each $\theta_0 \in \Theta$, there exists a neighborhood $U_i(\theta_0)$ of θ_0 and a nonnegative random variable $D_i(\theta_0)$ with $E(D_i(\theta_0)) < \infty$ such that $|\dot{L}_t(\theta)_i| \le D_i(\theta_0)$ for all $\theta \in U_i(\theta_0)$.

Here $\dot{L}_t(\theta)_i$ denotes the ith component of $\dot{L}_t(\theta)$. Let

$$\ell_t(\theta) = \log L_t(\theta)\,. \tag{3.1.4}$$

This is the log likelihood function. The condition (R1) implies that the score vector

$$\dot{\ell}_t(\theta) = \frac{\partial \log L_t(\theta)}{\partial \theta} = \frac{\dot{L}_t(\theta)}{L_t(\theta)} \tag{3.1.5}$$

has finite expectation for all $\theta \in \Theta$ and for all $t \geq 0$. Furthermore,

$$\frac{\partial}{\partial\theta_i}\int_A L_t(\theta)dP = \int_A \dot{L}_t(\theta)dP \tag{3.1.6}$$

for all $A \in \mathcal{F}$ and $t \geq 0$. Note that $\{\dot{\ell}_t(\theta), \mathcal{F}_t, t \geq 0\}$ is a martingale. This can be checked in the following way. Let $A \in \mathcal{F}_s$ and $s \leq t$. Then

$$\begin{aligned}
\int_A \dot{\ell}_t(\theta)_i dP_\theta &= \int_A \dot{L}_t(\theta)_i dP \\
&= \frac{\partial}{\partial\theta_i}\int_A L_t(\theta)dP \\
&= \frac{\partial}{\partial\theta_i}\int_A L_s(\theta)dP \\
&= \int_A \dot{L}_s(\theta)_i dP^s \\
&= \int_A \dot{\ell}_s(\theta)_i dP_\theta^s \\
&= \int_A \dot{\ell}_s(\theta)_i dP_\theta
\end{aligned}$$

and hence, $E[\dot{\ell}_t(\theta)|\mathcal{F}_s] = \dot{\ell}_s(\theta)$ whenever $s \leq t$. It is also easy to see that $E[\dot{\ell}_t(\theta)] = 0$ from the above calculations.

The **quadratic variation** $J_t(\theta)$ of the score vector martingale $\{\dot{\ell}_t(\theta)\}$ viz.

$$J_t(\theta) = \left[\dot{\ell}(\theta)\right]_t \tag{3.1.7}$$

may be interpreted as a type of Fisher information. It is known that

$$\sum_{j=1}^{2^m} \left[\Delta_j^{(m)} \dot{\ell}_t(\theta)\right] \left[\Delta_j^{(m)} \dot{\ell}_t(\theta)\right]^T \xrightarrow{p} J_t(\theta) \tag{3.1.8}$$

in P_θ-measure as $m \to \infty$ where

$$\Delta_j^{(m)} \dot{\ell}_t(\theta) = \dot{\ell}_{jt2^{-m}}(\theta) - \dot{\ell}_{(j-1)t2^{-m}}(\theta) \tag{3.1.9}$$

and the convergence is uniform for t in a compact interval (cf. [11, Theorem I-4-47]). Here α^T denotes the transpose of a vector α. In view of this result, we interpret the process $J_t(\theta)$ as the **incremental observed information.** Relation (3.1.8) implies that $J_t(\theta)$ can be calculated approximately from the observed likelihood process. Note that $J_t(\theta)$ depends on $L_s(\theta),\ s \le t$ and not just $L_t(\theta)$ alone as in the classical i.i.d. case. An alternate way of computing $J_t(\theta)$ is given by the formula (cf. [11, Theorem I-4-52]),

$$J_t(\theta) = \left\langle \dot{\ell}^{(c)}(\theta) \right\rangle_t + \sum_{s \le t} \left[\Delta \dot{\ell}_s(\theta)\right] \left[\Delta \dot{\ell}_s(\theta)\right]^T \tag{3.1.10}$$

where $\dot{\ell}_t^{(c)}(\theta)$ denotes the continuous martingale component of $\dot{\ell}_t(\theta)$ and $\Delta \dot{\ell}_t(\theta) = \dot{\ell}_t(\theta) - \dot{\ell}_{t-}(\theta)$. If $\dot{\ell}_t(\theta)$ has no continuous martingale component, then the formula (3.1.10) simplifies to

$$J_t(\theta) = \sum_{s \le t} \left[\Delta \dot{\ell}_s(\theta)\right] \left[\Delta \dot{\ell}_s(\theta)\right]^T . \tag{3.1.11}$$

In particular, if the time parameter t is discrete, then

$$J_t(\theta) = \sum_{j=1}^{t} \left[\dot{\ell}_j(\theta) - \dot{\ell}_{j-1}(\theta)\right] \left[\dot{\ell}_j(\theta) - \dot{\ell}_{j-1}(\theta)\right]^T . \tag{3.1.12}$$

We further observe that if the ith component of $\dot{\ell}_t(\theta)$ is continuous in t while the jth component is a purely discontinuous martingale, then $J_t(\theta)_{ij} = 0$ from (3.1.10).

In addition to the condition (R1), we assume that

(R2) for all $t \geq 0$, $\theta \in \Theta$ and $1 \leq i, j \leq k$,

$$E\left(\left|\dot{L}_t(\theta)_i \dot{L}_t(\theta)_j\right| / L_t(\theta)\right) < \infty .$$

Under the conditions (R1) and (R2), the score vector $\dot{\ell}_t(\theta)$ is a square integrable martingale under P_θ and $E[\dot{\ell}_t(\theta)]^2 < \infty$ for all $t < \infty$.

Let us now consider the **quadratic characteristic**

$$I_t(\theta) = \langle \dot{\ell}(\theta) \rangle_t \tag{3.1.13}$$

of the score vector martingale. The process $I_t(\theta)$ exists under the condition (R2). It is known that

$$I_t(\theta) = \lim_{m\to\infty} \sum_{j=1}^{2^m} E_\theta\left(\left[\Delta_j^{(m)}\dot{\ell}_t(\theta)\right]\left[\Delta_j^{(m)}\dot{\ell}_t(\theta)\right]^T \middle| \mathcal{F}_{(j-1)t2^{-m}}\right) \tag{3.1.14}$$

where the convergence is weakly in L_1 as was proved by Rao [17]. Here $\Delta_j^{(m)}\dot{\ell}_t(\theta)$ is as defined in (3.1.9). It is known that the particular partition chosen for $[0, t]$ is immaterial in (3.1.14) as well as in (3.1.8). Any sequence of partitions with mesh size tending to zero can be used. Note that $I_t(\theta)$ also depends on the process $\{L_s(\theta),\ s \leq t\}$. The process $I_t(\theta)$ is predictable and of locally integrable variation. In particular,

$$E_\theta\left(\left|I_t(\theta)_{ij}\right|\right) < \infty \tag{3.1.15}$$

for all $1 \leq i,\ j \leq n,\ t \geq 0$, and $\theta \in \Theta$. In the discrete time case,

$$I_t(\theta) = \sum_{j=1}^{t} E_\theta\left(\left[\dot{\ell}_j(\theta) - \dot{\ell}_{j-1}(\theta)\right]\left[\dot{\ell}_j(\theta) - \dot{\ell}_{j-1}(\theta)\right]^T \middle| \mathcal{F}_{j-1}\right) . \tag{3.1.16}$$

The process $I_t(\theta)$ is called **incremental expected information.** This concept of information has been used in [4] in their discussion on the maximum likelihood estimation for discrete time stochastic processes.

The diagonal elements of $I_t(\theta)$ as well as those of $J_t(\theta)$ are increasing processes. Furthermore the matrix valued processes $I_t(\theta)$ and $J_t(\theta)$ are nonnegative definite and increasing in the natural order for nonnegative definite matrices.

Let

$$i_t(\theta) = E_\theta\left(\dot{\ell}_t(\theta)\dot{\ell}_t(\theta)\right)^T . \qquad (3.1.17)$$

The term $i_t(\theta)$ is called the **expected information.** This corresponds to the classical Fisher information for independent and identically distributed random variables. Since the score vector is a square integrable martingale under (R2), it is known that $J_t(\theta)$ is a process of locally integrable variation with $I_t(\theta)$ as its predictable compensator (cf. [11, Proposition I-4-50]). In particular, each component of $J_t(\theta) - I_t(\theta)$ is a zero-mean martingale. Hence

$$i_t(\theta) = E_\theta\left(J_t(\theta)\right) = E_\theta\left(I_t(\theta)\right) . \qquad (3.1.18)$$

REMARK 3.1 If $J_t(\theta)_{ij}$ is predictable, for instance if it is continuous, then $I_t(\theta)_{ij} = J_t(\theta)_{ij}$. In general $J_t(\theta)$ and $I_t(\theta)$ may not be even asymptotically equivalent as $t \to \infty$. If

$$E_\theta\left(\sup_{t\geq 0}\left|\Delta\dot{\ell}_t(\theta)_i\right|^2\right) < \infty , \qquad (3.1.19)$$

then it follows from the strong law of large numbers due to Lepingle [14] that

$$\frac{J_t(\theta)_{ii}}{I_t(\theta)_{ii}} \to 1 \text{ a.s. } [P_\theta] \text{ on the set } [I_\infty(\theta)_{ii} = \infty] . \qquad (3.1.20)$$

This follows from the observation $\Delta J_t(\theta)_{ii} = [\Delta\dot{\ell}_t(\theta)_i]^2$. Note that $[I_\infty(\theta)_{ii} = \infty] = [J_\infty(\theta)_{ii} = \infty]$ a.s. $[P_\theta]$. The general asymptotic equivalence follows from results in [9, Theorem 2.23] for the discrete time processes and from [18, p. 284] for the continuous time processes under (3.1.19). A result in [15] is also useful for checking (3.1.20). ∎

Let us now assume that the following condition holds:

(R3) (i) $L_t(\theta)$ is twice continuously differentiable with respect to θ, and

(ii) the class of random matrices $\{\ddot{L}_t(\theta),\ \theta \in \Theta\}$ is locally dominated integrable under P.

Note that, under (R3) (i),

$$j_t(\theta) = -\ddot{\ell}_t(\theta) \qquad (3.1.21)$$

exists. The process $j_t(\theta)$ is called the **observed information.** It is easy to check that

$$j_t(\theta) = \dot{\ell}_t(\theta)\dot{\ell}_t(\theta)^T - \frac{\ddot{L}_t(\theta)}{L_t(\theta)} \tag{3.1.22}$$

and hence, under (R3) (ii),

$$\begin{aligned} E_\theta\left(j_t(\theta)\right) &= i_t(\theta) - E_\theta\left(\frac{\ddot{L}_t(\theta)}{L_t(\theta)}\right) \\ &= i_t(\theta) - \int \ddot{L}_t(\theta)dP \\ &= i_t(\theta) - \frac{\partial^2}{\partial\theta^2}\int L_t(\theta)dP \\ &= i_t(\theta)\,. \end{aligned} \tag{3.1.23}$$

It can be checked that each component of $\frac{\ddot{L}_t(\theta)}{L_t(\theta)}$ is a zero-mean martingale under P_θ, and hence the diagonal elements of $j_t(\theta)$ are submartingales under P_θ from (3.1.22). The unique increasing predictable process in the Doob–Meyer decomposition of $j_t(\theta)_{ii}$ is $I_t(\theta)_{ii}$ from (3.1.22). Furthermore, each component of $j_t(\theta) - I_t(\theta)$ is a martingale, and hence so are the components of $j_t(\theta) - J_t(\theta)$.

The asymptotic equivalence of the two information quantities $j_t(\theta)_{ii}$ and $I_t(\theta)_{ii}$ under P_θ follows from the fact that their difference $M_t^{(i)}(\theta)$ is a martingale using results in [15] under some conditions on the martingales $\{M_t^{(i)}(\theta)\}$. For instance if $M_t^{(i)}(\theta)$ is a square integrable martingale with quadratic characteristic $< M^{(i)}(\theta) >_t$, then the asymptotic equivalence follows provided $L_\infty(\theta) = \infty$ and

$$\int_0^\infty [1 + I_t(\theta_{ii})]^{-2} d < M^{(i)}(\theta) >_t < \infty \text{ a.s. } [P_\theta]\,.$$

Stochastic models where the ratio between $j_t(\theta)$ and $i_t(\theta)$ converges to a random matrix are usually referred to as non-ergodic models (cf. [4, 5]). It is important to note that the incremental informations $J_t(\theta)$ and $I_t(\theta)$ depend on the way the actual observed experiment E_t is embedded into an indexed set of experiments $\{E_s : s \geq 0\}$.

REMARK 3.2 Suppose that two components of θ (say) θ_i and θ_j are observed to be orthogonal for all t, that is $j_t(\theta)_{ij} = 0$. Relation (3.1.22) implies that $\dot{\ell}_t(\theta)_i \dot{\ell}_t(\theta)_j$ is a martingale, and hence, the quadratic characteristic $I_t(\theta)_{ij} = 0$. Thus, orthogonality of θ_i and θ_j implies that $j_t(\theta)_{ij} = I_t(\theta)_{ij} = i_t(\theta)_{ij} = 0$. However, $J_t(\theta)_{ij}$ need not be zero even when there is orthogonality. ∎

3.2 Examples

Example 3.1

Consider the autoregressive process

$$X_k = \theta X_{k-1} + \varepsilon_k, \quad k = 1, 2, \ldots \tag{3.2.1}$$

where $\theta \in R$, $x_0 = 0$ and ε_k, $k \geq 1$ are i.i.d. standard normal random variables. Then the score function corresponding to X_k, $1 \leq k \leq t$ is

$$\dot{\ell}_t(\theta) = \sum_{k=1}^{t} X_{k-1} (X_k - \theta X_{k-1}) \ , \tag{3.2.2}$$

which can be checked to be a martingale. It can also be seen that the incremental observed information is

$$J_t(\theta) = \sum_{k=1}^{t} X_{k-1}^2 (X_k - \theta X_{k-1})^2 \tag{3.2.3}$$

and the incremental expected information is

$$I_t(\theta) = \sum_{k=1}^{t} X_{k-1}^2 \tag{3.2.4}$$

for all $\theta \in \Theta$. Note that $J_t(\theta)$ depends on θ where as $I_t(\theta)$ does not. Let $I_t = I_t(\theta)$. Suppose that $|\theta| > 1$, the so called explosive case. Then it is known that there exists a random variable Y such that

$$\theta^{-t} X_t \to Y \text{ a.s. as } t \to \infty \tag{3.2.5}$$

and it follows that

$$\theta^{-2t} I_t(\theta) \to Y^2 / \left(\theta^2 - 1\right) \text{ a.s. as } t \to \infty \tag{3.2.6}$$

(cf. [2]). These results follow from the martingale convergence theorem and the Toeplitz lemma, respectively. Furthermore,

$$\theta^{-2t} J_t(\theta) \to Y^2 U_\infty \text{ a.s. as } t \to \infty \tag{3.2.7}$$

where

$$U_\infty = \lim_{t\to\infty} U_t \text{ a.s. } U_t = \sum_{j=1}^{t} \theta^{-2(t-j+1)} \varepsilon_j^2 . \tag{3.2.8}$$

This follows again from the Toeplitz lemma. In order to show that U_t converges a.s., we consider $\hat{U}_t = U_t - E(U_t)$, which is a supermartingale with bounded second moments, and hence, converges a.s. by the supermartingale convergence theorem. The limiting variable is nondegenerate with mean $(\theta^2 - 1)^{-1}$. Hence, from (3.2.7) and (3.2.6), it follows that

$$\frac{J_t(\theta)}{I_t(\theta)} \to U_\infty \left(\theta^2 - 1\right) \text{ a.s. as } t \to \infty . \tag{3.2.9}$$

This in particular shows that the incremental expected information and the incremental observed information need not be equivalent even as $t \to \infty$.

It follows from (3.2.2) that $j_t = I_t$ given by (3.2.4). Furthermore,

$$i_t(\theta) = \begin{cases} t(1-\theta^2)^{-1} & \text{if } |\theta| < 1 \\ \frac{1}{2}t^2 & \text{if } |\theta| = 1 \\ \theta^{2t}\left(\theta^2 - 1\right)^{-2} & \text{if } |\theta| > 1 . \end{cases} \tag{3.2.10}$$

Hence, for $|\theta| < 1$ (stable case),

$$\frac{j_t}{i_t(\theta)} \xrightarrow{p} 1 \text{ as } t \to \infty , \tag{3.2.11}$$

and for $|\theta| > 1$ (explosive case),

$$\frac{j_t}{i_t(\theta)} \to W \text{ a.s. as } t \to \infty \tag{3.2.12}$$

where $W \sim \chi_1^2$ indicating nonergodic behavior (cf. [5]). If $|\theta| = 1$, then $\frac{j_t}{i_t(\theta)}$ converges in law but not in probability [2]. Asymptotic equivalence of j_t and J_t for $|\theta| < 1$ can be obtained using [9, Theorem 2.23]. □

Example 3.2

Note that a pure birth process is a counting process $\{N_t\}$ with intensity λN_{t-}. Let $N_0 = 1$. The score function of the process continuously observed over $[0, t]$ is

$$\dot{\ell}_t(\lambda) = \lambda^{-1}(N_t - 1) - S_t \tag{3.2.13}$$

where

$$S_t = \int_0^t N_s ds \tag{3.2.14}$$

is the total time lived in the population before t. It can be checked that $\{\dot{\ell}_t(\lambda)\}$ is a martingale,

$$J_t(\lambda) = j_t(\lambda) = \lambda^{-2}(N_t - 1), \tag{3.2.15}$$

$$I_t(\lambda) = \lambda^{-1} S_t \tag{3.2.16}$$

and

$$i_t(\lambda) = \left(e^{\lambda t} - 1\right)/\lambda^2. \tag{3.2.17}$$

Here I_t and j_t are asymptotically equivalent but

$$\frac{j_t(\lambda)}{i_t(\lambda)} \to W \text{ a.s. as } t \to \infty$$

where $W \sim \frac{1}{2}\chi_2^2$ indicating the nonergodic behavior (cf. [12]). □

Example 3.3

Let X be a random field on R_+^2 solving the stochastic differential equation

$$dX_z = \theta a_z(X)dz + dW_z,\ z \in R_+^2 \tag{3.2.18}$$

with $X_{(x,0)} = X_{(0,y)} = 0$ for all $x, y \in R_+$. Here W is a Wiener random field and $a_z(\cdot)$ is a functional such that, for $z = (x, y)$, the functional depends only on the behaviour of the random field X in the domain $[0, x] \times [0, y]$. Suppose $\theta \in \Theta$ open in R containing zero. Consider the family of experiments where X is observed in the windows $K_t = [0, t] \times [0, t]$. Under the condition

$$P_\theta\left(\int_{K_t} a_z^2(x)dz < \infty\right) = 1,\ \theta \in \Theta,\ t \geq 0\,, \tag{3.2.19}$$

the likelihood function for the observation over K_t exists and is given by

$$\frac{dP_\theta^t}{dP_0^t} = \exp\left\{\theta \int_{K_t} a_z(X)dX_z - \frac{1}{2}\theta^2 \int_{K_t} a_z^2(X)dz\right\} \tag{3.2.20}$$

and the score function is

$$\dot{\ell}_t(\theta) = \int_{K_t} a_z(X)dX_z - \theta \int_{K_t} a_z^2(X)dz$$

$$= \int_{K_t} a_z(X)dW_z\,. \tag{3.2.21}$$

It is known that $\{\dot{\ell}_t(\theta)\}$ is a martingale. Since $\dot{\ell}_t(\theta)$ is continuous, it follows that $J_t(\theta)$ and $I_t(\theta)$ are identical and it can be checked that

$$J_t(\theta) = I_t(\theta) = j_t(\theta) = \int_{K_t} a_z^2(X)dz\,. \tag{3.2.22}$$

For a discussion on the diffusion fields, see Cairoli and Walsh [6] and Wong and Zakai [20]. Prakasa Rao [16] gives some background information for inference for diffusion fields. □

Example 3.4

Let X be the nonhomogeneous Poisson process with the intensity $\exp(\alpha + \beta t)$, $(\alpha, \beta) \in R^2$. The score vector, when X is observed over $[0, t]$, is

$$\dot{\ell}_t(\alpha, \beta) = \begin{bmatrix} X_t - \int_0^t \exp(\alpha + \beta s)ds \\ \sum_{j=1}^{X_t} \tau_j - \int_0^t s\exp(\alpha + \beta s)ds \end{bmatrix} \tag{3.2.23}$$

where $\tau_1, \tau_2, \ldots$ are the jump times of X. It can be checked that $\dot{\ell}_t(\alpha, \beta)$ is a square integrable vector martingale. The incremental observed information is

$$J_t(\alpha, \beta) = \begin{bmatrix} X_t & \sum_{j=1}^{X_t} \tau_j \\ \sum_{j=1}^{X_t} \tau_j & \sum_{j=1}^{X_t} \tau_j^2 \end{bmatrix} \tag{3.2.24}$$

and the incremental expected information is

$$I_t(\alpha, \beta) = \begin{bmatrix} \int_0^t e^{\alpha+\beta s}ds & \int_0^t se^{\alpha+\beta s}ds \\ \int_0^t se^{\alpha+\beta s}ds & \int_0^t s^2e^{\alpha+\beta s}ds \end{bmatrix}. \tag{3.2.25}$$

Since $J_t(\alpha, \beta)_{22} - I_t(\alpha, \beta)_{22}$ is a square integrable martingale with the quadratic characteristic

$$\int_0^t s^4 \exp(\alpha + \beta s)ds ,$$

it follows from [15] that

$$\frac{J_t(\alpha, \beta)_{22}}{I_t(\alpha, \beta)_{22}} \to 1 \text{ a.s. as } t \to \infty . \tag{3.2.26}$$

The law of large numbers for martingales due to Lepingle [14] shows that

$$\frac{J_t(\alpha,\beta)_{11}}{I_t(\alpha,\beta)_{11}} \to 1 \text{ a.s. as } t \to \infty\,. \tag{3.2.27}$$

In this example, the information on the two parameters α and β increase at different rates. It can be shown that

$$i_t(\alpha,\beta) = j_t(\alpha,\beta) = I_t(\alpha,\beta)$$

where $I_t(\alpha,\beta)$ is as given by (3.2.25). □

Example 3.5

Consider the solution $\{X_t\}$ of the stochastic differential equation

$$dX_t = (\theta_1 + \theta_2 X_t)\,dt + dW_t + dZ_t, \quad X_0 = x_0 \tag{3.2.28}$$

where $\{W_t\}$ is a standard Wiener process and

$$Z_t = \sum_{i=1}^{N_t} Y_i \tag{3.2.29}$$

is a compound Poisson process. Here N_t is a Poisson process with parameter $\lambda > 0$ and the Y_i's are independent exponentially distributed random variables with mean $\frac{1}{\mu} > 0$. Suppose that (θ_1, θ_2) varies over R^2. It can be checked that (θ_1, θ_2), λ, and μ are orthogonal parameters. Suppose the process X is observed over $[0, t]$. Then the incremental observed information $J_t(\theta_1, \theta_2, \lambda, \mu)$ is the 4×4 matrix given by

$$\begin{bmatrix} t & \int_0^t X_s ds & 0 & 0 \\ \int_0^t X_s ds & \int_0^t X_s^2 ds & 0 & 0 \\ 0 & 0 & \lambda^{-2} N_t & -\lambda^{-1}\sum_{i=1}^{N_t}\left(Y_i - \frac{1}{\mu}\right) \\ 0 & 0 & -\lambda^{-1}\sum_{i=1}^{N_t}\left(Y_i - \frac{1}{\mu}\right) & \sum_{i=1}^{N_t}\left(Y_i - \frac{1}{\mu}\right)^2 \end{bmatrix}.$$

The observed information $j_t(\theta_1, \theta_2, \lambda, \mu)$ is given by

$$\begin{bmatrix} t & \int_0^t X_s ds & 0 & 0 \\ \int_0^t X_s ds & \int_0^t X_s^2 ds & 0 & 0 \\ 0 & 0 & \lambda^{-2} N_t & 0 \\ 0 & 0 & 0 & \mu^{-2} N_t \end{bmatrix},$$

the incremental expected information $I_t(\theta_1, \theta_2, \lambda, \mu)$ is

$$\begin{bmatrix} t & \int_0^t X_s ds & 0 & 0 \\ \int_0^t X_s ds & \int_0^t X_s^2 ds & 0 & 0 \\ 0 & 0 & \lambda^{-1} t & 0 \\ 0 & 0 & 0 & \mu^{-2} \lambda t \end{bmatrix}$$

and the expected information $i_t(\theta_1, \theta_2, \lambda, \mu)$ is

$$\begin{bmatrix} t & f_1(t; \theta_1, \theta_2, \lambda, \mu) & 0 & 0 \\ f_1(t; \theta_1, \theta_2, \lambda, \mu) & f_2(t; \theta_1, \theta_2, \lambda, \mu) & 0 & 0 \\ 0 & 0 & \lambda^{-1} t & 0 \\ 0 & 0 & 0 & \mu^{-2} \lambda t \end{bmatrix}$$

where

$$f_1(t; \theta_1, \theta_2, \lambda, \mu) = \begin{cases} \theta_2^{-1}(x_0 + k)\left(e^{\theta_2 t} - 1\right) - kt & \text{if } \theta_2 \neq 0 \\ \frac{1}{2}\left(\theta_1 + \frac{\lambda}{\mu}\right) t^2 & \text{if } \theta_2 = 0 \end{cases}$$

and

$$f_2(t; \theta_1, \theta_2, \lambda, \mu) = \begin{cases} \frac{1}{2}\theta_2^{-1}\left[(x_0 + k)^2 + v\right]\left(e^{2\theta_2 t} - 1\right) & \\ \quad -2\theta_2^{-1} k (x_0 + k)\left(e^{\theta_2 t} - 1\right) & \\ \quad +\left(k^2 - v\right) t & \text{if } \theta_2 \neq 0 \\ \frac{1}{2}\left(1 + \lambda\mu^{-2}\right) t^2 + \frac{1}{3}\left(\theta_1 + \lambda\mu^{-1}\right)^2 t^3 & \text{if } \theta_2 = 0 \end{cases}$$

with $k = \theta_2^{-1}(\theta_1 + \lambda\mu^{-1})$ and $v = \frac{1}{2}\theta_2^{-1}(1 + 2\lambda\mu^{-2})$.

It can be checked that all the four types of information are asymptotically equivalent expect when $\theta_2 > 0$. In that case,

$$\frac{j_t(\theta)_{22}}{i_t(\theta)_{22}} \to \frac{\left\{\int_0^\infty e^{-\theta_2 s} dW_s + \int_0^\infty e^{-\theta_2 s} d\bar{Z}_s + x_0 + k\right\}^2}{(x_0 + k)^2 + \nu}$$

almost surely $[P_\theta]$ as $t \to \infty$. Here $\bar{Z}_s = Z_s - \frac{\lambda}{\mu} s$ is a martingale. The limiting random information matrices J_t, I_t, and j_t are asymptotically equivalent for all values of θ_2. The diagonal elements of the information matrices tend to infinity at different rates when $\theta_2 \geq 0$. □

3.3 Asymptotic Likelihood Theory for a Class of Exponential Families of Semimartingales

Let us consider a family of semimartingales discussed in Section 2.3 with the likelihood function

$$L_t(\theta, k) = H_t^\theta \exp\left((Y_k - 1) * (\mu - \nu)_t + \left(\log\left(Y_k\right) - Y_k + 1\right) * \mu_t\right) \tag{3.3.1}$$

where

$$H_t^\theta = \exp\left[\sum_{i=1}^n \theta_i Z^{(i)} \bullet X_t^{(c)} - \frac{1}{2} \sum_{i,j=1}^n \theta_i \theta_j \left(Z^{(i)} Z^{(j)}\right) \circ \beta_t\right] \tag{3.3.2}$$

and

$$Y_k(\omega, s, x) = \psi(\omega, s, x) \exp\left(\sum_{i=1}^{\ell} k_i Q_i(s, x)\right) = \psi \exp\left(k' Q\right). \tag{3.3.3}$$

See the relations (2.2.24) and (2.2.25). For other details, see Section 2.2. Note that the score function is the vector of the derivatives of the log-likelihood function. The log-likelihood function has the form

$$\ell_t(\theta, k) = \theta^T \left(Z \bullet X_t^{(c)}\right) - \frac{1}{2}\theta^T \left[ZZ^T \circ \beta_t)\right]\theta + k^T B_t - g(k, t) \quad (3.3.4)$$

where

$$B_t = \left(B_t^{(1)}, \dots, B_t^{(\ell)}\right)^T ,$$

$$B_t^{(i)} = Q_i I_A * (\mu - \nu)_t + Q_i I_{A^C} * \mu_t, \; 1 \leq i \leq \ell , \quad (3.3.5)$$

$$g(k, t) = \left(\psi e^{k^T Q} - 1 - \log \psi - k^T Q\right) I_A * \nu_t$$

$$+ \left(\psi e^{k^T Q} - 1\right) I_{A^C} * \nu_t . \quad (3.3.6)$$

Here A is a fixed set in $\mathcal{B}_{[0,\infty]} \otimes \mathcal{E}$ with $\mathcal{E}$ as the Borel σ-algebra on $[-1, 1]$ satisfying, for all $t > 0$,

$$\left.\begin{array}{l}
\text{(a)} \;\; |Q_i|\, I_A * (\mu - \nu)_t < \infty \text{ a.s. } [P], 1 \leq i \leq \ell ; \\
\text{(b)} \;\; |Q_i|\, I_{A^c} * \mu_t < \infty \text{ a.s. } [P], 1 \leq i \leq \ell ; \\
\text{(c)} \;\; \left|\log \psi + k^T Q - \psi e^{k^T Q} + 1\right| I_A * \nu_t \\
\qquad\qquad < \infty \text{ a.s. } [P], k \in \boldsymbol{K} , \\
\text{and} \\
\text{(d)} \;\; \left|\psi e^{k^T Q} - 1\right| I_{A^c} * \nu_t < \infty \text{ a.s. } [P], k \in \boldsymbol{K} .
\end{array}\right\} \quad (3.3.7)$$

We have deleted the terms independent of θ and k in $\ell_t(\theta, k)$, as it will not affect the likelihood theory calculations. The parameters θ and k are orthogonal in the sense that the likelihood function factors into two terms, one dependent on θ and the other on k. Let $\ell_t(\theta) = \ell_t^{(1)}(\theta) + \ell_t^{(2)}(k)$. Then

$$\dot{\ell}_t^{(1)}(\theta) = \left(\frac{\partial \ell_t}{\partial \theta_1}, \dots, \frac{\partial \ell_t}{\partial \theta_n}\right)^T = Z \bullet X_t^{(c)} - \left[\left(ZZ^T\right) \circ \beta_T\right]\theta . \quad (3.3.8)$$

The matrix $(ZZ^T) \circ \beta_T$ is symmetric and nonnegative definite. If it is positive definite, then

$$\hat{\theta}_t = \left[\left(ZZ^T\right) \circ \beta_t\right]^{-1} \left(Z \bullet X_t^{(c)}\right) \qquad (3.3.9)$$

is a solution of the likelihood equation, and if $\hat{\theta}_t \in \Theta$, then this is the unique MLE.

Under $P_{\theta,\boldsymbol{k}}$, the process $X_t^{(c)} - \theta^T(Z \circ \beta_t)$ is a locally square integrable martingale, and hence each component of (3.3.8) is a locally square integrable martingale such that

$$\left\langle \frac{\partial \ell_t}{\partial \theta_i}, \frac{\partial \ell_t}{\partial \theta_j} \right\rangle = \left(Z^{(i)} Z^{(j)}\right) \circ \beta_t, \; 1 \leq i, \; j \leq n\,. \qquad (3.3.10)$$

Thus, $I_t = (ZZ^T) \circ \beta_t$ is the incremental expected information (see Section 3.1). Further note that

$$\frac{\partial^2 \ell_t}{\partial \theta_i \partial \theta_j} = -\left(Z^{(i)} Z^{(j)}\right) \circ \beta_t, \; 1 \leq i, \; j \leq n \qquad (3.3.11)$$

and it is the observed Fisher information matrix. The relation (3.3.11) implies that the elements of the matrix I_t are finite with probability one under $P_{\theta,\boldsymbol{k}}$.

We now briefly consider results about the consistency and the asymptotic normality of the maximum likelihood estimator $\hat{\theta}_t$ given by (3.3.9) for the case $n = 1$. Note that

$$\begin{aligned} \hat{\theta}_t &= \left[\left(ZZ^T\right) \circ \beta_t\right]^{-1} \left(Z \bullet X_t^{(c)}\right) \\ &= \left[\left(ZZ^T\right) \circ \beta_t\right]^{-1} \left(\dot{\ell}_t^{(1)}(\theta) + \left[\left(ZZ^T\right) \circ \beta_t\right]\theta\right) \\ &= \theta + I_t^{-1} \dot{\ell}_t^{(1)}(\theta)\,, \end{aligned} \qquad (3.3.12)$$

and the last term tends to θ a.s. $[P_{\theta,\boldsymbol{k}}]$ on $[I_\infty = \infty]$ by Lepingle's strong law of large numbers [14]. Suppose that the following conditions hold:

(i) $i_t(\theta, \boldsymbol{k}) = E_{\theta,\boldsymbol{k}}(I_t) < \infty, \; t > 0\,,$ (3.3.13)

(ii) $i_t(\theta, \boldsymbol{k}) \to \infty$ as $t \to \infty$, and (3.3.14)

$$\text{(iii)} \quad \frac{I_t}{i_t(\theta, \boldsymbol{k})} \to \eta^2 \text{ in } P_{\theta,\boldsymbol{k}} - \text{ probability as } t \to \infty \tag{3.3.15}$$

where $\eta^2 \geq 0$ is possibly random.

Applying a central limit theorem for martingales (see Chapter 1 or [8]), it follows that

$$[i_t(\theta, \boldsymbol{k})]^{1/2} \left(\hat{\theta}_t - \theta\right) = [i_t(\theta, \boldsymbol{k})]^{-1/2} \, \dot{\ell}_t^{(1)}(\theta) \xrightarrow{\mathcal{L}} W \,, \tag{3.3.16}$$

and

$$I_t^{1/2} \left(\hat{\theta}_t - \theta\right) = I_t^{-1/2} \dot{\ell}_t^{(1)}(\theta) \xrightarrow{\mathcal{L}} N(0, 1) \tag{3.3.17}$$

where W has the distribution of the normal variance mixture with mixing distribution of η^2.

Let

$$D_t = \frac{L_t(\theta_0, \boldsymbol{k})}{L_t\left(\hat{\theta}_t, \boldsymbol{k}\right)} \tag{3.3.18}$$

be the likelihood ratio statistic for testing the hypothesis $H_0 : \theta = \theta_0$. Note that D_t does not depend on $\boldsymbol{k}$ by assumption. Then it follows from (3.3.17) that, under $P_{\theta_0,\boldsymbol{k}}$,

$$-2 \log D_t \xrightarrow{\mathcal{L}} \chi_1^2 \text{ as } t \to \infty \,.$$

Suppose that the interchange of the differentiation with respect to $\boldsymbol{k}$ and the integration with respect to ν are possible in the definition of $g(\boldsymbol{k}, t)$ given by (3.3.6) Further if

$$Q_i \left(\psi e^{\boldsymbol{k}^T \boldsymbol{Q}} - 1\right) I(|x| \leq 1) \quad \text{and} \quad Q_i \psi e^{\boldsymbol{k}^T \boldsymbol{Q}} I(|x| > 1); \, 1 \leq i \leq k$$

are locally dominated integrable with respect to ν, then it can be checked that

$$\dot{\ell}_t^{(2)}(\boldsymbol{k}) = \left(\frac{\partial \ell_t}{\partial k_1}, \ldots, \frac{\partial \ell_t}{\partial k_\ell}\right)^T = \boldsymbol{Q} \star \left(\mu - \psi e^{\boldsymbol{k}^T \boldsymbol{Q}} \circ \nu\right)_t \,. \tag{3.3.19}$$

Suppose $Q_i \in \zeta_{loc}(\mu, P_{\theta,\boldsymbol{k}})$, $1 \leq i \leq \ell$. Then each component of (3.3.19) is a purely discontinuous local martingale under $P_{\theta,\boldsymbol{k}}$ and

$$\left[\frac{\partial \ell_t}{\partial k_i}, \frac{\partial \ell_t}{\partial k_j}\right] = \left(Q_i Q_j\right) * \mu, \; 1 \leq i, \; j \leq \ell \,. \tag{3.3.20}$$

Since $\frac{\partial \ell_t}{\partial \theta_i}$ is continuous and $\frac{\partial \ell_t}{\partial k_j}$ is a purely discontinuous local martingale, it follows that

$$\left[\frac{\partial \ell_t}{\partial \theta_i}, \frac{\partial \ell_t}{\partial k_j}\right] = 0, \ 1 \le i \le n, \ 1 \le j \le \ell . \tag{3.3.21}$$

If $Q_i \in \zeta^2_{loc}(\mu, P_{\theta,\boldsymbol{k}})$, $1 \le i \le \ell$, then the components of (3.3.19) are locally square integrable,

$$\left\langle \frac{\partial \ell_t}{\partial k_i}, \frac{\partial \ell_t}{\partial k_j} \right\rangle = (Q_i Q_j) * \left(\psi e^{\boldsymbol{k}^T Q} \circ \nu\right), \ 1 \le i, \ j \le \ell , \tag{3.3.22}$$

and

$$\left\langle \frac{\partial \ell_t}{\partial \theta_i}, \frac{\partial \ell_t}{\partial k_j} \right\rangle = 0, \ 1 \le i \le n, \ 1 \le j \le \ell . \tag{3.3.23}$$

The last result follows from the orthogonality of θ_i and k_j.

The quadratic variation matrix $\{(Q_i Q_j) * \mu\}$ is the incremental observed information. The incremental expected information need not exist in general. Under an additional condition on the interchange of differentiation with respect to $\boldsymbol{k}$ and integration with respect to ν, the incremental expected information exists and is equal to the observed information; that is,

$$-\frac{\partial^2 \ell_t}{\partial k_i \partial k_j} = Q_i Q_j * \left(\psi e^{\boldsymbol{k}^T Q} \circ \nu\right)_t , \ 1 \le i, \ j \le \ell . \tag{3.3.24}$$

Example 3.6

Let us consider the exponential family studied in Section 2.2. In addition to the conditions (R) assumed there, suppose further that the following conditions hold:

(R$\star$)

(i) $1 * \nu_t < \infty$, a.s. $[P]$, $t > 0$,

(ii) $|\log \psi| I_A * \nu_t < \infty$, a.s. $[P]$, $t > 0$, and

(iii) $|Q_i| \, I_A * \nu_t < \infty$, a.s. $[P]$, $t > 0$, $1 \le i \le \ell$.

Then the likelihood function (2.2.26) can be written in the form

$$L_t(\theta, \boldsymbol{k}) = q_t Z_t^\theta \exp\left(\boldsymbol{k}^T (Q * \nu_t) - h(\boldsymbol{k}, t)\right) \tag{3.3.25}$$

where

$$q_t = a_t \exp\left((\log \psi) I_A * \nu_t + 1 * \nu_t\right) ,$$

$$h(k,t) = \psi e^{k^T Q} * \nu_t, \quad Q * \mu_t = (Q_1 * \mu_t, \ldots, Q_\ell * \mu_t)^T .$$

If $\ell = 1$ and $Q_i \equiv 1$, then

$$h(k,t) = e^k (\psi * \nu_t)$$

and $Q_1 * \mu_t$ is the number of jumps made by X before time t. The families $\{P_{\theta,k}\}$ include the exponential families of counting processes (cf. Example 2.2).

Suppose $n = 0$ and $\psi \equiv 1$. Then the log-likelihood function is given by

$$\ell_t(k) = k\,(1 * \mu_t) - e^k\,(1 * \nu_t) . \tag{3.3.26}$$

Here the parameter of interest is $\lambda = \exp(k)$. It is easy to see that

$$\dot{\ell}_t(k) = 1 * \mu_t - e^k\,(1 * \nu_t) ,$$

and

$$-\ddot{\ell}_t(k) = e^k\,(1 * \nu_t) .$$

Hence, the maximum likelihood estimate of k is given by

$$\exp\left(\hat{k}_t\right) = \frac{1 * \mu_t}{1 * \nu_t} .$$

Furthermore, $\dot{\ell}_t(k)$ is a locally square integrable martingale with $\langle \dot{\ell}(k) \rangle_t = e^k(1 * \nu_t)$ under P_k. Hence, by the Lepingle's law of large numbers,

$$\hat{\lambda}_t = \lambda + \frac{\lambda\left[(1 * \mu_t) - e^k * \nu_t\right]}{e^k * \nu_t} \to \lambda \text{ a.s. as } t \to \infty$$

on $[1 * \nu_\infty = \infty]$ under P_k. Suppose that

$$\left[\dot{\ell}(k)\right]_t = 1 * \nu_t \to \infty \text{ a.s. as } t \to \infty$$

and there exists a nonrandom function $c_t > 0$ and a nonnegative random variable η^2 such that

$$(1 * \mu_t)/c_t \to \eta^2 \text{ in } P_k - \text{probability as } t \to \infty .$$

Then it follows from the central limit theorem for local martingales (cf. Chapter 1) that, under P_k,

$$\left[\dot{\ell}(\lambda)\right]_t^{1/2}\left(\hat{\lambda}_t - \lambda\right) \xrightarrow{\mathcal{L}} N(0, \lambda) \text{ as } t \to \infty ,$$

and

$$\left[\dot{\ell}(k)\right]_t^{-1} \dot{\ell}_t(k)^2 \xrightarrow{\mathcal{L}} \chi_1^2 \text{ as } t \to \infty .$$

□

For the Example 2.2 with intensity λH_t, $\lambda > 0$ for some predictable process H_t, it can be checked that $d\nu = dH_t dt$ and $1 * \mu_t$ is equal to the counting process itself.

3.4 Asymptotic Likelihood Theory for General Processes

We now consider the asymptotic likelihood theory for general processes. We assume that the following conditions holds with the notation in Section 3.1:
(M1) For each $\theta \in \Theta \subset R^k$, the following hold under P_θ as $t \to \infty$;

$$\text{(i)} \quad i_t(\theta)_{jj} \to \infty, \ 1 \le j \le k ; \tag{3.4.1}$$

$$\text{(ii)} \quad i_t(\theta)^{-1/2} E_\theta\left(\sup_{s \le t}\left|\Delta i_s(\theta)_j\right|\right) \to 0, \ 1 \le j \le k ; \tag{3.4.2}$$

$$\text{(iii)} \quad D_t(\theta)^{-1/2} J_t(\theta) D_t(\theta)^{-1/2} \xrightarrow{p} W(\theta) \tag{3.4.3}$$

where $W(\theta)$ is a random nonnegative definite matrix and

$$D_t(\theta) = \text{diag } (i_t(\theta)_{11}, \ldots, i_t(\theta)_{kk}) \tag{3.4.4}$$

and

$$\text{(iv)} \quad D_t(\theta)^{-1/2} i_t(\theta) D_t(\theta)^{-1/2} \to H(\theta) \tag{3.4.5}$$

where $H(\theta)$ is a positive definite matrix.

REMARK 3.3 If $W(\theta)$ is nonrandom, then the model is called ergodic. Otherwise it is said to be nonergodic. ∎

The following conclusion about the weak convergence of the process $\dot{\ell}_t(\theta)$ as $t \to \infty$, under P_θ, follows from the central limit theorem for vector martingales (cf. [19]). See Chapter 1.

THEOREM 3.1
Under the condition $(M1)$,

$$(i) \quad D_t(\theta)^{(-1/2)}\dot{\ell}_t(\theta) \xrightarrow{\mathcal{L}} Z \ (stably), \tag{3.4.6}$$

where Z *is the normal variance mixture with the characteristic function* $\psi(u) = E_\theta\{\exp(-\frac{1}{2}u^T W(\theta)u)\}$, $u^T = (u_1, \ldots, u_n)$;
(ii) conditionally on the event $[\det(W(\theta)) > 0]$,

$$\left[D_t(\theta)^{1/2} J_t(\theta)^{-1} D_t(\theta)^{1/2}\right]^{1/2} D_t(\theta)^{-1/2}\dot{\ell}_t(\theta)$$

$$\xrightarrow{\mathcal{L}} N_k(0, I_k) \ (mixing) \tag{3.4.7}$$

where I_k *is the identity matrix of order k, and*

$$(iii) \quad \dot{\ell}_t(\theta)^T J_t(\theta)\dot{\ell}_t(\theta) \xrightarrow{\mathcal{L}} \chi_k^2 \ (mixing) \tag{3.4.8}$$

as $t \to \infty$.

REMARK 3.4 For the definition of mixing and stable convergence, see Chapter 1. If the diagonal elements of the matrix $i_t(\theta)$ tend to infinity at the same rate, then (3.4.7) simplifies to

$$J_t(\theta)^{-1/2}\dot{\ell}_t(\theta) \xrightarrow{\mathcal{L}} N_k(0, I_k) \ \text{(mixing)} \tag{3.4.9}$$

under P_θ as $t \to \infty$ conditionally on the event $[\det W(\theta) > 0]$. This is proved by replacing $i_t(\theta)_{jj}$, $2 \le j \le k$ by $i_t(\theta)_{11}$ in (M1). Relation (3.4.5) holds trivially whenever $i_t(\theta)$ is a diagonal matrix and in particular for the one-dimensional parameter θ. ∎

Let us now further assume that the following condition holds.
($M2$) Let

$$C_\theta = [\det W(\theta) > 0] . \tag{3.4.10}$$

Suppose that, for each $\theta \in \Theta$, the following holds under P_θ almost surely on C_θ:

(i) a solution $\hat{\theta}_t$ of the likelihood equation

$$\dot{\ell}_t(\theta) = 0 \tag{3.4.11}$$

exists for t sufficiently large and $\hat{\theta}_t \xrightarrow{p} \theta$ as $t \to \infty$;

(ii) there exists a random nonnegative definite matrix $V(\theta)$ with $\det(V(\theta)) > 0$ on C_θ such that for any sequence of k-dimensional random vectors a_t lying on the line segment connecting θ and $\hat{\theta}_t$,

$$D_t(\theta)^{-1/2} j_t(a_t) D_t(\theta)^{-1/2} \xrightarrow{p} V(\theta) \text{ as } t \to \infty . \tag{3.4.12}$$

Applying the Taylor's expansion, we have

$$\dot{\ell}_t(\theta) = j_t\left(\tilde{\theta}_t\right)\left(\hat{\theta}_t - \theta\right) \tag{3.4.13}$$

where $\tilde{\theta}_t$ is a convex combination of $\hat{\theta}_t$ and θ. Using the relations (3.4.6), (3.4.7), and (3.4.12), we find that, conditionally on $C_\theta = [\det W(\theta) > 0]$,

$$\left[D_t(\theta)^{1/2} J_t(\theta)^{-1} D_t(\theta)^{1/2}\right]^{1/2} D_t(\theta)^{-1/2} j_t\left(\tilde{\theta}_t\right)\left(\hat{\theta}_t - \theta\right)$$

$$\xrightarrow{\mathcal{L}} N_k(0, I_k) \text{ (mixing)} \tag{3.4.14}$$

weakly as $t \to \infty$ under P_θ. Since $\hat{\theta}_t$ is consistent, if we replace $\tilde{\theta}_t$ by θ, this result holds. Relation (3.4.14) proves the asymptotic normality of the maximum likelihood estimator under random norming. One can give alternate sufficient conditions ensuring the asymptotic normality of the maximum likelihood estimator. We refer the reader to [3].

We now give some sufficient conditions for the existence of a consistent maximum likelihood estimator. Let

$$G_t(\theta) = J_t(\theta) - j_t(\theta) . \tag{3.4.15}$$

It is easy to see that $\{G_t(\theta)\}$ is a P_θ-martingale as was stated in Section 3.1.

For a $k \times k$ matrix $A = ((a_{ij}))$, let $\|A\|^2 = \sum_{i=1}^k \sum_{j=1}^k a_{ij}^2$. Let $C_\theta = [\det W(\theta) > 0]$ where $W(\theta)$ is as defined earlier.

THEOREM 3.2

Suppose the conditions (M1) holds. In addition suppose that the following conditions hold for every $\alpha > 0$, under P_θ-measure, as $t \to \infty$;

$$(i) \quad \sup_{\tilde{\theta} \in S_t^{(\alpha)}} \left\| D_t(\theta)^{-1/2} G_t\left(\tilde{\theta}\right) D_t(\theta)^{-1/2} \right\| \xrightarrow{p} 0 \ \text{on } C_\theta , \tag{3.4.16}$$

and

$$(ii) \quad \sup_{\tilde{\theta} \in S_t^{(\alpha)}} \left\| D_t(\theta)^{-1/2} J_t\left(\tilde{\theta}\right) D_t(\theta)^{-1/2} - W(\theta) \right\| \xrightarrow{p} 0 \ \text{on } C_\theta ; \tag{3.4.17}$$

where

$$S_t^{(\alpha)} = \left\{\tilde{\theta} : \left\| W(\theta)^{1/2} D_t(\theta)^{1/2} \left(\tilde{\theta} - \theta\right) \right\| \le \alpha \right\} .$$

Then, under P_θ, there exists an estimator $\hat{\theta}_t$ on C_θ that is a solution of the likelihood equation

$$\dot{\ell}_t\left(\bar{\theta}\right) = 0 \tag{3.4.18}$$

with a probability tending to $P_\theta(C_\theta)$ as $t \to \infty$. Furthermore, $\hat{\theta}_t \xrightarrow{p} \theta$ on C_θ and for any sequence of random vectors $\mathbf{a}_t$ lying on the line segment joining θ and $\hat{\theta}_t$,

$$D_t(\theta)^{-1/2} J_t(\mathbf{a}_t) D_t(\theta)^{-1/2} \xrightarrow{p} W(\theta) \ \text{on } C_\theta \tag{3.4.19}$$

and

$$D_t(\theta)^{-1/2} j_t(\mathbf{a}_t) D_t(\theta)^{-1/2} \xrightarrow{p} W(\theta) \ \text{on } C_\theta \tag{3.4.20}$$

as $t \to \infty$.

PROOF If the likelihood equation (3.4.18) has no solution, let $\hat{\theta}_t = \infty$. Otherwise, choose the solution closest to θ in the sense that $\hat{\theta}_t \in S_t^{(n)}$ while there are no solutions in $S_t^{(m)}$ for $1 \leq m \leq n-1$. Let

$$E_{t,c} = \left[\hat{\theta}_t \in S_t^{(c)}\right]. \tag{3.4.21}$$

Consider the Taylor's expansion

$$\dot{\ell}_t\left(\bar{\theta}\right) = \dot{\ell}_t(\theta) + \left[G_t\left(\tilde{\theta}\right) - J_t\left(\tilde{\theta}\right)\right]\left(\bar{\theta} - \theta\right) \tag{3.4.22}$$

where $\tilde{\theta}$ is chosen such that $\|W(\theta)^{1/2}D_t(\theta)^{1/2}(\tilde{\theta} - \theta)\| = c$. Note that the vector $\tilde{\theta} \in S_t^{(c)}$. Fix $\varepsilon > 0$. Since $W(\theta)^{-1/2}D_t(\theta)^{-1/2}\dot{\ell}_t(\theta)$ converges weakly by (3.4.6)(i), there exists $k > 0$ and $t_1 > 0$ such that the event

$$A_t = \left\{\left\|W(\theta)^{-1/2}D_t(\theta)^{-1/2}\dot{\ell}_t(\theta)\right\| \leq k\right\} \cap C_\theta \tag{3.4.23}$$

has probability greater than $P_\theta(C_\theta) - \varepsilon$ for $t > t_1$. Fix $\delta > 0$ and choose c large so that $ck + \delta - c^2 < 0$. Relation (3.4.22) implies that, for $\bar{\theta} \in \partial S_t^{(c)}$,

$$\begin{aligned}\left(\bar{\theta} - \theta\right)^T \dot{\ell}_t\left(\bar{\theta}\right) &= \left(\bar{\theta} - \theta\right)^T D_t(\theta)^{-1/2}W(\theta)^{1/2}W(\theta)^{-1/2}D_t(\theta)^{-1/2}\dot{\ell}_t(\theta) \\ &+ \left(\bar{\theta} - \theta\right)^T D_t(\theta)^{1/2}D_t(\theta)^{-1/2}\left[G_t\left(\tilde{\theta}\right)\right. \\ &\left. - J_t\left(\tilde{\theta}\right)\right] D_t(\theta)^{-1/2}D_t(\theta)^{1/2}\left(\bar{\theta} - \theta\right)\end{aligned} \tag{3.4.24}$$

for some $\tilde{\theta} \in S_t^{(c)}$. Let U_t denote the supremum over $\bar{\theta} \in \partial S_t^{(c)}$ of the absolute value of the difference between $-c^2$ and the last term on the right side of (3.4.24). It follows from (3.4.16) and (3.4.17) that there exists $t_2 > 0$ such that

$$P_\theta\left(B_t\right) \geq P_\theta\left(C_\theta\right) - \varepsilon \text{ for } t > t_2$$

where

$$B_t = [U_t < \delta] \cap C_\theta .$$

On $A_t \cap B_t$, we have $(\bar{\theta}-\theta)^T \dot{\ell}_t(\bar{\theta}) \leq 0$ for all $\bar{\theta} \in \partial S_t^{(c)}$. Hence, the likelihood equation has at least one solution in $S_t^{(c)}$ since $\dot{\ell}_t(\bar{\theta})$ is a continuous function of $\bar{\theta}$ by Brouwer's fixed point theorem (cf. [1]). Note that $A_t \cap B_t \subset E_{t,c}$. Since $P_\theta(A_t \cap B_t) \geq P_\theta(C_\theta) - 2\varepsilon$ for $t > \max(t_1, t_2)$, it follows that

$$P_\theta\left(\hat{\theta}_t < \infty\right) \to P_\theta\left(C_\theta\right) \text{ as } t \to \infty .$$

Furthermore, $\|\hat{\theta}_t - \theta\| \leq \frac{c}{\lambda^-(t)}$ on $E_{t,c}$ where $\lambda^-(t)$ is the smallest eigenvalue of $W(\theta)^{1/2} D_t(\theta)^{1/2}$. Since $\frac{c}{\lambda^-(t)} \xrightarrow{p} 0$ on C_θ, it follows that $\hat{\theta}_t \xrightarrow{p} \theta$ on C_θ. Similarly one can check that (3.4.19) and (3.4.20) follow from (3.4.16) and (3.4.17). ∎

3.5 Exercises

Problem 1: Consider a conditional exponential family of Markov processes [7]. This is a p-dimensional, time-homogeneous Markov chain with one-step transition density (with respect to a dominating measure on R^p) of the exponential form

$$f(y;\theta|x) = h(x,y)\exp\left(\theta^T \boldsymbol{m}(y,x) - \beta(\theta;x)\right)$$

where $\theta \in \Theta \subset R^k$, $\boldsymbol{m}(y,x)$ is a k-dimensional vector and $\beta(\theta;x)$ is one-dimensional. Suppose $X_0 = x_0$ and the Markov chain is observed at the times $i = 1,\ldots,t$. Obtain the information quantities $I_t(\theta)$, $j_t(\theta)$ and $J_t(\theta)$.

Problem 2: Suppose the parameter θ is one-dimensional and the family is a conditionally additive exponential family as in Problem 1 (cf. [10]). This means that there exist functions $\gamma(\theta)$ and $r(x)$ such that $\beta(\theta,x) = \gamma(\theta)r(x)$. An example of such a family is the Gaussian autoregressive model given by

$$X_k = \theta X_{k-1} + Z_k$$

where $\theta \in R$, $X_0 = 0$ and $Z_i, i \geq 1$ are i.i.d. $N(0,1)$. Here $m(y,x) = yx$, $r(x) = x^2$ and $\gamma(\theta) = \frac{1}{2}\theta^2$. Suppose $|\theta| < 1$. Discuss the asymptotic properties of an MLE $\hat{\theta}_t$ based on $X_1,\ldots,X_t$.

Problem 3: Let $\{X_t\}$ be a stochastic process solving the stochastic differential equation

$$dX_t = \theta X_t dt + dW_t, \ \ X_0 = 1$$

where W_t is a Wiener process. Obtain the score function and the information quantities $i_t(\theta)$ and $J_t(\theta)$. Show that they are independent of θ.

Problem 4: Let $P_{\theta,M}$ be a probability measure under which the process $\{X_t\}$ is a solution of the stochastic differential equation

$$dX_t = \theta X_t dt + dM_t, X_0 = 1$$

where $\{M_t\}$ is a square integrable martingale with the quadratic characteristic $\langle M\rangle t = \int_0^t A_s ds$ for some predictable process A. Prove that if M is a continuous process, then a sufficient condition for the consistency of the maximum likelihood estimator of θ based on an observation of X over $[0, t]$ is that $\langle M\rangle_\infty = \infty$ a.s. Obtain the information quantities j_t and J_t.

Problem 5: Let $P_{\theta,\alpha,\sigma}$ be a probability measure under which $\{X_t\}$ solves the equation

$$dX_t = \theta X_t dt + \sigma \tanh(\alpha X_t)\, dW_t$$

where $\alpha > 0$ and $\sigma > 0$. Find the information quantity $J_t(\theta)$ and prove that

$$e^{-2\theta t} J_t(\theta) \to \frac{1}{2}\theta^{-1}\sigma^2 U^2 \text{ a.s. as } \quad t \to \infty$$

where $U = \lim e^{-\theta t} X_t$ a.s. and $E_{\theta,\alpha,\sigma}(U) = 1$. Show that the maximum likelihood estimator is strongly consistent on $C_{\theta,\alpha,\sigma} = [U > 0]$.

Problem 6: Let $\{X_t\}$ be the solution of the stochastic differential equation

$$dX_t = \theta X_t dt + \sigma \tanh\left(e^{-\theta t} X_t\right) dW_t\,.$$

Show that

$$e^{-2\theta t}(\theta) J_t \to \frac{1}{2}\theta^{-1}\sigma^2 U^2 \tanh^2(U) \text{ a.s. as } t \to \infty$$

where $U = \lim e^{-\theta t} X_t$ a.s.

Problem 7: Consider the stochastic differential equation

$$dX_t = \theta X_t dt + \sigma \tanh\left(e^{-t} X_t\right) dW_t\,.$$

Show that J_t is independent of θ. Further prove that, if $\theta > 1$, then

$$e^{-2\theta t} J_t \text{ a.s. } \rightarrow \frac{1}{2}\theta^{-1}\sigma^2 U^2 \text{ a.s. as } t \rightarrow \infty$$

and, for $\theta = 1$,

$$e^{-2\theta t} J_t \rightarrow \frac{1}{2}\sigma^2 U^2 \tanh^2(U) \text{ a.s. as } t \rightarrow \infty$$

where $U = \lim_{t\to\infty} e^{-t} X_t$ a.s. Further show that if $\frac{1}{2} < \theta < 1$, then

$$e^{2(1-2\theta)t} J_t \rightarrow (1-2\theta)^{-1}\sigma^2 U^4 \text{ a.s. as } t \rightarrow \infty$$

and if $\theta = 1/2$, then

$$t^{-1} J_t \text{ a.s. } \rightarrow \sigma^2 U^4 \text{ a.s. as } t \rightarrow \infty .$$

Further show that

$$e^{2(1-2\theta)t} X_t^2 \left[\tanh\left(e^{-t} X_t\right)\right]^2 \rightarrow U^4 \text{a.s. as } t \rightarrow \infty$$

for $\theta < 1$ which implies that J_t does not tend to infinity for $\theta < \frac{1}{2}$.

References

[1] Aitchinson, J. and Silvey, S.D. (1958) Maximum-likelihood estimation of the parameters subject to restraints, *Ann. Math. Statist.*, **29**, 813–828.

[2] Anderson, T.W. (1959) On asymptotic distributions of estimates of parameters of stochastic difference equations, *Ann. Math. Statist.*, **30**, 676–687.

[3] Barndorff–Nielsen, O.E. and Sorensen, M. (1994) A review of some aspects of symptotic likelihood theory for stochastic processes, *Internat. Statist. Rev.*, **62**, 133–165.

[4] Basawa, I.V. and Prakasa Rao, B.L.S. (1980) *Statistical Inference for Stochastic Processes,* Academic Press, London.

[5] Basawa, I.V. and Scott, D.J. (1983) *Asymptotic Optimal Inference for Non-ergodic Models,* Lecture Notes in Statistics, **17**, Springer, Heidelberg.

[6] Cairoli, R. and Walsh, J.B. (1975) Stochastic integrals in the plane, *Acta. Math.,* **134**, 111–183.

[7] Feigin, P.D. (1981) Conditional exponential families and a representation theorem for asymptotic inference, *Ann. Statist.,* **9**, 597–603.

[8] Feigin, P.D. (1985) Stable convergence of semimartingales, *Stoch. Proc. Appl.,* **19**, 125–134.

[9] Hall, P. and Heyde, C. (1980) *Martingale Limit Theory and Its Applications,* Academic Press, New York.

[10] Heyde, C. and Feigin, P.D. (1975) On efficiency and exponential families in stochastic process estimation, *Statistical Distributions in Scientific Work,* Vol. 2 (Ed. G.P. Patil et al.), Dordrecht, Holland, 227–240.

[11] Jacod, J. and Shiryayev, A.N. (1987) *Limit Theorems for Stochastic Processes,* Springer, Heidelberg.

[12] Keiding, N. (1974) Estimation in the birth process, *Biometrika,* **61**, 71–80.

[13] Kuchler, U. and Sorensen, M. (1989) Exponential families of stochastic processes: a unifying semimartingale approach, *Internat. Statst. Rev.,* **57**, 123–144.

[14] Lepingle, D. (1978) Sur le comportment asymptotique des martingales locales, *Seminaire de Probabilites XII,* Lecture Notes in Math., No. **649**, Springer, Berlin, 148–161.

[15] Liptser, R.S. (1980) A strong law of large numbers for local martingales, *Stochastics,* **3**, 217–228.

[16] Prakasa Rao, B.L.S. (1984) On Bayes estimation for diffusion fields, *Statistics: Applications and New Directions,* (Eds. J.K. Ghosh and J. Roy), Statistical Publishing Society, Calcutta.

[17] Rao, K.M. (1968) On decomposition theorems of Meyer, *Math. Scand.*, **24**, 66–78.

[18] Rebolledo, R. (1980) Central limit theorems for local martingales, *Z. Warsch. verw Gebiete,* **51**, 269–286.

[19] Sorensen, M. (1991) Likelihood methods for diffusions with jumps, *Statistical Inference in Stochastic Processes,* (Eds. N.U. Prabhu and I.V. Basawa), 67–105.

[20] Wong, E. and Zakai, M. (1977) Likelihood ratios and transformation of probability associated with two-parameter Wiener processes, Z. *Warsch. verw Gebiete,* **40**, 283–308.

Chapter 4

Asymptotic Likelihood Theory for Diffusion Processes with Jumps

4.1 Introduction

We now develop likelihood methods for inference about parameters that determine the drift and jump mechanism of a diffusion process with jumps following Sorensen [24, 25]. We assume that a *continuously* observed sample path is available. There are various situations where stochastic modeling through diffusion with jumps is found to be appropriate. Applications of such models include general stock price model in [1, 2], modeling soil moisture [21], hydrology [6], etc. The theory we will discuss here generalizes results for ordinary diffusions discussed in [4] and [19]. An extensive discussion is given in [22] on parametric as well as nonparametric inference for diffusion type processes with methods and applications, in both the cases when complete realizations of the sample paths of the process are available in the ideal case as well as when sampled data of the process is available. We will not go into these aspects here. A typical example of a diffusion with jumps is the solution of the stochastic differential equation

$$dX_t = \mu(X_t)\,dt + \sigma(X_t)\,dW_t + \delta(X_t)\,dZ_t$$

where $X_0 = x_0$, W_t is a Wiener process and Z_t is a compound Poisson process.

Diffusions with Jumps

Let us recall some concepts from Chapter 1.

Let Ω be the set of all functions $\omega : R_+ \rightarrow R^d$ that are right-continuous with left limits (RCLL). Let $\mathcal{F}_t = \bigcap_{s>t} \sigma\{\omega(u) : u \leq s\}$. Then $\{\mathcal{F}_t\}$ is increasing and right-continuous. Let $\mathcal{F} = \vee_t \mathcal{F}_t$. The space $(\Omega, \mathcal{F}, \{\mathcal{F}_t\})$ is called the *canonical space.*

Let $(\Omega', \mathcal{F}', \{\mathcal{F}'_t\}, P')$ be a stochastic basis. In other words $\{\mathcal{F}'_t\}$ is a filtration and P' is a probability measures on the measurable space $(\Omega', \mathcal{F}')$. Suppose that this stochastic basis is endowed with an m-dimensional Wiener process W and a class of homogeneous Poisson random measures $\{p^\theta : \theta \in \Theta\}$ on $R_+ \times R^k$. We assume that $\Theta \subset R^n$, Θ nonempty.

DEFINITION 4.1 *An integer valued random measure p on $R_+ \times R^k$ is called a Poisson measure if the following conditions are satisfied: (i) for $q(A) = E(p(A))$, $A \in \mathcal{B}(R_+) \otimes \mathcal{B}(R^k)$, $q(\cdot)$ is σ-finite and $q(\{t\} \times R^k) = 0$ for all $t \geq 0$, (ii) for all $s \geq 0$ and all $A \in \mathcal{B}(R_+) \otimes \mathcal{B}(R^k)$ with $A \subseteq (s, \infty) \times R^k$, the random variable $p(A)$ is independent of $\mathcal{F}'_s$.*

The measure q is called the *intensity measure* of p. Since p^θ is homogeneous, its intensity measure is given by

$$q^\theta(dt, dy) = F_\theta(dy)dt \tag{4.1.1}$$

where F_θ is a σ-finite measure on $(R^k, \mathcal{B}(R^k))$.

An important special case is provided by the random measure

$$p(\omega : dt, dy) = \sum_s I\left(\Delta Y_s(\omega) \neq 0\right) \varepsilon_{(s, \Delta Y_s(\omega))}(dt, dy) \tag{4.1.2}$$

where ε_a is the Dirac measure at a and Y is a k-dimensional compound Poisson process with intensity λ. If the probability distribution function of the jumps is G, then

$$q(dt, dy) = \lambda G(dy)dt \ .$$

Suppose that for each $\theta \in \Theta$, the stochastic differential equation (SDE)

$$dY_t = \beta(\theta; t, Y_t)\, dt + \gamma(t, Y_{t-})\, dW_t$$

$$+ \int_{R^k} \delta(\theta; t, Y_{t-}, z) \left(p^\theta(dt, dz) - q^\theta(dt, dz)\right) ,$$

$$Y_0 = x_0 \tag{4.1.3}$$

has a solution $\{Y_t^\theta\}$. Here x_0 is a nonrandom d-dimensional vector and

$$\gamma : R_+ \times R^d \to R^d \times R^m ,$$

$$\beta(\theta; .,.) : R_+ \times R^d \to R^d ,$$

and

$$\delta(\theta; \cdot, \cdot, \cdot) : R_+ \times R^d \times R^k \to R^d$$

are Borel measurable functions. We assume that the local Lipschitz conditions and linear growth conditions hold on the coefficients in the SDE which ensure that there exist a unique solution of (4.1.3). See Chapter 1 (cf. [14, Theorem III-2.32]).

By a solution of (4.1.3), we mean an $\{\mathcal{F}_t'\}$-adapted d-dimensional stochastic process $\{Y_t\}$ that is RCLL such that

$$Y_t^\theta = x_0 + \int_0^t \beta(\theta; s, Y_s)\, ds + \int_0^t \gamma(s, Y_{s-})\, dW_s$$

$$+ \int_0^t \int_{R^k} \delta(\theta; s, Y_{s-}, z)\left(p^\theta(ds, dz) - q^\theta(ds, dz)\right) .$$

We assume that the integrals in the above equation exist. It can be noted that if the process corresponding to p^θ makes a jump of size z at time t, then Y^θ will make a jump of size $\delta(\theta; t, Y_{t-}, z)$ at the same time.

The problem we will discuss is the estimation of the parameter θ based on the observation of the process $\{Y_t^\theta\}$ over $[0, T]$. The solution $\{Y_t^\theta\}$ of (4.1.3) induces a probability measure P_θ on the canonical space $(\Omega, \mathcal{F})$ by the mapping $\omega' \in \Omega \to Y^\theta(\omega') \in \Omega$. Under P_θ, the canonical process $X_t(\omega) = \omega(t),\ \omega \in \Omega$ is a semimartingale starting at $X_0 = x_0$ with the local characteristics (B, C, ν) given by

$$B_t(\theta) = \int_0^t b(\theta; s, X_s)\, ds , \tag{4.1.4}$$

$$C_t = \int_0^t c(s, X_s)\, ds \,, \tag{4.1.5}$$

and

$$\nu(\theta; dy, dt) = K_t(\theta; X_{t-}, dy)\, dt \,. \tag{4.1.6}$$

Here the kernel K_t is defined by

$$K_t(\theta; x, A) = \int_{R^k} I_{A-\{0\}}(\delta(\theta; t, x, y)) F_\theta(dy) \tag{4.1.7}$$

for all Borel subsets A of R^d,

$$b(\theta; t, x) = \beta(\theta; t, x) - \int_{\|y\|>1} y\, K_t(\theta; x, dy) \tag{4.1.8}$$

and

$$c(t, x) = \gamma(t, x)\gamma(t, x)^T \,. \tag{4.1.9}$$

The process X is a **special semimartingale** (cf. [7]). See Chapter 1. Note that the integral in (4.1.8) exists.

Hereafter, we will assume that the data are an observation of the canonical process. We will further assume that a parametric model for the canonical process is given by the local characteristics (B, C, ν) defined by (4.1.4)–(4.1.6) where

$$b(\theta; ., .) : R_+ \times R^d \to R^d$$

and

$$c(., .) : R_+ \times R^d \to R^d \times R^d$$

are Borel measurable functions such that the matrix $c(t, x)$ is symmetric and nonnegative definite and $K_t(\theta; x, dy)$ is a Borel measurable transition kernel from $R_+ \times R^d$ into R^d with $K_t(\theta; x, \{0\}) = 0$. A semimartingale with such local characteristics is called a *diffusion process with jumps.*

The assumption that the diffusion coefficient γ does not depend on θ is needed for otherwise the restriction of P_θ and P_ϕ to $\mathcal{F}_t$ for $\theta \neq \phi$ would be singular for all $t > 0$ (see Section 4.2). Since we assumed that a continuous path is available, we can estimate γ, when it is not known, from the fact that under P_θ, the matrix

$$\sum_{j=1}^{2^n} \left(X_{js2^{-n}} - X_{(j-1)s2^{-n}}\right)\left(X_{js2^{-n}} - X_{(j-1)s2^{-n}}\right)^T$$

$$-\sum_{u \le s} (\Delta X_u)(\Delta X_u)^T \xrightarrow{p} C_s \quad \text{as} \quad n \to \infty$$

as $n \to \infty$ uniformly for $s \le t$.

4.2 Absolute Continuity for Measures Generated by Diffusions with Jumps

In the following, we will assume that the condition (C1) given below is satisfied by the triplet (b, c, K).

Condition (C1)

(i) b is bounded on $[0, n] \times \{x : \|x\| \le n\}$ for all $n \ge 1$,

(ii) c is continuous on $R_+ \times R^d$ and positive definite, and

(iii) the function

$$(t, x) \to \int_A \left(|y|^2 \wedge 1\right) K_t(dx, dy)$$

is continuous on $R_+ \times R^d$ for all Borel sets $A \subset R^d$.

We assume that, for each $\theta \in \Theta$, the triplet $(b(\theta), c, K(\theta))$ satisfies the condition (C1). It is known that this assumption implies that, for each $\theta \in \Theta$, there exists a unique probability measure P_θ on $(\Omega, \mathcal{F})$ such that the canonical process under P_θ is a diffusion with jumps with the local characteristics $(B(\theta), C, \nu(\theta))$ given by (4.1.4)–(4.1.6). Under P_θ, the process

$$X_t^{(c)}(\theta) = X_t - X_0 - \sum_{s \le t} \Delta X_s I\left(\|\Delta X_s\| > 1\right) - B_t(\theta)$$

$$- \int_0^t \int_{\|x\| \le 1} (\mu - \nu(\theta))(dx, ds) \qquad (4.2.1)$$

is a continuous local martingale. Here the random measure μ is defined by

$$\mu(\omega; dx, dt) = \sum_s I\left(\Delta X_s(\omega) \neq 0\right) \varepsilon_{(x, \Delta X_s(\omega))}(dt, dx) . \qquad (4.2.2)$$

We will further assume that for some fixed $\theta_0 \in \Theta$,

$$\text{(C2)} \qquad K_t(\theta; , x, dy) = Y(\theta; t, x, y) K_t\left(\theta_\theta; x, dy\right), \theta \in \Theta \qquad (4.2.3)$$

where $(t, x, y) \to Y(\theta; t, x, y)$ is a strictly positive function from $R_+ \times R^d \times R^d$ into R_+.

We now discuss sufficient conditions for $P_\theta^{(t)} \simeq P_{\theta_0}^{(t)}$, where $P_\theta^{(t)}$ denotes the restriction of P_θ to $\mathcal{F}_t$, and obtain an expression for the likelihood function $L_t(\theta) = \frac{dP_\theta^t}{dP_{\theta_0}^t}$.

For each $\theta \in \Theta$, define the d-dimensional process $v(\theta)$ by

$$v_t(\theta) = b\left(\theta; t, X_{t-}\right) - b\left(\theta_0; t, X_{t-}\right)$$

$$- \int_{\|y\| \le 1} y\left[Y\left(\theta; t, X_{t-}, y\right) - 1\right] K_t\left(\theta_0 : X_{t-}, dy\right) \qquad (4.2.4)$$

and another process $\alpha_t(\theta) = c_t^{-1} v_t(\theta)$, where $c_t = c(t, X_{t-})$. If the integral in (4.2.4) diverges, we set v_t and α_t equal to $+\infty$. Furthermore, define a nonnegative increasing real process $\{A_t(\theta)\}$ by

$$A_t(\theta) = \int_0^t \alpha_s(\theta)^T c_s \alpha_s(\theta) ds$$

$$+\int_0^t \int_{R^d} [Y(\theta; s, X_s, y) - 1]\, I_{[y>2]} K_s(\theta_0; X_s, dy)\, ds$$

$$+\int_0^t \int_{R^d} [Y(\theta; s, X_s, y) - 1]^2\, I_{[y\leq 2]} K_s(\theta_0; X_s, dy)\, ds\,,$$

$$(4.2.5)$$

which may take the value infinity. This process is continuous except possibly at a point where it jumps to infinity.

THEOREM 4.1
Suppose that

$$P_\theta\,(A_t(\theta) < \infty) = P_{\theta_0}\,(A_t(\theta) < \infty) = 1\,.$$

Then

$$P_\theta^t \simeq P_{\theta_0}^t$$

and

$$\frac{dP_\theta^t}{dP_{\theta_0}^t} = \exp\left\{\int_0^t v_s(\theta)^T c_s^{-1} dX_s^{(c)}(\theta_0) - \frac{1}{2}\int_0^t v_s(\theta)^T c_s^{-1} v_s(\theta) ds\right.$$

$$+\int_0^t \int_{R^d} \left[Y(\theta; s, X_{s-}, y) - 1\right](\mu - \nu(\theta_0))(dy, ds)$$

$$\left.\int_0^t \int_{R^d} \left[\log(Y(\theta; s, X_{s-}, y)) - Y(\theta; s, X_{s-}, y) + 1\right]\mu(dy, ds)\right\}$$

$$(4.2.6)$$

where $X^{(c)}(\theta_0)$ is the continuous P_{θ_0}-martingale defined by (4.2.1) and μ is the random measure defined by (4.2.2).

REMARK 4.1 This result follows from Theorem 4.2 and Theorem 4.5 (b) in [13] (cf. [14, Theorem III-5.34]). See Chapter 1. We omit the details. ∎

Special case: Suppose that the σ-finite measure F_θ in (4.1.1) has a density $f(\theta; y)$ with respect to the Lebesgue measure on R^k and that $\{F_\theta, \theta \in \Theta\}$ are equivalent. Further suppose that $k = d$ and that, for each (θ, t, x), the function

$$z \to \delta(\theta; t, x, z)$$

has a continuously differentiable inverse $y \to \phi(\theta; t, x, y)$ with Jacobian $|J_\phi(\theta; t, x, y)| \neq 0$ for all $y \in \text{Im}(\delta)$. Under these assumptions,

$$K_t(\theta; x, A) = \int_{A \cap \text{Im}(\delta)} f(\theta; \phi(\theta; t, x, y)) \left|J_\phi(\theta; t, x, y)\right| dy$$

and (4.2.3) holds with

$$Y(\theta; t, x, y) = \frac{f(\theta; \phi(\theta; t, x, y)) \left|J_\phi(\theta; t, x, y)\right|}{f(\theta_0; \phi(\theta_0; t, x, y)) \left|J_\phi(\theta_0; t, x, y)\right|} .$$

Suppose further that the Poisson measure p^θ is generated by a compound Poisson process with intensely $\lambda(\theta)$. If the distribution of the jumps has density $g(\theta; y)$ with respect to the Lebesgue measure on R^d, then $f(\theta; y) = \lambda(\theta) g(\theta; y)$ and F_θ is a bounded measure. Note that the process $A(\theta)$ defined by (4.2.5) satisfies the relation

$$A_t(\theta) \leq t\left(\lambda(\theta) + 2\lambda(\theta_0)\right) + \int_0^t v_s(\theta)^T c_s^{-1} v_s(\theta) ds . \tag{4.2.7}$$

This inequality simplifies the conditions for the existence of likelihood ratio and we have the following theorem as a corollary to Theorem 4.1.

THEOREM 4.2

Let F_θ, defined in (4.1.1), be given by the relation $F_\theta(dy) = \lambda(\theta)g(\theta; y)dy$ for some probability density function $g(\theta; y)$. Let $d = k$ and assume the conditions on δ as given above. Then

$$P_\theta\left(\int_0^t \alpha_s(\theta)^T c_s \alpha_s(\theta) ds < \infty\right) = P_{\theta_0}\left(\int_0^t \alpha_s(\theta)^T c_s \alpha_s(\theta) ds < \infty\right) = 1$$

implies that

$$P_\theta^t \simeq P_{\theta_0}^t$$

and

$$\frac{dP^t}{dP_{\theta_0}^t} = \exp\left[\int_0^t v_s(\theta)^T c_s^{-1} dX_s^{(c)}(\theta_0) - \frac{1}{2}\int_0^T v_s(\theta)^T c_s^{-1} v_s(\theta) ds\right.$$

$$\left. + \sum_{s \in S_t} \log\left(Y(\theta; s, X_{s-}, \Delta X_s)\right) + \left(\lambda(\theta_0) - \lambda(\theta)\right)t\right]$$

where $S_t = \{s \le t : \Delta X_s \ne 0\}$.

REMARK 4.2 Deleting the terms independent of θ, we note that the log-likelihood function can be written in the form

$$\ell_t(\theta) = \int_0^t v_s(\theta)^T c_s^{-1} dX_s^{(c)}(\theta_0) - \frac{1}{2}\int_0^T v_s(\theta)^T c_s^{-1} v_s(\theta) ds$$

$$+ N_t \log(\lambda(\theta)) - t\lambda(\theta)$$

$$+ \sum_{s \in S_t} \log\left[g\left(\theta; \phi\left(\theta; s, X_{s-}, \Delta X_s\right)\right) \left|J_\phi\left(\theta; s, X_{s-}, \Delta X_s\right)\right|\right]$$

where N_t is the Poisson process counting the number of jumps before time t. Observe that the log-likelihood function is a sum of three components: a diffusion part, a Poisson process part, and a jump-size part. ∎

4.3 Score Vector and Information Matrix

Assume that the conditions for the validity of Theorem 4.1 hold. Then the log-likelihood function is given by

$$\ell_t(\theta) = \int_0^t v_s(\theta)^T c_s^{-1} dX_s^{(c)}(\theta_0) - \frac{1}{2}\int_0^t v_s(\theta)^T c_s^{-1} v_s(\theta) ds$$

$$+ \int_0^t \int_{R^d} \left[Y(\theta; s, X_{s-}, y) - 1\right]\left(\mu - \nu_{\theta_0}\right)(dy, ds)$$

$$+ \int_0^t \int_{R^d} \left[\log\left(Y(\theta; s, X_{s-}, y) - Y(\theta; s, X_{s-}, y) + 1\right]\mu(dy, ds)\,. \tag{4.3.1}$$

(For convenience, we have denoted $\nu(\theta)$ by ν_θ here). Suppose that $\ell_t(\theta)$ is differentiable with respect to θ and that the differentiation with respect to θ and the stochastic integration can be interchanged. Conditions for the justification of such a procedure are discussed in Appendix C (cf. [12, 15]). Then

$$\dot{\ell}_t(\theta) = \int_0^t \dot{v}_s(\theta)^T c_s^{-1} dX_s^{(c)}(\theta_0) - \int_0^t \dot{v}_s(\theta)^T c_s^{-1} v_s(\theta) ds$$

$$+ \int_0^t \int_{R^d} \dot{Y}(\theta : s, X_{s-}, y)\left(\mu - \nu_{\theta_0}\right)(dy, ds)$$

$$+ \int_0^t \int_{R^d} \left[\dot{H}(\theta; s, X_{s-} y) - \dot{Y}(\theta; s, X_{s-}, y)\right]\mu(dy, ds) \tag{4.3.2}$$

where the (i, j)th element of the matrix $\dot{v}_s(\theta)$ is given by $\dot{v}_s(\theta)_{ij} = \frac{\partial v_s(\theta)_i}{\partial \theta_j}$, the ith element of the vector $\dot{Y}(\theta, s, X_{s-}y)$ is given by $\dot{Y}(\theta, s, x, y)_i = \frac{\partial Y(\theta; s,, x, y)}{\partial \theta_i}$, and $H(\theta; s, X_{s-}, y) = \log Y(\theta; s, x_{s-}, y)$.

Suppose that

$$P_\theta\left(\int_0^t\int_{R^d} \left|\dot{Y}(\theta; s, X_s, y)\right| \nu_{\theta_0}(dy, ds) < \infty\right) = 1$$

for all $i = 1, \ldots, n, \quad \theta \in \Theta$ and $t \geq 0$. Then

$$\dot{\ell}_t(\theta) = \int_0^t \dot{v}_s(\theta)^T c_s^{-1} dX_s^{(c)}(\theta)$$

$$+ \int_0^t\int_{R^d} \dot{H}(\theta; s, X_{s-}, y)\,(\mu - \nu_\theta)(dy, ds) \tag{4.3.3}$$

where, under P_θ,

$$X_t^{(c)}(\theta) = X_t^{(c)}(\theta_0) - \int_0^t v_s(\theta)\, ds$$

is a continuous local martingale with the quadratic characteristic C given by (4.1.5). The score vector $\dot{\ell}_t(\theta)$ is, under P_θ, a locally square integrable martingale provided the quadratic characteristic

$$\langle\dot{\ell}(\theta)\rangle_t = \int_0^t \dot{v}_s(\theta)^T c_s^{-1} \dot{v}_s(\theta) ds$$

$$+ \int_0^t\int_{R^d} \dot{H}(\theta; s, X_{s-}, y)\, \dot{H}(\theta; s, X_{s-}, y)^T \nu_\theta(dy, ds)$$

(4.3.4)

is almost surely finite for all t. Furthermore, if $E_\theta(\langle\dot{\ell}(\theta)\rangle_t)$ is finite for all $t \geq 0$, then the score vector $\dot{\ell}_t(\theta)$ is a square integrable martingale under P_θ. Note that the quadratic variation of $\dot{\ell}_t(\theta)$ is

$$\left[\dot{\ell}(\theta)\right]_t = \int_0^t \dot{v}_s(\theta)^T c_s^{-1} \dot{v}_s(\theta) ds$$

$$+ \int_0^t \int_{R^d} \dot{H}(\theta; s, X_{s-}, y)\, \dot{H}(\theta; s, X_{s-}, y)^T \mu(dy, ds) . \tag{4.3.5}$$

Recall that the matrix $I_t(\theta) = \langle\dot{\ell}(\theta)\rangle_t$ is the *incremental expected information* while $J_t(\theta) = [\dot{\ell}(\theta)]_t$ is the *incremental observed information.* The difference $J_t(\theta) - I_t(\theta)$ is a matrix of local P_θ-martingales. If $E_\theta(I_t(\theta)) < \infty$ for all $t \geq 0$, then the difference is a matrix of zero-mean martingales.

Suppose the log-likelihood function is twice differentiable and that we can interchange differentiation with respect to θ and stochastic integration. Then

$$j_t(\theta) \equiv -\ddot{\ell}_t(\theta)$$

$$= J_t(\theta) - \int_0^t \ddot{v}_s(\theta)^T c_s^{-1} dX_s^{(c)}(\theta)$$

$$- \int_0^t \int_{R^d} \frac{\ddot{Y}(\theta; s, X_{s-}, y)}{Y(\theta; s, X_{s-}, y)} (\mu - \nu_\theta)(dy, ds)$$

$$= I_t(\theta) - \int_0^t \ddot{v}_s(\theta)^T c_s^{-1} dX_s^{(c)}(\theta)$$

$$- \int_0^t \int_{R^d} \ddot{H}(\theta; s, X_{s-} y) (\mu - \nu_\theta)(dy, ds) \qquad (4.3.6)$$

provided

$$P_\theta \left(\int_0^t \int_{R^d} \left| \ddot{Y}(\theta; s, X_{s-}, y)_{ij} \right| \nu_{\theta_0}(dy, ds) < \infty \right) = 1$$

for all $1 \le i, j \le n, \theta \in \Theta$, and $t \ge 0$. The $(i, (k, \ell))$th element of the $d \times n^2$ matrix $\ddot{v}_s(\theta)$ is given by $\ddot{v}_s(\theta)_{i,(k,\ell)} = \frac{\partial^2 v_s(\theta)_i}{\partial\theta_k \partial\theta_\ell}$ and the (i, j)th element of the $n \times n$ matrix $\ddot{Y}$ is given by $\ddot{Y}(\theta; t, x, y) = \frac{\partial^2 Y(\theta; t, x, y)}{\partial\theta_i \partial\theta_j}$. If, moreover,

$$E_\theta \left(\int_0^t \ddot{v}_s(\theta)^T c_s^{-1} \ddot{v}_s(\theta) ds \right) \infty ,$$

and

$$E_\theta \left(\int_0^t \int_{R^d} \ddot{Y}(\theta; s, X_{s-}, y) \nu_{\theta_0}(dy, ds) \right) < \infty ,$$

for all $t > 0$, then the difference $j_t(\theta) - J_t(\theta)$ is a matrix of zero-mean martingales under P_θ. Hence,

$$i_t(\theta) = E_\theta(j_t(\theta)) = E_\theta(I_t(\theta)) = E_\theta(J_t(\theta))$$

provided $E_\theta(I_t(\theta)) < \infty$. Note that $\dot{\ell}_t(\theta) = E_\theta(\dot{\ell}_t(\theta)\dot{\ell}_t(\theta)^T)$ is the *expected information.*

THEOREM 4.3

Suppose that all the diagonal elements of $i_t(\theta)$ *tend to infinity as* $t \to \infty$ *and that, as* $t \to \infty$,

$$i_t(\theta)_{jj}^{-1/2} E_\theta \left\{ \sup_{s \le t} \left| \dot{H}(\theta; s, X_{s-}, \Delta X_s) I(\Delta X_s \neq 0) \right| \right\}$$

$$\to 0, \ 1 \le j \le n, \tag{4.3.7}$$

$$D_t(\theta)^{-1/2} J_t(\theta) D_t(\theta)^{-1/2} \xrightarrow{p} \eta^2(\theta) \ \text{under} \ P_\theta, \tag{4.3.8}$$

and

$$D_t(\theta)^{-1/2} i_t(\theta) D_t(\theta)^{-1/2} \to \Sigma(\theta) \tag{4.3.9}$$

where $D_t(\theta) = \text{diag}\,(i_t(\theta)_{11}, \ldots, i_t(\theta)_{nn})$, $\eta^2(\theta)$ *is a random nonnegative definite matrix, and* $\sum(\theta)$ *is a nonrandom positive definite matrix. Then*

$$D_t(\theta)^{-1/2} \dot{\ell}_t(\theta) \xrightarrow{\mathcal{L}} Z \ \text{(stably)} \tag{4.3.10}$$

and conditionally on det $[(\eta^2(\theta)) > 0]$,

$$\left[D_t(\theta)^{1/2} J_t(\theta)^{-1} D_t(\theta)^{1/2} \right]^{1/2} D_t(\theta)^{-1/2} \dot{\ell}_t(\theta) \xrightarrow{\mathcal{L}} N(0, I_n) \ \text{(mixing)} \tag{4.3.11}$$

and

$$\dot{\ell}_t(\theta)^T J_t(\theta)^{-1} \dot{\ell}_t(\theta) \to \chi_n^2 \ \text{(mixing)} \tag{4.3.12}$$

where the distribution of Z *is a mixture of the normal distributions and has the characteristic function*

$$\phi(u) = E_\theta \left(\exp\left(-\frac{1}{2} u^T \eta^2(\theta) u \right) \right), \ u = (u_1, \ldots, u_n)^T \tag{4.3.13}$$

and I_n *is the identity matrix of order n.*

REMARK 4.3 (i) Since

$$\Delta \dot{\ell}_t(\theta) = \dot{H}(\theta; t, X_{t-}, \Delta X_t) I(\Delta X_t \neq 0), \tag{4.3.14}$$

the conditions of the central limit theorem for square integrable martingales are satisfied for $\dot{\ell}_t(\theta)$ and the result follows as in Theorem 3.1. We omit the details.

(ii) If $i_t(\theta)$ is a diagonal matrix and the diagonal elements tend to infinity at the same rate, then it follows again from the central limit theorem that

$$J_t(\theta)^{-1/2}\dot{\ell}_t(\theta) \xrightarrow{\mathcal{L}} N(0, I_n) \text{ (mixing)} \tag{4.3.15}$$

conditionally on the event $[\det(\eta^2(\theta)) > 0]$. ∎

4.4 Asymptotic Likelihood Theory for Diffusion Processes with Jumps

Consistency

We now show that, under the condition stated earlier, there exists a consistent solution of the likelihood equation. Let $\lambda^+(A)$ and $\lambda^-(A)$ denote the largest and smallest eigenvalues of a nonnegative definite matrix A.

THEOREM 4.4

Suppose that, under P_{θ_0}, the following conditions hold almost surely on the set $\Gamma_{\theta_0} = [\lambda^-(I_t(\theta_0)) \to \infty]$.

(D1) There exists an increasing nonnegative function h such that

$$\int_{\varepsilon}^{\infty} h(x)^{-2} dx < \infty$$

for some $\varepsilon > 0$ such that

$$\limsup_{t\to\infty} \frac{h\left(\lambda^+(I_t(\theta_0))\right)}{\lambda^-(I_t(\theta_0))} < \infty .$$

(D2) Let

$$G_t(\theta) = \int_0^t \ddot{v}_s^T(\theta) c_s^{-1} dX_s^{(c)}(\theta)$$

$$+ \int_0^t \int_{R^d} \ddot{H}(\theta; s, X_{s-}, y)(\mu - \nu_\theta)(dy, ds) .$$

Suppose that, for any $\varepsilon(t)$, such that $\varepsilon(t) \to 0$ as $t \to \infty$,

$$\sup_{\|\theta-\theta_0\| \le \varepsilon(t)} \left\| I_t(\theta_0)^{-1} G_t(\theta) \right\| \to 0$$

and

$$\sup_{\|\theta-\theta_0\| \le \varepsilon(t)} \left\| I_t(\theta_0)^{-1} I_t(\theta) - \boldsymbol{I}_n \right\| \to 0$$

as $t \to \infty$ where $\boldsymbol{I}_n$ is the identity matrix of order n.

Then, under P_{θ_0}, it holds, almost surely on the set Γ_{θ_0}, that the likelihood equation

$$\dot{\ell}_t(\theta) = 0 \tag{4.4.1}$$

has a solution $\hat{\theta}_t$ for t sufficiently large and $\hat{\theta}_t \xrightarrow{p} \theta_0$ as $t \to \infty$.

PROOF Applying the Taylor's expansion, we get that

$$\dot{\ell}_t(\theta) = \dot{\ell}_t(\theta_0) + \ddot{\ell}_t(\bar{\theta})(\theta - \theta_0)$$

$$= \dot{\ell}_t(\theta_0) + \left[G_t(\bar{\theta}) - I_t(\bar{\theta})\right](\theta - \theta_0) \tag{4.4.2}$$

where $\|\bar{\theta} - \theta_0\| \le \|\theta - \theta_0\|$. Note that

$$\left\| I_t(\theta_0)^{-1} \dot{\ell}_t(\theta_0) \right\| \le \lambda^-(I_t(\theta_0))^{-1} \|\dot{\ell}_t(\theta_0)\|$$

$$\le \lambda^-(I_t(\theta_0))^{-1} \sum_i |\dot{\ell}_t(\theta_0)_i|$$

$$\leq \left[\frac{h\left(\lambda^{+}\left(I_t\left(\theta_0\right)\right)\right)}{\lambda^{-}\left(I_t\left(\theta_0\right)\right)}\right] \sum_i \frac{\dot{\ell}_t\left(\theta_0\right)_i}{h\left(\langle \dot{\ell}\left(\theta_0\right)_i \rangle_t\right)} .$$

By Lepingle's law of large numbers (cf. [18]), it follows that

$$\frac{\dot{\ell}_t\left(\theta_0\right)_i}{h\left(\langle \dot{\ell}\left(\theta_0\right)_i \rangle_t\right)} \rightarrow 0 \text{ a.s. } \left[P_{\theta_0}\right] \text{ on } \Gamma_{\theta_0} .$$

Hence,

$$I_t\left(\theta_0\right)^{-1} \dot{\ell}_t\left(\theta_0\right) \rightarrow 0 \text{ as } t \rightarrow \infty \text{ a.s. } \left[P_{\theta_0}\right] \text{ on } \Gamma_{\theta_0} .$$

Relation (4.4.2) implies that

$$\begin{aligned}(\theta-\theta_0)^T I_t\left(\theta_0\right)^{-1} \dot{\ell}_t(\theta) = {} & (\theta-\theta_0)^T I_t\left(\theta_0\right)^{-1} \dot{\ell}_t\left(\theta_0\right) \\ & + (\theta-\theta_0)^T I_t\left(\theta_0\right)^{-1} \\ & \left\{G_t\left(\bar{\theta}\right) - I_t\left(\bar{\theta}\right)\right\}(\theta-\theta_0) .\end{aligned}$$

In view of the condition (D2) of the theorem, it follows that for sufficiently large t, we have, P_{θ_0}-almost surely on Γ_{θ_0}, that

$$\begin{aligned}(\theta-\theta_0)^T I_t\left(\theta_0\right)^{-1} \dot{\ell}_t(\theta) \leq 0 \text{ for all } \theta \text{ such that } & \|\theta-\theta_0\| \\ & = \left\| I_t\left(\theta_0\right)^{-1} \dot{\ell}_t\left(\theta_0\right)\right\|^{1/2} .\end{aligned}$$

Since $\dot{\ell}_t(\theta)$ is a continuous function of θ, we conclude that the equation $I_t(\theta_0)^{-1}\dot{\ell}_t(\theta) = 0$, and hence the equation (4.4.1) has at least one solution in the set $\{\theta : \|\theta-\theta_0\| \leq \|I_T(\theta_0)^{-1}\dot{\ell}_t(\theta_0)\|^{1/2}\}$. ▮

REMARK 4.4 The solution $\hat{\theta}_t$ in the above theorem gives a local maximum of the likelihood function. ▮

Limiting Distribution

THEOREM 4.5

Suppose that for t large, there exists a solution $\hat{\theta}_t$ of the likelihood equation $\dot{\ell}_t(\theta) = 0$ such that $\hat{\theta}_t \to \theta$ in probability under P_θ as $t \to \infty$. Further suppose that the conditions stated in Theorem 4.4 hold. In addition assume that

$$D_t(\theta)^{-1/2} j_t\left(\tilde{\theta}_t\right) D_t(\theta)^{-1/2} \xrightarrow{p} \eta^2(\theta) \text{ under } P_\theta \text{ as } t \to \infty \tag{4.4.3}$$

for any sequence of n-dimensional random vectors $\tilde{\theta}_t$ lying on the line segment connecting θ and $\hat{\theta}_t$. Then, conditionally on the event [det $(\eta^2(\theta)) > 0$],

$$D_t(\theta)^{1/2}\left(\hat{\theta}_t - \theta\right) \xrightarrow{\mathcal{L}} \eta(\theta)^{-2} Z \text{ (stably)}, \tag{4.4.4}$$

$$\left[D_t(\theta)^{-1/2} j_t\left(\hat{\theta}_t\right) D_t(\theta)^{-1/2}\right]^{1/2} D_t(\theta)^{1/2}\left(\hat{\theta}_t - \theta\right) \to N\left(0, I_n\right) \text{ (mixing)}, \tag{4.4.5}$$

and

$$2\left(\ell_t\left(\hat{\theta}_t\right) - \ell_t(\theta)\right) \xrightarrow{\mathcal{L}} \chi_n^2 \text{ (mixing)} \tag{4.4.6}$$

under P_θ as $t \to \infty$.

PROOF Expanding via Taylor's expansion, we see that

$$\dot{\ell}_t(\theta) = -\ddot{\ell}_t\left(\tilde{\theta}_t\right)\left(\hat{\theta}_t - \theta\right)$$

for some $\tilde{\theta}_t$ such that $\|\theta - \tilde{\theta}_t\| \leq \|\theta - \hat{\theta}_t\|$ since $\dot{\ell}(\hat{\theta}_t) = 0$. Hence,

$$\dot{\ell}_t(\theta) = j_t\left(\tilde{\theta}_t\right)\left(\hat{\theta}_t - \theta\right).$$

Similarly

$$\ell_t\left(\hat{\theta}_t\right) - \ell_t(\theta) = \frac{1}{2}\left(\hat{\theta}_t - \theta\right) j_t\left(\tilde{\theta}_t\right)\left(\hat{\theta}_t - \theta\right).$$

The theorem follows from the above expansions and Theorem 4.4. Relation (4.4.5) follows from (4.3.11). ∎

As a corollay to Theorem 4.5, we have the following theorem giving the asymptotic distribution of $\hat{\theta}_t$ under random norming.

THEOREM 4.6
Suppose that the condition of Theorem 4.4 holds and the condition (4.4.3) holds. Further suppose that the components of $D_t(\theta)$ tend to infinity at the same rate. Then

$$j_t\left(\hat{\theta}_t\right)^{1/2}\left(\hat{\theta}_t-\theta\right)\xrightarrow{\mathcal{L}} N\left(0, I_n\right)\ (\text{mixing}) \tag{4.4.7}$$

under P_θ as $t\to\infty$, conditionally on the event $[det\ (\eta^2(\theta))>0]$.

This result follows from the central limit theorem for martingales.

4.5 Asymptotic Likelihood Theory for a Special Class of Exponential Families

Let us now consider the special case of exponential families. Suppose the parameter space consists of two parts θ and λ of dimensions n_1 and n_2, respectively, that vary independently. Let us further assume that the process v_t defined by (4.2.4) depends on θ only and the process Y defined by (4.2.3) depends on λ only. Then the likelihood function (4.2.6) splits into a product of two factors one depending on θ and the other depending on λ. Here the parameters θ and λ are orthogonal, and all the information matrices discussed in the Section 4.3 are block-diagonal, that is, they split in two blocks of size $n_1\times n_1$ and $n_2\times n_2$, respectively, in the diagonal with zeros outside and the two blocks depend on θ and λ, respectively.

Let us consider models for which

$$v_t(\theta)=a_t+D_t\theta \tag{4.5.1}$$

where a_t is a d-dimensional stochastic process while D_t is a $d\times n_1$ matrix of stochastic processes. Let us further suppose that the likelihood function exists.

Then the part of the log-likelihood function depending on θ is given by

$$\ell_t^{(1)}(\theta) = \theta^T H_t - \frac{1}{2}\theta^T \int_0^t D_s^T c_s^{-1} D_s ds\, \theta \qquad (4.5.2)$$

where

$$H_t = \int_0^t D_s^T c_s^{-1} dX_s^{(c)}(\theta_0, \lambda_0) - \int_0^t D_s^T c_s^{-1} a_s ds\,. \qquad (4.5.3)$$

We assume that (θ_0, λ_0) is the parameter value corresponding to the dominating measure. The score vector is given by

$$\dot{\ell}_t^{(1)}(\theta) = H_t - \left(\int_0^t D_s^T c_s^{-1} D_s d_s \right) \theta \qquad (4.5.4)$$

from (4.5.2) and the observed information is

$$j_t^{(1)} = -\ddot{\ell}_t^{(1)}(\theta) = \int_0^t D_s^T c_s^{-1} D_s ds\,. \qquad (4.5.5)$$

Note that the score vector defined by (4.5.4) is a locally square integrable continuous martingale with the quadratic characteristic and the quadratic variation equal to $j_t^{(1)}$. Hence, the three types of information I_t, J_t and j_t coincide in this case as far as the parameter θ is concerned. We further observe that this block of information matrix does not depend on the parameters. The matrix $j_t^{(1)}$ is nonnegative definite. It is strictly positive definite almost surely provided the n_1 d-dimensional processes given by the columns of D_t are almost surely linearly independent; that is, if for any fixed $t > 0$, the set $\{s \leq t : D_s x = 0\}$ where x is a n_1-dimensional vector has a strictly positive Lebesgue measure, then $x = 0$. Hence, if the columns of D_t are almost surely linearly independent, then the likelihood function has almost surely a unique maximum at

$$\hat{\theta}_t = j_t^{(1)^{-1}} H_t\,. \qquad (4.5.6)$$

If $\hat{\theta}_t$ belongs to the parameter set, then it is the unique maximum likelihood estimator and we have the following result giving the asymptotic behavior of $\hat{\theta}_t$.

THEOREM 4.7
Suppose that $i_t^{(1)}(\theta;\lambda) = E_{\theta,\lambda}(j_t^{(1)}) < \infty$ for all t and its diagonal elements tend to infinity as $t \to \infty$. Furthermore, assume that, as $t \to \infty$

$$D_t(\theta,\lambda)^{-1/2} j_t^{(1)} D_t(\theta,\lambda)^{-1/2} \xrightarrow{P} \eta^2(\theta,\lambda) \text{ under } P_{\theta,\lambda} \tag{4.5.7}$$

and

$$D_t(\theta,\lambda)^{-1/2} i_t^{(1)}(\theta,\lambda) D_t(\theta,\lambda)^{-1/2} \to \Sigma(\theta,\lambda) \tag{4.5.8}$$

where $D_t(\theta,\lambda)=\text{diag}\,(i_t^{(1)}(\theta,\lambda)_{11},\ldots,i_t^{(1)}(\theta,\lambda)_{n_1 n_1})$, $\eta^2(\theta,\lambda)$ is a random nonnegative definite matrix and $\sum(\theta,\lambda)$ is a nonrandom positive definite matrix. Then, under $P_{\theta,\lambda}$, conditionally on $[\det \eta^2(\theta,\lambda) > 0]$,

$$\hat{\theta}_t \xrightarrow{P} \theta \text{ as } t \to \infty\,, \tag{4.5.9}$$

$$\left[D_t(\theta,\lambda)^{-1/2} j_t^{(1)} D_t(\theta,\lambda)^{-1/2}\right]^{1/2} D_t(\theta,\lambda)^{1/2}\left(\hat{\theta}_t - \theta\right) \xrightarrow{\mathcal{L}} N\left(0, I_{n_1}\right) \tag{4.5.10}$$

and

$$2\left(\ell_t^{(1)}\left(\hat{\theta}_t\right) - \ell_t^{(1)}(\theta)\right) \xrightarrow{\mathcal{L}} \chi^2_{n_1}\,. \tag{4.5.11}$$

If the diagonal elements of $D_t(\theta,\lambda)$ tend to infinity at the same rate, then

$$j_t^{(1)1/2}\left(\hat{\theta}_t - \theta\right) \xrightarrow{\mathcal{L}} N\left(0, I_{n_1}\right) \tag{4.5.12}$$

under $P_{\theta,\lambda}$ as $t \to \infty$.

PROOF It is easy to check from the relations (4.5.4), (4.5.5), and (4.5.6) that

$$j_t^{(1)}\left(\hat{\theta}_t - \theta\right) = \dot{\ell}_t^{(1)}(\theta)\,.$$

Since $\dot{\ell}_t^{(1)}(\theta)$ is a continuous square integrable martingale with the quadratic variation $j_t^{(1)}$, the results (4.5.10) and (4.5.12) follow from the central limit theorem for martingales. It is clear that the relation (4.5.9) on consistency is a consequence of (4.5.10) and the property (4.5.11) follows from the relation

$$2\left(\ell_t^{(1)}\left(\hat{\theta}_t\right) - \ell_t^{(1)}(\theta)\right) = \left(\hat{\theta}_t - \theta\right)^T j_t^{(1)}\left(\hat{\theta}_t - \theta\right). \tag{4.5.13}$$

▮

REMARK 4.5 Sufficient conditions under which the full likelihood function has an exponential structure are discussed in [23]. Conditions on Y in the equation (4.2.3) implying the exponential structure of the part of the likelihood function that depends on λ are given in [16]. ▮

4.6 Examples

Example 4.1

Consider the stochastic differential equation

$$dX_t = \theta X_t dt + \sigma dW_t + dN_t, t \geq 0, X_0 = x_0 \tag{4.6.1}$$

where W_t is a standard Wiener process and N_t is a Poisson process with intensity λ. Suppose $\sigma > 0$ is *known* and the parameters $\theta \in R$ and $\lambda > 0$ are to be estimated. In this case, the likelihood function can be computed and the maximum likelihood estimates of (θ, λ) depending on an observation of the process $\{X_s, 0 \leq s \leq t\}$ are

$$\hat{\theta}_t = \int_0^t X_{s-} dX_s^{(c)} / \int_0^t X_s^2 ds, \tag{4.6.2}$$

and

$$\hat{\lambda}_t = N_t / t \tag{4.6.3}$$

where

$$X_t^{(c)} = X_t - \sum_{s \le t} \Delta X_s \,. \tag{4.6.4}$$

Note that N_t is the number of jumps before time t. Here $\Delta X_s = X_s - X_{s-}$ as before.

Suppose the model (4.6.1) is replaced by the model

$$dX_t = (\theta X_t + \lambda)\, dt + \sigma dW_t, t \ge 0, X_0 = x_0 \,. \tag{4.6.5}$$

This model is obtained by replacing N_t by λt, which is its compensator. In this case,

$$\tilde{\theta}_t = \frac{t \int_0^t X_s dX_s - (X_t - x_0) \int_0^t X_s ds}{t \int_0^t X_s^2 ds} \tag{4.6.6}$$

and

$$\hat{\lambda}_t = \frac{X_t - x_0}{t} \,. \tag{4.6.7}$$

□

Example 4.2

Consider the stochastic differential equation

$$dX_t = \theta dt + dW_t + dN_t \tag{4.6.8}$$

where W_t and N_t are as in the Example 4.1. The problem is to estimate θ and λ. Here the maximum likelihood estimates of θ and λ are

$$\hat{\theta}_t = \frac{X_t^{(c)}}{t}, \hat{\lambda}_t = \frac{N_t}{t} \tag{4.6.9}$$

where $X_t^{(c)} = X_t - \sum_{s \le t} \Delta X_s$ and $\Delta X_s = X_s - X_{s-}$.

If we consider the model

$$dX_t = (\theta + \lambda)dt + dW_t \tag{4.6.10}$$

obtained by replacing N_t by λt, then we obtain the maximum likelihood estimate of $\mu = \theta + \lambda$ to be

$$\hat{\mu}_t = \frac{X_t}{t} \,. \tag{4.6.11}$$

Note that λ and θ are not individually estimable in this model. □

Example 4.3
Consider the compound Poisson process

$$X_t = \sum_{i=1}^{N_t} Y_i \tag{4.6.12}$$

where N_t is a Poisson process with intensity λ and Y_i are i.i.d. exponential random variables independent of $\{N_t\}$ with mean θ. The problem is to estimate θ and λ. The maximum likelihood estimates of θ and λ are

$$\hat{\theta}_t = \frac{X_t}{N_t} \text{ and } \hat{\lambda}_t = \frac{N_t}{t}. \tag{4.6.13}$$

If we replace N_t by λ_t, then the model

$$X_t = \sum_{i=1}^{\lambda t} Y_i$$

leads to the equation

$$X_t = \lambda t \theta$$

by the method of moments, which gives the estimate for $\mu = \lambda\theta$ to be

$$\hat{\mu} = \frac{X_t}{t}, \tag{4.6.14}$$

which is also the maximum likelihood estimate of μ. □

Example 4.4
Consider a stochastic differential equation of the form

$$dX_t = \alpha(\theta; t, X_t)\, dt + \gamma(t, X_{t-}) dW_t$$

$$+ \delta\left(\theta; t, X_{t-}, \Delta Z_t^\theta\right) dZ_t^\theta, t \geq 0, X_0 = x_0. \tag{4.6.15}$$

where $\{Z_t^\theta\}$ is a compound Poisson process and $\{W_t\}$ is a Wiener process.

Suppose that

$$\alpha(\theta; t, X_t) = \alpha(t, X_t) + \gamma(t, X_t)\pi_t\theta$$

where π_t is a *known* nonrandom regular $d \times d$ matrix, $\alpha : R_+ \times R^d \to R^d$ and $\gamma : R_+ \times R^d \to R^d \times R^d$ is nonsingular. Suppose further that the parameter consists of two parts θ and λ where the drift of the diffusion part of the process depends on the n_1-dimensional vector θ only and the jump mechanism depends on the vector λ only. Further suppose that $n_1 = d = m$. It can be checked that

$$\ell_t^{(1)}(\theta) = \theta^T \int_0^t \pi_s^T \gamma(s, X_{s-})^{-1} dX_s^{(c)}(\theta_o) - \frac{1}{2}\theta^T \int_0^t \pi_s^T \pi_s ds\, \theta \quad (4.6.16)$$

for any fixed θ_0. Note that

$$\int_0^t \pi_s^T \gamma(s, X_{s-})^{-1} dX_s^{(c)}(\theta_o)$$

$$= \int_0^t \pi_s^T \gamma(s, X_{s-})^{-1} dX_s - \int_0^t \pi_s^T \gamma(s, X_s)^{-1} \alpha(s, X_s) ds$$

$$- \sum_{s \le t} \pi_s^T \gamma(s, X_{s-}) \Delta X_s$$

where

$$X_t^{(c)}(\theta) = X_t - X_0 - \sum_{s \le t} \Delta X_s - \int_0^t \alpha(\theta; s, X_s) ds\,,$$

and the latter is a continuous P_θ-martingale. Relation (4.6.2) implies that the MLE is given by

$$\hat{\theta}_t = \left[\int_0^t \pi_s^T \pi_s ds\right]^{-1} \left\{\int_0^t \pi_s^T \gamma(s, X_{s-})^{-1} dX_s^{(c)}(\theta_o)\right\}. \quad (4.6.17)$$

This can be written, under P_θ-measure, as

$$\hat{\theta}_t = \left[\int_0^t \pi_s^T \pi_s ds\right]^{-1} \left\{\int_0^t \pi_s^T \gamma(s, X_{s-})^{-1} dX_s^{(c)}(\theta)\right.$$

$$+ \int_0^t \pi_s^T \gamma(s, X_{s-})^{-1} \gamma(s, X_{s-}) \pi_s ds \, \theta \Big\}$$

$$= \left[\int_0^t \pi_s^T \pi_s ds\right]^{-1} \int_0^T \pi_s^T dW_s + \theta . \tag{4.6.18}$$

Note that, under P_θ,

$$\hat{\theta}_t \simeq N\left(\theta, \left[\int_0^t \pi_s^T \pi_s ds\right]^{-1}\right) \tag{4.6.19}$$

exactly since

$$\int_0^t \pi_s dW_s \simeq N\left(0, \int_0^t \pi_s^T \pi_s ds\right) .$$

In order to test the hypothesis $H_0 : \theta = \theta_0$ against the alternative $H_1 : \theta \neq \theta_0$, one can use the log-likelihood ratio statistic (Wald's test statistic) in the form

$$2\left(\ell_t^{(1)}\left(\hat{\theta}_t\right) - \ell_t^{(1)}(\theta_0)\right) = \left(\hat{\theta}_t - \theta_0\right)^T j_t^{(1)}\left(\hat{\theta}_t - \theta_0\right) , \tag{4.6.20}$$

and this will have an exact $\chi^2_{n_1}$ under P_{θ_0}. The test statistic follows a noncentral $\chi^2_{n_1}$ with noncentrality parameter $(\theta - \theta_0)^T (\int_0^t \pi_s^T \pi_s ds)^{-1} (\theta - \theta_0)$ under P_θ. □

Example 4.5 **(Model for security prices)**
In the study of capital markets, security prices play an important role. Firms use them to guide their investment decisions. The allocation of an investor's funds across securities is based on these prices. Stochastic models for security prices in continuous times have been discussed by Merton [20], and Black and Scholes [5] for processes with continuous sample paths, and Harrison and Pliska [10, 11], and Aase [1] for processes that allow jumps. Consider a particular case of a stochastic differential equation (4.6.15) of the type

$$dX_t = \theta X_t dt + \sigma X_{t-} dW_t + X_{t-} dZ_t \tag{4.6.21}$$

where Z_t is a compound Poisson process, that is,

$$Z_t = \sum_{i=1}^{N_t} \varepsilon_i \tag{4.6.22}$$

where N_t is a Poisson process with parameter λ and $\varepsilon_i, i \geq 1$ are i.i.d. random variables. We assume that σ is known. If this is unknown, one can estimate it by using the fact that, under P_θ,

$$\sum_{i=1}^{2^n} \left(X_{jt2^{-n}} - X_{(j-1)t2^{-n}}\right)^2 - \sum_{u \leq t} (\Delta X_u)^2 \to \sigma^2 \int_0^t X_{s-}^2 ds \text{ a.s.}$$

as $n \to \infty$ uniformly in t for any finite interval [0,T]. We will assume that $\sigma = 1$. The equation (4.6.21) has a unique solution given by

$$X_t = x_o \exp\left[\left(\theta - \frac{1}{2}\right)t + W_t\right] \prod_{i=1}^{N_t} (1 + \varepsilon_i) \tag{4.6.23}$$

by the Doléans–Dade exponential formula. We assume that $\varepsilon_i > -1$ always implying that $X_t > 0$ always.

It can be checked that the maximum likelihood estimator of θ is given by

$$\begin{aligned} \hat{\theta}_t &= t^{-1} \int_0^t X_{s-}^{-1}\, dX_s^{(c)}(\theta o) \\ &= t^{-1} \log\left(\frac{X_t}{x_o}\right) + \frac{1}{2} - t^{-1} \sum_{s \leq t} \left(\frac{X_s}{X_{s-}}\right) \end{aligned} \tag{4.6.24}$$

and $\hat{\theta}_t \simeq N(\theta, \frac{1}{t})$. □

REMARK 4.6 If ε_1 belongs to a regular one-parameter exponential family with density with respect to the Lebesgue measure with mean μ, one can study the problem of estimation of (θ, λ, μ) (cf. [25]). ∎

Example 4.6 **(Model for soil moisture)**

The dynamics of water within the unsaturated zone of the soil represents an important component in the overall spectrum of processes that constitute the hydrological cycle. The significance of soil water movement stems from the fact that the processes associated with it such as infiltration, redistribution, drainage, and evapotranspiration (ET) determine the amount and timing of flows at the earth's surface. In particular, surface run off to stream channels as well as subsurface flow to the ground water system and eventually to the channel network are determined by soil moisture. There are several fundamental phenomena which produce the dynamic behavior of soil moisture. The random depth of cumulative infiltration, which is a function of both the random storm characteristics (intensity and duration) and the hydraulic properties of the soil, is responsible for wetting the soil. In addition the random rate of storm occurrence or time between the storms, the climatically controlled ET rate, the type of vegetation, and the soil hydraulic properties combine to produce variation of soil moisture between storms. Mtundu and Koch [21] developed a model to study the dynamics of water within the unsaturated root zone of the soil, and these are represented by a pair of stochastic differential equations, one representing the "surplus" state of the soil moisture and the other the "deficit" state. The inputs to the model are the climatically controlled random infiltration events and ET which are modeled as a compound Poisson process and a Wiener process, respectively. The stochastic differential equation considered by them is of the type

$$dX_t = (\theta_1 + \theta_2 X)\,dt + \sigma\,dW_t + dZ_t \tag{4.6.25}$$

where Z_t is a compound Poisson process, that is,

$$Z_t = \sum_{i=1}^{N_t} \varepsilon_i$$

where N_t is a Poisson process with parameter λ and ε_i and i.i.d. random variables with ε_i having a one parameter exponential family with density with respect to the Lebesgue measure with mean μ. Let $T(\cdot)$ be the canonical statistic (cf [3]). The problem is to estimate $(\theta_1, \theta_2, \lambda, \mu)$. For the dominating measure, we chose $(\theta_1, \theta_2, \lambda, \mu) = (0, 0, 1, \mu_0)$. Equation (4.6.25) has a unique solution given by

$$X_t = x_0 e^{\theta_2 t} + \frac{\theta_1}{\theta_2}\left(e^{\theta_2 t} - 1\right) + \sigma \int e^{\theta_2(t-s)}\,dW_s$$

$$+\int_0^t e^{\theta_2(t-s)} dZ_s \quad \text{if} \quad \theta_2 \neq 0$$

$$= \theta_1 t + \sigma W_t + Z_t \quad \text{if} \quad \theta_2 = 0 \tag{4.6.26}$$

by the Ito's formula. It can be checked that

$$\hat{\theta_{1t}} = \frac{X_1^{(c)}(\theta_{10}, \theta_{20}) \int_0^t X_s^2 ds - \int_0^t X_s ds \int_0^t X_{s-} dX_s^{(c)}(\theta_{10}, \theta_{20})}{t \int_0^t X_s^2 ds - \left(\int_0^t X_s ds\right)^2} \tag{4.6.27}$$

and

$$\hat{\theta_{2t}} = \frac{t \int_0^t X_{s-} dX_s^{(c)}(\theta_{10}, \theta_{20}) - X_t^{(c)}(\theta_{10}, \theta_{20}) \int_0^t X_s ds}{t \int_0^t X_s^2 ds - \left(\int_o^t X_s ds\right)^2} \tag{4.6.28}$$

where

$$X_t^{(c)}(\theta_{10}, \theta_{20}) = X_t - \sum_{s \le t} \Delta X_s \,. \tag{4.6.29}$$

By the Ito's formula,

$$\int X_{s-} dX_s^{(c)}(\theta_{10}, \theta_{20})$$

$$= \frac{1}{2}\left[X_t^2 - x_0^2 - \sigma^2 t - \sum_{s \le t}\left(X_s^2 - X_{s-}^2\right)\right]. \tag{4.6.30}$$

The MLE's of $\hat{X}_t$ and $\hat{\mu}_t$ are given by

$$\hat{\lambda}_t = \frac{N_t}{t} \tag{4.6.31}$$

and

$$\hat{\mu}_t = \frac{1}{N_t} \sum_{s \in S_t} T(\Delta X_s) \tag{4.6.32}$$

where S_t is the set of jump times. The incremental observed information matrix $J_t = [\dot{\ell}(\theta)]_t$ equals

$$\begin{bmatrix} \sigma^{-2}t & \sigma^{-2}\int_0^t X_s ds & 0 & 0 \\ \sigma^{-2}\int_0^t X_s ds & \sigma^{-2}\int_0^t X_s^2 ds & 0 & 0 \\ 0 & 0 & \lambda^{-2}N_t & \lambda^{-1}J_t^{(1)}(\mu) \\ 0 & 0 & \lambda^{-1}J_t^{(1)}(\mu) & J_t^{(2)}(\mu) \end{bmatrix} \tag{4.6.33}$$

where

$$J_t^{(1)}(\mu) = (V(\mu))^{-1} \sum_{s \in S_t} (T(\Delta X_s) - \mu) \ , \tag{4.6.34}$$

$$J_t^{(2)}(\mu) = (V(\mu))^{-2} \sum_{s \in S_t} (T(\Delta X_s) - \mu)^2 \ , \tag{4.6.35}$$

and

$$V(\mu) = V_\mu (T(\varepsilon_1)) \ . \tag{4.6.36}$$

See [3] for the definition of $V(\mu)$.

Let us suppose that $\sigma^2 = 1$ for simplicity. If $\theta_2 \neq 0$, then

$$E(X_t) = (x_0 + k) e^{\theta_2 t} - k \tag{4.6.37}$$

and

$$E\left(X_t^2\right) = \left((x_0 + k)^2 + \nu\right) e^{2\theta_2 t} - 2k(x_0 + k) e^{\theta_2 t} + \left(k^2 - \nu\right) \tag{4.6.38}$$

where $k = (\theta_1 + \lambda m)/\theta_2$, $\nu = \frac{1}{2}(1 + \lambda m^{(2)})/\theta^2$, $m = E_\mu(\varepsilon_1)$, and $m^{(2)} = E_\mu(\varepsilon_1^2)$. This follows from the representation (4.6.26) and using the fact that

$Z_t - \lambda mt$ is a martingale with the quadratic characteristic $\lambda m^{(2)}(t)$. Relations (4.6.37) and (4.6.38) prove that

$$E\left(\int_0^t X_s ds\right) = \theta_2^{-1}(x_0 + k)\left(e^{\theta_2 t} - 1\right) - kt \tag{4.6.39}$$

and

$$E\left(\int_0^t X_s^2 ds\right) = \frac{1}{2}\theta_2^{-1}\left((x_0 + k)^2 + \nu\right)\left(e^{2\theta_2 t} - 1\right) \tag{4.6.40}$$

$$- 2\theta_2^{-1}k(x_0 + k)\left(e^{\theta_2 t} - 1\right) + \left(k^2 - \nu\right)t\,.$$

If $\theta_2 = 0$, then the representation (4.6.26) gives

$$E\left(\int_0^t X_s ds\right) = \frac{1}{2}(\theta_1 + \lambda m)t^2 \tag{4.6.41}$$

and

$$E\left(\int_0^t X_s^2 ds\right) = \frac{1}{2}\left(1 + \lambda m^{(2)}\right)t^2 + \frac{1}{3}(\theta_1 + \lambda m)^2 + 3\,. \tag{4.6.42}$$

The upper left corner of the expected information matrix i_t can be obtained using (4.6.39) to (4.6.42). The lower right corner will be $\begin{bmatrix} t\lambda^{-1} & 0 \\ 0 & t\lambda V(\mu)^{-1} \end{bmatrix}$.

Suppose $\theta_2 < 0$. It can be seen that the expected information matrix i_t satisfies the condition

$$t^{-1}i_t \to \Sigma \text{ as } t \to \infty \tag{4.6.43}$$

where

$$\Sigma = \begin{bmatrix} 1 & -k & 0 & 0 \\ -k & (k^2 - \nu) & 0 & 0 \\ 0 & 0 & \lambda^{-1} & 0 \\ 0 & 0 & 0 & \lambda V(\mu)^{-1} \end{bmatrix}. \tag{4.6.44}$$

Since $\nu < 0$ for $\theta_2 < 0$, the matrix Σ is positive definite. By the ergodic theorem,

$$t^{-1}\int_0^t X_s ds \to -k \text{ as } t \to \infty \text{ a.s.} \tag{4.6.45}$$

and

$$t^{-1}\int_0^t X_s^2 ds \to k^2 - \nu \text{ as } t \to \infty \text{ a.s.} \tag{4.6.46}$$

for $\theta_2 < 0$. Relations (4.6.45) and (4.6.46) together with the law of large numbers show that

$$t^{-1} J_t \to \Sigma \text{ a.s. as } t \to \infty . \tag{4.6.47}$$

It can be checked that the score function and the maximum likelihood estimator are asymptotically normal after proper normalization following Theorem 4.3 and Theorem 4.6. Note that

$$\begin{aligned}\hat{\theta}_{1t} - \theta_1 &= \frac{X_t^{(c)}(\theta_1,\theta_2)\int_0^t X_s^2 ds - \int_0^t X_s ds \int_0^t X_{s-} dX_s^{(c)}(\theta_1,\theta_2)}{t\int_0^t X_s^2 ds - \left(\int_0^t X_s ds\right)^2} \\ &= \frac{t^{-1}X_t^{(c)}(\theta_1,\theta_2)\, t^{-1}\int_0^t X_s^2 ds - t^{-1}\int_0^t X_s ds\, t^{-1}\int_0^t X_{s-} dX_s^{(c)}(\theta_1,\theta_2)}{t^{-1}\int_0^t X_s^2 ds - \left(t^{-1}\int_0^t X_s ds\right)^2} .\end{aligned} \tag{4.6.48}$$

It follows from (4.6.45), (4.6.46) and the law of large numbers due to Lepingle [18] that $\hat{\theta}_{1t} - \theta_1 \to 0$ a.s. since $X_t^{(c)}(\theta_1, \theta_2)$ is a martingale with the quadratic characteristic $\int_0^t X_s^2 ds \simeq t$. Similar arguments prove the strong consistency of $\hat{\theta}_{2t}$ to θ_2 using the relation

$$\hat{\theta}_{2t} - \theta_2 = \frac{t\int_0^t X_{s-} dX_s^{(c)}(\theta_1, \theta_2) - X_t^{(c)}(\theta_1, \theta_2)\int_0^t X_s ds}{t\int_0^t X_s^2 ds - \left(\int_0^t X_s ds\right)^2}. \tag{4.6.49}$$

Suppose $\theta_2 > 0$. Then the components of i_t increase to infinity at different rates. In this case, we apply Theorem 4.5 to obtain the simultaneous asymptotic normality of MLEs. We can chose $D_t(\theta) = \text{diag}\,(t, e^{2\theta_2 t}, t, t)$ in Theorem 4.5. Let

$$\begin{aligned} H_t &= e^{-\theta_2 t} X_t - x_0 - k\left(1 - e^{-\theta_2 t}\right) \\ &= \int_0^t e^{-\theta_2 s} dW_s + \int_0^t e^{-\theta_2 s} d\bar{Z}_s . \end{aligned}$$

Here $\bar{Z}_t = Z_t - \lambda m t$. Then H_t is a martingale and $E(H_t^2) = \nu(1 - e^{-2\theta_2 t}) \leq \nu$. Hence, by the martingale convergence theorem, $e^{-\theta_2 t} X_t$ converges a.s. to a random variable $H_\infty + x_0 + k$ and $e^{-2\theta_2 t} X_t^2$ converges almost surely to $(H_\infty + x_0 + k)^2$. Applying the integral version of the Toeplitz lemma, we have

$$e^{-\theta_2 t}\int_0^t X_s ds \to \frac{H_\infty + x_0 + k}{\theta_2}$$

and

$$e^{-2\theta_2 t}\int_0^t X_s^2 ds \to \frac{\frac{1}{2}(H_\infty + x_0 + k)^2}{\theta_2}$$

almost surely as $t \to \infty$. It can now be checked that the condition (4.4.3) holds for $\eta^2 = \text{diag}\,(1, \frac{\frac{1}{2}(H_\infty + x_0 + k)^2}{\theta_2}, \lambda^{-1}, \lambda V(\mu)^{-1})$. We now leave it to the reader to prove the strong consistency and the asymptotic normality of the MLE (cf. [25]) under some conditions on the density of ε_1. $\square$

REMARK 4.7 For a review of stochastic differential equations with applications in hydrology, see [6]. ∎

4.7 Exercises

Problem 1: (Model for dynamics of population growth) Consider the stochastic differential equation

$$dX_t = \theta_1 X_t \left(1 - \frac{X_t}{K}\right) dt + \sigma X_{t-} dW_t - X_{t-} dZ_t$$

where Z_t is a compound Poisson process. Suppose the probability distribution of the jumps of Z_t are concentrated in the interval (0, 1). This model is used for studying the dynamics of a population that grows logistically between disasters. A disaster when it occurs removes a random fraction of the population. The parameter K is the carrying capacity of the environment (cf. [8, 9]). Show that the solution of the equation is given by

$$X_t = x_0 \exp\left(\left(\theta_1 - \frac{1}{2}\sigma^2\right) t - \theta_1 K^{-1} \int_0^t X_s ds + \sigma W_t + \sum_{i=1}^{N_t} \log(1 - \varepsilon_i)\right)$$

using the Doléans–Dade exponential formula. Let $\theta_2 = -\theta_1 K^{-1}$. Suppose ε_1 has the beta-distribution with parameter k_1, k_2. Obtain the MLEs of θ_1, θ_2, k_1, and k_2 [25].

Problem 2: Consider the one-dimensional stochastic differential equation

$$dY_t = \beta(\theta; t, Y_t) dt + \gamma(t, Y_{t-}) dW_t$$

$$+ \int_R \delta(\theta; t, Y_{t-}, z) \left(p^\theta(dt, dz) - q^\theta(dt, dz)\right),$$

$$Y_0 = x_0$$

with

$$\tilde{\beta}\left(\tilde{\theta}_1, \theta_2; t, x\right) = \theta_1 + \theta_2 x\,, \quad \gamma = 1\,, \quad \delta(\theta; t, x, z) = z\,,$$

and

$$q^{\alpha}(dt, dz) = z^{-1} e^{-\alpha z} dt dz\,.$$

Here $\theta = (\theta_1, \theta_2, \alpha) \in R \times R \times R_+$. Study the MLE of the parameter α [25].

References

[1] Aase, K.K. (1986) New option pricing formulas when the stock market is a combined continuous point process, *Technical Report*, No. 60, CCREMS, MIT, Cambridge, MA.

[2] Aase, K.K. (1986) Probabilistic solutions of option pricing and corporate security valuation, *Technical Report*, No. 63, CCREMS, MIT, Cambridge, MA.

[3] Barndorff–Nielsen, O. (1978) *Information and Exponential Families*, Wiley, Chicester.

[4] Basawa, I.V. and Prakasa Rao, B.L.S. (1980) *Statistical Inference for Stochastic Processes*, Academic Press, New York.

[5] Black, F. and Scholes, M. (1973) The pricing of options and corporate liabilities, *Journal of Political Economy*, **81**, 637–654.

[6] Bodo, B.A., Thompson, M.E., and Unny, T.E. (1987) A review on stochastic differential equations for applications in hydrology, *Stochastic Hydrol. Hydraul.*, **1**, 81–100.

[7] Elliott, R.J. (1982) *Stochasic Calculus and Applications*, Springer, New York.

[8] Guttorp, P. and Kulperger, R. (1984) Statistical inference for some Volterra population processes in a random environment, *Can. J. Statist.*, **12**, 289–302.

[9] Hanson, F.B. and Tuckwell, H.C. (1981) Logistic growth with random density independent disasters, *Theor. Pop. Biol.*, **19**, 1–18.

[10] Harrison, J.M. and Pliska, S.R. (1981) Martingales and stochastic integrals in the theory of continuous trading, *Stoch. Proc. Appl.*, **11**, 215–260.

[11] Harrison, J.M. and Pliska, S.R. (1983) A stochastic calculus model of continuous trading: complete markets, *Stoch. Proc. Appl.*, **15**, 313–316.

[12] Hutton, J.E. and Nelson, P.I. (1984) Interchanging the order of differentiation and stochastic integration, *Stoch. Proc. Appl.*, **18**, 371–377.

[13] Jacod, J. and Memin, J. (1976) Caracteristiques locales et conditions de cotinuite absolue pour les martingales, *Z. Warsch. verw Gebiete*, **35**, 1–37.

[14] Jacod, J. and Shiryayev, A.N. (1987) *Limit Theorems for Stochastic Processes,* Springer, Heidelberg.

[15] Karandikar, R.L. (1983) Interchanging the order of stochastic integration and ordinary differentiation, *Sankhya Ser. A*, **45**, 120–124.

[16] Küchler, U. and Sorensen, M. (1989) Exponential families of stochastic processes: a unifying semimartingale approach, *Internat. Statist. Rev.*, **57**, 123–144.

[17] Kutoyants, Y.A. (1984) *Parameter Estimation for Stochastic Processes,* (Trans. and Ed. B.L.S. Prakasa Rao), Heldermann, Berlin.

[18] Lepingle, D. (1978) Sur le comportement asymptotique des martingales locales, *Lecture Notes in Mathematics,* **649**, (Eds. C. Dellacherie, P.A. Meyer and M. Weil), Springer, Berlin, 148–161.

[19] Liptser, R.S. and Shiryayev, A.N. (1977) *Statistics of Random Processes: General Theory,* Springer, New York.

[20] Merton, R.C. (1971) Optimum consumption and portfolio rules in a continuous-time model, *Journal of Economic Theory,* **3**, 373–413.

[21] Mtundu, N.D. and Koch, R.W. (1987) A stochastic differential equation approach to soil moisture, *Stochastic Hydrol. Hydraul.*, **1**, 101–116.

[22] Prakasa Rao, B.L.S. (1999) *Statistical Inference for Diffusion Type Processes,* Arnold, London.

[23] Skokan, V. (1984) A contribution to maximum likelihood estimation in processes given by certain types of stochastic differential equations, *Asymptotic Statistics,* **2**, (Eds. P. Mandl and M. Huskova), Elsevier, Amsterdam.

[24] Sorensen, M. (1989) A note on the existence of a consistent maximum likelihood estimator for diffusions with jumps, *Markov Processes and Control Theory,* (Eds. H. Langer and V. Nollau), Akademie-Verlag, Berlin, 229–234.

[25] Sorensen, M. (1991) Likelihood methods for diffusions with jumps, *Statistical Inference in Stochastic Proceses,* (Eds. N.U. Prabhu and I.V. Basawa), Marcel Dekker, New York, 67–105.

Chapter 5

Quasi Likelihood and Semimartingales

5.1 Quasi Likelihood and Discrete Time Processes

Let $\{Y_1, Y_2, \ldots\}$ be a discrete time real-valued stochastic process. Suppose that the joint distribution F of $(Y_1, \ldots Y_n)$ belongs to a family $\mathcal{F}$ of probability distributions on R^n. Let $\theta = \theta(F)$, $F \in \mathcal{F}$ be a real-valued parameter. Let h_i be a real-valued function of $Y_1, \ldots, Y_i$ and θ such that

$$E_{i-1,F}\left[h_i\left(Y_1, \ldots, Y_i; \theta(F)\right)\right] = 0,\ 1 \le i \le n,\ F \in \mathcal{F} \tag{5.1.1}$$

where $E_{i-1,F}$ denotes the conditional expectation of h_i given $Y_1, \ldots, Y_{i-1}$. For simplicity, we write hereafter

$$E_{i-1,F} \equiv E_{i-1} \text{ and } E_{0,F} \equiv E_F \equiv E\ .$$

An example of h_i is

$$h_i = Y_i - E_{i-1}\left(Y_i\right)\ . \tag{5.1.2}$$

It is easy to see that

$$E\left(h_i h_j\right) = 0,\ i \ne j\ . \tag{5.1.3}$$

Hence the functions $h_1, h_2, \ldots$ are uncorrelated.

DEFINITION 5.1 *Any real-valued function g of the random variables $Y_1, Y_2, \ldots, Y_n$ and the parameter θ is called an estimating function and it is called an unbiased estimating function if*

$$E_F\left[g\left(Y_1, \ldots, Y_n;\ \theta(F)\right)\right] = 0,\ F \in \mathcal{F}\ . \tag{5.1.4}$$

Among all the unbiased estimating functions g, an estimating function g^* is said to be *optimal* if

$$\frac{E_F\left[g^2\left(Y_1,\ldots,Y_n;\ \theta(F)\right)\right]}{\left\{E_F\left(\left[\frac{\partial g(Y_1,\ldots,Y_n;\ \theta(F))}{\partial\theta}\right]_{\theta=\theta(F)}\right)\right\}^2} \tag{5.1.5}$$

is minimized for all $F \in \mathcal{F}$ at $g = g^*$ minimization being done over the class of g which are differentiable with respect to θ and for which (5.1.5) is well defined.

Let us consider the estimating functions of the form

$$g = \sum_{i=1}^{n} h_i f_{i-1} \tag{5.1.6}$$

where the functions h_i, $1 \le i \le n$ are as defined earlier and f_{i-1} is a function of the random variable $Y_1, \ldots, Y_{i-1}$ and the parameter θ for $i = 1, \ldots, n$. It is clear that

$$E(g) = 0 \tag{5.1.7}$$

from (5.1.1). Let $\mathcal{L}$ be the class of all estimating functions g defined by (5.1.6) for a given set $\{h_i\}$.

THEOREM 5.1

In the class $\mathcal{L}$ of estimating functions, the function g^* minimizing (5.1.5) is

$$g^* = \sum_{i=1}^{n} h_i f_{i-1}^* \tag{5.1.8}$$

where

$$f_{i-1}^* = \frac{E_{i-1}\left(\frac{\partial h_i}{\partial\theta}\right)}{E_{i-1}\left(h_i^2\right)}. \tag{5.1.9}$$

PROOF Note that

$$E\left(g^2\right) = E\left(\sum_{i=1}^{n} h_i f_{i-1}\right)^2$$

$$= E\left(\sum_{i=1}^{n} f_{i-1}^2 E_{i-1}\left(h_i^2\right)\right) \tag{5.1.10}$$

from (5.1.3) and

$$\left\{E\left(\frac{\partial g}{\partial \theta}\right)\right\}^2 = \left\{E\left(\sum_{i=1}^{n}\left(\frac{\partial h_i}{\partial \theta} f_{i-1} + h_i \frac{\partial f_{i-1}}{\partial \theta}\right)\right)\right\}^2$$

$$= \left\{E\left[\sum_{i=1}^{n}\left(E_{i-1}\left(\frac{\partial h_i}{\partial \theta}\right) f_{i-1} + \frac{\partial f_{i-1}}{\partial \theta} E_{i-1}(h_i)\right)\right]\right\}^2$$

$$= \left(E\left[\sum_{i=1}^{n} E_{i-1}\left(\frac{\partial h_i}{\partial \theta}\right) f_{i-1}\right]\right)^2 \tag{5.1.11}$$

since $E_{i-1}(h_i) = 0$ by (5.1.1). An application of the Cauchy–Schwartz inequality implies that

$$\frac{\left[E\left(\frac{\partial g}{\partial \theta}\right)\right]^2}{E\left(g^2\right)} = \frac{(E[B])^2}{E\left(A^2\right)} \le E\left(\frac{B^2}{A^2}\right)$$

where

$$B = \sum_{i=1}^{n} f_{i-1} E_{i-1}\left(\frac{\partial h_i}{\partial \theta}\right)$$

and

$$A = \left\{\sum_{i=1}^{n} f_{i-1}^2 E_{i-1}\left(h_i^2\right)\right\}^{1/2}.$$

Note that $\frac{B^2}{A^2}$ is maximized when $B = \sum_{i=1}^{n} f_{i-1}^{*} E_{i-1}\left(\frac{\partial h_i}{\partial \theta}\right)$ and $A =$

$\{\sum_{i=1}^{n} f_{i-1}^{*2} E_{i-1}(h_i^2)\}^{1/2}$, and in such an event

$$E\left(\frac{B^2}{A^2}\right) = \frac{\{E(B)\}^2}{E\left(A^2\right)} .$$

This proves the optimality of g^* defined by (5.1.8). ∎

5.2 Quasi Likelihood and Continuous Time Processes

Let $\{X_t, 0 \leq t \leq T\}$ be a stochastic process defined on a probability space $(\Omega, \mathcal{B}, P_\theta), \theta \in \Theta$ open in R^n. We assume further that $(\Omega, \mathcal{B}, P_\theta)$ is a complete probability space. Let $\{\mathcal{B}_t, t \geq 0\}$ be a filtration satisfying the usual conditions $\mathcal{B}_s \subset \mathcal{B}_t \subseteq \mathcal{B}$ for $s \leq t$, $\mathcal{B}_0$ is augmented by sets of measure zero of $\mathcal{B}$ and $\mathcal{B}_t = \mathcal{B}_{t+}$ where $\mathcal{B}_{t+} = \cup_{s>t}\mathcal{B}_s$.

Consider the class ζ of zero mean square integrable estimating functions $G_t(\theta) = G_t(\{X_s, 0 \leq s \leq t\}, \theta)$ for which $E_\theta(G_t(\theta)) = 0$ for each $P_\theta \in \mathcal{P}$ with index θ. Let $M \subset \zeta$ be the class of estimating functions $\{G_s(\theta), \mathcal{B}_s, 0 \leq s \leq t\}$ that are martingales for each $P_\theta \in \mathcal{P}$ with index θ. Suppose the process $\{X_s\}$ is observed over $[0, t]$. Estimators θ_t^* of θ can be obtained by solving the estimating equation

$$G_t\left(\theta_t^*\right) = 0 .$$

Let us further restrict attention to the subclass ζ_1 of ζ of the estimating functions $G_t(\theta)$ which are almost surely differentiable with respect to the components of θ and for which the matrices

$$E_\theta \dot{G}_t(\theta) = \left(\left(E_\theta \frac{\partial G_{ti}(\theta)}{\partial \theta_j}\right)\right)$$

and

$$E_\theta\left(G_t(\theta) G_t^T(\theta)\right)$$

are nonsingular. Here the expectations are taken with respect to P_θ, $\dot{G}_t(\theta)$ denotes the gradient of $G_t(\theta)$ with respect to θ, and α^T denotes the transpose of a vector α.

DEFINITION 5.2 *An estimating function G_t^* is said to be optimal if*

$$\left(E\dot{G}_t\right)^{-1}\left(EG_tG_t^T\right)\left(\left(E\dot{G}_t\right)^{-1}\right)^T - \left(E\dot{G}_t^*\right)^{-1}E\left(G_t^*G_t^{*T}\right)\left(\left(E\dot{G}_t^*\right)^{-1}\right)^T$$

is nonnegative definite for all $G_t \in \zeta_1$, $\theta \in \Theta$, and $P_\theta \in \mathcal{P}$. (Here A^T denotes the transpose of matrix A.) Such an optimal estimating function G_t^* is called the quasi-score function and the corresponding equation $G_t^* = 0$ is the quasi-likelihood equation. A solution of the quasilikelihood equation is called a quasi-likelihood estimator.

5.3 Quasi Likelihood and Special Semimartingale

Let $\{X_t\}$ be a d-dimensional continuous time stochastic process defined on a probability space $(\Omega, \mathcal{F}, P)$ with a right-continuous filtration. We assume that $\{X_t\}$ is RCLL and adapted to a filtration on the space $(\Omega, \mathcal{F}, P)$ generated by $\{X_t\}$. Suppose the probability distribution of $\{X_t\}$ depends on a parameter $\theta \in \Theta \subset R^n$ and that the process X_t has the representation

$$X_t = X_0 + \int_0^t f_s(\theta)d\lambda_s + m_t(\theta), \quad t \geq 0 . \tag{5.3.1}$$

Here $\{\lambda_t\}$ is a real nondecreasing right-continuous predictable process with $\lambda_0 = 0$, $\{f_t\}$ is a d-dimensional predictable process, and $\{m_t(\theta)\}$ is an RCLL local martingale with $m_0(\theta) = 0$. The representation (5.3.1) implies that $\{X_t\}$ is a special semimartingale (cf. [11, p. 30]). See Chapter 1. Recall that any local martingale can be decomposed uniquely into a continuous local martingale and a purely discontinuous local martingale. Let the decomposition of $m_t(\theta)$ be given by

$$m_t(\theta) = m_t^{(c)}(\theta) + m_t^{(d)}(\theta) . \tag{5.3.2}$$

A continuous local martingale is always locally square integrable. Hence, the quadratic characteristic of $m_t^{(c)}(\theta)$ exists. It has the form

$$\left\langle m^{(c)}(\theta)\right\rangle_t = \int_0^t a_s(\theta)d\lambda_s \tag{5.3.3}$$

where $a_s(\theta)$ is a symmetric nonnegative definite $d \times d$ matrix of predictable processes (cf. [12, Proposition II 2-9]). Let μ be the random measure on $[0, \infty) \times R^d$ given by the jumps of $\{X_t\}$, that is,

$$\mu(dt, dx) = \sum_s I\,(\Delta X_s \neq 0)\, \varepsilon_{(s,\Delta X_s)}(dt, dx)$$

where ε_a is the Dirac measure at a. Let $\nu(\theta)$ be the predictable compensator of μ. Then

$$m_t^{(d)}(\theta) = \int_0^t \int_{R^d} x(\mu - \nu(\theta))(dx, ds)\,.$$

The existence of this integral follows from the fact that $\{X_t\}$ is a special semimartingale [12, Proposition II-2.9]. The random measure $\nu(\theta)$ has the form

$$\nu(\theta; dx, dt) = Y(\theta; x, t)K_t(dx)d\lambda_t \tag{5.3.4}$$

where all the terms in (5.3.4) depend implicitly on ω, the element of the basic probability space. For fixed (ω, t), $Y > 0$ is a Borel measurable function and $K_t(\cdot)$ is a positive measure on $(R^d, \mathcal{B}_d)$. For fixed x and for fixed $B \in \mathcal{B}_d$, $Y(\theta; x, t)$ and $K_t(B)$ are predictable processes. Note that we have assumed that K is independent of θ. The representation (5.3.4) follows from [12, Proposition II-2-9]. See Section 4.2.

In order to define the quasi-score function, we assume that

$$\int_{R^d} |x_i|\, Y(\theta; x, t)K_t(dx) < \infty \tag{5.3.5}$$

almost surely for all $t > 0$ and $1 \le i \le d$. We further assume that

$$g_t(\theta) = f_t(\theta) - \int_{R^d} xY(\theta; x, t)\, K_t(dx)\,, \tag{5.3.6}$$

and $Y(\theta; x, t)$ are continuously differentiable with respect to θ for all x and t and for almost all ω, and that for all $\theta \in \Theta$ and $1 \le i, j \le n$,

$$\int_0^t \left| \left((\dot{g}_s(\theta))^T\, a_s^+(\theta)\, (\dot{g}_s(\theta)) \right)_{ij} \right| d\lambda_s < \infty$$

and

$$\int_0^t \int_{R^d} \left(\dot{H}(\theta; x, s) \dot{H}(\theta; x, s)^T \right)_{ij} \nu(\theta; dx, ds) < \infty$$

almost surely. Here $a_s^+(\theta)$ is the Moore–Penrose inverse of $a_s(\theta)$ and $H(\theta; x, s) = \log(Y(\theta; x, s))$. In addition,

$$\dot{g}_s(\theta)_{ij} = \frac{\partial g_s(\theta)_i}{\partial \theta_j}$$

and

$$\dot{H}(\theta; x, s)_i = \frac{\partial H(\theta; x, s)}{\partial \theta_i}\,.$$

We define the *quasi-score function* based on a sample path of the process $\{X_s\}$ observed over $[0, t]$ to be

$$Q_t(\theta) = \int_0^t \dot{g}_s(\theta)^T a_s^+(\theta) dm_s^{(c)}(\theta)$$

$$+ \int_0^t \int_{R^d} \dot{H}(\theta; x, s)(\mu - \nu(\theta))(dx, ds)\,. \tag{5.3.7}$$

In the notation for stochastic integrals with respect to semimartingales, we write

$$Q_t(\theta) = \dot{g}(\theta)^T a^+(\theta) \bullet m_s^{(c)}(\theta)_t + \dot{H}(\theta) * (\mu - \nu(\theta))_t \,.$$

The stochastic integrals in (5.3.7) exist under the assumptions stated above. If the likelihood function exists, then the true score function will be equal to the function $Q_t(\theta)$ given by (5.3.7) under some assumptions provided $\{X_t\}$ is quasi-left-continuous. If $\{X_t\}$ is not quasi-left-continuous, then there will exist fixed time points at which the processes will jump with a positive probability. If, for all θ, the process jumps with probability one at these points, then (5.3.7) can still be used and $Q_t(\theta)$ equals the true score function if it exists. Otherwise, the time-points at which the probability of a jump is less than one are known from the specified model and the jumps that take place should be treated separately. The rest of the jumps should be used to construct the quasi-score function given by (5.3.7). An example of a nonquasi-left-continuous process is obtained when a discrete time process is treated as a continuous time process by setting $X_t = X_{[t]}$. Here the process jumps with probability one at all the jump times.

Optimality

We now prove that the estimating function $Q_t(\theta)$ given by (5.3.7) is optimal according to the criterion discussed earlier within the class of estimating functions of the form

$$G_t(\theta) = \int_0^t \alpha_s(\theta) dm_s^{(c)}(\theta) + \int_0^t \int_{R^d} \beta(\theta; x, s)(\mu - \nu(\theta))(dx, ds) \qquad (5.3.8)$$

where $\{\alpha_t(\theta)\}$ is an $n \times d$ matrix of predictable processes and $\beta(\theta; x, s)$ is an n-dimensional vector of predictable random functions from $R^d \times [0, \infty)$ to R. Note that $\{G_t(\theta)\}$ is a local vector martingale when θ is the true parameter. Furthermore, every n-dimensional local vector martingale $\{M_t(\theta)\}$ has the decomposition

$$M_t(\theta) = \int_0^t \alpha_s(\theta) dm_s^{(c)}(\theta) + \int_0^t \int_{R^d} \beta(\theta; x, s)(\mu - \nu(\theta))(dx, ds) + N_t(\theta)$$

where $\{N_t(\theta)\}$ is a vector of local martingales orthogonal to the observed process $\{X_t\}$ in the sense that

$$\left\langle N^{(c)}(\theta)_i, m^{(c)}(\theta)_j \right\rangle = 0, t \geq 0, 1 \leq i \leq n, 1 \leq j \leq d,$$

$\{N_t^{(c)}(\theta)\}$ is the continuous martingale component of $\{N_t(\theta)\}$, and

$$E\left\{\int_0^\infty \int_{R^d} \Delta N_s(\theta)_i \mu(dx, ds)\right\} = 0, 1 \leq i \leq n.$$

Here $\Delta N_s(\theta)_i = N_s(\theta)_i - N_{s-}(\theta)_i$. This can be seen from the martingale representation theory (cf. [12, Lemma II-4-24]). See Chapter 1. Hence, the estimating function of the type (5.3.8) gives the most general form of interest here. A result in [12, Section III-4] gives sufficient conditions for $N_t = 0$ for $t \geq 0$, in which case (5.3.8) gives all the estimating functions of interest. If

$$\int_0^t \left|\left(\alpha_s(\theta) a_s(\theta) \alpha_s(\theta)^T\right)_{ij}\right| d\lambda_s < \infty,$$

and

$$\int_0^t \int_{R^d} \left|\left(\beta(\theta; x, s)\beta(\theta; x, s)^T\right)_{ij}\right| \nu(\theta; dx, ds) < \infty$$

for $t > 0$, $1 \leq i, j \leq n$, then $\{G_t(\theta)\}$ is a locally square integrable martingale under P_θ with the quadratic characteristic

$$\langle G(\theta)\rangle_t = \int_0^t \alpha_s(\theta) a_s(\theta) \alpha_s(\theta)^T d\lambda_s$$

$$+ \int_0^t \int_{R^d} \beta(\theta; x, s)\beta(\theta; x, s)^T \nu(\theta; dx, ds). \qquad (5.3.9)$$

This follows from the fact that the first integral in (5.3.8) is a continuous local martingale and the second is a purely discontinuous martingale. Since $Q_t(\theta)$ has the same properties as $G_t(\theta)$, it follows that

$$\langle G(\theta), Q(\theta)\rangle_t = \int_0^t \alpha_s(\theta)\dot{g}_s(\theta)d\lambda_s$$

$$+ \int_0^t \int_{R^d} \beta(\theta; x, s)\dot{H}(\theta; x, s)^T \nu(\theta; dx, ds) \quad (5.3.10)$$

provided $a_t(\theta)$ is almost surely positive definite for all $t > 0$.

We now assume sufficient conditions which ensure that the estimating functions (5.3.8) are zero-mean martingales and are differentiable with respect to θ.

Let ζ be the class of estimating functions $G_t(\theta)$ of the type (5.3.8) such that, for all $t > 0$,

(C1) $E_\theta < G(\theta) >_t < \infty$,

(C2) $G_t(\theta)$ is differentiable with respect to θ and

$$\dot{G}_t(\theta) = \int_0^t \dot{\alpha}_s(\theta)dm_s^{(c)}(\theta) + \int_0^t \int_{R^d} \dot{\beta}(\theta; x, s)(\mu - \nu(\theta))(dx, ds)$$

$$- \int_0^t \alpha_s(\theta)\dot{g}_s(\theta)d\lambda_s$$

$$- \int_0^t \int_{R^d} \beta(\theta; x, s)\dot{Y}(\theta; x, s)^T K_s(dx)d\lambda_s\,,$$

(C3) $E_\theta(\int_0^t |\dot{\alpha}_s(\theta)a_s(\theta)\dot{\alpha}_s(\theta)^T_{i,j,k,l}|d\lambda_s < \infty, 1 \le i, j, k, l \le n,$

(C4) $E_\theta(\int_0^t |\dot\beta(\theta, x, s)_{ij}\dot\beta(\theta, x; s)_{kl}|\nu(\theta, dx, ds) < \infty, 1 \le i, j, k, l \le n,$ and

(C5) the matrices $\langle G(\theta)\rangle_t$ and $\langle G(\theta), Q(\theta) >_t$ are almost surely nonsingular and have nonsingular expectations.

If $\{G_t(\theta)\} \in \zeta$, then the conditions (C3)–(C5) imply that $\{\dot G_t(\theta) - \bar G_t(\theta)\}$ is a zero-mean square integrable martingale where

$$\bar G_t(\theta) = -\int_0^t \alpha_s(\theta)\dot g_s(\theta)d\lambda_s$$

$$-\int_0^t\int_{R^d} \beta(\theta; x, s)\dot Y(\theta; d, s)^T K_s(dx)d\lambda_s . \qquad (5.3.11)$$

Hence

$$E_\theta\left(\bar G_t(\theta)\right) = E_\theta\left(\dot G_t(\theta)\right) .$$

The following theorem shows that $Q_t(\theta)$ is an optimal quasi-score function in the sense of the criterion given above.

THEOREM 5.2

Suppose that $Q_t(\theta) \in \zeta$ and that $a_t(\theta)$ is almost surely strictly positive definite for all $t > 0$. Then

$$E_\theta\left(Q_t(\theta)Q_t(\theta)^T\right) - E_\theta\left(\dot G_t(\theta)\right)^T\left(E_\theta\left(G_t(\theta)G_t(\theta)^T\right)\right)^{-1} E_\theta\left(\dot G_t(\theta)\right) \qquad (5.3.12)$$

is nonnegative definite and

$$\langle Q(\theta)\rangle_t - \bar G_t(\theta)^T\langle G(\theta)\rangle_t^{-1}\bar G_t(\theta) \qquad (5.3.13)$$

is almost surely nonnegative definite for any $\{G_t(\theta)\} \in \zeta$.

PROOF Note that

$$\langle G(\theta), Q(\theta)\rangle_t = -\bar{G}_t(\theta)$$

by (5.3.10) and hence

$$E_\theta\left(G_t(\theta)G_t(\theta)^T\right) = -E_\theta\left(\bar{G}_t(\theta)\right) = -E_\theta\left(\dot{G}_t(\theta)\right) .$$

Note that

$$\langle Q(\theta)\rangle_t = -\bar{Q}_t(\theta) \text{ and } E_\theta\left(Q_t(\theta)Q_t(\theta)^T\right) = -E_\theta\left(\dot{Q}_t(\theta)\right) .$$

Let $G_t = (G_{1t}, \ldots, G_{nt})^T$ and $Q_t = (Q_{1t}, \ldots, Q_{nt})^T$ and $Z^T = (G_{1t}, \ldots, G_{nt}, Q_{1t}, \ldots, Q_{nt})$.

Consider the partitioned matrix

$$D = \begin{pmatrix} E_\theta G_t G_t^T & \vdots\ E_\theta G_t Q_t^T \\ (E_\theta G_t Q_t^T)^T & \vdots\ E_\theta Q_t Q_t^T \end{pmatrix}$$

$$= \begin{pmatrix} E_\theta G_t G_t^T & \vdots\ -E_\theta \dot{G}_t \\ (-E_\theta \dot{G}_t)^T & \vdots\ E_\theta Q_t Q_t^T \end{pmatrix}$$

and the partitioned matrix

$$C = \begin{pmatrix} \langle G\rangle_t & \vdots\ \langle G, Q\rangle_t \\ \langle G, Q\rangle_t & \vdots\ \langle Q\rangle_t \end{pmatrix}$$

$$= \begin{pmatrix} \langle G\rangle_t & \vdots\ -\bar{G}_t \\ -\bar{G}_t^T & \vdots\ \langle Q\rangle_t \end{pmatrix} .$$

Note that D is the expected value of C. Now C and D are nonnegative definite since for any arbitrary $2n$-dimensional vector $\boldsymbol{u}$,

$$\boldsymbol{u}^T D \boldsymbol{u} = \boldsymbol{u}^T E_\theta(ZZ^T)\boldsymbol{u}$$

$$= E_\theta \left(u^T Z Z^T u\right)$$

$$= E_\theta \left(Z^T u\right)^2 \geq 0$$

while almost surely

$$u^T C u = u^T \langle Z, Z \rangle_t u = \langle u^T Z, u^T Z \rangle_t$$

$$= \langle u^T Z \rangle_t \geq 0 \, .$$

Hence, by [19, p. 266],

$$E_\theta G_t G_t^T - \left(E_\theta \dot{G}_t\right) \left(E_\theta Q_t Q_t^T\right)^{-1} \left(E_\theta \dot{G}_t\right)^T$$

is nonnegative definite and

$$\langle G \rangle_t - \bar{G}_t \langle Q \rangle_t^{-1} \bar{G}_t^T$$

is nonnegative definite almost surely. Since all the matrices involved here are almost surely nonsingular, we obtain by inversion that

$$E_\theta Q_t Q_t^T - \left(E_\theta \dot{G}_t\right)^T \left(E_\theta G_t G_t^T\right)^{-1} \left(E_\theta \dot{G}_t\right)$$

is nonnegative definite and

$$\langle Q \rangle_t - \bar{G}_t^T \langle G \rangle_t^{-1} \bar{G}_t$$

is nonnegative definite almost surely. Since $E_\theta Q_t Q_t^T = -E_\theta \dot{Q}_t$ and $\langle Q \rangle_t = -\bar{Q}_t$, both $(E_\theta \dot{G}_t)^T (E_\theta G_t G_t^T)^{-1} (E_\theta \dot{G}_t)$ and $\bar{G}_t^T \langle G \rangle_t^{-1} \bar{G}_t$ are maximized in the partial order for nonnegative definite matrices by choosing $G_t = Q_t$. Thus the estimating function $Q_t(\theta)$ defined by (5.3.8) is optimal. ∎

REMARK 5.1 The estimating function $Q_t(\theta)$ defined by (5.3.8) is a local martingale. If the likelihood function exists, then $Q_t(\theta)$ is equal to the true score function under some regularity conditions. If the local martingale $m(\theta)$ is continuous, then $Q_t(\theta)$ reduces to the quasi-score function proposed by Hutton and Nelson [10]. ∎

Asymptotic Properties

Suppose that the quadratic characteristic of $\{Q_t(\theta)\}$ given by

$$\langle Q(\theta)\rangle_t = \left(\dot{g}(\theta)^T a^+(\theta)\dot{g}(\theta)\right) \bullet \lambda_t + \left(\dot{H}(\theta)\dot{H}(\theta)^T\right) * \nu(\theta)_t \tag{5.3.14}$$

has finite expectation for all $t \geq 0$. This implies that $\{Q_t(\theta)\}$ is a square integral martingale. Let

$$B_t(\theta) = E_\theta\left(\langle Q(\theta)\rangle_t\right) , \tag{5.3.15}$$

and the quadratic variation of $\{Q_t(\theta)\}$ is given by

$$[Q(\theta)]_t = \left(\dot{g}(\theta)^T a^+(\theta)\dot{g}(\theta)\right) \bullet \lambda_t + \left(\dot{H}(\theta)\dot{H}(\theta)^T\right) * \mu_t . \tag{5.3.16}$$

Let

$$I_t(\theta) = \operatorname{diag}\left(\langle Q(\theta)_1\rangle_t, \ldots, \langle Q(\theta)_n\rangle_t\right) . \tag{5.3.17}$$

where $Q(\theta)_i$ denotes the ith component of $Q(\theta)$ and

$$D_t(\theta) = E_\theta\left(I_t(\theta)\right) . \tag{5.3.18}$$

The following theorem is an immediate consequence of the central limit theorem for square integrable vector martingales stated in Chapter 1. Note that

$$\Delta Q_t(\theta) = \dot{H}\left(\theta; t, \Delta X_t\right) I\left(\Delta X_t \neq 0\right) .$$

THEOREM 5.3
Suppose that, as $t \to \infty$,

(i) $D_t(\theta)_{ii} \to \infty,\ 1 \leq i \leq n$,

(ii) $D_t(\theta)_{ii}^{-1/2} E_\theta\left\{\sup_{s\leq t}\left|\dot{H}\left(\theta; s, \Delta X_s\right)_i\right| I\left(\Delta X_s \neq 0\right)\right\} \to 0,\ 1 \leq i \leq n$,

(iii) $D_t(\theta)^{-1/2}[Q(\theta)]_t D_t(\theta)^{-1/2} \xrightarrow{p} W(\theta)$ *where* $W(\theta)$ *is a random non-negative definite matrix, and (iv)* $D_t(\theta)^{-1/2}B_t(\theta)D_t(\theta)^{-1/2} \to \sum(\theta)$

where $\Sigma(\theta)$ *is a positive definite matrix. Then, as* $t \to \infty$,

$$D_t(\theta)^{-1/2}Q_t(\theta) \overset{\mathcal{L}}{\to} Z \text{ (stably)} \tag{5.3.19}$$

and conditionally on $[det\ (W(\theta)) > 0]$

$$(D_t(\theta)^{1/2}[Q(\theta)]_t^{-1}D_t(\theta)^{1/2})^{1/2}$$

$$D_t(\theta)^{-1/2}Q_t(\theta) \overset{\mathcal{L}}{\to} N(0, I_n) \ \ (mixing)\,. \tag{5.3.20}$$

Here the distribution of Z *is the normal mixture and has the characteristic function*

$$\phi(u) = E_\theta\left(\exp\left(-\frac{1}{2}u^T W(\theta)u\right)\right), u = (u_1, \ldots, u_n)^T \ .$$

Existence and Consistency of the Quasi Likelihood Estimator

Suppose that $Q_t(\theta)$ is differentiable with respect to θ and that conditions (C1)–(C5) hold for $G_t(\theta) = Q_t(\theta)$. Further suppose that the exchangeability of differentiation and stochastic integration is allowed (cf. [9, 13]). See Appendix B. Then

$$\dot{Q}_t(\theta) = M_t^{(1)}(\theta) - \langle Q(\theta)\rangle_t \tag{5.3.21}$$

where

$$M_t^{(1)}(\theta) = \dot{\alpha}(\theta) \bullet m^{(c)}(\theta)_t + \ddot{H}(\theta) * (\mu - \nu(\theta))_t \tag{5.3.22}$$

is a local matrix martingale with parameter θ. Here

$$\alpha_t(\theta) = \dot{g}_t(\theta)^T a_t^+(\theta)$$

and $\dot{\alpha}_t(\theta)$ is the $n \times n$ matrix of d-dimensional row vectors with $\dot{\alpha}_t(\theta)_{ij} = \frac{\partial \alpha_t(\theta)_i}{\partial \theta_j}$ and $\alpha_t(\theta)_i$ is the ith row of $\alpha_t(\theta)$. Another representation for $\dot{Q}_t(\theta)$ is given by

$$\dot{Q}_t(\theta) = M_t^{(2)}(\theta) - [Q(\theta)]_t \tag{5.3.23}$$

where $M_t^{(2)}$ is a local martingale with parameter θ given by

$$M_t^{(2)}(\theta) = \dot{\alpha}(\theta) \bullet m^{(c)}(\theta)_t + \left(\frac{\ddot{Y}(\theta)}{Y(\theta)}\right) * (\mu - \nu(\theta))_t \,.$$

THEOREM 5.4

Suppose θ_0 is the true value of the parameter θ and that the following conditions hold for this parameter value almost surely on $\Gamma_{\theta_0} = [\langle Q(\theta_0)_i\rangle_t \to \infty$ as $t \to \infty, 1 \le i \le n]$:

$$\left[\frac{\dot{\alpha}(\theta_0)_{ij}\, a(\theta_0)\, \dot{\alpha}(\theta_0)_{ij}^T}{\left(1 + \langle Q(\theta_0)_i\rangle\right)^2}\right] \bullet \lambda_\infty$$

$$+ \left[\frac{\ddot{H}(\theta_0)_{ij}^2}{\left(1 + \langle Q(\theta_0)_i\rangle\right)\left(1 + \langle Q(\theta_0)_j\rangle + |\ddot{H}(\theta_0)_{ij}|\right.}\right] * \nu(\theta_0)_\infty$$

$$< \infty, \; 1 \le i, j \le n\,; \tag{5.3.24}$$

$$\sup_{\theta \in R_t} \left\| I_t(\theta_0)^{-1} \langle Q(\theta)\rangle_t - V(\theta_0)\right\| \to 0 \text{ as } t \to \infty\,, \tag{5.3.25}$$

where $V(\theta_0)$ is a nonnegative definite random matrix with det $(V(\theta_0)) > 0$ almost surely on Γ_{θ_0} and where, for $\delta > 0$,

$$R_t = \left\{\theta : \left\| V(\theta_0)^{1/2}(\theta - \theta_0)\right\| \le \left\| V(\theta_0)^{-1/2} I_t(\theta_0)^{-1} Q_t(\theta_0)\right\|^{1-\delta}\right\}$$

and

$$\sup_{\theta \in R_t} \left\| I_t(\theta_0)^{-1} \left\{M_t^{(1)}(\theta_0) - M_t^{(1)}(\theta)\right\}\right\| \to 0 \text{ as } t \to \infty\,. \tag{5.3.26}$$

Then, almost surely on Γ_{θ_0}, the quasi-likelihood equation

$$Q_t(\theta) = 0 \tag{5.3.27}$$

has a solution $\hat{\theta}_t$ for t large enough satisfying

$$\hat{\theta}_t \to \theta_0 \text{ as } t \to \infty .$$

PROOF Let $\theta \in \partial R_t$ where ∂R_t denotes the boundary of R_t. Applying the Taylor's expansion about θ_0, we have

$$Q_t(\theta) = Q_t(\theta_0) + \dot{Q}_t(\bar{\theta})(\theta - \theta_0) \tag{5.3.28}$$

where $\|\bar{\theta} - \theta_0| \leq \|\theta - \theta_0|$. In view of (5.3.21), we have

$$\begin{aligned}
&(\theta - \theta_0)^T I_t(\theta_0)^{-1} Q_t(\theta) \\
&\quad = (\theta - \theta_0)^T I_t(\theta_0)^{-1} Q_t(\theta_0) \\
&\qquad + (\theta - \theta_0)^T I_t(\theta_0)^{-1} \left\{ M_t^{(1)}(\bar{\theta}) - < Q(\bar{\theta}) >_t \right\} (\theta - \theta_0)
\end{aligned}$$

for some $\bar{\theta} \in R_t$. By the strong law of large numbers for local martingales due to Liptser [15] (see Chapter 1), it follows that

$$\sup_{\theta \in R_t} \left\| I_t(\theta_0)^{-1} M_t^{(1)}(\theta) \right\| \to 0 \text{ as } t \to \infty \tag{5.3.29}$$

almost surely on Γ_{θ_0}. Furthermore, the law of large numbers for locally square integrable martingales due to Lepingle [14] (see Chapter 1) implies that

$$I_t(\theta_0)^{-1} Q_t(\theta_0) \to 0 \text{ as } t \to \infty \tag{5.3.30}$$

almost surely on Γ_{θ_0}. Then, it follows from (5.3.25) that for t sufficiently large

$$(\theta - \theta_0)^T I_t(\theta_0)^{-1} Q_t(\theta) \leq 0$$

for all $\theta \in \partial R_t$. Since $Q_t(\theta)$ is a continuous function of θ, it follows that the equation

$$I_t(\theta_0)^{-1} Q_t(\theta) = 0 \tag{5.3.31}$$

has at least one solution in R_t, and hence, $Q_t(\theta) = 0$ has at least one solution in R_t. ∎

REMARK 5.2 It is sufficient if $\langle Q(\theta)\rangle_t$ is finite for all t, which implies that $Q(\theta)$ is locally square integrable in Theorem 5.4 and $Q(\theta)$ need not be square integrable. The quasi-likelihood function has a local maximum at $\hat{\theta}_t$ specified in Theorem 5.4. If the elements of $I_t(\theta)$ tend to infinity at different rates, an alternate set of sufficient conditions for the existence and the consistency of a quasi-likelihood estimator have been given in [20] following [22] (cf. [17]). ∎

Asymptotic Normality of the Quasi Likelihood Estimator

THEOREM 5.5

Let θ_0 be the true parameter. Suppose the following condition hold:

(R1) *The conditions for Theorem 5.3 are satisfied,*

(R2) *There exists an estimator $\hat{\theta}_t$ of θ on C_{θ_0} that is a solution of $Q_t(\theta) = 0$ where $C_{\theta_0} = [det(W(\theta_0)) > 0]$ and $W(\theta)$ is as defined in Theorem 5.3,*

(R3) $-D_t(\theta_0)^{-1/2}\dot{Q}_t(a_t)D_t(\theta_0)^{-1/2} \xrightarrow{p} V(\theta_0)$ *as $t \to \infty$ for any sequence of random vectors a_t satisfying $\|a_t - \theta_0\| \le \|\hat{\theta}_t - \theta_0\|$ where $V(\theta_0)$ is a random nonnegative definite matrix such that $C_{\theta_0} \subset [det(V(\theta_0)) > 0]$, and*

(R4) $D_t(\theta_0)^{-1/2}[Q(\hat{\theta}_t)]_t D_t(\theta_0)^{-1/2} \xrightarrow{p} W(\theta_0)$ *as $t \to \infty$.*

Then, as $t \to \infty$,

$$\left[D_t\left(\theta_0\right)^{1/2}\left[Q\left(\hat{\theta}_t\right)\right]_t^{-1} D_t\left(\theta_0\right)^{1/2}\right]^{1/2}$$

$$D_t\left(\theta_0\right)^{-1/2} \dot{Q}_t\left(\theta_t\right)\left(\hat{\theta}_t - \theta_0\right) \xrightarrow{\mathcal{L}} N\left(0, I_n\right) \tag{5.3.32}$$

conditionally on the event $[\det(W(\theta_0)) > 0]$.

PROOF This result follows from (5.3.19) and the expansion

$$Q_t(\theta) = Q_t\left(\theta_0\right) + \dot{Q}_t\left(\bar{\theta}\right)\left(\theta - \theta_0\right)$$

where $\|\bar{\theta} - \theta_0\| \leq \|\theta - \theta_0\|$. ∎

REMARK 5.3 For more discussion and examples on the estimating functions for stochastic processes, see [16] and [3] in [18]. ∎

5.4 Quasi Likelihood and Partially Specified Counting Processes

Let $(\Omega, \mathcal{F})$ be a measurable space. Let X be a counting process and Y be an arbitrary d-dimensional process defined on the interval $[0, \infty)$. We interpret Y as a "covariate" process. Let $\{\mathcal{F}_t, t \geq 0\}$ be the filtration generated by X and Y. Let Θ be a one-dimensional parameter space. For each $\tau \in \Theta$, let a_τ be a predictable process. Suppose that there exists a probability distribution P on $\mathcal{F}$ such that X has the Doob–Meyer decomposition of the form

$$X(t) = M_\tau(t) + \int_0^t a_\tau(s)ds, t \geq 0 \tag{5.4.1}$$

where M_τ is a martingale and a_τ is the intensity process. The underlying probability measure P on $\mathcal{F}$ is, in general, not uniquely determined by the intensity process a_τ. Let $\mathcal{P}_\tau$ be the family of all probability measures for which the intensity process of X is a_τ. Note that the model (5.4.1) is partially specified in the above sense [6].

Let $\{f_\tau\}$ be a predictable process. Fix $t > 0$. Suppose that (under appropriate regularity conditions)

$$\int \int_0^t f_\tau dM_\tau dP = 0, \quad P \in \mathcal{P}_\tau, \tau \in \Theta s\,. \tag{5.4.2}$$

Then we can consider the function

$$\tau \to \int_0^t f_\tau dM_\tau \,.$$

as an estimating function or quasi-score function and the corresponding estimating equation is

$$\int_0^t f_\tau dM_\tau = 0 \,. \tag{5.4.3}$$

For simplicity, let us call $\tau \to f_\tau$ as an *estimating function.*

Fix $\theta \in \Theta$ and $P \in \mathcal{P}_\theta$. Let $b_t \to \infty$ as $t \to \infty$ and $V(b_t)$ be the set of all real-valued predictable processes v such that there exists a positive number $\|v\|$ with

$$b_t^{-2} \int_0^t v^2 a_\theta ds = \|v\|^2 + o_p(1) \,. \tag{5.4.4}$$

This defines a seminorm on $V(b_t)$.

DEFINITION 5.3 *Two processes $v \in V(b_t)$ and $w \in V(c_t)$ are said to be compatible if there exists a real number (v, w) such that*

$$b_t^{-1} c_t^{-1} \int_0^t vw \; a_\theta ds = \langle v, w \rangle + o_p(1) \,. \tag{5.4.5}$$

DEFINITION 5.4 *An estimating function $\tau \to f_\tau$ is called b_t-continuous at θ if f_θ is in $V(b_t)$ and*

$$b_t^{-2} \int_0^t (f_\tau - f_\theta)^2 \; a_\theta ds = o_p\left((\tau - \theta)^2\right) \,. \tag{5.4.6}$$

DEFINITION 5.5 *An estimating function $\tau \to f_\tau$ is called b_θ-differentiable* *at θ with the derivative d_θ if d_θ is in $V(b_t)$ and*

$$b_t^{-2}\int_0^t (f_\tau - f_\theta - (\tau-\theta)d_\theta)^2\, a_\theta ds = o_p\left((\tau-\theta)^2\right). \tag{5.4.7}$$

Here d_θ is determined only up to $\|\cdot\|$-equivalence.

DEFINITION 5.6 *An estimator $\hat{\theta}_t$ is said to be c_t-consistent if $c_t \to \infty$ and*

$$c_t\left(\hat{\theta}_t - \theta\right) = O_p(1). \tag{5.4.8}$$

THEOREM 5.6
Suppose that the function $\tau \to a_\theta^{-1}a_\tau$ is c_t-differentiable at θ with the derivative k_θ and the function $\tau \to f_\tau$ is b_t-continuous at θ. Assume that f_θ and k_θ are compatible. Then any c_t-consistent solution $\hat{\theta}_t$ of the estimating equation (5.4.3) admits a stochastic representation of the form

$$c_t\left(\hat{\theta}_t - \theta\right) = c_t\left(\int_0^t f_\theta k_\theta\, a_\theta ds\right)^{-1}\int_0^t f_\theta dM_\theta + o_p(1)$$

$$= b_t^{-1}\langle f_\theta, k_\theta\rangle^{-1}\int_0^t f_\theta dM_\theta + o_p(1). \tag{5.4.9}$$

If, in addition,

$$b_t^{-2}\int_0^t f_\theta^2 a_\theta I\left(|f_\theta| > \varepsilon b_t\right) a_\theta ds = o_p(1),\ \ \varepsilon > 0 \tag{5.4.10}$$

then

$$c_t\left(\hat{\theta}_t - \theta\right) \xrightarrow{\mathcal{L}} N\left(0, \langle f_\theta, k_\theta\rangle^{-2}\|f_\theta\|^2\right) \text{ as } t \to \infty. \tag{5.4.11}$$

PROOF Let r_t be defined by the relation

$$a_{\hat{\theta}_t} = a_\theta \left(1 + \left(\hat{\theta}_t - \theta\right) k_\theta + r_t\right) .$$

The estimating equation (5.4.3) implies, following (5.4.1), that

$$0 = \int_0^t f_{\hat{\theta}_t} dM_{\hat{\theta}_t} = \int_0^t f_\theta dM_\theta - \left(\hat{\theta}_t - \theta\right) \int_0^t f_\theta k_\theta a_\theta ds + R_t \tag{5.4.12}$$

where

$$R_t = \int_0^t \left(f_{\hat{\theta}_t} - f_\theta\right) dM_\theta - \int_0^t \left(f_\theta + f_{\hat{\theta}_t} - f_\theta\right) r_t a_\theta ds$$

$$- \left(\hat{\theta}_t - \theta\right) \int_0^t \left(f_{\hat{\theta}_t} - f_\theta\right) k_\theta a_\theta ds . \tag{5.4.13}$$

Since $\tau \to a_\theta^{-1} a_\tau$ is c_t -differentiable and $\hat{\theta}_t$ is c_t-consistent, it follows that

$$\int_0^t r_t^2 a_\theta ds = o_p(1) .$$

Since $\tau \to f_\tau$ is b_t-continuous,

$$b_t^{-2} \int_0^t \left(f_{\hat{\theta}_t} - f_\theta\right)^2 a_\theta ds = o_p(1) .$$

Using these relations, it can be checked that $b_t^{-1} R_t = o_p(1)$. But

$$b_t^{-2} \left\langle \int_0^{\cdot} f_\theta dM_\theta \right\rangle_t = b_t^{-2} \int_0^t f_\theta^2 a_\theta ds$$

$$= \|f_\theta\|^2 + o_p(1)\,.$$

Hence R_t is $o_p(1)$ in (5.4.12). Furthermore, the processes f_θ and k_θ are compatible; that is,

$$b_\tau^{-1}c_t^{-1}\int_0^t f_\theta k_\theta a_\theta ds = \langle f_\theta, k_\theta\rangle + o_p(1)\,.$$

The stochastic representation (5.4.9) follows now from (5.4.12). In view of the Lindeberg condition (5.4.10) the property of the asymptotic normality given in (5.4.11) follows from the central limit theorem for stochastic integrals with respect to martingales (cf. [12, p. 405, Theorem 5.4]). ∎

REMARK 5.4 The convergence rate of $\hat{\theta}_t$ is c_t in Theorem 5.6 irrespective of the rate b_t associated with the estimating function. Furthermore, the stochastic representation given by (5.4.9) is valid for c_t-consistent estimators $\hat{\theta}_t$. Additional stronger global conditions are needed to ensure the existence of a consistent solution without a rate. See Theorem 5.4. By the Cauchy–Schwarz inequality, we have

$$\langle f_\theta, k_\theta\rangle^{-2}\,\|f_\theta\|^2 \ge \|k_\theta\|^{-2} = \langle k_\theta, k_\theta\rangle^{-2}\,\|k_\theta\|^2\,. \tag{5.4.14}$$

Hence if $\tau \to k_\tau$ is continuous at θ, then we can apply Theorem 5.6 for $f_\tau = k_\tau$ and the estimating function $\tau \to k_\tau$ leads to an estimator that is asymptotically efficient among all the solutions of the estimating equations. ∎

Optimal Estimating Equation

Suppose that the mapping $\tau \to a_\tau(s)$ is P-a.s. differentiable with the derivative $a'_\tau(s)$. Let

$$G_t(\tau) = \int_0^t f_\tau dM_\tau$$

and $\bar{G}(\theta)$ be the compensator of the derivative $G'(\theta) = \frac{\partial G(\theta)}{\partial \theta}$ of $G(\theta)$. Then

$$\langle G(\theta)\rangle_t = \int_0^t f_\theta^2 a_\theta ds ,$$

$$G_t'(\theta) = \int_0^t f_\theta' dM_\theta - \int_0^t a_\theta' ds$$

and

$$\bar{G}_t(\theta) = -\int_0^t f_\theta a_\theta' ds .$$

Suppose further that $\tau \to a_\tau$ is pointwise differentiable and that $a_\theta^{-1} a_\theta'$ is a c_t-derivative of $\tau \to a_\theta^{-1} a_\tau$ at θ. Then it can be checked that under some conditions,

$$c_t^2 \bar{G}_t(\theta)^{-2} \langle G(\theta)\rangle_t = \left\langle f_\theta, a_\theta^{-1} a_\theta' \right\rangle \| f_\theta \|^2 + o_p(1) .$$

Consider the estimating function

$$G_t^*(\tau) = \int_0^t a_\tau^{-1} a_\theta' ds$$

based on the map $\tau \to a_\tau^{-1} a_\theta'$.

Note that

$$\left\langle f_\theta, a_\theta^{-1} a_\theta' \right\rangle^2 \| f_\theta \|^2 \geq \left\| a_\theta^{-1} a_\theta' \right\|^{-2}$$

$$= \left\langle a_\theta^{-1} a_\theta', a_\theta^{-1} a_\theta' \right\rangle^{-2} \left\| a_\theta^{-1} a_\theta' \right\|^2$$

from (5.4.14). It can be shown that

$$\left(\int G_t'(\theta)dP\right)^{-2}\int G_t^2(\theta)dP \geq \left(\int G_t^{*'}(\theta)dP\right)^{-2}\int G_t^{*^2}(\theta)dP$$

(cf. [23]), which shows that the mapping $\tau \rightarrow a_\tau^{-1}a_\tau'$ is optimal in the sense of the criterion for optimality discussed in Section 5.2. We omit the details. Hence

$$G_t^*(\tau) = \int_0^t a_\tau^{-1}a_\tau' dM_\tau$$

is an optimal quasi-score function and the equation

$$G_t^*(\tau) = 0$$

gives the quasi-likelihood estimator.

REMARK 5.5 Greenwood and Wefelmeyer [7] show that the quasi likelihood estimator discussed above is not only asymptotically efficient within the class of solutions of estimating equations, but is also asymptotically efficient within a large class of partial likelihood estimators as discussed later in Chapter 7. We omit the details. ∎

5.5 Examples

Example 5.1

Consider the multivariate counting process $X_t = (X_{1t}, \ldots, X_{pt})^T$ such that each X_{it} is of the form

$$X_{it} = \theta_i \int_0^t J_i(s)ds + m_{it} \tag{5.5.1}$$

with multiplicative intensity $\Lambda_i(t) = \theta_i J_i(t)$, $J_i(t) > 0$ almost surely being a predictable process and m_{it} a square integrable martingale (cf. [1, 2]). Note that

$$\langle m_i \rangle_t = A_{it} = \theta_i \int_0^t J_i(s)ds$$

is a nondecreasing compensator and $\langle m_i, m_j \rangle_t = 0$ for $i \neq j$ [1, Theorem 3.2]. Note that

$$X_t = \left(\int_0^t J(s)ds \right) \theta + m_t ,$$

where

$$J(s) = \text{diag}\left(J_1(s), \ldots, J_p(s)\right), \theta^T = \left(\theta_1, \ldots, \theta_p\right) ,$$

$$m_t = \left(m_{1t}, \ldots, m_{pt}\right)^T$$

so that

$$f_t(\theta) = J(t)\theta$$

in (5.3.1). Here

$$\langle m \rangle_t = \text{diag}\left(A_{1t}, \ldots, A_{pt}\right)^T$$

and

$$a_t(\theta) = \text{diag}\left(\theta_1 J_1(t), \ldots, \theta_p J_p(t)\right) .$$

The quasi-likelihood estimator $\hat{\theta}_t$ is given by

$$X_t = \hat{\theta}_t \int_0^t J(s)ds ,$$

which is also the maximum likelihood estimator (cf. [1]). □

In the above example and the earlier examples discussed in Section 4.6, one can compute the likelihood function and obtain the maximum likelihood estimators, which are interalia also quasi likelihood estimators. Let us now consider an example where it seems to be difficult to compute the likelihood, and the quasi likelihood approach is possibly applicable.

Example 5.2
Consider the one-dimensional process

$$dX_t = d_t(\theta; X)dt + \gamma_t(\theta; X)dW_t + \delta_t(\theta; X)dZ_t^\theta, t > 0 \tag{5.5.2}$$

for some $\theta \in \Theta \subset R^n$. Here $d_t(\theta; \cdot)$, $\gamma_t(\theta; \cdot)$, and $\delta_t(\theta; \cdot)$ are predictable functionals. The processes $\{W_t\}$ and $\{Z_t^\theta\}$ are the Wiener process and a marked counting process

$$Z_t^\theta = \sum_{i=1}^{N_t} Y_i , \tag{5.5.3}$$

respectively, where $\{N_t\}$ is a counting process with intensity $\{\alpha_t(\theta)\}$, and $\{Y_i\}$ are independent identically distributed random variables with a distribution F_θ independent of $\{N_t\}$. We assume that $F_\theta(dy)$ has a density $p(\theta; y)$ for all $t > 0$, $\gamma_t(\theta; X) \neq 0$ and $\delta_t(\theta, X) \neq 0$ a.s., and $E(\alpha_t(\theta)) < \infty$. In the notation of (5.3.1),

$$\lambda_t = t , \quad m_t^{(c)}(\theta) = \int_0^t \gamma_s(\theta; X)dW_s , \quad a_t(\theta) = \gamma_t(\theta, X)^2 ,$$

$$f_t(\theta) = d_t(\theta; X) + \alpha_t(\theta)\delta_t(\theta; x) \int_{-\infty}^{\infty} y\, p(\theta; y)dy ,$$

$$g_t(\theta) = d_t(\theta; X) ,$$

$$Y(\theta; x, t) = p(\theta; x/\delta_t(\theta; x))\alpha_t(\theta)/ |\delta_t(\theta; x)| ,$$

and

$$K_t(dx) = dx .$$

Equation (5.5.2) implies that

$$\int_0^t \gamma_s(\theta; X)dW_s = X_t^{(c)} - \int_0^t d_s(\theta; x)\, ds$$

where

$$X_t^{(c)} = X_t - \sum_{s \le t} \Delta X_s .$$

It can be checked that the quasi-score function $Q_t(\theta)$ is given by

$$Q_t(\theta) = \int_0^t \dot{d}_s(\theta; X)\gamma_s(\theta; X)^{-2} dX_s^{(c)}$$

$$- \int_0^t \dot{d}_s(\theta; X) d_s(\theta; X)\gamma_s(\theta; X)^{-2} ds$$

$$+ \sum_{i=1}^{N_t} \frac{\dot{\alpha}_{\tau_i}(\theta)}{\alpha_{\tau_i}(\theta)} - \int_0^t \dot{\alpha}_s(\theta) ds$$

$$+ \sum_{s \le t} \frac{d}{d\theta} \log \left(p\left(\theta; \Delta X_s / \delta_s(\theta; X)\right) / |\delta_s(\theta; X)|\right) I\left(\Delta X_s \ne 0\right)$$

where $\tau_1, \tau_2, \ldots$ are the jump times for the process X. Note that there are only finitely many jumps in $[0, t]$. It can be seen that the quasi-score function here is the sum of the three score functions based on the diffusion type process $\{X_t^{(c)}\}$, the counting process $\{N_t\}$, and the jumps had these been independent. The problem of estimation of the parameter θ can be done by following the quasi likelihood approach. $\square$

5.6 Exercises

Problem 1: Consider the SDE

$$dX_t = \theta\, dt + \theta\, dm_t, \quad t \ge 0$$

where $\{m_t\}$ is a locally square integrable martingale with $\langle m \rangle_t = t$ a.s. Obtain the quasi likelihood estimator for θ based on $X(t_i), 0 \le i \le n, \; t_0 = 0, \; t_n =$

T [10].

Problem 2: Let $\{m_t\}$ be a locally square integrable martingale with $\langle m \rangle_t = t$ a.s. Consider the model

$$dX_t = \left(\theta_1\theta_2 + \theta_1\sqrt{ct^{c-1}}\right) dt + dm_t, \ t \geq 0$$

where $\theta_1 > 0, \theta_2 > 0$ are unknown and $\frac{1}{2} < c < 1$ with c known. Show that there exists strongly consistent quasi likelihood estimators for θ_1 and θ_2 [10].

Problem 3: Consider a Galton–Watson branching process $\{Z_0 = 1, Z_1, \ldots Z_t, \ldots\}$ with $EZ_1 = \theta$ and $Var Z_1 = \sigma^2 < \infty$. Let $X_t = \sum_{i=1}^{t} Z_i$ and $m_t(\theta) = \sum_{i=1}^{t}(Z_i - \theta Z_{i-1})$. Show that the optimal quasi-score function is given by

$$\sigma^{-2}\sum_{i=1}^{t}(Z_i - \theta Z_{i-1})$$

and obtain the quasi likelihood estimator for θ [5].

Problem 4: Suppose $\{X_k, \mathcal{F}_k\}$ is a discrete time stochastic process with $E_\theta(X_k|\mathcal{F}_{k-1}) = \theta Y_k$ and $Var_\theta(X_k|\mathcal{F}_{k-1}) = g(\theta)Y_k$ where $\{Y_k\}$ is a sequence of random variables such that Y_k is $\mathcal{F}_{k-1}$-measurable. Obtain the quasi likelihood estimator for θ that is strongly consistent [10].

Problem 5: Let $\{X_t = (X_t^{(1)}, X_t^{(2)}), t \geq 0\}$ be a transient 2-dimensional linear birth and death process with instantaneous birth and death rates $b_{11}, b_{12}, b_{21}, b_{22}$ and $d_{11}, d_{12} = 0 = d_{21}, d_{22}$, respectively. Let B_t^{ij} = Number of births of type i particles to a parent of type j up to time t, and $D_t^{(i)}$ = Number of deaths of type i particles up to time t. Let $Y_t = (B_t^{(11)}, B_t^{(12)}, D_t^{(1)}, B_t^{(21)}, B_t^{(22)}, D_t^{(2)})^T$. Show that $\{Y_t\}$ is a semimartingale of the form (5.3.1). Let $S_T^{(i)} = \int_0^T X_t^{(i)} dt, \ i = 1, 2$. Show that

$$\hat{\theta}_T = \left(\frac{B_T^{(1,1)}}{S_T^{(1)}}, \frac{B_T^{(1,2)}}{S_T^{(2)}}, \frac{D_T^{(1)}}{S_T^{(1)}}, \frac{B_T^{(2,1)}}{S_T^{(1)}}, \frac{B_T^{(2,2)}}{S_T^{(2)}}, \frac{D_T^{(2)}}{S_T^{(2)}}\right)$$

is a strongly consistent quasi likelihood estimator for

$$\theta = (b_{11}, b_{12}, d_{11}, b_{21}, b_{22}, d_{22}) \ .$$

[8, 10].

Problem 6: Consider a process $\{X_t\}$ satisfying the stochastic differential equation

$$dX_t = (\theta_1 + \theta_2 t)\, dt + dW_t, \ \ t \geq 0, \ \ X_0 = 0 \, .$$

Obtain the maximum likelihood estimator of (θ_1, θ_2). Is it strongly consistent [10]?

Problem 7: Consider the process $\{X_t\}$ satisfying

$$dX_t = k_t \theta dt + b_t dW_t, \ \ t \geq 0, X_0 = 0$$

where $\{k_t\}$ and $\{b_t\}$ are predictable. Find sufficient conditions for the strong consistency of maximum likelihood estimator for θ [10].

References

[1] Aalen, O.O. (1978) Nonparametric inference for a family of counting processes, *Ann. Statist.*, **6**, 701–726.

[2] Andersen, P.K., Borgan, Φ., Gill, R.D., and Keiding, N. (1993) *Statistical Methods for Counting Processes,* Springer, New York.

[3] Bhat, B.R. (1996) Tests based on estimating functions, *Stochastic Processes and Statistical Inference,* (Eds. B.L.S. Prakasa Rao and B.R. Bhat), New Age International, New Delhi, 20–38.

[4] Godambe, V.P. (1985) The foundations of finite sample estimation in stochastic processes, *Biometrika,* **72**, 419–428.

[5] Godambe, V.P. and Heyde, C.C. (1987) Quasi-likelihood and optimal estimation, *Internat. Statist. Rev.*, **55**, 231–244.

[6] Greenwood, P. (1988) Partially specified semimartingale experiments, *Statistical Inference from Stochastic Processes,* (Ed. N.U. Prabhu), *Contemporary Math.,* **80**, 1–17.

[7] Greenwood, P. and Wefelmeyer, W. (1989) On optimal estimating functions for partially specified counting process models, preprint, No. 123, University of Cologne.

[8] Hutton, J.E. (1980) The recurrence and transience of two-dimensional linear birth and death processes, *Adv. Appl. Prob.,* **12**, 615–639.

[9] Hutton, J.E. and Nelson, P.I. (1984) Interchanging the order of differentiation and stochastic integration, *Stoch. Proc. Appl.,* **18**, 371–377.

[10] Hutton, J.E. and Nelson, P.I. (1986) Quasi-likelihood estimation for semimartingales, *Stoch. Proc. Appl.,* **22**, 245–257.

[11] Jacod, J. (1979) Calcul stochastique et problemes de martingales, *Lecture Notes in Mathematics,* No. 714, Springer, Berlin.

[12] Jacod, J. and Shiryayev, A.N. (1987) *Limit Theorems for Stochastic Processes,* Springer, Heidelberg.

[13] Karandikar, R.L. (1983) Interchanging the order of stochastic integration and ordinary differentiation, *Sankhya Ser. A,* **45**, 120–124.

[14] Lepingle, D. (1978) Sur le comportment asymptotique des martingales locales, *Seminaire de Probabilites XII,* (Eds. P.A. Meyer and M. Weil), *Lecture Notes in Mathematics,* No. 649, Springer, Berlin, 148–161.

[15] Liptser, R.S. (1980) A strong law of large numbers for local martingales, *Stochastics,* **3**, 217–228.

[16] Naik–Nimbalkar, U.V. (1996) Estimating functions for stochastic processes, *Stochastic Processes and Statistical Inference,* (Eds. B.L.S. Prakasa Rao and B.R. Bhat), New Age International, New Delhi, 52–72.

[17] Prakasa Rao, B.L.S. (1987) *Asymptotic Theory of Statistical Inference,* Wiley, New York.

[18] Prakasa Rao, B.L.S. and Bhat, B.R. (1996) *Stochastic Processes and Statistical Inference,* New Age International, New Delhi.

[19] Rao, C.R. (1965) *Linear Statistical Inference and Its Applications,* Wiley, New York.

[20] Sorensen, M. (1988) Some asymptotic properties of quasilikelihood estimators for semimartingales, *Proc. Fourth Prague Symp. on Asymptotic Statistics.*

[21] Sorensen, M. (1990) On quasilikelihood for semimartingales, *Stoch. Proc. Appl.,* **35**, 331–346.

[22] Sweeting, T.J. (1980) Uniform asymptotic normality of the maximum likelihood estimator, *Ann. Statist.,* **8**, 1375–1380. (Correction ibid. **10**, 320).

[23] Thavaneswaran, A. and Thompson, M.E. (1986) Optimal estimation for semimartingales, *J. Appl. Prob.,* **23**, 409–417.

[24] Wedderburn, R.W.M. (1974) Quasi-likelihood functions, generalized linear models, and the Gauss-Newton method, *Biometrika,* **61**, 439–447.

Chapter 6

Local Asymptotic Behavior of Semimartingale Experiments

6.1 Local Asymptotic Mixed Normality

Let $(\Omega, \mathcal{F})$ be a measurable space and $\{P_\theta, \theta \in \Theta\}$, $\Theta \subset R^k$ be a family of probability measures on $(\Omega, \mathcal{F})$ such that, under each $P_\theta, \theta \in \Theta$, $\boldsymbol{X}$ is a semimartingale with respect to the filtration $\{\mathcal{F}_t, t \geq 0\}$ generated by $\boldsymbol{X}$. Observe that the statistical experiment $(\Omega, \mathcal{F}_t, \{P_\theta|\mathcal{F}_t; \theta \in \Theta\})$ corresponds to the observation of the semimartingale $\boldsymbol{X}$ up to time t. Here, $P_\theta|\mathcal{F}_t$ denotes the probability measure P_θ restricted to the σ-algebra $\mathcal{F}_t$ contained in $\mathcal{F}$.

DEFINITION 6.1 *The net of experiments $(\Omega, \mathcal{F}_t, \{P_\theta|\mathcal{F}_t : \theta \in \Theta\})$ is said to be locally asymptotic mixed normal (LAMN) at $\theta \in int(\Theta)$ as $t \to \infty$ if there exists symmetric positive definite $k \times k$ matrices $\boldsymbol{\delta}_t$ with $\boldsymbol{\delta}_t \to 0$ and adapted processes $\{\boldsymbol{\Lambda}_t, t \geq 0\}$ and $\{\boldsymbol{G}_t, t \geq 0\}$ (possibly depending on θ) with values in R^k and the set of nonnegative definite $k \times k$ matrices, respectively, such that, for every $\boldsymbol{u} \in R^k$,*

$$\log \frac{dP_{\theta+\boldsymbol{\delta}_t \boldsymbol{u}}|\mathcal{F}_t}{dP_{\theta|\mathcal{F}_t}} - \boldsymbol{u}^T \boldsymbol{\Lambda}_t + \frac{1}{2}\boldsymbol{u}^T \boldsymbol{G}_t \boldsymbol{u} \to 0 \ \text{ in } P_\theta\text{-probability}$$

and

$$(\boldsymbol{\Lambda}_t, \boldsymbol{G}_t) \xrightarrow{\mathcal{L}} \left(\boldsymbol{G}^{1/2}\boldsymbol{Z}, \boldsymbol{G}\right) \quad \text{as } t \to \infty$$

where $\boldsymbol{G}$ and $\boldsymbol{Z}$ are independent random vectors with $\boldsymbol{Z}$ distributed as $N_k(0, \boldsymbol{I}_k)$. Here $\xrightarrow{\mathcal{L}}$ denotes the convergence in law.

REMARK 6.1 In this section, we assume that the process δ_t takes values in the class of diagonal matrices even though it is important to allow a general nonnegative definite matrix as the value of δ_t. If G is a constant matrix, then the class of experiments is said to be *locally asymptotic normal* (LAN). ∎

Let the process $X = (X_t, t \geq 0)$ be an R^d-valued RCLL process on $(\Omega, \mathcal{F})$ starting at $X_0 = x$. Assume that X is quasi-left-continuous. Here, $\{\mathcal{F}_t, t \geq 0\}$ is the right continuous filtration generated by X, that is, $\mathcal{F}_t = \cap_{s>t}\mathcal{F}_s^\circ$ and $\mathcal{F}_t^\circ = \sigma\{X_s : s \leq t\}$. Further suppose that $\mathcal{F} = \vee_t \mathcal{F}_t$. Suppose X is a P_θ-semimartingale for every $\theta \in \Theta \subset R^k$ with the triplet of predictable characteristics $(B(\theta), C, \nu(\theta))$ relative to a fixed Borel-measurable truncation function $h : R^d \to R^d$. The function $B(\theta)$ is the drift characteristic of X and $C(\theta) = \langle X^{(c)}(\theta)\rangle$ is the quadratic characteristic where $X^{(c)}(\theta)$ is the continuous martingale component of X. The function $\nu(\theta)$ is the compensator of the random measure μ on $R_+ \times R^d$ associated with the jumps of X by

$$\mu = \sum_{t \geq 0} \varepsilon_{(t, \Delta X_t)} I\{\Delta X_t \neq 0\}$$

with $\Delta X_t = X_t - X_{t-}$ and $\Delta X_0 = 0$. We assume that $C(\theta)$ does not depend on θ and denote $C(\theta)$ by C hereafter.

Recall that

$$H \bullet Z_t \text{ denotes } \int_0^t H_s dZ_s \,,$$

$$J \circ F \text{ denotes } \int_0^t J_s dF_s \,,$$

and

$$U * \eta_t \text{ denotes } \int_0^t U(s, x)\eta(ds, dx) \,,$$

which are the integral processes of H, J, and U with respect to the semimartingale Z, increasing process F and the random measure η, respectively. Let M_k denote the set of symmetric nonnegative definite $k \times k$ matrices and $\mathcal{P}$ denote the predictable σ-algebra on $\Omega \times R_+$.

Likelihood Ratio Process

Let

$$L_t(\theta, \tau) = \frac{dP_\tau|\mathcal{F}_t}{dP_\theta|\mathcal{F}_t} \text{ and } \log L_t(\theta, \tau) = \ell_t(\theta, \tau) .$$

The process $\{L_t(\theta, \tau), t \geq 0\}$ for any fixed θ, τ is called the *likelihood ratio process*. The condition that X is quasi-left-continuous under P_θ implies that $\nu(\theta, \{t\} \times R^d) = 0$ for any $\theta \in \Theta$ up to P_θ-indistinguishability (cf. [11, p. 70]). There exists a continuous adapted increasing process F with $F_0 = 0$ and a predictable M_d-valued process $\mathbf{c}$ such that $C = \mathbf{c} \circ F$ and, for every pair $\theta, \tau \in \Theta$, there exists a $\mathcal{P} \times \mathcal{B}_d$-measurable function $Y(\theta, \tau) : \Omega \times R_+ \times R^d \to R_+$ and a predictable R^d-valued process $\boldsymbol{t}(\theta, \tau)$ satisfying P_τ-a.s.:

$$|(Y(\theta, \tau) - 1)h| * \nu(\theta)_t < \infty , \tag{6.1.1}$$

$$\| \mathbf{c}\, \boldsymbol{t}(\theta, \tau)\| \circ F_t < \infty, \ \left\|\mathbf{c}^{1/2}\boldsymbol{t}(\theta, \tau)\right\|^2 \circ F_t < \infty, \ t \geq 0 , \tag{6.1.2}$$

$$B(\tau) = B(\theta) + (\mathbf{c}\, \boldsymbol{t}(\theta, \tau)) \circ F + (Y(\theta, \tau) - 1)h * \nu(\theta) ,$$

and

$$\nu(\tau, dt, dx) = Y(\theta, \tau, t, x)\, \nu(\theta, dt, dx) . \tag{6.1.3}$$

Here, $h(x) = xI(|x| \leq 1)$. This result follows from the Girsanov's theorem for semimartingales (cf. Theorem 1.85, Chapter 1, [11, p. 159]).

Suppose that the following condition holds:

(C0) all the P_θ-local martingales have the representation property (for the definition, see Section "Diffusion Processes" in Chapter 1) relative to the semimartingale X.

Then, P_θ-a.s.,

$$(Y(\theta, \tau)^{1/2} - 1)^2 * \nu(\theta)_t < \infty, \ t \geq 0 \tag{6.1.4}$$

and the representation

$$\begin{aligned} \log L(\theta, \tau) = {} & \boldsymbol{t}(\theta, \tau) \bullet \boldsymbol{X}^{(c)}(\theta) - \frac{1}{2} \left\|\mathbf{c}^{1/2}\boldsymbol{t}(\theta, \tau)\right\|^2 \circ F \\ & + (Y(\theta, \tau) - 1) * (\mu - \nu(\theta)) \\ & + (\log Y(\theta, \tau) - Y(\theta, \tau) + 1) * \mu \end{aligned} \tag{6.1.5}$$

holds following [11, p. 176 and 182] using the fact $L_0(\theta, \tau) = 1$ (cf. [17]).

Regularity Conditions

Let $\theta \in int(\Theta)$ and $\delta_{i,t} = \delta_{i,t}(\theta) > 0$ and $\delta_{i,t} \to 0$ as $t \to \infty,\ 1 \leq i \leq k$. Let $\boldsymbol{\delta}_t = \boldsymbol{\delta}_t(\theta)$ denote the $k \times k$ diagonal matrix with entries $\delta_{i,t}$ in the diagonal. Suppose the following conditions hold.

(C1) There exists a predictable process $\dot{\boldsymbol{t}}(\theta)$ with values in the set of all $k \times d$ matrices such that

$$\left\|\dot{\boldsymbol{\beta}}(\theta)\boldsymbol{c}\dot{\boldsymbol{\beta}}(\theta)^T\right\| \circ F_t < \infty \text{ a.s } [P_\theta] \text{ for every } t \geq 0,$$

and

$$\left\|\boldsymbol{c}^{1/2}(\boldsymbol{\beta}(\theta, \theta + \boldsymbol{\delta}_t \boldsymbol{u}) - \dot{\boldsymbol{\beta}}(\theta)^T \boldsymbol{\delta}_t \boldsymbol{u})\right\|^2 = o_{P_\theta}(1), \boldsymbol{u} \in R^k .$$

(C2) There exists a $\mathcal{P} \times \mathcal{B}_d$-measurable function

$$\dot{\boldsymbol{Y}}(\theta) : \Omega \times R_+ \times R^d \to R^k \text{ such that}$$

$\|\dot{\boldsymbol{Y}}(\theta)\|^2 * \nu(\theta)_t < \infty\ P_\theta -$ a.s. for every $t \geq 0$ and every τ in a neighborhood of θ and

$$\left[Y(\theta, \theta + \boldsymbol{\delta}_t \boldsymbol{u})^{1/2} - 1 - \boldsymbol{u}^T \boldsymbol{\delta}_t \dot{\boldsymbol{Y}}(\theta)\right]^2 * \nu(\theta)_t = o_{P_\theta}(1), \boldsymbol{u} \in R^k .$$

Consider a P_θ-local martingale $\dot{\ell}(\theta)$ defined by

$$\dot{\ell}(\theta) = \dot{\boldsymbol{\beta}}(\theta) \bullet \boldsymbol{X}^{(c)}(\theta) + \dot{\boldsymbol{Y}}(\theta) * (\mu - \nu(\theta)) . \tag{6.1.6}$$

Its quadratic variation is given by

$$[\dot{\ell}(\theta)] = \left(\dot{\boldsymbol{\beta}}(\theta)\boldsymbol{c}\dot{\boldsymbol{\beta}}(\theta)^T\right) \circ F + \dot{\boldsymbol{Y}}(\theta)\dot{\boldsymbol{Y}}(\theta)^T * \mu . \tag{6.1.7}$$

Note that the quadratic characteristic is

$$\langle\dot{\ell}(\theta)\rangle = \left(\dot{\boldsymbol{\beta}}(\theta)\boldsymbol{c}\dot{\boldsymbol{\beta}}(\theta)^T\right) \circ F + \dot{\boldsymbol{Y}}(\theta)\dot{\boldsymbol{Y}}(\theta)^T * \nu(\theta) . \tag{6.1.8}$$

In addition to the conditions (C1) and (C2), we assume that the following conditions hold.

(C3) There exists a M_k-valued random matrix $G(\theta)$ such that

$$\delta_t \left[\dot{\ell}(\theta)\right]_t \delta_t^T = G(\theta) + o_{P_\theta}(1) .$$

The jumps of $\dot{\ell}(\theta)$ are given by

$$\Delta\dot{\ell}(\theta)_{i,t} = \dot{Y}_i\,(\theta, t, \Delta X_t)\ I\,(\Delta X_t \neq 0)\ .$$

Suppose that the following additional conditions hold:

(C4) (i) $\delta_{i,t} E_\theta\{\sup_{s\leq t}|\dot{Y}_i(\theta, s, \Delta X_s)I(\Delta X_s \neq 0)\} = o(1),\ 1 \leq i \leq k,$

(ii) $\delta_{i,t}\sup_{s\leq t}|\dot{Y}_i(\theta, s, \Delta X_s)|I(\Delta X_s \neq 0) = o_{P_\theta}(1)$, for $1 \leq i \leq k$;

(C5) $E_\theta[\dot{\ell}(\theta)]_t < \infty$ for every $t \geq 0$ and there exists a positive definite $k \times k$ matrix $\sum(\theta)$ such that

$$\delta_t E_\theta \left[\dot{\ell}(\theta)\right]_t \delta_t^T = \sum(\theta) + o(1) .$$

THEOREM 6.1
Suppose the conditions (C1) to (C5) hold. Then, for every $u \in R^k$,

$$\log L_t\,(\theta, \theta + \delta_t u) \;=\; u^T \Lambda_t(\theta) - \frac{1}{2} u^T G_t(\theta) u + o_{P_\theta}(1)\,, \quad (6.1.9)$$

$$(\Lambda_t(\theta), G_t(\theta)) \xrightarrow{\mathcal{L}} K_\theta\ (P_\theta - \textit{stably})\ \textit{as}\ t \to \infty \quad (6.1.10)$$

where

$$\Lambda_t(\theta) = \delta_t \dot{\ell}_t(\theta),\ G_t(\theta) = \delta_t \left[\dot{\ell}(\theta)\right]_t \delta_t^T$$

and

$$K_\theta(\omega) = N(0, G(\theta, \omega)) \otimes \varepsilon G_{(\theta,\omega)}\ . \quad (6.1.11)$$

In particular,

$$(\Lambda_t(\theta), G_t(\theta)) \xrightarrow{\mathcal{L}} \left(G^{1/2}(\theta) Z, G(\theta)\right)\quad \textit{as}\ t \to \infty \quad (6.1.12)$$

under P_θ where $G(\theta)$ are Z and independent and $\mathcal{L}(Z) = N(0, I_k)$. (Here $\mathcal{L}(Z)$ denotes the probability distribution of Z).

PROOF Relations (6.1.5), (6.1.6), and (6.1.7) imply that

$$\log \quad L_t(\theta, \theta + \delta_t u) - u^T \Lambda_t + \frac{1}{2} u^T G_t u$$

$$= \beta^T(\theta, \theta + \delta_t u) \bullet X^{(c)}(\theta)_t - \frac{1}{2} \left\| c^{1/2} \beta(\theta, \theta + \delta_t u) \right\|^2 \circ F_t$$

$$+ (Y(\theta, \theta + \delta_t u) - 1) * (\mu - \nu(\theta))_t$$

$$+ (\log Y(\theta, \theta + \delta_t u) - Y(\theta, \theta + \delta_t u) + 1) * \mu_t$$

$$- u^T \delta_t \dot{\ell}_t(\theta) + \frac{1}{2} u^T \delta_t \left[\dot{\ell}(\theta)\right]_t \delta_t^T u$$

$$= \beta^T(\theta, \theta + \delta_t u) \bullet X^{(c)}(\theta)_t - \frac{1}{2} \left\| c^{1/2} \beta(\theta, \theta + \delta_t u) \right\|^2 \circ F_t$$

$$+ (Y(\theta, \theta + \delta_t u) - 1) * (\mu - \nu(\theta))_t$$

$$+ (\log Y(\theta, \theta + \delta_t u) - Y(\theta, \theta + \delta_t u) + 1) * \mu_t$$

$$- u^T \delta_t \dot{\beta}(\theta) \bullet X^{(c)}(\theta)_t - u^T \delta_t \dot{Y}(\theta) * (\mu - \nu(\theta))_t$$

$$+ \frac{1}{2} u^T \delta_t \left\{ \left(\dot{\beta}(\theta) c \dot{\iota}(\theta)^T\right) \circ F_t + \left(\dot{Y}(\theta) \dot{Y}(\theta)^T\right) \star \mu_t \right\} \delta_t^T u$$

$$= \left(\beta^T(\theta, \theta + \delta_t u) - u^T \delta_t \dot{\beta}\right) \bullet X^{(c)}(\theta)_t$$

$$- \frac{1}{2} \left(\left\| c^{1/2} \beta(\theta, \theta + \delta_t u) \right\|^2 - \left\| c^{1/2} \dot{\beta}^T \delta_t u \right\|^2 \right) \circ F_t$$

$$+ \left(Y(\theta, \theta + \delta_t u) - 1 - u^T \delta_t \dot{Y}\right) * (\mu - \nu)_t$$

$$+\left(\log Y(\theta,\theta+\delta_t u) - Y(\theta,\theta+\delta_t u) + 1 + \frac{1}{2}\left(u^T\delta_t\dot{Y}\right)^2\right) * \mu_t \,. \tag{6.1.13}$$

Step 1: Observe that

$$\left\langle\left(\boldsymbol{\beta}^T(\theta,\theta+\delta_t u) - u^T\delta_t\dot{\boldsymbol{\beta}}\right) * \boldsymbol{X}^{(c)}(\theta)\right\rangle_t$$

$$= \left\|c^{1/2}\left(\boldsymbol{\beta}(\theta,\theta+\delta_t u) - \dot{\boldsymbol{\beta}}^T\delta_t u\right)\right|^2 \circ F_t$$

$$= o_{P_\theta}(1) \tag{6.1.14}$$

by (C1). Hence, it follows that

$$\left(\boldsymbol{\beta}^T(\theta,\theta+\delta_t u) - u^T\delta_t\dot{\boldsymbol{\beta}}\right) \bullet \boldsymbol{X}_t^{(c)} = o_{P_\theta}(1) \tag{6.1.15}$$

by Lenglart's inequality (cf. Theorem 1.33). The condition (C3) implies that

$$\left\|c^{1/2}\dot{\boldsymbol{\beta}}^T\delta_t u\right\|^2 \circ F_t = O_{P_\theta}(1)\,. \tag{6.1.16}$$

It can be checked that

$$\left(\left\|c^{1/2}\boldsymbol{\beta}(\theta,\theta+\delta_t u)\right\|^2 - \left\|c^{1/2}\dot{\boldsymbol{\beta}}^T\delta_t u\right\|^2\right) \circ F_t\Big)$$

$$\leq 2\left(\left\|c^{1/2}\dot{\boldsymbol{\beta}}^T\delta_t u\right\|^2 \circ F_t\right)^{1/2}\left(\left\|c^{1/2}\boldsymbol{R}^t\right\|^2 \circ F_t\right)^{1/2}$$

$$+\left\|c^{1/2}\boldsymbol{R}^t\right\|^2 \circ F_t \tag{6.1.17}$$

where

$$\boldsymbol{R}^t = \boldsymbol{\beta}(\theta,\theta+\delta_t u) - \dot{\boldsymbol{\beta}}^T\delta_t u$$

by the relation

$$x^T \sum x - y^T \sum y = (x-y)^T \sum (x-y) + 2y^T \sum (x-y) ,$$

which in turn shows by the Cauchy–Schwarz inequality,

$$\left| x^T \sum x - y^T \sum y \right| \leq \left\| \Sigma^{1/2} (x-y) \right\|^2 + 2 \left\| \Sigma^{1/2} y \right\| \left\| \Sigma^{1/2} (x-y) \right\| .$$

Note that the terms on the right-hand side of (6.1.17) are $o_{P_\theta}(1)$ by (6.1.14), (6.1.15), and (6.1.16). Hence, the first two terms on the right-hand side of (6.1.13) tend to zero in P_θ-probability.

Step 2: Note that

$$\left(Y(\theta, \theta + \delta_t u) - 1 - u^T \delta_t \dot{Y} \right) * (\mu - \nu)$$

is a locally square integrable martingale under (C2). Furthermore,

$$\left\langle \left(Y(\theta, \theta + \delta_t u) - 1 - u^T \delta_t \dot{Y} \right) * \mu - \nu \right\rangle$$

$$= \left(Y(\theta, \theta + \delta_t u) - 1 - u^T \delta_t \dot{Y} \right)^2 * \nu_t$$

$$= o_{P_\theta}(1) \tag{6.1.18}$$

by (C2). Hence, by the Lenglart's inequality (cf. Theorem 1.33), it follows that

$$Y(\theta, \theta + \delta_t u) - 1 - u^T \delta_t \dot{Y} * (\mu - \nu)_t = o_{P_\theta}(1) . \tag{6.1.19}$$

It remains to prove that the last term on the right-hand side of (6.1.13) is $o_{P_\theta}(1)$ to conclude (6.1.9). Let

$$f(x) = \log(1+x) - x + \frac{1}{2}x^2, \ x > -1 . \tag{6.1.20}$$

Then

$$\left(\log Y(\theta, \theta + \delta_t u) - Y(\theta, \theta + \delta_t u) + 1 + \frac{1}{2} \left(u^T \delta_t \dot{Y} \right)^2 \right) * \mu_t$$

$$= f\,(Y\,(\theta,\theta+\delta_t u)-1) * \mu_t$$

$$-\frac{1}{2}\left[(Y\,(\theta,\theta+\delta_t u)-1)^2-\left(u^T\delta_t\dot{Y}\right)^2\right] * \mu_t\,. \qquad (6.1.21)$$

Observe that $(Y(\theta,\theta+\delta_t u)-1-u^T\delta_t\dot{Y})^2 * \mu$ is a P_θ-locally integrable increasing process and is dominated by its compensator in the sense of Lenglart [11, p. 35] and the compensator is

$$\left(Y\,(\theta,\theta+\delta_t u)-1-u^T\delta_t\dot{Y}\right)^2 * \nu\,.$$

Condition (C2) implies that

$$\left(Y\,(\theta,\theta+\delta_t u)-1-u^T\delta_t Y\right)^2 * \nu_t = o_{P_\theta}(1)\,. \qquad (6.1.22)$$

Hence, by the Lenglart's inequality (cf. Theorem 1.33), it follows that

$$\left(Y\,(\theta,\theta+\delta_t u)-1-u^T\delta_t\dot{Y}\right)^2 * \mu_t = o_{P_\theta}(1)\,. \qquad (6.1.23)$$

Condition (C5) and the equation (6.1.7) show that

$$\left(u^T\delta_t\dot{Y}\right)^2 * \mu_t = O_{P_\theta}(1)\,. \qquad (6.1.24)$$

Relations (6.1.23) and (6.1.24) prove that

$$(Y\,(\theta,\theta+\delta_t u)-1)^2 * \mu_t = O_{P_\theta}(1)\,. \qquad (6.1.25)$$

Let $\varepsilon > 0$. We will now prove that

$$f\,(Y\,(\theta,\theta+\delta_t u)-1)\,I\,(|Y\,(\theta,\theta+\delta_t u))-1|>\varepsilon) * \mu_t = o_{P_\theta}(1)\,. \qquad (6.1.26)$$

Note that for $a > 0$ and $0 < b < 1$.

$$\{|f\,(Y\,(\theta,\theta+\delta_t u)-1)\,I\,\{|Y\,(\theta,\theta+\delta_t u)-1|>\varepsilon\} * \mu_t| > a\}$$

$$\subset \{I\{|Y(\theta,\theta+\delta_t u)-1|>\varepsilon\} * \mu_t > b\} \qquad (6.1.27)$$

(since $|f(x)| = |\log(1+x) - x + \frac{1}{2}x^2| \le \min\left(\frac{|x|^3}{6}, |x|^2\right)$, it is sufficient to prove that

$$I\{|Y(\theta, \theta + \boldsymbol{\delta}_t \boldsymbol{u}) - 1| > \varepsilon\} * \mu_t = o_{P_\theta}(1) \tag{6.1.28}$$

to establish (6.1.26). Condition (C4) (ii) implies that

$$\sup_{s \le t} \left| \boldsymbol{u}^T \boldsymbol{\delta}_t \dot{\boldsymbol{Y}}(\theta, s, \Delta \boldsymbol{X}_s) \right| I(\Delta \boldsymbol{X}_s \ne 0) = o_{P_\theta}(1) . \tag{6.1.29}$$

Note that for $0 < a < \varepsilon^2$,

$$\left\{ \left(\boldsymbol{u}^T \boldsymbol{\delta}_t \dot{\boldsymbol{Y}} \right)^2 I\left\{ \left| \boldsymbol{u}^T \boldsymbol{\delta}_t \dot{\boldsymbol{Y}} \right| > \varepsilon \right\} * \mu_t > a \right\}$$

$$= \left\{ \sup_{s \le t} \left| \boldsymbol{u}^T \boldsymbol{\delta}_t \dot{\boldsymbol{Y}}(\theta, s, \Delta \boldsymbol{X}_s) \right| I\{\Delta \boldsymbol{X}_s \ne 0\} > \varepsilon \right\} . \tag{6.1.30}$$

Hence the equation (6.1.29) holds if and only if

$$\left(\boldsymbol{u}^T \boldsymbol{\delta}_t \dot{\boldsymbol{Y}} \right)^2 I\left\{ \left| \boldsymbol{u}^T \boldsymbol{\delta}_t \dot{\boldsymbol{Y}} \right| > \varepsilon \right\} * \mu_t = o_{P_\theta}(1) . \tag{6.1.31}$$

Let

$$r^t = Y(\theta, \theta + \boldsymbol{\delta}_t \boldsymbol{u}) - 1 - \boldsymbol{u}^T \boldsymbol{\delta}_t \dot{\boldsymbol{Y}} .$$

Then

$$I\{|Y(\theta, \theta + \boldsymbol{\delta}_t \boldsymbol{u}) - 1| > \varepsilon\} * \mu_t$$

$$\le I\left\{ \left| \boldsymbol{u}^T \boldsymbol{\delta}_t \dot{\boldsymbol{Y}} \right| > \varepsilon/2 \right\} * \mu_t + I\left\{ \left| r^t \right| > \varepsilon/2 \right\} * \mu_t$$

$$\le 4\varepsilon^{-2} \left(\boldsymbol{u}^T \boldsymbol{\delta}_t \dot{\boldsymbol{Y}} \right)^2 I\left\{ \left| \boldsymbol{u}^T \boldsymbol{\delta}_t \dot{\boldsymbol{Y}} \right| > \varepsilon/2 \right\} * \mu_t$$

$$+ 4\varepsilon^{-2} \left(r^t \right)^2 * \mu_t$$

$$= o_{P_\theta}(1) \tag{6.1.32}$$

by (6.1.31) and (6.1.23). Hence, the relation (6.1.26) holds. Note that

$$|f(x)| \le 2|x|^3 \text{ for } 0 < |x| \le 1/2 .$$

Therefore, for $0 \le \varepsilon \le 1/2$,

$$\begin{aligned} &|f(Y(\theta, \theta + \delta_t u) - 1) I\{|Y(\theta, \theta + \delta_t u)| \le \varepsilon\} * \mu_t| \\ &\le 2|Y(\theta, \theta + \delta_t u) - 1|^3 I\{|Y(\theta, \theta + \delta_t u)| \le \varepsilon\} * \mu_t \\ &\le 2\varepsilon\ (Y(\theta, \theta + \delta_t u) - 1)^2 * \mu_t \\ &= 2\varepsilon\ O_{P_\theta}(1) . \end{aligned} \tag{6.1.33}$$

Relations (6.1.26) and (6.1.33) prove that

$$f(Y(\theta, \theta + \delta_t u) - 1) * \mu_t = o_{P_\theta}(1) . \tag{6.1.34}$$

This completes the proof of (6.1.9) in view of results in Step 1 and equation (6.1.13). Applying a central limit theorem for multivariate martingales (see Chapter 1), one can obtain (6.1.10) and (6.1.11) in view of (C3) to (C5). ∎

REMARK 6.2 An estimator $\hat{\theta}_t$ of θ is said to be *regular* at θ if

$$\mathcal{L}\left(\delta_t^{-1}\left(\hat{\theta}_t - \theta\right) - u.\, G_t(\theta) \middle| P_{\theta+\delta_t u}\right) \xrightarrow{\mathcal{D}} (S, G) \tag{6.1.35}$$

for every $u \in R^k$. Here $\mathcal{L}(G) = \mathcal{L}(G)(\theta)|P_\theta)$. Suppose that $\det(G) > 0$ a.s. Then the conditional distribution $\mathcal{L}(S|G = g)$ is $\mathcal{L}(G)$-a.s. equal to $N(0, g^{-1})$ convoluted with a probability measure on R^k (cf. [3, 12]). A regular estimator is "asymptotically most concentrated" if $\mathcal{L}(S|g) = N(0, g^{-1/2})$ $\mathcal{L}(G)$-a.s. In other words $\mathcal{L}(S) = \mathcal{L}(G^{-1/2}Z)$ where Z is independent of G and Z is $N_k(0, I_k)$. ∎

An estimator $\hat{\theta}_t$ of θ is called *efficient* at θ if

$$\mathcal{L}\left(\delta_t^{-1}\left(\hat{\theta}_t - \theta\right)\middle| P_\theta\right) \xrightarrow{\mathcal{D}} G^{-1/2}Z \text{ as } t \to \infty . \tag{6.1.36}$$

Notice that the estimator satisfying

$$\delta_t^{-1}\left(\hat{\theta}_t - \theta\right) = G_t(\theta)^{-1}\mathfrak{Z}_t(\theta) + o_{P_\theta}(1) \tag{6.1.37}$$

are regular and efficient at θ. Recall that $G_t(\theta) = \delta_t[\dot{\ell}(\theta)]_t\delta_t^T$. In the light of (6.1.10) of Theorem 6.1, it follows that

$$\delta_t^{-1}\left(\hat{\theta}_t - \theta\right) \xrightarrow{\mathcal{L}} N\left(0, G(\theta, \omega)^{-1}\right) \; P_\theta - \text{ stably} . \tag{6.1.38}$$

In fact, relation (6.1.10) implies that,

$$G_t(\theta)^{-1/2}\mathfrak{Z}_t(\theta) \xrightarrow{\mathcal{L}} N(0, I_k) \; P_\theta - \text{ mixing} . \tag{6.1.39}$$

and we obtain that

$$G_t(\theta)^{1/2}\delta_t^{-1}\left(\hat{\theta}_t - \theta\right) \xrightarrow{\mathcal{L}} N(0, I_k) \quad (P_\theta\text{-mixing}) \tag{6.1.40}$$

as $t \to \infty$ for all estimators $\hat{\theta}_t$ satisfying (6.1.37).

If $\delta_{1,t} = \cdots = \delta_{k,t}$, then it follows that

$$\left[\dot{\ell}(\theta)\right]_t^{1/2}\left(\hat{\theta}_t - \theta\right) \xrightarrow{\mathcal{L}} N(0, I_k) \quad (P_\theta\text{-mixing}) \text{ as } t \to \infty . \tag{6.1.41}$$

proving the asymptotic normality of $\hat{\theta}_t$ under random norming.

6.2 Local Asymptotic Quadraticity

Let $X = \{X_t, t \geq 0\}$ be a R^d-valued process on a measurable space $(\Omega, \mathcal{F})$. Consider a statistical experiment $(\Omega, \mathcal{F}, \{P_\theta, \theta \in \Theta\})$, $\Theta \subset R^k$ open equipped with a filtration $\{\mathcal{F}_t, t \geq 0\}$ generated by X such that, under P_θ, X is a semimartingale with the drift characteristic $B(\theta)$, the quadratic characteristic C and the jump characteristic $\nu(\theta)$. The model $\mathcal{E}_t = (\Omega, \mathcal{F}_t, (P_\theta|\mathcal{F}_t, \theta \in \Theta))$ corresponds to the observation of the process X up to time t. It is assumed that the quadratic characteristic C is independent of θ. Suppose further that $\nu(\theta, \{t\} \times R^d) = 0$ for every $\theta \in \Theta$. Let

$$\ell_t(\theta, \tau) = \frac{dP_\tau \big|_{\mathcal{F}_t}}{dP_\theta \big|_{\mathcal{F}_t}} . \tag{6.2.1}$$

Following the notation and results given in Section 6.1, we have

$$\ell_t(\theta, \tau) = \int_0^t \beta_s(\theta, \tau)^T d\boldsymbol{X}_s^{(c)}(\theta) - \frac{1}{2}\int_0^t \left\| c_s^{1/2} \beta_s(\theta, \tau) \right\|^2 dF_s$$

$$+ \int_0^t \int (Y(\theta, \tau) - 1) d(\mu - \nu(\theta))$$

$$+ \int_0^t \int (\log Y(\theta, \tau) - Y(\theta, \tau) + 1) d\mu . \tag{6.2.2}$$

DEFINITION 6.2 *The net of experiments $\mathcal{E}_t$ is said to be locally asymptotically quadratic (LAQ) at $\theta \in \Theta$ as $t \to \infty$ if there exists symmetric positive definite $k \times k$ matrices $\boldsymbol{\delta}_t(\theta)$ with $\boldsymbol{\delta}_t(\theta) \to 0$ and adapted process $\{\boldsymbol{S}_t(\theta), t \geq 0\}$ and $\{\boldsymbol{G}_t(\theta), t \geq 0\}$ with values in R^k and M_k (the space of symmetric nonnegative definite $k \times k$ matrices), respectively, such that*

$$(i) \qquad \ell_t(\theta, \theta + \boldsymbol{\delta}_t \boldsymbol{u}_t) = \boldsymbol{u}_t^T \boldsymbol{S}_t(\theta) - \frac{1}{2} \boldsymbol{u}_t^T \boldsymbol{G}_t(\theta) \boldsymbol{u}_t + o_{P_\theta}(1) \tag{6.2.3}$$

for every eventually bounded net $\{\boldsymbol{u}_t\}$ in R^k,

$$(ii) \qquad (\boldsymbol{S}_t(\theta), \boldsymbol{G}_t(\theta)) \xrightarrow{\mathcal{L}} (\boldsymbol{S}(\theta), \boldsymbol{G}(\theta)) , \tag{6.2.4}$$

under P_θ, where $\boldsymbol{G}(\theta)$ is a positive definite matrix with positive probability, and

$$(iii) \qquad E_\theta \left[\exp\left(\boldsymbol{u}^T \boldsymbol{S}(\theta) - \frac{1}{2} \boldsymbol{u}^T \boldsymbol{G}(\theta) \boldsymbol{u} \right) \right] = 1, \ \boldsymbol{u} \in R^k . \tag{6.2.5}$$

REMARK 6.3 In the LAMN case discussed in Section 6.1,

$$\mathcal{L}(S(\theta), G(\theta)) = \mathcal{L}\left(G(\theta)^{1/2} Z, G(\theta)\right)$$

where $G(\theta)$ and Z are independent and Z is $N_k(0, I_k)$. It is easy to see that (6.2.5) holds automatically in this case. ∎

Let us assume that the following regularity conditions hold.

(D1) **Differentiability of the drift characteristic:** There exists a predictable process $\dot{t} = \dot{t}(\theta)$ with values in the set of all $k \times d$ matrices such that

$$\int_0^t \left\| \dot{\beta}_s c_s \dot{\beta}_s^T \right\| dF_s < \infty \text{ a.s. } [P_\theta] \text{ for every } t \geq 0,$$

and

$$\int_0^t \left\| c_s^{1/2} \left(\beta_s (\theta, \theta + \delta_t u_t) - \dot{\beta}_s^T \delta_t u_t \right) \right\|^2 dF_s = o_{P_\theta}(1)$$

for every eventually bounded net $\{u_t\}$ in R^k.

(D2) L^Θ**-differentiability of the jump characteristic:** There exists a $\mathcal{P} \times \mathcal{B}_d$-measurable function $\dot{Y} = \dot{Y}(\theta) : \Omega \times R_+ \times R^d \to R^k$ such that

$$\int_0^t \int \|\dot{Y}\|^2 d\nu(\theta) < \infty \text{ a.s. } [P_\theta],\ t \geq 0,$$

and

$$\int_0^t \int \left(Y(\theta, \theta + \delta_t u_t)^{1/2} - 1 - \frac{1}{2} u_t^T \delta_t \dot{Y} \right)^2 d\nu(\theta) = o_{P_\theta}(1)$$

for every eventually bounded net $\{u_t\}$ in R^k.

(D3) **Lindeberg condition:** For every $\varepsilon > 0$,

$$\int_0^t \int \|\delta_t \dot{Y}\|^2 I\left(\|\delta_t \dot{Y}\| > \varepsilon\right) d\nu(\theta) = o_{P_\theta}(1)\,.$$

Define a R^k-valued P_θ-local martingale $\dot{\ell}(\theta)$ by

$$\dot{\ell}_t(\theta) = \int_0^t \dot{\beta}_s(\theta) d\boldsymbol{X}_s^{(c)}(\theta) + \int_0^t \int \dot{Y}(\theta) d(\mu - \nu(\theta))\,. \quad (6.2.6)$$

Then $\dot{\ell}$ is a component wise locally square integrable under P_θ with $\dot{\ell}_0(\theta) = 0$. This can be considered as a score function and the process

$$\langle \dot{\ell}(\theta) \rangle_t = \int_0^t \dot{\beta}_s(\theta) c_s \dot{\beta}_s(\theta)_T dF_s + \int_0^t \int \dot{Y}(\theta) \dot{Y}(\theta)^T d\nu(\theta)\,. \quad (6.2.7)$$

as an information.

(D4) Suppose that

$$\delta_t \langle \dot{\ell}(\theta) \rangle_t \delta_t^T = O_{P_\theta}(1)\,.$$

THEOREM 6.2

Under the conditions (D1)–(D4), the stochastic representation given by (6.2.3) holds with

$$(\boldsymbol{S}_t(\theta), \boldsymbol{G}_t(\theta)) = \left(\delta_t \dot{\ell}_t(\theta), \delta_t \langle \dot{\ell}(\theta) \rangle_t \delta_t^T\right)\,. \quad (6.2.8)$$

Proof of this theorem is analogous to that given in Theorem 6.1. (cf. [17, 18]). We omit the details.

For $t \geq 0$, let $\sigma^t = \sigma_\theta^t : R_+ \to R_+$ be a continuous nondecreasing function with $\sigma^t(0) = 0$ and $\sigma^t(1) = t$. Let

$$\boldsymbol{M}_s^t = \delta_t \dot{\ell}_{\sigma^t(s)}(\theta)\,. \quad (6.2.9)$$

Then $\boldsymbol{M}^t = \boldsymbol{M}^t(\theta)$ is a locally square integrable local martingale under P_θ with respect to the filtration $\{\mathcal{F}_{\sigma^t(s)}, s \geq 0\}$ (cf. [16, Chapter 4.7]).

Suppose the following condition holds.

(D5) Suppose that

$$\mathcal{L}\left(M^t \,\middle|\, P_\theta\right) \xrightarrow{\mathcal{D}} M \tag{6.2.10}$$

(here $\xrightarrow{\mathcal{D}}$ denotes the weak convergence in $D([0,1], R^k)$ equipped with the Skorokhod topology) where $M = M(\theta)$ is a continuous local martingale with

(i) $M_0 = 0$,

(ii) $E\left[\exp\left(u^T M_1 - \frac{1}{2}u^T \langle M\rangle_1 u\right)\right] = 1,\ u \in R^k$, and (6.2.11)

(iii) $\langle M\rangle_1$ is positive definite with positive probability.

THEOREM 6.3
Under the conditions (D1)–(D3) and (D5), the net of experiments of $\mathcal{E}_t$ is LAQ at θ with rate δ_t,

$$(S_T(\theta), G_t(\theta)) = \left(\delta_t \dot{\ell}_t(\theta), \delta_t \langle \dot{\ell}(\theta)\rangle_t \delta_t\right),$$

and

$$(S_t(\theta), G_t(\theta)) \xrightarrow{\mathcal{L}} (S(\theta), G(\theta))$$

under P_θ where

$$(S(\theta), G(\theta)) = (M_1, \langle M\rangle_1). \tag{6.2.12}$$

PROOF Note that

$$M_1^t = \delta_t \dot{\ell}_{\sigma^t(1)}(\theta) = \delta_t \dot{\ell}_t(\theta). \tag{6.2.13}$$

Assumption (D5) implies that $\mathcal{L}(M_1^t | P_\theta) \xrightarrow{\mathcal{D}} M_1$. Hence,

$$\mathcal{L}\left(M_1^t, [M^t]_1 \,\middle|\, P_\theta\right) \xrightarrow{\mathcal{D}} (M_1, \langle M\rangle_1) \tag{6.2.14}$$

by the convergence theorem for semimartingales (cf. [11, p. 341]), where

$$[M^t]_s = \delta_t \left[\dot{\ell}\right]_{\sigma^t(s)} \delta_t^T,$$

and

$$[\dot{\ell}]_t = \int_0^t \dot{t}_s c_s \dot{t}_s^T dF_s + \int_0^t \int \dot{Y}\dot{Y}^T d\mu, \dot{\ell} = \dot{\ell}(\theta)\,.$$

This can be seen in the following way. Note that

$$\Delta M_s^t = \delta_t \dot{Y}\left(\sigma^t(s), \Delta X_{\sigma^t(s)}\right) I\left(\Delta X_{\sigma^t(s)} \neq 0\right)\,.$$

Hence, the jump measure μ^t of M^t satisfies the equation

$$\int_0^t \int I(A) d\mu^t = \int_0^{\sigma^t(s)} \int I\left(\delta_t \dot{Y} \in A\right) d\mu$$

for every $A \in \mathcal{B}_k$ with A not containing $\{0\}$. A similar equation holds for the compensator ν^t of μ^t under P_θ, namely,

$$\int \int_0^t I(A) d\nu^t = \int_0^{\sigma^t(s)} \int I\left(\delta_t \dot{Y} \in A\right) d\nu$$

where $\nu = \nu(\theta)$. Therefore the drift characteristic B^t of M^t relative to the truncation function $h : R^k \to R^k, h(x) = xI(\|x\| \leq 1)$ is given by

$$B_s^t = -\int_0^{\sigma^t(s)} \delta_t \dot{Y} I\left(\|\delta_t \dot{Y}\| > 1\right) d\nu$$

(cf. [11]). This in turn implies that the variation of the ith component B_i^t of B_t satisfies the inequality

$$\mathrm{Var} B_{is}^t \leq \int_0^{\sigma^t(s)} \int \|\delta_t \dot{Y}\| \, I\left(\|\delta_t \dot{Y}\| > 1\right) d\nu$$

$$\leq \int_0^t \|\delta_t \dot{Y}\|^2 I\left(\|\delta_t \dot{Y}\| > 1\right) d\nu$$

$$= o_{P_\theta}(1)$$

by (D3). The condition (D4) implies that

$$(M^t, [M^t]) \xrightarrow{\mathcal{L}} (M, \langle M \rangle)$$

under P_θ by the convergence theorem for semimartingales. This proves (6.2.14). We shall now show that

$$[M^t]_1 - \langle M^t \rangle_1 = o_{P_\theta}(1) \text{ as } t \to \infty . \tag{6.2.15}$$

Note that for the (i, j)th component,

$$\left| [M^t]_{ij,1} - \langle M^t \rangle_{ij,1} \right|$$

$$= \left| \int_0^t \int (\delta_t \dot{Y})_i (\delta_t \dot{Y})_j \, d(\mu - \nu) \right|$$

$$\leq \int_0^t \int \left| \delta_t \dot{Y})_i (\delta_t \dot{Y})_j \right| I \{ \| \delta_t \dot{Y} \| > \varepsilon \} \, d\mu$$

$$+ \int_0^t \int \left| (\delta_t \dot{Y})_i (\delta_t \dot{Y})_j \right| I \{ \| \delta_t \dot{Y} \| > \varepsilon \} \, d\nu$$

$$+ \left| \int_0^t \int (\delta_t \dot{Y})_i \left(\delta_t \dot{Y})_j \right) I \{ \| \delta_t \dot{Y} \| \leq \varepsilon \} \, d(\mu - \nu) \right|$$

for every $\varepsilon > 0$. Observe that the second term on the right-hand side of the above inequality tends to zero in P_θ-measure by (D3), and hence, the first term tends to zero in P_θ-measure by the Lenglart's inequality (cf. Theorem 1.33).

Furthermore,

$$\left\langle \int_0 \int (\delta_t \dot{Y})_i (\delta_t \dot{Y})_j I\left(\|\delta_t \dot{Y}\| \leq \varepsilon\right) d(\mu - \nu) \right\rangle_t$$

$$= \int_0^t \int (\delta_t \dot{Y})_i^2 (\delta_t \dot{Y})_j^2 I\left(\|\delta_t \dot{Y}\| \leq \varepsilon\right) d\nu$$

$$\leq \varepsilon^2 \int_0^t \int \left|(\delta_t \dot{Y})_i (\delta_t \dot{Y})_j\right| d\nu$$

$$= \varepsilon^2 \, O_{P_\theta}(1) \, .$$

This proves (6.2.15). Theorem 6.3 is now a consequence of Theorem 6.2. ∎

Limiting Distribution

Suppose the conditions (D1)–(D3) hold and the net $\mathcal{E}_t$ is LAQ at every point $\theta \in \Theta$ with $\delta_t(\theta)$ as the norming factor and $(S_t(\theta), G_t(\theta))$ defined by (6.2.8). Suppose further that $G(\theta)$ is positive definite a.s. It follows that

$$(S_t(\theta), G_t(\theta)) \xrightarrow{\mathcal{L}} (S(\theta), G(\theta))$$

under P_θ and

$$E_\theta \left[\exp\left(u^T S(\theta) - \frac{1}{2} u^T G(\theta) u \right) \right] = 1, u \in R^k \, .$$

It follows from Le Cam's first lemma (cf. Appendix F; [21, p. 103]; [15, p. 88]), that the probability measure $\{P_{\theta+\delta_t(\theta)u_t} | \mathcal{F}_t\}$ and $\{P_\theta | \mathcal{F}_t\}$ are contiguous.

Let us consider the estimator $\hat{\theta}_t : (\Omega, \mathcal{F}_t) \to (\bar{\Theta}, \mathcal{B}^k \cap \bar{\Theta})$ having the representation

$$\delta_t(\theta)^{-1} \left(\hat{\theta}_t - \theta \right) = G_t(\theta)^{-1} S_t(\theta) + o_{P_\theta}(1)$$

$$= \delta_t(\theta)^{-1} \langle \dot{\ell}(\theta) \rangle_t^{-1} \dot{\ell}_t(\theta) + o_{P_\theta}(1) \tag{6.2.16}$$

for every $\theta \in \Theta$ by Theorem 6.2. By contiguity and the Le Cam's third lemma (cf. Appendix F; [21, p. 104]; [15, p. 88]), it follows that for every net $\{u_t\}$ in R^k with $u_t \to u$ as $t \to \infty$, and for every $\theta \in \Theta$,

$$\mathcal{L}\left(\delta_t(\theta)^{-1}\left(\hat{\theta}_t - \theta - \delta_t(\theta)u_t\right) \Big| P_{\theta+\delta_t(\theta)u_t}\right) \xrightarrow{\mathcal{D}} \lambda_\theta(u) \tag{6.2.17}$$

and

$$\mathcal{L}\left(G_t(\theta)^{1/2}\delta_t(\theta)^{-1}\left(\hat{\theta}_t - \theta - \delta_t(\theta)u_t\right) \Big| P_{\theta+\delta_t(\theta)u_t}\right) \xrightarrow{\mathcal{D}} \rho_\theta(u) \tag{6.2.18}$$

where

$$\lambda_\theta(u) = \mathcal{L}\left(G(\theta)^{-1}S(\theta) - u \Big| Q_u\right), \tag{6.2.19}$$

$$\rho_\theta(u) = \mathcal{L}\left(G(\theta)^{-1/2}S(\theta) - G(\theta)^{1/2}u \Big| Q_u\right), \tag{6.2.20}$$

and

$$\frac{dQ_u}{dQ_o} = \exp\left(u^T S(\theta) - \frac{1}{2}u^T G(\theta) u\right). \tag{6.2.21}$$

Note that $\{Q_u, u \in R^k\}$ are connected with the limiting experiment, which is quadratic.

Under the conditions stated in Theorem 6.3,

$$(S(\theta), G(\theta) = (M_1, \langle M \rangle_1) \tag{6.2.22}$$

for a continuous local martingale $M = M(\theta)$ satisfying (6.2.11) defined on $(\Omega', \mathcal{F}', Q_0)$ equipped with a right-continuous filtration $\{\mathcal{F}'_t\}$. Suppose M_t is of the form

$$M_t = \int_0^t f(V_s)\, dV_s \tag{6.2.23}$$

where V is a continuous Gaussian R^m-martingale, $V_0 = 0$,

$$\langle V\rangle_t = \int_0^t D_s \, dA_s$$

with D M_m-valued, A continuous nondecreasing, $A_0 = 0$ with both A and D non-random, and $f : R^m \to R^{k\times m}$ is a measurable function such that

$$\int_0^t \left\| f(V_s) D_s f(V_s)^T \right\| dA_s < \infty \quad Q_0 - \text{a.s.}, \quad 0 \le t \le 1 .$$

Then the limiting distributions $\lambda_\theta(u)$ and $\rho_\theta(u)$ in (6.2.19) and (6.2.20) can be obtained by the following result.

THEOREM 6.4

Under the assumptions stated above, for every $u \in R^k$,

$$\lambda_\theta(u) \overset{\mathcal{D}}{=} \langle H\rangle_1^{-1} H_1 \quad \textit{and} \quad \rho_\theta(u) \overset{\mathcal{D}}{=} \langle H\rangle_1^{-1/2} H_1 \tag{6.2.24}$$

where

$$H_1 = \int_0^1 f(Z_s)\, d\hat{V}_s .$$

($X \overset{\mathcal{D}}{=} Y$ indicates that X and Y have the same distribution.) Here $\hat{V}$ is a continuous Gaussian martingale with $\langle \hat{V}\rangle = \langle V\rangle$ and Z satisfies the stochastic differential equation

$$dZ_t = D_t f(Z_t)^T u \, dA_t + d\hat{V}_t, \; Z_0 = 0, 0 \le t \le 1 \tag{6.2.25}$$

and

$$\langle H\rangle_t = \int_0^t f(Z_s) D_s f(Z_s)^T \, dA_s \tag{6.2.26}$$

(H depends on θ and u).

PROOF Theorem 6.3 and the relation (6.2.17) show that

$$\lambda_\theta(u) = \mathcal{L}\left(\langle M\rangle_1^{-1}(M_1 - \langle M\rangle_1 u) | Q_u\right)$$

where

$$\frac{dQ_u\big|\mathcal{F}_t^1}{dQ_0\big|\mathcal{F}_t^1} = \exp\left(u^T M_t - \frac{1}{2}u^T \langle M\rangle_t u\right), 0 \le t \le 1 .$$

By the Girsanov's theorem, it follows that the process $\hat{V}$ with components

$$\hat{V}_{i,t} = V_{i,t} - \left\langle V_i, u^T M\right\rangle_t, 0 \le t \le 1$$

is a continuous Gaussian martingale under Q_u with

$$\left\langle \hat{V}\right\rangle_t = \langle V\rangle_t = \int_0^t D_s dA_s .$$

Since $\langle V_i, M_j\rangle_t$ is the (i, j)-th component of the matrix $\int_0^t D_s f(V_s)^T dA_s$, we obtain that

$$\hat{V}_t = V_t - \int_0^t D_s f(V_s)^T u \, dA_s .$$

But

$$\langle M\rangle_t = \int_0^t f(V_s) D_s f(V_s)^T dA_s \text{ and } M_t = \int_0^t f(V_s) dV_s .$$

Hence,

$$M_t - \langle M\rangle_t u = \int_0^t f(V_s) dV_s .$$

Let $\boldsymbol{Z} = \boldsymbol{V}$. Then $\lambda_\theta(\boldsymbol{u}) = \mathcal{L}((\boldsymbol{H}_1)^{-1}\boldsymbol{H}_1|\boldsymbol{Q}\boldsymbol{u})$ and $\boldsymbol{Z}$ satisfies (6.2.25). It is easy to see that $\rho_\theta(\boldsymbol{u}) = \mathcal{L}((\boldsymbol{H}_1)^{-1/2}\boldsymbol{H}_1|\boldsymbol{Q}\boldsymbol{u})$. ▮

6.3 Local Asymptotic Infinite Divisibility

Let $(\Omega, \mathcal{F})$ be measurable space and $\{P_\theta, \theta \in \Theta\}, \Theta \subset R$ be a family of probability measures defined on $(\Omega, \mathcal{F})$. Let $X \equiv \{X_t, t \geq 0\}$ be a real-valued stochastic process defined on $(\Omega, \mathcal{F})$ such that under each $P_\theta, \theta \in \Theta, X$ is a semimartingale with respect to the filtration $\{\mathcal{F}_t, t \geq 0\}$ generated by X. The experiment $(\Omega, \mathcal{F}_t, \{P_\theta|\mathcal{F}_t, \theta \in \Theta\})$ corresponds to the observation of the semimartingale X up to time t. Suppose Θ is open.

DEFINITION 6.3 *The net of experiments $(\Omega, \mathcal{F}_t, \{P_\theta|\mathcal{F}_t, \theta \in \Theta\})$ is said to be locally asymptotically infinitely divisible (LAID) at θ as $t \to \infty$ if there exists a continuous function $\delta_t \equiv \delta_t(\theta)$ decreasing to zero as $t \to \infty$ such that, for every $u \in R$,*

$$\log d P_{\theta+\delta_t u}\left|\frac{\mathcal{F}_t}{d P_\theta}\right|\mathcal{F}_t - uA_t + Q(u) \xrightarrow{p} 0 \ \text{ in } P_\theta\text{-probability} \tag{6.3.1}$$

where $Q(u)$ is a nonnegative function and $\{A_{t_1} t \geq 0\}$ is an adapted process possibly depending on θ such that

$$A_t \xrightarrow{\mathcal{L}} \mathcal{K}(0, K_\theta) \ \text{ as } t \to \infty \tag{6.3.2}$$

under P_θ where $\mathcal{K}(0, K_\theta)$ is the infinitely divisible distribution with K_θ as its Kolmogorov function with $K_\theta(+\infty) = 1$.

REMARK 6.4 It is well known that any infinitely divisible distribution with finite variance is uniquely determined by its mean and a nondecreasing bounded function $K(y)$ with $K(-\infty) = 0$. This function K is called the *Kolmogorov function* and the infinitely divisible distribution with the mean a and the Kolmogorov function K is denoted by $\mathcal{K}(a, K)$. ▮

If $\{A_t\}$ in (6.3.1) is asymptotically $N(0, 1)$ and $Q(u) = u^2/2$, then it follows that the family of statistical experiments is locally asymptotically normal (LAN) at θ.

Let us recall that the semimartingale $X^\theta = \{X_t^\theta\}$ admits the canonical representation

$$X_t^\theta = X_0 + \alpha_t^\theta + m_t^\theta + \int_0^t \int_{|x|>1} x d\mu + \int_0^t \int_{|x|\leq 1} x d(\mu - \nu(\theta)) \tag{6.3.3}$$

where $\mu = \mu(dt, dx)$ is the random measure of the jumps of the process X^θ, $\nu(\theta) = \nu(\theta)(dt, dx)$ is its compensator relative to $\{\mathcal{F}_t, t \geq 0\}$, $\alpha^\theta = (\alpha_t^\theta, \mathcal{F}_t)$ is the predictable process of locally bounded variation, $m^\theta = (m_\theta^t, \mathcal{F}_t)$ is a continuous local martingale with the quadratic characteristics $\langle m^\theta \rangle = (\langle m^\theta \rangle_t, \mathcal{F}_t)$. The triple $(\alpha^\theta, \langle m^\theta \rangle, \nu(\theta))$ is the triplet of predictable characteristics of the semimartingale X^θ.

Assume that the compensators of the semimartingales $\{X^\theta, \theta \in \Theta\}$ are continuous, that is, $\nu(\theta(\{t\}, R - \{0\}) = 0$ for all $t > 0$. Further assume that $P^{\theta_1} \overset{loc}{\leq} P^{\theta_2}$ for all $\theta_1, \theta_2 \in \Theta$. Let P_t^θ be the probability measure generated by $\{X_s^\theta, 0 \leq s \leq t\}$. Define

$$\rho\,(x; \theta_1, \theta_2) = \frac{dP_0^{\theta_2}}{dP_0^{\theta_1}} \,. \tag{6.3.4}$$

We assume that all the processes $\{X^\theta, \theta \in \Theta\}$ have the same quadratic characteristic $\langle m^\theta \rangle$, hereafter denoted by $\langle m \rangle$ and the compensators $\nu(\theta_1)$ and $\nu(\theta_2)$ are mutually absolutely continuous for $\theta_1, \theta_2 \in \Theta$. Define

$$\frac{d\nu(\theta_2)}{d\nu\,(\theta_1)}(t, x) = Y\,(t, x; \theta_1, \theta_2)\ . \tag{6.3.5}$$

Furthermore, there exists a predictable process $\beta(\theta_1, \theta_2)$ such that

$$\int_0^t \beta_s\,(\theta_1, \theta_2)\, d\langle m \rangle_s = \alpha_t^{\theta_2} - \alpha_t^{\theta_1}$$

$$-\int_0^t \int_{|x|\le 1} x\,(Y(s,x;\theta_1,\theta_2)-1)\,\nu(\theta_1)(ds,dx) \tag{6.3.6}$$

for every $t > 0$. Under the assumptions stated above,

$$L_t(\theta_1,\theta_2) \equiv Z_t = \frac{dP_t^{\theta_2}}{dP_t^{\theta_1}} > 0 \text{ a.s.} \tag{6.3.7}$$

for any $t \ge 0$. Let $\ell_t(\theta_1,\theta_2) = \log L_t(\theta_1,\theta_2)$. Then

$$\ell_t(\theta_1,\theta_2) = \log p(X_0;\theta_1,\theta_2) + \int_0^t \beta_s(\theta_1,\theta_2)\,dm_s^{\theta_1}$$

$$+ \int_0^t \int_{R-\{0\}} \log Y(s,x;\theta_1,\theta_2)\,d(\mu-\nu(\theta_1))$$

$$-\frac{1}{2}\int_0^t \beta_s^2(\theta_1,\theta_2)\,d\langle m\rangle_s$$

$$-\int_0^t \int_{R-\{0\}} (Y(s,x;\theta_1,\theta_2)-1$$

$$-\log Y(s,x;\theta_1,\theta_2))\,\nu(\theta_1)(ds,dx) \tag{6.3.8}$$

where the stochastic integrals are locally square integrable martingales. It is easy to see that this is the same as the representation (6.1.5) in the one-dimensional case.

Regularity Conditions

Let $\theta \in \Theta$. Suppose the following conditions hold for the family of semi-martingales $\{X^\theta, \theta \in \Theta\}$:

(E0) $\rho(X_0; \theta, \theta_1) \to 1$ in P_θ-probability as $\theta_1 \to \theta$,

(E1) the random functions $\beta_s(\theta, \theta_1)$ and $Y(s, x; \theta, \theta_1)$ are differentiable in θ_1 at the point $\theta_1 = \theta$ P^θ-a.s. and their derivatives

$$\left.\frac{\partial \beta_s(\theta, \theta_1)}{\partial \theta_1}\right|_{\theta_1=\theta} \equiv \dot{\beta}_s(\theta) \text{ and } \left.\frac{\partial Y(s, x; \theta, \theta_1)}{\partial \theta_1}\right|_{\theta_1=\theta} = \dot{Y}(s, x; \theta)$$

are such that for $t > 0$,

$$\int_0^t \dot{\beta}_s^2(\theta) d\langle m\rangle_s < \infty \text{ and } \int_0^t \int_{R-\{0\}} \dot{Y}^2(s, x; \theta)\nu(\theta)(ds, dx) < \infty$$

P^θ-a.s.

Further suppose there exists $\delta_t \downarrow 0$ such that,

(E2) as $N \to \infty$, uniformly for $t > 0$,

$$\left\{I(Y > N)Y(\theta, \theta + u\delta_t) - I\left(Y < \tfrac{1}{N}\right)\log Y(\theta, \theta + u\delta_t)\right\}$$

$$*\nu(\theta)_t \xrightarrow{P_\theta} 0,$$

(E3) there exists a nondecreasing function $K_\theta(\cdot)$ with $K_\theta(-\infty) = 0$, $K_\theta(+\infty) = 1$ such that, for any $s > 0$,

$$\delta_t^2\left\{I\left(\bar{R}_+\right)(y)\dot{\beta}^2 \circ \langle m\rangle_{st} + I\left(\dot{Y} < y\delta_t^{-1}\right)Y^2 * \nu(\theta)_{st}\right\}$$

$$\xrightarrow{P_\theta} sK_\theta(Y)$$

for $y \in M = C(K_\theta) \cup \{+\infty\}$ where $C(g)$ denotes the set of points of continuity of $g(\cdot)$ and $\bar{R}_+ = (0, \infty]$, and

(E4) for $y \in M$ and $s > 0$, $u \in R - \{0\}$,

$$\left\{I\left(u^{-1}\log Y < y\right)\log Y(\theta, \theta + u\delta_t)\right.$$

$$- u\delta_t I\left(\dot{Y} < y\delta_t^{-1}\right)\dot{Y}(\theta)\}^2 \star \nu(\theta)_{st}$$

$$+ \left(\beta\left(\theta, \theta + u\delta_t\right) - u\delta_t\dot{\beta}(\theta)\right)^2 \circ \langle m\rangle_{st} \xrightarrow{P_\theta} 0\,.$$

Let

$$\ell_s^{(t)} \equiv \ell_{st}\left(\theta, \theta + \delta u_t\right), t \geq 0 \tag{6.3.9}$$

for any fixed $s > 0$ where $\ell_t(.,\cdot)$ is as given by (6.3.8).

THEOREM 6.5

Suppose the family of semimartingales $\{X^\theta, \theta \in \Theta\}$ satisfy the conditions (E0) to (E4). Then the process $\{\ell_s^{(t)}\}$ has the representation

$$\ell_s^{(t)} = u\Delta_s^{(t)} - sQ(u) + R_s^{(t)}(u) \tag{6.3.10}$$

for any $u \in R$ where

$$(i)\;\; R_s^{(t)}(u) \xrightarrow{P_\theta} 0 \text{ for every } s > 0\,, \tag{6.3.11}$$

$$(ii)\;\; Q(u) = \int_{-\infty}^{\infty}\left(e^{uy} - 1 - uy\right)y^{-2}dK_\theta(y)\,, \tag{6.3.12}$$

(iii) $\Delta_s^{(t)} \xrightarrow{\mathcal{D}} \eta_s$ where η is a homogeneous process with independent increments, $\eta_0 = 0$ and $\eta_s \simeq \mathcal{K}(0, sK_\theta)$ for every $s > 0$. Here $\xrightarrow{\mathcal{D}}$ denotes the weak convergence of a family of processes in the space $(D, \mathcal{D})$ with the Skorokhod topology.

We prove a few lemmas before giving proof of the theorem. Let us introduce the notation

$$\Delta_s^{(t)} = \delta_t\left[\dot{\beta} \bullet m_{st} + \dot{Y} * (\mu - \nu(\theta))_{st}\right] \tag{6.3.13}$$

and

$$A_s^{(t)}(u) = \frac{1}{2}\beta^2 \circ \langle m\rangle_{st} + (Y - 1 - \log Y) * \nu(\theta)_{st}\,. \tag{6.3.14}$$

LEMMA 6.1

Suppose the family of semimartingales $\{X^\theta, \theta \in \Theta\}$ satisfies the conditions (E0) to (E3) and that (E4) holds for $y = \infty$. Then

$$\ell_s^{(t)} - u\Delta_s^{(t)} + A_s^{(t)}(u) \xrightarrow{P_\theta} 0 \;\; as\; t \to \infty . \tag{6.3.15}$$

PROOF It is easy to see from (6.3.8) that

$$\ell_s^{(t)} - u\Delta_s^{(t)} + A_s^{(t)}(u) = \log \rho\,(X_0; \theta, \theta + u\delta_t) + R_s^{(t)}(u) \tag{6.3.16}$$

where

$$R_s^{(t)}(u) = \left(\beta - u\delta_t\dot{\beta}\right) \bullet m_{st} + \left[\log Y - u\delta_t\dot{Y}\right] * (\mu - \nu(\theta))_{st} . \tag{6.3.17}$$

Condition (E0) implies that

$$\log \rho\,(X_0; \theta, \theta + u\delta_t) \xrightarrow{P_\theta} 0 \text{ as } t \to \infty . \tag{6.3.18}$$

Furthermore, for any $\varepsilon > 0$,

$$P_\theta\left(\left|R_s^{(t)}(u)\right| > \varepsilon\right) \le \frac{N}{\varepsilon^2} + P_\theta\left\{\left[\left(\beta - u\delta_t\dot{\beta}\right)^2 \circ \langle m\rangle_{st}\right.\right.$$

$$\left.\left. + \;\left(\log Y - u\delta_t\dot{Y}\right)^2 * \nu(\theta)_{st}\right] > N\right\} \tag{6.3.19}$$

for every $N > 0$ by the property of locally square integrable martingales or by the Lenglart's inequality. The inequality (6.3.19) together with the fact that (E4) holds for $y = \infty$ shows that

$$R_s^{(t)}(u) \xrightarrow{P_\theta} 0 \tag{6.3.20}$$

for every $u \in R$. This proves the lemma in view of (6.3.18). ∎

We now a state a limit theorem due to Taraskin [23]. Suppose that m is a continuous local martingale defined on a probability space $(\Omega, \mathcal{F}, P)$ with the quadratic characteristic $\langle m\rangle$ and suppose that an integer valued random measure

μ with the compensator ν is given. Let $H = \{H_t, t \geq 0\}$ be a stochastic process for which $H \bullet m_t$ is defined for any $t > 0$ and is a locally square integrable martingale. Further suppose that $U = (U(s,x))$ is a random function on $R_+ \times (R - \{0\})$ such that the stochastic integral $U * (\mu - \nu)_t$ is defined for any $t > 0$ and is a locally square integrable martingale. Let

$$J = H \bullet m + U * (\mu - \nu) . \tag{6.3.21}$$

THEOREM 6.6 **[22]**

Let $\nu(\{t\}, R - \{0\}) = 0$ for any $t > 0$ and suppose there exists a continuous function $\delta_t \downarrow \infty$ as $t \to \infty$ such that

$$\delta_t^2 \left\{ I\left(\bar{R}_+\right)(y) H^2 \circ \langle m \rangle_{st} + I\left(U < y\delta_t^{-1}\right) U^2 * \nu_{st} \right)\right\} \xrightarrow{P} s\ K(y) \tag{6.3.22}$$

for any $s > 0$, at each point of continuity of K and at $y = +\infty$, where $K(\cdot)$ is a nonrandom nondecreasing bounded function with $K(-\infty) = 0$. Then the family of processes $J^{(t)} = (J_{st}, s \geq 0)$ has the property

$$J^{(t)} \xrightarrow{\mathcal{D}} \eta \ \ as\ t \to \infty \tag{6.3.23}$$

where η is a homogeneous process with independent increments beginning at zero and η_s has the distribution $\mathcal{K}(o, sK)$. In particular,

$$\delta_t J_t \xrightarrow{\mathcal{L}} \mathcal{K}(0, K) \ \ as\ t \to \infty . \tag{6.3.24}$$

REMARK 6.5 In order that (6.3.24) holds, it is sufficient if the condition (6.3.22) is satisfied for $s = 1$. ∎

LEMMA 6.2

Under the conditions (E3) and (E4), for each $s > 0$ and $y \in R$,

$$F_t(y; s, u) \xrightarrow{P_\theta} s\ K_\theta(y) \ \ as\ t \to \infty \tag{6.3.25}$$

where

$$F_t(y; s, u) = u^{-2}\left\{I\left(\bar{R}_+\right)(y)\beta^2 \circ \langle m\rangle_{st} + I\left(u^{-1}\log Y < y\right)\log^2 Y * \nu(\theta)_{st}\right\}. \quad (6.3.26)$$

PROOF Condition (E3) implies that

$$K_t(y; s) = \delta_t^2\left\{I\left(\bar{R}_+\right)(y)\dot{\beta}^2 \circ \langle m\rangle_{st} + I\left(\dot{Y} < y\delta_t^{-1}\right)\dot{Y}^2 * \nu(\theta)_{st}\right\}$$

$$\xrightarrow{P_\theta} s\ K_\theta(y) \text{ as } t \to \infty. \quad (6.3.27)$$

Hence, it is sufficient to prove that,

$$F_t(y; s, u) - K_t(y; s) \xrightarrow{P_\theta} 0 \text{ as } t \to \infty. \quad (6.3.28)$$

Note that

$$\begin{aligned}
&F_t(y; s, u) - K_t(y; s) \\
&= u^{-2}\left\{\left[I\left(u^{-1}\log Y < y\right)\log Y - u\delta_t I\left(\dot{Y} < y\delta_t^{-1}\right)\dot{Y}\right]^2 * \nu(\theta)_{st}\right. \\
&\left. + I\left(\bar{R}_+\right)(y)\left(\beta - u\delta_t\dot{\beta}\right)^2 \circ \langle m\rangle_{st}\right\} \\
&+ 2u^{-1}\left\{I\left(\bar{R}_+\right)(y)\delta_t\dot{\beta}\left(\beta - u\delta_t\dot{\beta}\right) \circ \langle m\rangle_{st}\right. \\
&+ I\left(\dot{Y} < y\delta_t^{-1}\right)\delta_t\dot{Y}\left[I\left(u^{-1}\log Y < y\right)\log Y\right. \\
&\left.\left. - u\delta_t I\left(\dot{Y} < y\delta_t^{-1}\right)\dot{Y}\right] * \nu(\theta)_{st}\right\}. \quad (6.3.29)
\end{aligned}$$

The first term on the right-hand side (6.3.29) tends to zero in P_θ-probability by (E4). Let B_t be the second term on the right-hand side of (6.3.29). Then

$$\left|\frac{uB_t}{2}\right| \le \left|\dot{\beta}^2\delta_t^2 \circ \langle m\rangle_{st}\right|^{1/2} \left|(\beta - u\delta_t\dot{\beta})^2 \circ \langle m\rangle_{st}\right|^{1/2}$$

$$+ \left|\delta_t^2 \circ \dot{Y}^2 * \nu(\theta)_{st}\right|^{1/2} \left\{\left[I\left(u^{-1}\log Y < y\right)\log Y\right.\right.$$

$$\left.\left. - u\delta_t I\left(\dot{Y} < y\delta_t^{-1}\right)\dot{Y}\right]^2 * \nu(\theta)_{st}\right\}^{1/2} \tag{6.3.30}$$

by the Cauchy–Schwartz inequality. The factors in the first term in the summands on the right-hand side of the relation (6.3.30) are bounded by (E1) and the factors in the second term converge to zero in P_θ-probability by (E4). Hence, $B_t \overset{P_\theta}{\to} 0$ and the lemma is proved. ∎

LEMMA 6.3

Suppose the conditions (E0) to (E4) hold. Then

$$A_s^{(t)}(u) \overset{P_\theta}{\to} s\ Q(u) \quad as\ t \to \infty \tag{6.3.31}$$

where $A_s^t(u)$ is as defined by (6.3.14) and $Q(u)$ is given by (6.3.12).

PROOF Let

$$F_t^*(y; s, u) = u^{-2} I\left(u^{-1}\log Y < y\right)\log^2 Y * \nu(\theta)_{st}\,. \tag{6.3.32}$$

Then

$$(Y - 1 - \log Y) * \nu(\theta)_{st} = \int_{-\infty}^{\infty} \left(e^{uy} - 1 - uy\right) y^{-2} dF_t^*(y; s, u) \tag{6.3.33}$$

and hence,

$$A_s^{(t)}(u) = \int_{-\infty}^{\infty} \left(e^{uy} - 1 - uy\right) y^{-2} dF_t(y; s, u) \tag{6.3.34}$$

from the definition of $F_t(y; s, u)$ given in (6.3.26) and (6.3.12). The integrand in the above integral can be assumed to be continuous by redefining the value of the integrand at zero as

$$\lim_{y \to 0} \frac{e^{uy} - 1 - uy}{y^2} .$$

Furthermore, for any $s > 0$ and $y \in R$,

$$F_t(y; s, u) \xrightarrow{P_\theta} s\, K_\theta(y) \text{ as } t \to \infty \tag{6.3.35}$$

by Lemma 6.2. Finally, the family of integrals is stochastically uniformly integrable in the sense that given $\varepsilon > 0$, there exists $0 < \delta < 1$ and $N_{\varepsilon,\delta} \geq 1$ such that for $N > N_{\varepsilon,\delta}$,

$$P\left[\int_{|y|>N} \left|\left(e^{uy} - 1 - uy\right) y^{-2}\right| dF_t(y; s, u) > \varepsilon\right] \leq \delta$$

for all $t > 0$. This can be seen from the relation

$$\int_{|y|>N} \left(e^{uy} - 1 - uy\right) y^{-2} dF_t(y; s, u)$$

$$= I\left(\left|u^{-1} \log Y\right| > N\right) (Y - 1 - \log Y) * \nu(\theta)_{st}$$

$$\leq \left\{I\left(Y > e^{|u|N}\right) Y - I\left(Y < e^{-|u|N}\right) \log Y\right\} * \nu(\theta)_{st}$$

for any $N > 0$ and the last term tends to zero as $N \to 0$ uniformly for $t > 0$ by (E2). Hence,

$$A_s^t(u) = \int_{-\infty}^{\infty} \left(e^{uy} - 1 - uy\right) y^{-2} dF_t(y; s, u) \xrightarrow{P_\theta} s$$

$$\int_{-\infty}^{\infty} \left(e^{uy} - 1 - uy\right) y^{-2} dK_\theta(y)$$

as $t \to \infty$ from the Lemma 6.2 and the stochastic dominated convergence theorem given later in this section. ∎

PROOF (Proof of Theorem 6.5) Lemmas 6.1 and 6.3 prove that

$$R_s^{(t)}(u) = \ell_s^{(t)} - u\Delta_s^{(t)} + sQ(u) \xrightarrow{P_\theta} 0 \text{ as } t \to \infty .$$

Condition (E3) implies that, following Theorem 6.6 for stochastic integrals,

$$\ell_s^{(t)} \xrightarrow{\mathcal{D}} \eta_s \text{ as } t \to \infty$$

where η is a homogeneous process with independent increments with $\eta_0 = 0$ and $\mathcal{L}(\eta | P_\theta) = \mathcal{K}(0, sK_\theta)$. This completes the proof of the theorem. ∎

It is easy to see from Theorem 6.5 that the family of statistical experiments $(\Omega, \mathcal{F}, \{P_\theta, \theta \in \Theta\})$ corresponding to the semimartingales $\{X^\theta, \theta \in \Theta\}$ is local asymptotic infinite divisible (LAID) at θ under the conditions (E0) to (E4). However, it is easy to check that a weaker set of conditions is sufficient for LAID. The following theorem holds. We omit the details.

THEOREM 6.7

Suppose the conditions (E0) to (E2) hold and in addition (E3) and (E4) are satisfied for $s = 1$. Then the family of statistical experiments $(\Omega, \mathcal{F}, \{P_\theta, \theta \in \Theta\})$ corresponding to the semimartingales $\{X^\theta, \theta \in \Theta\}$ is LAID at θ. In particular if the function $K_\theta(\cdot)$ in (E3) is $N(0, 1)$, then the class of experiments is locally asymptotic normal (LAN).

A Stochastic Dominated Convergence Theorem

Let $(\Omega, \mathcal{F}, P)$ be a probability space. Suppose that for each $t \in T$, there exists a nondecreasing random function $F_t(y, \omega)$ on R such that $F_t(-\infty, \omega) = 0$ and $F_t(+\infty, \omega) < \infty$ a.s. A function $g(y)$, $y \in R$ is said to be *stochastically uniformly integrable* with respect to $\{F_t, t \in T\}$ if, for any $\varepsilon > 0$ and $0 < \delta < 1$, there exists $N_{\varepsilon,\delta}$ such that for $N > N_{\varepsilon,\delta}$,

$$P\left\{\int_{|Y|>N} |g(y)|dF_t(y) > \varepsilon\right\} \leq \delta \text{ for all } t \in T .$$

Suppose the function F_0 possesses the same properties as the family F_t, $t \in T$.

THEOREM 6.8

If $g(\cdot)$ is continuous on R and stochastically uniformly integrable with respect to $\{F_t, t \in T\}$ and if

$$F_t(y) \xrightarrow{P} F_0(y), \quad y \in S \cup \{+\infty\}$$

as $t \to t_0$, where S is dense in R, then

$$\int_{-\infty}^{\infty} g(y)dF_t(y) \xrightarrow{P} \int_{-\infty}^{\infty} g(y)dF_0(y) \quad as\ t \to t_0 .$$

Proof is left as an exercise for the reader.

6.4 Local Asymptotic Normality (Infinite Dimensional Parameter Case)

We have seen earlier that the semimartingales are characterized by their triplets of predictable characteristics. As we have noted earlier, there is often a natural way to choose a basic parameter which describes a family of stochastic processes. Examples are the intensity function for a family of counting processes and the drift function for a family of diffusion processes with common diffusion coefficient. Here the local parameterization gives rise to an infinite dimensional version of the local asymptotic normality under some conditions.

Let Θ be an arbitrary parameter space possibly infinite dimensional. For $n \geq 1$, let $\{P_{n,\theta}, \theta \in \Theta\}$ be a family of probability measures on a measurable space $(\Omega_n, \mathcal{F}_n)$. Let us call this *global model.*

Fix $\theta \in \Theta$. A *local model* at θ is introduced in the following way. Let $\boldsymbol{H}$ be a separable Hilbert space with the inner product $(\cdot, \cdot)$ and the norm $\|\cdot\|$. Let $\boldsymbol{V} \subset \boldsymbol{H}$ be a cone called the *local parameter space.* Assume that, for each $v \in \boldsymbol{V}$, there exists a sequence θ_{nv} in Θ such that the *local model*

$$\{P_{n,\theta_{nv}}, v \in \boldsymbol{V}\}$$

has the following LAN property: there exists a process Z_n indexed by $\boldsymbol{V}$ such that, for $v \in \boldsymbol{V}$,

$$\log \frac{dP_{n,\theta_n v}}{dP_{n,\theta}} = Z_n(v) - \frac{1}{2}\|v\|^2 + o_{P_{n,\theta}}(1) \tag{6.4.1}$$

and

$$Z_n(v) \xrightarrow{\mathcal{L}} N\left(0, \|v\|^2\right) \text{ as } n \to \infty \text{ under } \left\{P_{n,\theta}\right\}.$$

(We write $\mathcal{L}(Z_n(v)|P_{n,\theta}) \to N(0, \|v\|^2)$ as $n \to \infty$. Here $\mathcal{L}(X|P)$ denotes the distribution of X under the probability measure P and convergence is in the sense of weak convergence.)

REMARK 6.6 In applications, the sequence θ_{nv} converges to θ at a rate δ_n^{-1} and δ_n is a localizing sequence (normalization) chosen so that $Z_n(v)$ has a nondegenerate limiting distribution. The norm $\|v\|$ represents the effort involved to distinguish between θ and θ_{nv}. The inner product $(\cdot, \cdot)$ in $\boldsymbol{V}$ is called the *acuity.* ▮

DEFINITION 6.4 *Let $k : \Theta \to R$ be a functional defined on Θ. The function $k(\cdot)$ is said to be differentiable at θ with gradient $g \in \boldsymbol{H}$ if there exists a sequence $\{\delta_n\} \to \infty$ such that*

$$\delta_n \left(k\left(\theta_{nv}\right) - k(\theta)\right) \to (v, g) \quad as\ n \to \infty$$

for all $v \in \boldsymbol{V}$.

REMARK 6.7 Since V is a subset of H, the gradient g is not uniquely determined by the values of the linear functional $(\cdot, g)$ on V. If $h - g \perp V$, then $(v, h - g) = 0$ for all $v \in V$, and hence, h is also a gradient. The projection of g into the closed linear span of V is unique and it is called the *canonical gradient.* ▮

DEFINITION 6.5 *An estimator $\hat{k}_n : \Omega_n \to \mathbb{R}$ of $\hat{k}(\theta)$ is called regular on V for $k(\cdot)$ at θ with limit R if*

$$\mathcal{L}\left(\delta_n\left(\hat{k}_n - k\left(\theta_{nv}\right)\right)\Big| P_{n\theta_n v}\right) \Rightarrow R \quad as \quad n \to \infty .$$

The following result is an analogue of the Hajek–Le Cam convolution theorem for the infinite dimensional parameter case.

THEOREM 6.9 **(Convolution theorem)**
Let $k(\cdot)$ be differentiable at θ with gradient $g \in V$. Let $\hat{k}_n$ be an estimator regular on V for $k(\cdot)$ at θ. Then there exists a probability measure S on $\mathbb{R}$ such that

$$\mathcal{L}\left(\left(Z_n(g), \delta_n\left(\hat{k}_n - k(\theta)\right) - Z_n(g)\right)\Big| P_{n\theta}\right) \Rightarrow N\left(0, \|g\|^2\right) \times S$$

where $\mu \times \nu$ denotes the product measure of μ and ν.

PROOF Let $T_n = \delta_n(\hat{k}_n - k(\theta))$. Since $k(\theta)$ is differentiable at θ and $\hat{k}_n$ is regular, it follows that

$$\mathcal{L}\left(T_n - (v, g)\,|P_{n\theta_n v}\right) \Rightarrow R \text{ as } n \to \infty .$$

It is sufficient to show that

$$\mathcal{L}\left(\left(Z_n(g), T_n - Z_n(g)\right) \mid P_{n,\theta}\right) \Rightarrow N\left(0, \|g\|^2\right) \times S \text{ as } n \to \infty .$$

Note that V is a cone. Consider the one-dimensional subset

$$\left\{P_{n,\theta_n, v(r)} : r \geq 0\right\}$$

with $v(r) = \frac{rg}{\|g\|^2}$ of the local model $\{P_{n,\theta_n v}, v \in V\}$.

Note that

$$(v(r), g) = r(g, g)/\|g\|^2 = r$$

and the sequence of distributions of T_n is regular in the sense that it is asymptotically shift equivariant, that is,

$$\mathcal{L}\left(T_n - r \,\middle|\, P_{n,\theta_n v(r)}\right) \Rightarrow R \text{ as } n \to \infty .$$

The main result now follows from the Bickel's proof of Hajek convolution theorem for the one-dimensional case (cf. [5, p. 35, Theorem 2.3]). We omit the details. ∎

REMARK 6.8 Suppose that the estimator $\hat{k}_n$ is regular and efficient in the sense that

$$\mathcal{L}\left(\delta_n\left(\hat{k}_n - k\left(\theta_n v\right)\right) \middle| P_{n,\theta_n v}\right) \Rightarrow N(0, \|g\|^2) \text{ as } n \to \infty$$

for all $v \in V$. Then the probability measure S in Theorem 6.9 is equal to the Dirac measure at zero and Theorem 6.9 implies that

$$\delta_n\left(\hat{k}_n - k(\theta)\right) = Z_n(g) + o_{P_{n,\theta}}(1) .$$

∎

DEFINITION 6.6 *An estimator $\hat{k}_n : \Omega_n \to \mathbb{R}$ of $k(\theta)$ is said to be asymptotically linear for $k(\theta)$ with the influence function $f \in H$*

$$\delta_n\left(\hat{k}_n - k(\theta)\right) = Z_n(f) + o_{P_{n,\theta}}(1) .$$

REMARK 6.9 Since Z_n is linear, it follows, by the Cramér–Wold technique, that

$$\mathcal{L}((Z_n(v), Z_n(f)|P_{n,\theta}) \Rightarrow N\left(\begin{pmatrix}0\\0\end{pmatrix}, \begin{pmatrix}\|v\|^2 & (v, f)\\ (v, f) & \|f\|^2\end{pmatrix}\right) \text{ as } n \to \infty .$$

Hence, by the Le Cam's third lemma (cf. Appendix F; [15, p. 468]; [21]), it follows that

$$\mathcal{L}\left(\delta_n\left(\hat{k}_n - k\left(\theta_{n v}\right)\right)\Big| P_{n,\theta_n v}\right) \Rightarrow N((v,(f-g)), \|f\|^2) \text{ as } n \to \infty\,,$$

and hence, an asymptotically linear estimator is regular if and only if its influence function is a gradient, that is, $f - g \perp V$. This implies that the regular and asymptotically efficient estimators are exactly the same as those that are asymptotically linear with the influence function equal to the canonical gradient. ∎

DEFINITION 6.7 *Let* $k : \Theta \to \mathbb{R}^p$. *The function* $k = (k_1, \ldots, k_p)'$ *is said to be differentiable at* θ *with gradient* $g = (g_1, \ldots, g_p)'$ *if the components* k_j *are differentiable with gradient* g_j *for* $1 \le j \le p$. *Let the components of* g *be canonical. An estimator* $\hat{k}_n = (\hat{k}_{n1}, \ldots, \hat{k}_{np})'$ *is said to be regular for* k *at* θ *with limit* R *if, for* v *in the linear span of* $g_1, \ldots, g_p$,

$$\mathcal{L}\left(\delta_n\left(\hat{k}_n - k\left(\theta_{n v}\right)\right)\Big| P_{n\theta_n v}\right) \Rightarrow R \text{ as } n \to \infty\,.$$

The following result is a multidimensional version of Theorem 6.9.

THEOREM 6.10 **(Convolution theorem)**
Let $k : \Theta \to \mathbb{R}^p$ *be differentiable at* θ *with gradient* $g \in V^p$. *Let* $\hat{k}_n$ *be regular for* k *at* θ *with limit* R. *Then there exists a probability measure* S *such that*

$$\mathcal{L}\left(\left(Z_n(g), \delta_n\left(\hat{k}_n - k(\theta)\right) - Z_n(g)\right)\Big| P_{n\theta}\right)$$

$$\Rightarrow N\left(0, (g, g')\right) \times S$$

where (g, g') *denotes the matrix with the acuity inner product* (g_i, g_j) *as the* (i, j)*-th element.*

PROOF Fix $b \in \mathbb{R}^p$. Define $q(r) = b'r$ for $r \in \mathbb{R}^p$. Then $b'k$ is differentiable at θ with the gradient $b'g$. Furthermore, $b'\hat{k}_n$ is regular for $b'k$ at θ with limit $R \circ q$. By the convolution Theorem 6.9, it follows that there exists a probability measure S_b on $\mathbb{R}$ such that

$$\mathcal{L}\left(\left(b'Z_n(g), \delta_n b'\left(\hat{k}_n - k(\theta)\right) - b'Z_n(g)\right)\right)$$

$$\Rightarrow N\left(0, b'\left(g, g'\right) b\right) \times S_b \text{ as } n \to \infty .$$

Since $\delta_n(\hat{k}_n - k(\theta))$ and $Z_n(g)$ are tight, there exists a probability measure S on $\mathbb{R}^p$ such that for some subsequence

$$\mathcal{L}\left(\delta_n\left(\hat{k}_n - k(\theta)\right) - Z_n(g)\right) \Rightarrow S \text{ as } n \to \infty .$$

Hence,

$$S_b = S \circ q ,$$

and

$$N\left(0, b'\left(g, g'\right) b\right) \times S_b = N\left(0, \left(g, g'\right) \times S \circ q .$$

In particular $R \circ q = N(0, (g, g') \times S \circ q$. Hence, S is independent of the subsequence. The result now follows from the Cramér–Wold technique. ∎

6.5 Multiplicative Models and Asymptotic Variance Bounds

We now consider an example of a LAN family with an infinite dimensional parameter.

For $n \geq 1$, let $\{P_{n\theta} : \theta \in \Theta\}$ be a family of probability measures on a measurable space $(\Omega_n, \mathcal{F}_n)$. Let $\{X_n\}$ be a semimartingale. Suppose that its Doob–Meyer decomposition with respect to $P_{n\theta}$ and the filtration $\{\mathcal{F}_{nt}, t \geq 0\}$ generated by X_n is

$$X_n = M_{n\theta} + A_{n\theta} \tag{6.5.1}$$

where $M_{n\theta}$ is a martingale and $A_{n\theta}$ is a predictable process called the compensator. Examples of such models are as follows.

(i) Let X_n be a jump process. Then the compensator is of the form

$$A_{n\theta}(t) = \int_0^t \int x \; \nu_{n\theta}(ds, dx) \tag{6.5.2}$$

where $\upsilon_{n\theta}$ is the compensator of the random jump measure of X_n with respect to $\{\mathcal{F}_{nt}, t \geq 0\}$ and $P_{n\theta}$. The sequence of models is called *multiplicative* if

$$\nu_{n\theta}(ds, dx) = Y_n(s, x)\theta(ds, dx) . \tag{6.5.3}$$

For a counting process,

$$\theta(ds, dx) = \theta\,(ds, d\varepsilon_1) . \tag{6.5.4}$$

Here the mass is concentrated on the jumps of size 1 and ε_1 is the measure degenerate at 1. For a discrete time parameter process

$$\theta(ds, dx) = \theta\left(\sum_{i=1}^{\infty} d\delta_{t_i}, dx\right) ,$$

that is, the mass is concentrated at the discrete time set $\{t_i, i \geq 1\}$.

(ii) Let X_n be a diffusion process. In this case, the compensator is of the form

$$A_{n\theta}(t) = \int_0^t b_{n\theta}(s) ds \, .$$

Here the model is said to be *multiplicative* if

$$b_{n\theta}(s) = Y_n(s)\theta(s) .$$

Following the remarks made earlier, we now calculate the asymptotic variance bounds in the case of counting processes with multiplicative intensity. Similar analysis can be applied for the diffusion type processes, point processes, and general discrete time processes. For details, see [8].

Counting Processes with Multiplicative Intensity

Let D denote the space of all functions defined on $[0, 1]$ with at most discontinuties of the first kind only endowed with the Skorokhod topology. We call D the Skorokhod space on $[0, 1]$. For every $n \geq 1$, let $(\Omega_n, \mathcal{F}_n)$ be a measurable space and $X_n : \Omega_n \to D$ be a counting process. Let $\{\mathcal{F}_{nt}, 0 \leq t \leq 1\}$ be the filtration generated by X_n. Let Θ be a set of positive functions contained in D. For $\theta \in \Theta$, let $P_{n\theta}$ be a probability measure on $(\Omega_n, \mathcal{F}_n)$ such that X_n has the Doob–Meyer decomposition with respect to the filtration $\{\mathcal{F}_{nt}, 0 \leq t \leq 1\}$ and $P_{n\theta}$ of the form

$$dX_n(s) = dM_{n\theta}(s) + Y_n(s)\theta(s)ds \,. \tag{6.5.5}$$

Let us consider local model corresponding to the global model given above. Fix $\theta \in \Theta$. Suppose there exists a bounded function $y(\cdot) : [0, 1] \to R$ such that

$$\delta_n^{-2} Y_n(s) \to y(s) \text{ uniformly in } s \in [0, 1] \text{ in } P_{n\theta}\text{-probability} \tag{6.5.6}$$

as $\delta_n \to \infty$. Let $L_2(\theta)$ denote the Hilbert space of square integrable functions with respect to the measure $\theta(s)ds$. Let $(\cdot, \cdot)_\theta$ and $\|\cdot\|_\theta$ denote the inner product and the norm in $L_2(\theta)$, respectively. Let $H = L_2(\theta)$ and $V \subset L_2(\theta)$ be a cone such that for $v \in V$ there exists a sequence θ_{nv} in Θ which is Hellinger differentiable at θ with the derivative v, that is,

$$\left\| \delta_n \left(\left(\frac{\theta_{nv}}{\theta} \right)^{1/2} - 1 \right) - \frac{1}{2} v \right\|_\theta \to 0 \text{ as } n \to \infty \,. \tag{6.5.7}$$

THEOREM 6.11
If (6.5.6) and (6.5.7) hold, then

$$\log \left(\frac{dP_{n,\theta_{nv}}}{dP_{n\theta}} \right)_{\mathcal{F}_{n1}} = \delta_n^{-1} \int_0^1 v \, dM_{n\theta} - \frac{1}{2} \|v\|_{y\theta}^2 + o_{P_{n\theta}}(1) \tag{6.5.8}$$

with

$$\delta_n^{-1} \int_0^1 v \, dM_{n\theta} \Rightarrow N\left(0, \|v\|_{y\theta}^2\right) \quad \text{as } n \to \infty \tag{6.5.9}$$

(here $L_2(y\theta)$ denotes the space of square integrable functions with respect to the measure $y(s)\theta(s)ds$ and $(\cdot,\cdot)_{y\theta}$ and $\|\cdot\|_{y\theta}$ the inner product and the norm, respectively, in the space $L_2(y\theta)$).

PROOF For the counting process, the log-likelihood ratio has the form

$$\log\left(\frac{dP_{n,\theta_{nv}}}{dP_{n,\theta}}\right)_{\mathcal{F}_{n1}} = \int_0^1 \log\left(\frac{\theta_{nv}}{\theta}\right) dM_{n\theta} - \int_0^1 (\theta_{nv}-\theta)\, Y_n ds\,. \tag{6.5.10}$$

Note that $\log x = 2\log(1+x^{1/2}-1)$. Hence, by the Taylor's expansion,

$$\log\left(\frac{dP_{n,\theta_{nv}}}{dP_{n,\theta}}\right)_{\mathcal{F}_{n1}} = 2\int_0^1 \left\{\left(\frac{\theta_{nv}}{\theta}\right)^{1/2} - 1\right\} dM_{n\theta}$$

$$-2\int_0^1 \left\{\left(\frac{\theta_{nv}}{\theta}\right)^{1/2} - 1\right\}^2 Y_n\theta ds + R_n \tag{6.5.11}$$

where

$$R_n = -2\int_0^1 \left(\left(\frac{\theta_{nv}}{\theta}\right)^{1/2} - 1\right) I\left\{\left|\left(\frac{\theta_{nv}}{\theta}\right)^{1/2} - 1\right| > \varepsilon\right\} dX_n$$

$$+\int_0^1 \left(\left(\frac{\theta_{nv}}{\theta}\right)^{1/2} - 1\right)^2 I\left\{\left|\left(\frac{\theta_{nv}}{\theta}\right)^{1/2} - 1\right| > \varepsilon\right\} dX_n$$

$$-\int_0^1 \left(\left(\frac{\theta_{nv}}{\theta}\right)^{1/2} - 1\right)^2 I\left\{\left|\left(\frac{\theta_{nv}}{\theta}\right)^{1/2} - 1\right| > \varepsilon\right\} dM_{n\theta}$$

$$+\int_0^1 r\left(\left(\frac{\theta_{nv}}{\theta}\right)^{1/2}-1\right) I\left\{\left|\left(\frac{\theta_{nv}}{\theta}\right)^{1/2}-1\right|\le\varepsilon\right\} dX_n$$

$$-\int_0^1 \left(\left(\frac{\theta_{nv}}{\theta}\right)^{1/2}-1\right)^2 I\left\{\left|\left(\frac{\theta_{nv}}{\theta}\right)^{1/2}-1\right|\le\varepsilon\right\} dM_{n\theta}$$

(6.5.12)

and

$$r(x)=\log(1+x)-x+\frac{x^2}{2}\,.$$

We shall prove later that $R_n = o_{P_{n,\theta}}(1)$. In order to complete the proof, it is sufficient to replace the expression $\left(\frac{\theta_{nv}}{\theta}\right)^{1/2}-1$ by $\frac{1}{2}\delta_n^{-1}v$ and Y_n by $\delta_n^2 y$ in (6.5.11). This is justified by the Hellinger differentiability (6.5.7) and the assumption (6.5.6). Note that $\delta_n^{-1}\int_0^1 v\,dM_{n\theta}$ is asymptotically normal because the limit of the corresponding quadratic variation processes is continuous in t.

Step 1: In order to show that the first three terms in R_n are of the order $o_{P_{n\theta}}(1)$, it is sufficient to check that the probability of a jump in $\left\{\left|\left(\frac{\theta_{nv}}{\theta}\right)^{1/2}-1\right|>\varepsilon\right\}$ is negligible, that is,

$$P_{n\theta}\left\{\int_0^1 I\left\{\left|\left(\frac{\theta_{nv}}{\theta}\right)^{1/2}-1\right|>\varepsilon\right\} dX_n\ge 1\right\}\to 0 \text{ as } n\to\infty\,.$$

Since $\delta_n^{-2}Y_n\to y$ uniformly in probability and y is bounded, it follows that

$$\int_0^1 I\left\{\left|\left(\frac{\theta_{nv}}{\theta}\right)^{1/2}-1\right|>\varepsilon\right\} Y_n\theta\,ds$$

$$\le\varepsilon^{-2}\int_0^1\left(\left(\frac{\theta_{nv}}{\theta}\right)^{1/2}-1\right)^2 I\left\{\left|\left(\frac{\theta_{nv}}{\theta}\right)^{1/2}-1\right|>\varepsilon\right\} Y_n\theta\,ds$$

$$\leq \frac{1}{4}\varepsilon^{-2}\left\{\begin{matrix}\sup y(s)\\ s\in[0,1]\end{matrix}\right\}\int_0^1 v^2 I\left\{\left|\left(\frac{\theta_{nv}}{\theta}\right)^{1/2}-1\right|>\varepsilon\right\}\theta ds + o_{P_{n,\theta}}(1)$$

$$= o_{P_{n,\theta}}(1)\,.$$

From the fact that $d\langle M_{n\theta}\rangle = Y_n\theta ds$ and the Lenglart's inequality, we have

$$\int_0^1 I\left\{\left|\left(\frac{\theta_{nv}}{\theta}\right)^{1/2}-1\right|>\varepsilon\right\}dX_n = o_{P_{n,\theta}}(1)\,.$$

This shows that the first two terms of R_n are $o_{P_{n\theta}}(1)$. The third term in R_n is $o_{P_{n,\theta}}(1)$ since it can be written in the form

$$\int_0^1 \left(\left(\frac{\theta_{nv}}{\theta}\right)^{1/2}-1\right)^2 I\left\{\left|\left(\frac{\theta_{nv}}{\theta}\right)^{1/2}-1\right|>\varepsilon\right\}(dX_n - Y_n\theta ds)\,.$$

Step 2: Choose a sequence $\varepsilon_n \to 0$ so that the first three terms of R_n are $o_{P_{n,\theta}}(1)$. Note that $r(x) \leq 2|x|^3$. Consider

$$\int_0^1 r\left(\left(\frac{\theta_{nv}}{\theta}\right)^{1/2}-1\right) I\left\{\left|\left(\frac{\theta_{nv}}{\theta}\right)^{1/2}-1\right|\leq\varepsilon_n\right\}dX_n$$

$$\leq 2\varepsilon_n\int_0^1\left(\left(\frac{\theta_{nv}}{\theta}\right)^{1/2}-1\right)^2 dX_n$$

$$= o_{P_{n,\theta}}(1)$$

by the Lenglart's inequality.

Step 3: The fifth term in R_n, with ε replaced by ε_n, is of order $o_{P_{n,\theta}}(1)$ because

$$\left\langle \int_0^1 \left(\left(\frac{\theta_{nv}}{\theta}\right)^{1/2} - 1\right)^2 I\left\{\left|\left(\frac{\theta_{nv}}{\theta}\right)^{1/2} - 1\right| \leq \varepsilon_n\right\} dM_{n\theta}\right\rangle$$

$$= \int_0^1 \left(\left(\frac{\theta_{nv}}{\theta}\right)^{1/2} - 1\right)^4 I\left\{\left|\left(\frac{\theta_{nv}}{\theta}\right)^{1/2} - 1\right| \leq \varepsilon_n\right\} Y_n(\theta)ds$$

$$\leq \varepsilon_n^2 \int_0^1 \left(\left(\frac{\theta_{nv}}{\theta}\right)^{1/2} - 1\right)^2 Y_n\theta ds = o_{P_{n,\theta}}(1) .$$

This completes the proof of Theorem 6.11. ∎

REMARK 6.10 As a consequence of Theorem 6.11, we have the LAN of the form (6.4.1) with

$$Z_n(v) = \delta_n^{-1} \int_0^1 v dM_{n\theta} .$$

Note that

$$Z_n(v) \xrightarrow{\mathcal{L}} N\left(o, \|v\|_{y\theta}^2\right) \text{ as } n \to \infty$$

under P_θ. Here the acuity $(\cdot, \cdot)_{y\theta}$ is different from the natural inner product $(\cdot, \cdot)_\theta$ on the local parameter space V. Recall that a functional $k : \Theta \to R$ is differentiable at θ with gradient $g \in L_2(y\theta)$ if

$$\delta_n \left(k\left(\theta_{nv}\right) - k(\theta)\right) \to (v, g)_{y\theta} \text{ as } n \to \infty \text{ for } v \in V .$$

In such a case the convolution theorem 6.4.1 holds and $\|g\|^2 = \|g\|_{y\theta}^2$ is the asymptotic variance bound, that is, the lower bound for the asymptotic variance. ∎

Example 6.1

(i) Let us consider the cumulative intensity function

$$\Lambda(t) = \int_0^t \theta(s)ds$$

where $\theta \in \Theta$ and Θ consists of all the intensity functions. Then the equation (6.5.5) can be written in the form

$$dX_n = dM_{n\theta} + Y_n d\Lambda \ .$$

Suppose that the function $y(s)$ given by (6.5.6) is bounded away from zero. The function $\Lambda(t)$ can be estimated by the Nelson–Aalen estimator $\hat{\Lambda}_n(t)$ defined by

$$d\hat{\Lambda}_n = J_n Y_n^{-1} dX_n$$

with $J_n = I(Y_n > 0)$. If $t_{n1} < \cdots < t_{nn}$ denote the jump times of X_n, then $\hat{\Lambda}_n(t)$ is given by

$$\hat{\Lambda}_n(t) = \sum_{t_{nj} \leq t} \left[Y\left(t_{nj}\right)\right]^{-1} .$$

We will now show that this estimator is asymptotically efficient in the sense that it achieves the asymptotic variance bound. Let us consider a more general function

$$k(\theta) = \int_0^1 h d\Lambda$$

where h is a bounded function. Note that $k(\theta)$ reduces to $\Lambda(t)$ if $h = I_{[0,t]}$. Suppose that $k(\cdot)$ is differentiable at θ in the sense that

$$\delta_n(k(\theta_{nv}) - k(\theta)) \to \int_0^1 vhd\lambda = \left(v, hy^{-1}\right)_{y\theta} .$$

Assume that $V = L_2(\theta)$. Then the gradient $g = hy^{-1}$ is canonical and the asymptotic variance bound for estimating k is

$$\|g\|^2_{y\theta} = \int_0^1 h^2 y^{-1} d\Lambda \,. \tag{6.5.13}$$

One can choose $\hat{k}_n = \int_0^1 hd\hat{\Lambda}_n$ as a possible estimator for $k(\theta) = \int_0^1 hd\Lambda$. Note that

$$\hat{k}_n - \int_0^1 hJ_n d\Lambda = \int_0^1 hJ_n Y_n^{-1} dM_{n\theta}$$

and hence

$$\delta_n\left(\hat{k}_n - k(\theta)\right) = \delta_n\left(\int_0^1 hd\hat{\Lambda}_n - \int hd\Lambda\right)$$

$$= \delta_n^{-1}\int_0^1 hy^{-1} dM_{n\theta} + o_{P_{n\theta}}(1)\,.$$

This proves that $\hat{k}_n$ is asymptotically linear with the influence function equal to the canonical gradient. Hence, $\hat{k}_n$ is asymptotically efficient.

If we take $h(s) = I_{[0,t]}(s)$, the functional $k(\theta)$ reduces to $s(t)$ and the estimator $\hat{k}_n$ to $\hat{\Lambda}_n(t)$. Hence, the Nelson–Aalen estimator is asymptotically efficient. (ii) Suppose Θ consists of all the intensity functions with $\Lambda(1) = 1$. The condition (6.5.7) restricts V to be given by

$$V = L_{2,0}(\theta) = \left\{v \in L_2(\theta) : \int_0^1 vd\Lambda = 0\right\}\,.$$

We have seen in (i) that the functional $k(\theta) = \int_0^1 hd\Lambda$ has the gradient $g = hy^{-1}$. The canonical gradient g_0 is obtained by projecting g into $L_{20}(\theta)$ with respect to the acuity $(\cdot,\cdot)_{y\theta}$. Note that

$$g_0 = \left(h - \int_0^1 h y^{-1} d\Lambda \Big/ \int_0^1 y^{-1} d\Lambda \right) y^{-1} .$$

Hence, the asymptotic variance bound for estimating k is

$$\|g_0\|^2_{y\theta} = \int_0^1 h^2 y^{-1} d\Lambda - \left(\int_0^1 h y^{-1} d\Lambda \right)^2 / \int_0^1 y^{-1} d\Lambda .$$

Note that this bound is smaller than the bound $\int_0^1 h^2 y^{-1} d\Lambda$ for the full family of all intensity functions. One can choose

$$k_n' = \int_0^1 h d\hat{\Lambda}_n / \hat{\Lambda}_n(1)$$

as an estimator for $k(\theta)$. Note that

$$\begin{aligned} \delta_n \left(k_n' - k(\theta) \right) &= \delta_n \left(\int_0^1 h d\hat{\Lambda}_n - \int_0^1 h d\Lambda \right) - \delta_n \left(\hat{\Lambda}_n(1) - 1 \right) \int_0^1 h d\hat{\Lambda}_n \\ &\quad + o_{P_{n,\theta}}(1) \\ &= \delta_n^{-1} \int_0^1 g_0 dM_n + o_{P_{n,\theta}}(1) . \end{aligned}$$

Therefore k_n' is asymptotically linear with the influence function equal to the canonical gradient. We leave it to the reader to check that

$$k_n'' = \left(2 - \hat{\Lambda}_n(1) \right) \int_0^1 h d\hat{\Lambda}_n$$

is an asymptotically efficient estimator for k.

(iii) Suppose Θ consists of all constant intensity functions. Then V also consists of all constant functions. Consider the functional

$$k(\theta) = \int_0^1 h d\Lambda$$

as discussed in (i). The canonical gradient g_0 is obtained by projecting $g = hy^{-1}$ into V with respect to $(\cdot, \cdot)_{y\theta}$. Hence,

$$g_0 = \frac{\int_0^1 h d\Lambda}{\int_0^1 y d\Lambda} = \frac{\int_0^1 h ds}{\int_0^1 y ds} .$$

Note that $\Lambda(t) = ct$ for some constant c. Hence, the asymptotic variance bound for estimating k is

$$\|g_0\|^2_{y\theta} = g_0^2 \int_0^1 y d\Lambda = \theta \left(\int_0^1 hds \right)^2 / \int_0^1 yds . \tag{6.5.14}$$

Here we have interpreted the intensity function $\theta(t)$ as a constant θ. By the Cauchy–Schwartz inequality, the bound in (6.5.14) is smaller than the bound obtained in (i) for the family of *all* intensity functions, namely,

$$\int_0^1 h^2 y^{-1} d\Lambda = \theta \int_0^1 h^2 y^{-1} ds . \tag{6.5.15}$$

One can choose

$$k_n''' = \hat{\Lambda}_n(1) \int_0^1 hds$$

as an estimator for $k(\theta)$ and check that k_n''' is asymptotically linear with the influence function

$$f = y^{-1}\int_0^1 h\,ds\,.$$

Hence its asymptotic variance is

$$\|f\|_{y\theta}^2 = \theta\left(\int_0^1 h\,ds\right)^2\left\{\int_0^1 y^{-1}ds\right\}.$$

By the Cauchy–Schwartz inequality, this bound is smaller than the bound (6.5.15) for the full family of intensity functions but larger than the bound (6.5.14) for the present model. Hence, k_n''' is not efficient. However, k_n''' is regular since its influence function f is a gradient because

$$(f - g, v)_{y\theta} = \theta v\left(\int_0^1 h\,ds - \int_0^1 y g_0\,ds\right) \quad \text{for } v \in V\,.$$

It can be shown that

$$k_n^* = \hat{\theta}_n \int_0^1 h\,ds$$

is asymptotically efficient where $\hat{\theta}_n$ is the maximum likelihood estimator of θ. This follows from the fact that, under some regularity conditions,

$$\mathcal{L}\left(\delta_n\left(\hat{\theta}_n - \theta\right)\middle| P_{n\theta}\right) \Rightarrow N\left(0, \frac{\theta}{\int_0^1 y\,ds}\right) \quad \text{as } n \to \infty\,.$$

(iv) **Right Censoring:**

Let $X_i \geq 0,\ 1 \leq i \leq n$ be i.i.d. life times with the density f and the distribution function F satisfying $F(1) < 1$. Let

$$\theta(s) = \frac{f(s)}{1 - F(s)}$$

be the hazard rate and

$$\Lambda(t) = \int_0^t \theta(s)ds$$

be the cumulative hazard rate. Let $Y_i \geq 0$, $1 \leq i \leq n$ be i.i.d. censoring times with a distribution function G satisfying $G(1) < 1$. The observations are $Z_i = \min(X_i, Y_i), \delta_i = I(X_i \leq Y_i)$, $1 \leq i \leq n$. Let H be the distribution function of Z_i. Then $(1-H) = (1-F)(1-G)$. Define $X_n(t) = \sum_{i=1}^{n} I(Z_i \leq t)$ and

$$X_n^{(1)}(t) = \sum_{i=1}^{n} \delta_i I(Z_i \leq t) .$$

The process $X_n^{(1)}$ has the Doob–Meyer decomposition given by

$$dX_n^1(s) = dM_{n\theta}(s) + (n - X_n(s-))\theta(s)ds .$$

Here, $Y_n(s) = n - X_n(s-)$ in the form (6.5.5). This is a multiplicative intensity model with hazard rate $\theta(\cdot)$ as the natural parameter. Note that

$$n^{-1}Y_n = n^{-1}(n - X_{n-}) \to 1 - H = y \text{ as } n \to \infty$$

where X_{n-} denotes the process $\{X_n(s-)\}$. Here, the acuity is

$$(\cdot, \cdot)_{y\theta} = (\cdot, \cdot)_{(1-H)\theta} .$$

Let $\mathcal{F}$ be the family of all absolutely continuous distribution functions F on $[0, \infty)$. Suppose we want to estimate $F(t)$ for $t \leq 1$. This can be considered as a functional on Θ, namely,

$$F(t) = 1 - \exp(-\Lambda(t)) .$$

Let the distribution function F have the hazard rate θ. Choose (θ_{nv}) Hellinger differentiable at θ with derivative v satisfying (6.5.7). Let F_{nv} be the corresponding distribution function. Then

$$n^{1/2}(F_{nv}(t) - F(t)) \to \exp(-\Lambda(t)) \int_0^t y d\Lambda$$

$$= (1 - F(t))\left(v, (1-H)^{-1}I[0,t]\right)_{1-H)\theta} .$$

Hence, the canonical gradient is

$$g(s) = (1 - F(t))(1 - H(s))^{-1}I_{[0,t]}(s) ,$$

and the asymptotic variance bound for estimating $F(t)$ is

$$\|g\|^2_{(1-H)\theta} = (1 - F(t))^2 \int_0^t (1-H)^{-1}(1-F)^{-1}dF$$

$$= (1 - F(t))^2 \int_0^t (1-F)^{-2}(1-G)^{-1}dF .$$

This shows that the Kaplan–Meier estimator (cf. [7]) for F is asymptotically efficient. Since the Kaplan–Meier estimator is stochastically equivalent to $1 - \exp(-\hat{\Lambda}_n(t))$ where $\hat{\Lambda}_n$ is the Nelson estimator, it also follows directly that the Kaplan–Meier estimator is asymptotically efficient. Note that Nelson estimator is a special case of the Nelson–Aalen estimator and

$$\hat{\Lambda}_n(t) = \sum_{j=1}^{X_n(t)} (n - j + 1)^{-1} .$$

The asymptotic variance bound is not affected by the restrictions on the family of censoring distributions. □

6.6 Exercises

Problem 1: (*Counting process*) Let X be a counting process with $X_0 = 0$ and a predictable intensity $\lambda(\theta)$ under a probability measure P_θ. Suppose the $\lambda(\theta) = \sum_{i=1}^k \theta_i\lambda_i$ where $\theta_i > 0$ and λ_i predictable R_+-valued process for

$1 \leq i \leq k$ so that $\lambda(\theta) > 0$ for all $\theta = (\theta_1, \cdots, \theta_k)$. Study the LAMN for the family under suitable conditions on the score function [17].

Problem 2: (*Two-dimensional Gaussian diffusion process*) The instantaneous axis of the earth's rotation is displaced with repect to the minor axis of the terrestrial ellipsoid. This displacement consists of a periodic part and a fluctuation part. The following process has been used as a model of the fluctuation part (cf. [1, 2]). Under P_θ, let $\boldsymbol{X} = (X_1, X_2)^T$ be the unique strong solution of the stochastic differential equation

$$d\boldsymbol{X}_t = \boldsymbol{A}(\theta)\boldsymbol{X}_t\, dt + d\boldsymbol{W}_t, \quad \boldsymbol{X}_0 = 0$$

where

$$A(\theta) = \begin{pmatrix} -\theta_1 & -\theta_2 \\ \theta_2 & -\theta_1 \end{pmatrix},$$

$\boldsymbol{W}_t$ is a 2-dimensional standard Wiener process and $\Theta = R^2$. Show that

$$\boldsymbol{X}_t = e^{\boldsymbol{A}(\theta)t} \int_0^t e^{-\boldsymbol{A}(\theta)s} d\boldsymbol{W}_s$$

where

$$e^{\boldsymbol{A}(\theta)t} = e^{-\theta_1 t} \begin{pmatrix} \cos\theta_2 t & -\sin\theta_2 t \\ \sin\theta_2 t & \cos\theta_2 t \end{pmatrix}.$$

Show that the family is LAMN at every θ with $\theta_1 < 0$, LAN at every θ with $\theta_1 > 0$ and that a singularity occurs for $\theta_1 = 0$ (cf. [6, 18]). Further prove that the maximum likelihood estimator (MLE) is regular and asymptotically efficient at every θ with $\theta_1 \neq 0$. Obtain the asymptotic distribution of the MLE under suitable norming for $\theta_1 \neq 0$ [17, 18].

Problem 3: (*Ornstein–Uhlenbeck process*) Let X_t be the unique solution of the stochastic differential equation

$$dX_t = -\theta A(t) X_t dt + dW_t, \quad X_0 = 0$$

under the probability measure P_θ where $A : R_+ \to R$ with

$$\int_0^t A(s)^2 ds < \infty, \quad t \geq 0.$$

Let $\Theta_0 = \{\theta : -\theta \int_0^t A(s)ds \to \infty\}$. Suppose that Θ_0 is nonempty. Show that the family is LAMN at every $\theta \in \Theta_0$. Show that the score function is given by

$$\dot{\ell}_t(\theta) = -\int_0^t A(s)X_s dW_s \,.$$

Let

$$f(\theta, t) = \exp\left(-\theta \int_0^t A(s)ds\right)$$

and Θ_1 be the set of $\theta \notin \Theta_0$ with $E_\theta \langle \dot{\ell}(\theta) \rangle_t \to \infty$ such that

$$A(t)^2 f(\theta, t)^4 \left(\int_0^t f(\theta, s)^{-2} ds\right)^2 = o\left(E_\theta \langle \dot{\ell}(\theta) \rangle_t\right) \,.$$

Further, suppose that A is continuous. Show that the family is LAN at every $\theta \in \Theta_1$ under the conditions stated above. Find the maximum likelihood estimator of θ and obtain its asymptotic distribution [17].

Problem 4: (*Diffusion with jumps*) Let $\{X_t, \quad t \geq 0\}$ be the unique solution of the stochastic differential equation

$$dX_t = (-\theta_1 X_t + \theta_2)\, dt + dN_t(\theta), X_0 = 0, \quad t \geq 0$$

where $\theta = (\theta_1, \theta_2, \theta_3) \in \Theta = R^2 \times \Theta_3$ with $\Theta_3 \subset R^{k-2}$ open and $\{N_t(\theta), \ t \geq 0\}$ is a process with independent stationary increaments with $N_0(\theta) = 0$ (cf. [4, 19]). Suppose the jump characteristic of the process X does not depend on θ_1 and θ_2. Under some conditions, show that the family is LAMN at every θ with $\theta_1 < 0$ provided a suitable jump condition is satisfied and it is LAN at every θ with $\theta_1 > 0$. If $\theta_1 = 0$ and $a(\theta) = \theta_2 + E_\theta N_1(\theta) \neq 0$, then prove that the family is LAN. Derive the maximum likelihood estimators for θ_1, θ_2 and θ_3 and study their asymptotic properties [17].

Problem 5: (*Continuation of Problem* 2) Show that the family of measures is LAQ for every θ with $\theta_1 = 0$ [18].

Problem 6: (*Continuation of Problem* 4) Show that the family of measures is LAQ for every $\theta \in \Theta$ with $\theta_1 = 0$ and $a(\theta) = \theta_2 + E_\theta N_1(\theta) = 0$ [18].

Problem 7: (*Point process*) Let $X = \{X_t, t \geq 0\}$ be a univariate point process with $X_0 = 0$ and a predictable intensity $\lambda(\theta) > 0$. Then the log-likelihood ratio process $\ell_t(\theta, \tau)$ is given by

$$\ell_t(\theta, \tau) = \int_0^t \log \frac{\lambda(\tau, s)}{\lambda(\theta, s)} dX_s - \int_0^t (\lambda(\tau, s) - \lambda(\theta, s)) ds .$$

(i) (*Self-Correcting model*) Suppose that

$$\lambda(\theta, t) = \exp(\theta_1 + \theta_2(t - X_{t-})), \quad \theta = (\theta_1, \theta_2) \in R^2$$

(cf. [9, 20]). Show that the family of measures is LAN at $\theta = (\theta_1, 0), \theta_1 \neq 0$ and LAQ at $\theta = (0, 0)$.

(ii) Suppose that

$$\lambda(\theta, t) = \theta_1 I_{(-\infty,0]}(t - X_{t-}) + \theta_2 I_{(0,\infty)}(t - X_{t-}) ,$$

$\theta = (\theta_1, \theta_2) \in (0, 1) \times (1, \infty) \cup \{(1, 1)\}$ (cf. [10]). Show that the family is LAN for $\theta \in (0, 1) \times (1, \infty)$ and is LAQ at $\theta = (1, 1)$ [10, 18].

Problem 8: Consider the stochastic differential equation

$$dX_t = \frac{\theta X_t}{1 + t} dt + dW_t, \quad X_0, \quad t \geq 0, \quad \theta \in R .$$

Show that the family of measures is LAN for every $\theta < 1/2$ is LAMN for $\theta > 1/2$ and is LAQ for $\theta = 1/2$ [17, 18].

Problem 9: (*Diffusion type processes*) Consider the stochastic differential equation

$$dX_t = a_t(X, \theta)dt + b_t(X)dW_t, \quad t \geq 0$$

where X_0 is a random variable whose distribution does not depend on the parameter θ. Show that, under suitable conditions, on the functions $a_t(.,.)$ and $b_t(.,)$, the family of measures is LAMN for every θ [2, 13, 23].

Problem 10: Let $\{X_t, \quad t \geq 0\}$ be a Markov process satisfying the equation

$$X_t = X_0 + \int_0^t a_s(X_s; \theta) ds + \int_0^t b_s(X_s) dW_s$$

$$+ \int_0^t \int_{|x|>1} x d\mu + \int_0^t \int_{|x|\leq 1} x d(\mu - \nu(\theta))$$

where X_0 is a random variable with distribution not depending on θ, $a_s(x, \theta)$ and $b_s(x)$ are known functions, $\mu(ds, dx)$ is the jump measure of the process X with the compensator

$$\nu(\theta)([0, t], \Gamma) = \int_0^t Q(s, X_s, \Gamma; \theta)\, ds .$$

Here Q is a function on $R_+ \times R \times \mathcal{B} \times \Theta$, $\mathcal{B}$ being the Borel σ-algebra of subsets of $E = R - \{0\}$. Suppose that the measures $P_t^{\theta i}$, $i = 1, 2$ corresponding to the processes X on $[0, t]$ are mutually absolutely continuous. Let Y and β be the solutions of the equations

$$Q(t, x, \Gamma; \theta_2) = \int_\Gamma Y(t, x, v; \theta_1, \theta_2)\, Q(t, x, dv; \theta_1)$$

for any $t \geq 0$, $x \in R$ and $\Gamma \in \mathcal{B}$ and

$$a_t(x, \theta_2) - a_t(x, \theta_1) = b_t^2(x) \beta(t, x; \theta_1, \theta_2)$$

$$+ \int_{|v|\geq 1} v\,(Y(t, x, v; \theta_1, \theta_2) - 1)\, Q(t, x, dv; \theta_1) .$$

Under suitable conditions, show that the family of measures is LAID at θ [23].

Problem 11: (*Process with independent increments*) Consider the family $\{X_t,\ t > 0\}$ of centered processes with independent increments without continuous component and with jump measures compensator

$$\nu(\theta)(ds, dx) = 4s\left(s^2 + s + \frac{1}{2}\right)^{-1} \theta^{-x} e^{2s-x} I(x > s) ds\ dx$$

where $\theta > 0$. Show that the family of measures is LAID at every $\theta > 0$ [23].

References

[1] Arato, M. (1982) *Linear Stochastic Systems with Constant Coefficients: A Statistical Approach,* Lecture Notes in Control and Information Sciences, **45**, Springer, Berlin.

[2] Basawa, I.V. and Prakasa Rao, B.L.S. (1980) *Statistical Inference for Stochastic Processes,* Academic Press, New York.

[3] Basawa, I.V. and Scott, D.J. (1983) *Asymptotic Optimal Inference for Non-Ergodic Models,* Lecture Notes in Statistics, **17**, Springer, Heidelberg.

[4] Cinlar, E. and Pinsky, M. (1971) A stochastic integral in storage theory, *Z. Warsch. verw Gebiete,* **17**, 227–240.

[5] Droste, W. and Wefelmeyer, W. (1984) On Hajek's convolution theorem, *Statistics and Decisions,* **2**, 131–144.

[6] Feigin, P. (1979) Some comments on a curious singularity, *J. Appl. Prob.,* **16**, 440–444.

[7] Fleming, T.R. and Harrington, D.P. (1991) *Counting Processes and Survival Analysis,* Wiley, New York.

[8] Greenwood, P.E. and Wefelmeyer, W. (1989) Efficient bounds for estimating functionals of stochastic processes, *Tech. Report,* 117, University of Cologne.

[9] Hayashi, T. (1988) Local asymptotic normality in self-correcting point processes, *Statistical Theory and Data Analysis* (Ed. K. Matusita), North Holland, Amsterdam, 551–558.

[10] Inagaki, N. and Hayashi, T. (1990) Parameter estimation for the simple self-correcting process, *Ann. Inst. Statist. Math.,* **42**, 89–98.

[11] Jacod, J. and Shiryayev, A.N. (1987) *Limit Theorems for Stochastic Processes,* Springer, Heidelberg.

[12] Jeganathan, P. (1982) On the asymptotic theory of estimation when the limit of the loglikelihood ratio is mixed normal, *Sankhya Ser. A,* **44**, 173–212.

[13] Kutoyants, Y.A. (1984) *Parameter Estimation for Stochastic Processes,* (Trans. and Ed. B.L.S. Prakasa Rao), Heldermann, Berlin.

[14] Le Cam, L. (1960) Locally asymptotically normal families of distributions, *Univ. California Publ. Statist.*, **3**, 37–98.

[15] Le Cam, L. (1986) *Asymptotic Methods in Statistical Decision Theory,* Springer, New York.

[16] Liptser, R.S. and Shiryayev, A.N. (1989) *The Theory of Martingales,* Kluwer, Dordrecht.

[17] Luschgy, H. (1992) Local asymptotic mixed normality for semimartingale experiments, *Probab. Theory Relat. Fields,* **92**, 151–176.

[18] Luschgy, H. (1994) Asymptotic inference for semimartingale models with singular parameter points, *J. Stat. Plan. Inf.*, **39**, 156–186.

[19] Moran, P.A.P. (1969) A theory of dams with continuous input and a general release rule, *J. Appl. Probab.*, **6**, 88–98.

[20] Ogata, Y. and Vere-Jones, D. (1984) Inference for earthquake models: a self-correcting model, *Stoch. Proc. Appl.*, **17**, 337–347.

[21] Prakasa Rao, B.L.S. (1987) *Asymptotic Theory of Statistical Inference,* Wiley, New York.

[22] Taraskin, A.F. (1984) Central limit theorem for stochastic integrals, *Teor. Sluch. Proc.*, **12**, 81–90 (in Russian).

[23] Taraskin, A.F. (1984) On the behavior of the likelihood ratio of semimartingales, *Theory. Probab. Appl.*, **29**, 452–464.

Chapter 7

Likelihood and Asymptotic Efficiency

7.1 Fully Specified Likelihood (Factorizable Models)

Let $(\Omega, \mathcal{F})$ be a measurable space and $\boldsymbol{X} = (X_1, X_2)$ be an RCLL process on the interval $[0, \infty)$. Let $\mathcal{F} = \{\mathcal{F}_s, s \geq 0\}$ be the filtration generated by $\boldsymbol{X}$. For simplicity, assume that $\mathcal{F}_0 = \phi$ and suppose that X_1 and X_2 are real-valued processes.

We now recall some notation. Let

$$\mu(\omega; dt, dx) = \sum_s I\left(\Delta \boldsymbol{X}_s(\omega) \neq 0\right) \varepsilon_{(s, \Delta \boldsymbol{X}_s(\omega))}(dt, dx) ,$$

where ε_a denotes the Dirac measure at the point a. This is the (random) jump measure of the process $\boldsymbol{X}$. Let us write

$$(X \bullet Y)_t(\omega) = \int_0^t X_s(\omega) dY_s(\omega) ,$$

and

$$(W * \mu)_t(\omega) = \int \int_0^t W(\omega; s, x) \mu(\omega; ds, dx) .$$

Let $\mathcal{P}$ be the predictable σ-algebra on $\Omega \times [0, \infty)$ and $\{P_\theta, \theta \in \Theta\}$ be a family of probability measures on $(\Omega, \mathcal{F})$. Fix $\theta \in \Theta$, let $\boldsymbol{X}$ be a semimartingale with the triplet of predictable local characteristics $T_\theta = (\boldsymbol{B}_\theta, \boldsymbol{C}_\theta, \nu_\theta)$ with respect to

P_θ and $\mathcal{F}$. Here $\boldsymbol{B}_\theta$ is the drift characteristic, $\boldsymbol{C}_\theta$ is the quadratic characteristic and ν_θ is the compensator of the jump measure μ_θ.

Fix $\theta \in \Theta$ and select a continuous adapted increasing process F and a predictable process $\boldsymbol{c}$ with values in the set of 2×2 symmetric matrices such that

$$C_{ij\theta} = c_{ij} \circ F .$$

Let $\tau \in \Theta$. Suppose that

$$P_\tau|\mathcal{F} \text{ and } P_\theta|\mathcal{F}_s \text{ are mutually absolutely continuous for } s \geq 0 . \tag{7.1.1}$$

We further assume that

$$\nu_\tau\left(\{s\} \times R^2\right) = \nu_\theta\left(\{s\} \times R^2\right), s \geq 0 . \tag{7.1.2}$$

In other words, we are avoiding the case when there are different fixed jump times for different parameters. This will allow the case of no common jumps as well as the discrete time parameter case. We further assume that

$$\nu_\delta\left([0, s], R^2\right) < \infty, \ s \geq 0 . \tag{7.1.3}$$

By the Girsanov's theorem (cf. [13, p. 159, Theorem 3.24]), the following relations hold P_τ-a.s. and hence also P_θ-a.s.

$$\boldsymbol{C}_\tau = \boldsymbol{C}_\theta = \boldsymbol{C} , \tag{7.1.4}$$

there exists a positive $\mathcal{P} \times R^2$-measurable process Y such that

$$\nu_\tau = Y \star \nu_\theta , \tag{7.1.5}$$

and there exists a two-dimensional predictable process $\boldsymbol{t}$ such that

$$B_{i\tau} = B_{i\theta} + (c_{i1}\beta_1 + c_{i2}\beta_2) \circ F , \tag{7.1.6}$$

$$\left(\boldsymbol{\beta c \beta}^T\right) \circ F_s < \infty, s \geq 0 , \tag{7.1.7}$$

and

$$\left(Y^{1/2} - 1\right)^2 * \nu_{\theta s} < \infty, s \geq 0 . \tag{7.1.8}$$

The log-likelihood ratio process $\ell_{\theta\tau}$ between τ and θ restricted to $\mathcal{F}_t$ is defined by

$$\ell_{\theta\tau}(t) = \log\left(\frac{dP_\tau|\mathcal{F}_t}{dP_\theta|\mathcal{F}_t}\right), t \geq 0 . \tag{7.1.9}$$

Let μ and $\boldsymbol{X}_\theta^{(c)}$ denote the jump measure and the continuous martingale part of $\boldsymbol{X}$ with respect to P_θ, respectively. Then the representation theorem for the likelihood ratio process gives

$$\begin{aligned}\ell_{\theta\tau} &= \boldsymbol{\beta} \bullet \boldsymbol{X}_\theta^{(c)} - \frac{1}{2}\left(\boldsymbol{\beta} c \boldsymbol{\beta}^T\right) \circ F \\ &\quad + \sum_{s \leq \cdot} \log\left(1 + \Delta(Y-1) * (\mu - \nu_\theta)_s\right) \\ &\quad + ((Y-1) * (\mu - \nu_\theta))^{(c)} .\end{aligned} \tag{7.1.10}$$

If ν_δ has a density in t, then

$$\begin{aligned}\ell_{\theta\tau} &= \boldsymbol{\beta} \bullet \boldsymbol{X}_\theta^{(c)} - \frac{1}{2}\left(\boldsymbol{\beta} c \boldsymbol{\beta}^T\right) \circ F \\ &\quad + (\log Y) * \mu - (Y-1) * \nu_\theta .\end{aligned} \tag{7.1.11}$$

Factorization of the Likelihood

Fix $\theta, \tau \in \Theta$. Assume that for $\delta \in \{\theta, \tau\}$, the compensator ν_δ of the jump measure of $\boldsymbol{X}$ has a density in the time variable and that the processes X_1 and X_2 have no common jumps. Then

$$\begin{aligned}\nu_\delta(ds, dx_1, dx_2) &= y_{1\delta}(s, x_1)\, \varepsilon_0(dx_2)\, ds dx_1 \\ &\quad + y_{2\delta}(s, x_2)\, \varepsilon_0(dx_1) ds dx_2\end{aligned} \tag{7.1.12}$$

and

$$\begin{aligned}Y(s, x_1, x_2) &= y_{1\theta}(s, x_1)^{-1}\, y_{1\tau}(s, x_1)\, I\{x_1 \neq 0\} \\ &\quad + y_{2\theta}(s, x_2)^{-1}\, y_{2\tau}(s, x_2)\, I\{x_2 \neq 0\} .\end{aligned} \tag{7.1.13}$$

Suppose that the quadratic characteristic $C_\theta = c \circ F$ has values in the set of the *diagonal* nonnegative matrices. Then the likelihood process (7.1.11) factors and

$$\ell_{\theta\tau} = \ell_{\theta\tau}^{(1)} + \ell_{\theta\tau}^{(2)} \tag{7.1.14}$$

with

$$\ell_{\theta\tau}^{(i)} = \beta_i \bullet X_{i\theta}^{(c)} - \frac{1}{2}\left(c_{ii}\beta_i^2\right) \circ F$$

$$+ \log\left(y_{i\theta}^{-1} y_{i\tau}\right) * \mu^i - \left(y_{i\theta}^{-1} y_{i\tau} - 1\right) * \nu_{i\theta} \tag{7.1.15}$$

and

$$\nu_{i\delta}(ds, dx_i) = y_{i\delta}(s, x_i)\, ds dx_i, \; i = 1, 2 \,. \tag{7.1.16}$$

Here X_i has the triplet of local characteristic $T_\delta^{(i)} = (B_{i\delta}, C_{i\delta}, \nu_{i\delta})$ and the triplet $T_\delta = (\boldsymbol{B}_\delta, \boldsymbol{C}_\delta, \nu_\delta)$ of $\boldsymbol{X}$ is determined by $T_\delta^{(1)}$ and $T_\delta^{(2)}$ with

$$\boldsymbol{B}_\delta = (B_{1\delta}, B_{2\delta}) \;; \tag{7.1.17}$$

$$\boldsymbol{C} = \operatorname{diag}(C_{ii}) \;; \tag{7.1.18}$$

$$\nu_\delta(ds, dx_1, dx_2) = \nu_{1\delta}(ds, dx_1)\, \varepsilon_o(dx_2)$$

$$+ \nu_{2\delta}(ds, dx_2)\, \varepsilon_o(dx_1) \,. \tag{7.1.19}$$

Hence, the likelihood $\exp(\ell_{\theta\tau})$ factors into the product of $\exp(\ell_{\theta\tau}^{(i)})$, $i = 1, 2$, and $\exp(\ell_{\theta\tau}^{(i)})$ is expressed in terms of $T_\theta^{(i)}$ only for $i = 1, 2$.

Local Asymptotic Normality

For each $\theta \in \Theta$, let the process $\boldsymbol{X} = (X_1, X_2)$ be a semimartingale with triplet $T_\theta = (\boldsymbol{B}_\theta, \boldsymbol{C}_\theta, \nu_\theta)$. We will now find conditions for the local asymptotic normality for the model and then obtain the asymptotic variance bound for estimators of functionals of the parameter θ.

Fix $\theta \in \Theta$. Let F be a continuous adapted increasing process and $\boldsymbol{C}$ a predictable process with values in the set of 2×2 nonnegative diagonal matrices such that

$$C_{ij}(\theta) = c_{ij} \circ F \,. \tag{7.1.20}$$

Let $b_{i\theta}$ be a predictable process such that

$$B_{i\theta} = (c_{ii} b_{i\theta}) \circ F \,. \tag{7.1.21}$$

Let ν_θ^i denote the compensator of the jump measure of X_i with respect to P_θ and $\mathcal{F}$, the filtration generated by $\boldsymbol{X} = (X_1, X_2)$.

We have seen earlier that the representation of the likelihood process can be given in terms of the triplet of predictable characteristics. This triplet can be replaced by a pair of drift densities $b = (b_1, b_2)'$ and the two-dimensional jump density y for factorizable models under the assumption stated earlier. Here $(\alpha_1, \alpha_2)'$ denotes transpose of (α_1, α_2). Under the condition (7.1.12), the jump density y is determined by a vector (y_1, y_2) of the one-dimensional jump densities. In order to investigate the local asymptotic normality of the model, we fix a pair (b, y) and choose paths $b_{itv_i}, t \geq 0$ tending to b_i and $y_{itw_i}, t \geq 0$ tending to y_i at a rate $r_t \to \infty$, that is,

$$r_t \left(b_{itv_i} - b_i\right) \to v_i, r_t \left(y_i^{-1/2} y_{itw_i}^{1/2} - 1\right) \to \frac{1}{2} w_i \text{ as } t \to \infty \,. \tag{7.1.22}$$

Let $v = (v_1, v_2)'$ and $w = (w_1, w_2)'$. The *local parameters* (v, w) indicate the possible directions in which the natural parameter (b, y) can extend in the model. The rate is chosen so that the log likelihood between (b_{tv}, y_{tw}) and (b, y) admits a stochastic expansion as $t \to \infty$ with a nondegenerate limit distribution.

For $i = 1, 2$, let V_i denote a cone of predictable processes $v : \Omega \times [0, \infty) \to R$ such that, for each pair $v, \bar{v} \in V_i$, there exists a real number $(v, \bar{v})_{ci}$ such that

$$r_t^{-2} (v c_{ii} \bar{v}) \circ F_t = (v, \bar{v})_{ci} + o_{P_\theta}(1) \,. \tag{7.1.23}$$

For $i = 1, 2$, let W_i denote a case of $\mathcal{P} \times R$-measurable processes $w : \Omega \times [0, \infty) \times R \to R$ such that, for each pair $w, \bar{w} \in w_i$, there exists a real number $(w, \bar{w})_i$ such that

$$r_t^{-2} (w \bar{w}) * \nu_{\theta t}^i = (w, \bar{w})_i + o_{P_\theta}(1) \,. \tag{7.1.24}$$

Let $\|v\|_{ci}^2 = (v, v)_{ci}$ and $\|w\|_i^2 = (w, w)_i$. We now introduce a *local model.* Let U be a cone. Assume that, for $u \in U$, there exists $\theta_{tu} \in \Theta, t \geq 0$ with the following properties:

(R1) (i) $P_{\theta_{tu}}|\mathcal{F}_s$ and $P_\theta|\mathcal{F}_s$ are mutually absolutely

$$\text{continuous}\ , s \geq 0\ , \tag{7.1.25}$$

(ii) $$\nu_{\theta_{tu}}([0,s], R \times R) < \infty,\ s \geq 0\ , \tag{7.1.26}$$

(iii) $$\nu_{\theta_{tu}}(ds, dx_1, dx_2) = y_{1\theta_{tu}}(s, x_1)\,\varepsilon_0(dx_2)\,ds dx_2$$

$$+\ y_{2\theta_{tu}}(s, x_2)\,\varepsilon_0(dx_1)\,ds dx_2\ , \tag{7.1.27}$$

(iv) $$B_{i\theta_{tu}} = \left(c_{ii} b_{i\theta_{tu}}\right) \circ F\ . \tag{7.1.28}$$

Recall that, by the Girsanov's theorem,

$$C_{\theta_{tu}} = C \text{ for } t, u \geq 0\ . \tag{7.1.29}$$

(R2) The drift densities $b_{i\theta_{tu}}$ are differentiable at θ with the derivative $b_i^{(1)}(u)$: $\Omega \times [0,\infty) \to R$ in V_i, that is,

$$r_t^{-2}(c_{ii}\left(r_t\left(\left(b_{i\theta_{tu}} - b_{i\theta}\right) - b_i^{(1)}(u)\right)^2\right) \circ F = o_{P_\theta}(1)\ , \tag{7.1.30}$$

and the derivative $b_i^{(1)}(u)$ is linear in u and satisfies the Lindeberg condition, that is, for every $\varepsilon > 0$,

$$r_t^{-2}\left(c_{ii} b_i^{(1)}(u)^2 I\left\{\left|b_i^{(1)}(u)\right| > \varepsilon r_t\right\}\right) \circ F_t = o_{P_\theta}(1)\ . \tag{7.1.31}$$

(R3) The jump densities $y_{i\theta_{tu}}$ are Hellinger differentiable at θ with the derivative $y_i^{(1)}(u) : \Omega \times [0,\infty) \times R \to R$ in W_i, that is,

$$r_t^{-2}\left(r_t\left(y_{i\theta}^{-1/2} y_{i\theta_{tu}}^{1/2} - 1\right) - \frac{1}{2} y_i^{(1)}(u)\right)^2 * \nu_{\theta t}^i = o_{P_\theta}(1)\ , \tag{7.1.32}$$

and the derivative $y_i^{(1)}(u)$ is linear in u and satisfies the Lindeberg condition, that is, for every $\varepsilon > 0$,

$$r_t^{-2}\left(y_i^{(1)}(u)^2 I\left\{\left|y_i^{(1)}(u)\right| > \varepsilon r_t\right\}\right) * \nu_{\theta k}^i = o_{P_\theta}(1)\,. \tag{7.1.33}$$

THEOREM 7.1

Suppose the conditions (R1) to (R3) hold. Then, for $u \in U$,

$$\ell_{\theta,\theta_{tu}}(t) = Z_t(u) - \frac{1}{2}\|u\|^2 + o_{P_\theta}(1)\,, \tag{7.1.34}$$

where

$$\mathcal{L}\left(Z_t(u)|P_\theta\right) \to N\left(0, \|u\|^2\right) \quad as\ t \to \infty\,, \tag{7.1.35}$$

$$Z_t(u) = r_t^{-1}\sum_{i=1}^{2}\left(b_i^{(1)}(u) \bullet X_{i\theta t}^{(c)} + y_i^{(u)}(u) * \left(\mu^i - \nu_\theta^i\right)_t\right)\,, \tag{7.1.36}$$

and

$$\|u\|^2 = \sum_{i=1}^{2}\left(\left\|b_i^{(1)}(u)\right\|_{ci}^2 + \left\|y_i^{(1)}(u)\right\|_i^2\right)\,. \tag{7.1.37}$$

For proof of Theorem 7.1, see [7].

Recall that a functional $k : \Theta \to R$ is said to be differentiable at θ with the gradient $g \in U$ if

$$r_t\left(k\left(\theta_{tu}\right) - k(\theta)\right) \to (u, g), u \in U\,. \tag{7.1.38}$$

and an estimator $\hat{k}_t$ is said to be *regular for k at θ with limit J* if

$$\mathcal{L}\left(r_t\left(\hat{k}_t - k\left(\theta_{tu}\right)\right)\Big| P_{\theta_{tu}}\right) \to J, u \in U\,. \tag{7.1.39}$$

It is sufficient if the regularity holds in the direction g, that is,

$$\mathcal{L}\left(r_t\left(\hat{k}_t - k\left(\theta_{t,\alpha,g}\right)\right)\Big| P_{\theta_{t,\alpha g}}\right) \Rightarrow J \text{ as } t \to \infty, \alpha \in R\,. \tag{7.1.40}$$

The following convolution theorem can be proved giving a characterization for the limiting distribution J.

THEOREM 7.2
Let the functional k be differentiable at θ with gradient $g \in U$. Let the estimator $\hat{k}$ be regular for k at θ. Then there exists a probability measure S such that

$$\mathcal{L}\left(r_t\left(\hat{k}_t - k(\theta)\right)\Big| P_\theta\right) \Rightarrow N\left(0, \|g\|^2\right) * S \tag{7.1.41}$$

where $\|g\|^2$ is as defined by (7.1.37). Here $*$ denotes the convolution operation.

For proof, see [7]. The quantity

$$\|g\|^2 = \sum_{i=1}^{2}\left(\left\|b_i^{(1)}(g)\right\|_{ci}^2 + \left\|y_i^{(1)}(g)\right\|_i^2\right) \tag{7.1.42}$$

is the asymptotic variance bound for the estimators of the functional k at θ. Recall that an estimator $\hat{k}_t$ of $k(\theta)$ is asymptotically efficient and regular for k at θ if and only if it is asymptotically linear with influence function equal to the gradient g, that is,

$$r_t\left(\hat{k}_t - k(\theta)\right) = Z_t(g) + o_{P_\theta}(1)\,. \tag{7.1.43}$$

7.2 Partially Specified Likelihood

Let $X = (X_1, X_2)$ be a stochastic process defined on (Ω, ζ) and Θ be an arbitrary set. For each $\theta \in \Theta$, let $T_\theta^{(1)} = (B_\theta^{(1)}, C_\theta^{(1)}, \nu_\theta^{(1)})$ be a triplet of predictable local characteristics such that there exists a unique probability measure P_θ on $(\Omega, \mathcal{F})$ such that X_1 is a semimartingale with respect to this probability measure and the filtration $\{\mathcal{F}_s, s \geq 0\}$ which is the filtration generated by X with predictable characteristics given by $T_\theta^{(1)}$. Let $X_{1\theta}^{(c)}$ denote the continuous martingale component of $X_{1\theta}$. The family $\{T_\theta^{(1)}\}, \theta \in \Theta$ is called a *partially specified model.* A partially specified likelihood process is expressed in terms of $T_\theta^{(1)}$. We now give a construction of such a process.

Let $\phi, \psi \in \Theta$. Suppose that

$$C_{\psi}^{(1)} = C_{\phi}^{(1)} = C^{(1)} . \tag{7.2.1}$$

Let F be a continuous adapted increasing process and c_1 a nonnegative predictable process such that

$$c^{(1)} = c_1 \circ F . \tag{7.2.2}$$

We further assume that

$$\nu_{\phi}^{(1)}(\{s\} \times R) = \nu_{\psi}^{(1)}(\{s\} \times R\}, s \geq 0 , \tag{7.2.3}$$

and that for $\xi \in \{\phi, \psi\}$,

$$\nu_{\xi}^{(1)}([0, s], R) < \infty, \ s \geq 0 . \tag{7.2.4}$$

Assume that there exists a positive predictable process Y_1 such that

$$\begin{aligned} &\text{(i)} \ \ \nu_{\psi}^{(1)} = Y_1 \star \nu_{\phi}^{(1)} , \\ &\text{(ii)} \ \ (Y_1^{1/2} - 1)^2 * \nu_{\phi s}^{(1)} < \infty, \ s \geq 0 , \end{aligned} \tag{7.2.5}$$

and the predictable process β_1 such that

$$\text{(iii)} \ \ B_{\psi}^{(1)} = B_{\phi}^{(1)} + (c_1 \beta_1) \circ F ,$$

$$\text{(iv)} \ \ \left(c_1 \beta_1^2\right) \circ F_s < \infty, \ s \geq 0 .$$

Let $\mu^{(1)}$ denote the jump measure of X_1. Then the partially specified log-likelihood process between ψ and ϕ is defined as

$$\ell_{\phi\psi}^{(1)} = \beta_1 \bullet X_{1\phi}^{(c)} - \frac{1}{2}\left(c_1\beta_1^2\right) \circ F$$
$$+ \sum_{s\leq \cdot} \log\left(1 + \Delta\left(Y_1 - 1\right) * \left(\mu^{(1)} - \nu_{\phi}^{(1)}\right)_s\right)$$
$$+ \left(\left(Y_1 - 1\right) * \mu^{(1)} - \nu_{\phi}^{(1)}\right)^{(c)} . \tag{7.2.6}$$

Assume that

$$\nu_{\xi}^{(1)}(ds, dx) = y_{1\xi}(s, x) ds dx . \tag{7.2.7}$$

Then

$$\ell_{\phi\psi}^{(1)} = \beta_1 \bullet X_{1\phi}^{(c)} - \frac{1}{2}\left(c_1\beta_1^2\right) \circ F$$
$$+ \log\left(y_{1\phi}^{-1} y_{1\psi}\right) * \mu^{(1)}$$
$$- \left(y_{1\phi}^{-1} y_{1\psi} - 1\right) * \nu_{\phi}^{(1)} , \tag{7.2.8}$$

and it is the partially specified likelihood process between ψ and ϕ. If $\mathcal{F}$ were generated by X_1, then this is the likelihood process between ψ and ϕ. Let us now consider a localized model.

Let $\phi \in \Phi$ and suppose P_ϕ is the unique probability measure associated with X_1 and $T_\phi^{(1)}$. Choose a continuous adapted increasing process F and a nonnegative predictable process c_1 such that

$$C_\phi^{(1)} = c_1 \circ F . \tag{7.2.9}$$

Let $b_{1\phi}$ be a predictable process such that

$$B_\phi^{(1)} = \left(c_1 b_{1\phi}\right) \circ F . \tag{7.2.10}$$

Let us introduce the cones V_1 and W_1 of process v and w satisfying (7.1.23) and (7.1.24), respectively, with P_θ replaced by P_ϕ and θ replaced by ϕ. Let U_1 be a cone such that, for $u \in U_1$, there exists $\phi_{tu} \in \Theta,\ t \geq 0$ satisfying the following conditions:

$$\text{(R1) (i)}\quad C^{(1)}_{\phi_{tu}} = C^{(1)}_{\phi}\,,$$

$$\text{(ii)}\quad \nu^{(1)}_{\phi_{tu}}([0,s],R) < \infty,\ s \geq 0\,;$$

$$\text{(iii)}\quad \nu^{(1)}_{\phi_{tu}}(ds,dx) = y_{1\phi_{tu}} ds dx$$

$$\text{(iv)}\quad B^{(1)}_{\phi_{tu}} = \left(c_1\, b_{1\phi_{tu}}\right) oF\,,$$

(R2) The drift density $b_{1\phi_{tu}}$ and the jump density $y_{1\phi_{tu}}$ admit derivatives $b_1^{(1)}(u)$ and $y_1^{(1)}(u)$ at ϕ in the sense of (7.1.30)–(7.1.33)) with P_θ replaced by P_ϕ and θ replaced by ϕ.

Let $\mu^{(1)}$ denote the jump measure of X_1. The following result gives sufficient conditions for the local asymptotic normality.

THEOREM 7.3
Under conditions (R1) and (R2) for $u \in U_1$,

$$\ell^{(1)}_{\phi\phi_{tu}t} = Z_t^{(1)}(u) - \frac{1}{2}\|u\|^2 + o_{P_\phi}(1)\,, \tag{7.2.11}$$

$$\mathcal{L}\left(Z_t^{(1)}(u)\middle| P_\phi\right) \Rightarrow N\left(0, \|u\|^2\right)\,, \tag{7.2.12}$$

as $t \to \infty$ where

$$Z_t^{(1)}(u) = r_t^{-1}\left(b_1^{(1)}(u) \bullet X^{(c)}_{1\phi_t} + y_1^{(1)}(u) * \left(\mu^{(1)} - \nu_\phi^{(1)}\right)_t\right)\,, \tag{7.2.13}$$

and

$$\|u\|^2 = \left\|b_1^{(1)}(u)\right\|^2_{c1} + \left\|y_1^{(1)}(u)\right\|^2\,. \tag{7.2.14}$$

We omit the proof. Let $(\cdot,\cdot)$ denote the inner product corresponding to the norm given by (7.2.14). We introduce a local model whose log-likelihood is $\ell^{(1)}_{\phi\phi_{tu}t}$ by the relation

$$d\bar{P}_{\phi_{tu}t} = \exp\left(\ell^{(1)}_{\phi\phi_{tu}t}\right) dP_\phi,\ u \in U_1\,. \tag{7.2.15}$$

Let $m : \Theta \to R$ be a functional. Suppose m is differentiable at $\phi \in \Phi$ with gradient $g_1 \in U_1$, that is,

$$r_t\left(m\left(\phi_{tu}\right) - m(\phi)\right) \to \left(u, g_1\right) \text{ as } t \to \infty \tag{7.2.16}$$

for $u \in U_1$. Let $\hat{m}_t$ be an estimator for m. Suppose that $\hat{m}_t$ is regular for m at ϕ with the limit J_1, that is,

$$\mathcal{L}\left(r_t\left(\hat{m}_t - m\left(\phi_{tu}\right)\right)\middle|\, \bar{P}_{\phi_{tut}}\right) \Rightarrow J_1 \text{ as } t \to \infty \tag{7.2.17}$$

for $u \in U_1$.

By the arguments similar to those given in the earlier section, it follows that we have a convolution theorem for the partially specified model; that is, there exists a probability measure S_1 such that

$$\mathcal{L}\left(r_t\left(\hat{m}_t - m(\theta)\right)\middle|\, P_\phi\right) \Rightarrow N\left(0, \|g_1\|^2\right) * S_1 \text{ as } t \to \infty \tag{7.2.18}$$

where

$$\|g_1\|^2 = \left\|b_1^{(1)}\left(g_1\right)\right\|_{c1}^2 + \left\|y_1^{(1)}\left(g_1\right)\right\|_1^2 \tag{7.2.19}$$

that gives an asymptotic variance bound for the estimators of the functional m at ϕ. Recall that an estimator $\hat{m}_t$ for m is asymptotically efficient and regular for m at ϕ if and only if it is asymptotically linear with the influence function equal to the gradient g_1, that is,

$$r_t\left(\hat{m}_t - m(\phi)\right) = Z_t^{(1)}\left(g_1\right) + o_P(1)\,. \tag{7.2.20}$$

Example 7.1

(Linear regression model) Consider the linear regression model

$$Y_{1n} = \beta Y_{2,n-1} + Z_n \tag{7.2.21}$$

where Z_n are i.i.d. random variables with the standard normal density f and Y_{2n} are random. Suppose we observe a two-dimensional discrete time process $\boldsymbol{X} = (X_1, X_2)$ defined by $\boldsymbol{X}_n = (X_{1n}, X_{2n})$ with the increments $\boldsymbol{Y}_n = (Y_{1n}, Y_{2n})$. Then for each $\beta \in R$, the conditional density of Y_{1n} given the σ-algebra $\mathcal{F}_{n-1}$ generated by $\boldsymbol{X}_{n-1}$ is

$$f\left(x_1 - \beta y_{2,n-1}\right)\,.$$

We can rephrase the model and put it in a semimartingale framework as follows. Let $X_0 = 0$ and let $\mathcal{F}_0$ be the trivial σ-field and extend the process X to continuous time by

$$X_s = X_n, \; \mathcal{F}_s = \mathcal{F}_n, \; n \le s < n+1 .$$

Let $\mathcal{F} = \{\mathcal{F}_s, s \ge 0\}$. Suppose that, for each $\beta \in R$ and the triplet $(0, 0, \nu_\beta^{(1)})$ with

$$\nu_\beta^{(1)}(ds, dx_1) = \sum_n \varepsilon_n(ds) y_{1n\beta}(x_1)\, dx_1 , \tag{7.2.22}$$

there exists a probability measure P_β on $\mathcal{F}_{\infty-}$ such that X_1 is a semimartingale with respect to P_β and the filtration $\mathcal{F}$ with predictable characteristics given by $(0, 0, \nu_\beta^{(1)})$. Let $\mathcal{P}_\beta$ denote the set of all such solutions. This is a partially specified model in the sense that we have specified the family of possible triplets $(0, 0, \nu_\beta^{(1)})$, $\beta \in R$ of X_1 but not the possible joint distributions of the process $X = (X_1, X_2)$. The partially specified log-likelihood between β' and β is

$$\ell_{\beta\beta'(t)}^{(1)} = \sum_{n \le t} \log\left(f\left(Y_{1n} - \beta Y_{2n-1}\right)^{-1} f\left(Y_{1n} - \beta' Y_{2,n-1}\right)\right)$$

$$= (\beta' - \beta) \sum_{n \le t} Z_n Y_{2,n-1} - \frac{1}{2}(\beta' - \beta)^2 \sum_{n \le t} Y_{2,n-1}^2 . \tag{7.2.23}$$

Fix $\beta \in R$. Assume that there exists a positive number I_β such that

$$r_t^{-2} \sum_{n \le t} Y_{2,n-1}^2 = I_\beta + o_P(1) , \tag{7.2.24}$$

and that the following Lindeberg condition holds for every $\varepsilon > 0$:

$$r_t^{-2} \sum_{n \le t} \int_{-\infty}^{\infty} \left(Y_{2,n-1} x\right)^2 I\left\{\left|Y_{2,n-1} x\right| > \varepsilon r_t\right\} f(x) dx = o_P(1) . \tag{7.2.25}$$

Introduce a local parameter $u \in R$ by

$$\beta_{tu} = \beta + r_t^{-1} u . \tag{7.2.26}$$

Then we obtain the local asymptotic normality of the partially specified log-likelihood (7.2.23) under the conditions (7.2.24) and (7.2.25). In other words,

$$\ell^{(1)}_{\beta\beta_{tu}t} = ur_t^{-1} \sum_{n \le t} Z_n Y_{2,n-1} - \frac{1}{2} u^2 I_\beta + o_P(1) \tag{7.2.27}$$

and

$$\mathcal{L}\left(r_t^{-1} \sum_{n \le t} Z_n Y_{2,n-1} \middle| P_\beta \right) \Rightarrow N\left(0, I_\beta\right) \text{ as } t \to \infty . \tag{7.2.28}$$

From the local asymptotic normality, we see that the asymptotic variance bound for the estimators for β is I_β^{-1}. An efficient estimator $\hat{\beta}_t$ can be obtained as a solution of the estimating equation based on the partially specified likelihood:

$$\sum_{n \le t} \left(Y_{1n} - \beta Y_{2,n-1}\right) Y_{2,n-1} = 0 . \tag{7.2.29}$$

In this case, the maximum likelihood estimator is the least squares estimator given by

$$\hat{\beta}_t = \left(\sum_{n \le t} Y_{2,n-1}^2 \right)^{-1} \sum_{n \le t} Y_{1n} Y_{2,n-1} \tag{7.2.30}$$

and

$$\hat{\beta}_t - \beta = \left(\sum_{n \le t} Y_{2,n-1}^2 \right)^{-1} \sum_{n \le t} Z_n Y_{2,n-1} . \tag{7.2.31}$$

From (7.2.24) it follows that

$$r_t \left(\hat{\beta}_t - \beta \right) = r_t^{-1} I_\beta^{-1} \sum_{n \le t} Z_n Y_{2,n-1} + o_P(1) , \tag{7.2.32}$$

which implies that

$$\mathcal{L}\left(r_t \left(\hat{\beta}_t - \beta \right) \middle| P_\beta \right) \Rightarrow N(0, I_\beta^{-1}) \text{ as } t \to \infty \tag{7.2.33}$$

from (7.2.28). Hence, $\hat{\beta}_t$ attains the asymptotic variance bound and it is asymptotically efficient.

This can also be seen from the methods given earlier. Note that

$$Z_t(u) = ur_t^{-1} \sum_{n \le t} Z_n Y_{2,n-1} , \tag{7.2.34}$$

with

$$\|u\|^2 = u^2 I_\beta, (u, v) = uI_\beta v \tag{7.2.35}$$

where $Z_t(u)$ and $\|u\|^2$ are as defined in Theorem 7.3 concerning the local asymptotic normality. Consider the functional $m(\beta) = \beta$. Then

$$r_t \left(m\left(\beta_{tu}\right) - m(\beta)\right) = u, \ u \in R .$$

A gradient $g_1 \in U_1 = R$ is determined by

$$u = (u, g_1) = uI_\beta g_1, u \in R ,$$

and hence $g_1 = I_\beta^{-1}$. Therefore, the relation (7.2.32) reduces to

$$r_t \left(\hat{\beta}_t - \beta\right) = Z_t(g_1) + o_P 0(1) . \tag{7.2.36}$$

This shows that $\hat{\beta}_t$ is regular and asymptotically efficient following the characterization of regular and efficient estimators given earlier. □

Example 7.2

(Markov chains in a random environment) Let $\boldsymbol{X} = (X, X_2)$ be a two-dimensional discrete time process with increments $\boldsymbol{Y}_n = (Y_{1n}, Y_{2n})$. Let Φ be the set of positive functions $f : [0, 1] \times [0, 1]^2 \to R$ with $f(\cdot|y)$ a probability density for each $y \in [0, 1]^2$. Suppose that, for $f \in \Phi$, there exists a unique probability measure P_f on $\mathcal{F}$ such that Y_1 is a Markov chain with the state space $[0, 1]$ and the transition distribution with density f, that is,

$$P\left(Y_{1n} \in dx | \boldsymbol{Y}_{n-1} = y\right) = f(x|y)dx . \tag{7.2.37}$$

Then the transition distribution depends on X_2 through $Y_{2,n-1}$. The compensator of the jump measure of X_1 with respect to the filtration generated by $\boldsymbol{X}$ is

$$\nu_f^{(1)}(ds, dx) = \sum_n \varepsilon_n(ds) f\left(x | \boldsymbol{Y}_{n-1}\right) dx . \tag{7.2.38}$$

This is a partially specified model with triplets $T_f^{(1)} = (0, 0, \nu_f^{(1)}),\ f \in \Phi$. The partially specified log likelihood between g and f is

$$\ell_{fgt}^{(1)} = \sum_{n \le t} \log \left(f\left(Y_{1n} | \boldsymbol{Y}_{n-1}\right)^{-1} g\left(Y_{1n} | \boldsymbol{Y}_{n-1}\right) \right) . \tag{7.2.39}$$

Fix $f \in \Phi$. Assume that $\boldsymbol{Y}$ is stationary under P_f with density π_f. Further suppose that π_f is bounded away from zero.

Let the local parameter space U_1 be the set of all bounded measurable function $u : [0, 1] \times [0, 1]^2 \to R$ such that

$$\int u(x|\boldsymbol{y}) f(x|\boldsymbol{y}) dx = 0,\ \ \boldsymbol{y} \in [0, 1]^2 . \tag{7.2.40}$$

Define

$$f_{tu}(x|\boldsymbol{y}) = \left(1 + t^{-1/2} u(x|\boldsymbol{y})\right) f(x|\boldsymbol{y}) . \tag{7.2.41}$$

Note that each function $f_{tu}(\cdot|\boldsymbol{y})$ is a probability density for each $y \in [0, 1]^2$ in view of the condition (7.2.41). By the ergodic theorem, for $u,\ v \in U_1$,

$$t^{-1} \sum_{n \le t} \int u\left(x\,|\boldsymbol{Y}_{n-1}\right) v\left(x\,|\boldsymbol{Y}_{n-1}\right) f\left(x\,|\boldsymbol{Y}_{n-1}\right) dx$$

$$= \int u(x|\boldsymbol{y}) v(x|\boldsymbol{y}) f(x|\boldsymbol{y}) \pi_f(\boldsymbol{y}) dx d\boldsymbol{y} + o_P(1) . \tag{7.2.42}$$

Let W_{d1} be the collection of all sequences such that

$$w(\omega; n, x) = u\left(x\,|\boldsymbol{Y}_{n-1}(\omega)\right), n \ge 1,\ \ u \in U_1 .$$

It can be checked that the sequence of conditional densities

$$y_{1mf_{tu}}(x) = f_{tu}\left(x\,|\boldsymbol{Y}_{n-1}\right),\ n \ge 1 \tag{7.2.43}$$

is Hellinger differentiable at f with the derivative $y_{1n}^{(1)}(u) \in W_{d1}$ defined by

$$y_{1n}^{(1)}(u)(x) = u\left(x\,|\boldsymbol{Y}_{n-1}\right),\ n \ge 1 ,$$

that is,

$$t^{-1}\sum_{n\le t}\int\left(t^{1/2}\left(y_{1n}(x)^{-1/2}y_{1nf_{tu}}(x)^{1/2}-1\right)-\frac{1}{2}y_{1n}^{(1)}(u)(x)\right)^2 y_{1n}(x)dx$$

$$= o_P(1)\,. \tag{7.2.44}$$

Note further that the derivative $y_{1n}^{(1)}(u)$ is linear in u and u being bounded, $y_{1n}^{(1)}(u)$ satisfies the Lindberg condition in the sense that

$$t^{-1}\sum_{n\le t}\int y_{1n}^{(1)}(u)(x)^2 I\left\{\left|y_{1n}^{(1)}(u)(x)\right| > \varepsilon t^{1/2}\right\} y_{1n}(x)dx = o_P(1)$$

$$\tag{7.2.45}$$

for all $\varepsilon > 0$. Conditions (7.2.44) and (7.2.45) imply the local asymptotic normality of the partially specified log likelihood (7.2.39). In fact

$$\text{(i)}\quad \ell_{ff_{tu}}^{(1)}(t) = Z_t(u) - \frac{1}{2}\|u\|^2 + o_P(1)\,, \tag{7.2.46}$$

$$\text{(ii)}\quad \mathcal{L}\left(Z_t(u)|\,P_f\right) \Rightarrow N\left(0,\,\|u\|^2\right)\ \text{ as } t\to\infty$$

with

$$\text{(i)}\qquad Z_t(u) = t^{-1/2}\sum_{n\le t}\int u\,(x\,|\mathbf{Y}_{n-1})\left(\mu^{(1)}(\{n\},dx) - f\,(x\,|\mathbf{Y}_{n-1})\,dx\right), \tag{7.2.47}$$

and

$$\text{(ii)}\qquad \|u\|^2 = \left\|y_1^{(1)}(u)\right\|_{d1}^2 = \int\int u^2(x|y)f(x|y)\pi_f(y)dx\,dy\,.$$

Let us consider the financial

$$m(f) = \int\int I_A(x)f(x|y)dx\,dy\,. \tag{7.2.48}$$

This is the probability that, starting from $\mathbf{Y}_{n-1}$, a randomly chosen point in $[0,1]^2$, $Y_{1n} \in A$ after one step of the Markov chain. We now compute the asymptotic variance bound for the estimator of m at f.

For $u \in U$,

$$t^{1/2}\left(m\left(f_{tu}\right) - m(f)\right) = \int\int I_A(x)u(x|y)f(x|y)dxdy\,. \qquad (7.2.49)$$

A gradient $g_1 \in U_1$ of m is determined by expressing the right-hand side of the above equation in terms of the acuity defined by the norm (7.2.47) (ii):

$$\int\int I_A(x)u(x|y)f(x|y)dxdy$$

$$\int\int u(x|y)g_1(x|y)f(x|y)\pi_f(y)dxdy, \quad u \in U_1\,. \qquad (7.2.50)$$

Using the condition (7.2.40), it follows that

$$g_1(x|y) = \pi_f(y)^{-1}\left(I_A(x) - \int I_A(x)f(x|y)dx\right)\,. \qquad (7.2.51)$$

Note that g_1 is bounded since π_f is bounded away from zero. Hence, $g_1 \in U_1$. From the convolution theorem and the relations (7.2.47) and (7.2.48), it follows that the asymptotic variance bound for estimator of m at f is

$$\|g_1\|^2 = \int \pi_f(y)^{-1}\int I_A(x)f(x|y)dx\left(1 - \int I_A(x)f(x|y)dx\right)dy\,.$$

(7.2.52)

□

Example 7.3 **(Diffusions with jumps and multiplicative drift)**
Consider a stochastic differential equation

$$dX_{1s} = \lambda X_{2s}ds + dW_s + dZ_s \qquad (7.2.53)$$

where W is the Wiener process and Z is a jump process such that the compensator of the jump measure has the density:

$$\nu_s^{(1)}(ds, dx) = y(s, x)dsdx \,. \tag{7.2.54}$$

This is a semiparametric partially specified model with triplets $T_{\lambda y}^{(1)} = (B_\lambda^{(1)}, C^{(1)}, \nu_y^{(1)})$ where

$$B_{\lambda t}^{(1)} = \lambda \int_0^t X_{2s}ds, \; C_t^{(1)} = t \,. \tag{7.2.55}$$

The partially specified log-likelihood ratio process between (λ', y') and (λ, y) is

$$\begin{aligned} \ell_{\lambda y, \lambda' y', t}^{(1)} &= (\lambda' - \lambda) \int_0^t X_{2s}dX_{1\lambda s}^{(c)} - \frac{1}{2}(\lambda' - \lambda)^2 t \\ &+ \log\left(y^{-1}y'\right) * \mu_t^{(1)} \\ &- \left(y^{-1}y' - 1\right) * \nu_{yt}^{(1)} \,. \end{aligned} \tag{7.2.56}$$

We can separate the jumps from the continuous part of X_1, and hence the problem of estimating λ from X_2 and X_1 is equivalent to the problem of estimating λ from X_2 and the continuous part of X_1.

From the general theory discussed earlier, it follows that, for $m(\lambda, y) = \lambda$, the asymptotic variance bound for known y coincides with the bound for unknown y. Hence, for the problem of asymptotically efficient estimation of λ, it is sufficient to determine the bound for known y for computing the asymptotic variance bound.

Fix λ and y. Suppose that there exists a positive number $I_{\lambda, y}$ such that

$$r_t^{-2} \int_0^t X_{2s}^2 ds = I_{\lambda y} + o_P(1) \,, \tag{7.2.57}$$

and that the following Lindeberg condition holds:

$$r_t^{-2} \int_0^t X_{2s}^2 I\{|X_{2s}| > \varepsilon r_t\} ds = o_P(1), \ \varepsilon > 0 . \tag{7.2.58}$$

Let us introduce a local parameter $u \in R$ by

$$\lambda_{tu} = \lambda + r_t^{-1} u . \tag{7.2.59}$$

Let V_1 be the collection of all processes

$$v(\omega : t) = uX_{2t}(\omega), \ u \in R ,$$

and $U_1 = R$. The drift density $b_{1\lambda_{tu}} = \lambda_{tu} X_2$ is differentiable at λ with the derivative $b_1^{(1)}(u) = uX_2 \in V_1$ in the sense of Hellinger differentiability. Theorem 7.3, implies the local asymptotic normality of the partially specified log-likelihood process between (λ_{tu}, u) and (λ, y) viz

$$\ell^{(1)}_{\lambda y, \lambda_{tu} y, t} = u r_t^{-1} X_2 \bullet X^{(c)}_{1\lambda t} - \frac{1}{2} u^2 I_{\lambda, y} + o_P(1) , \tag{7.2.60}$$

and

$$\mathcal{L}\left(r_t^{-1} X_2 \bullet X^{(c)}_{1\lambda t} \middle| P_{\lambda, y} \right) \Rightarrow N\left(0, I_{\lambda, y}\right) \text{ as } t \to \infty . \tag{7.2.61}$$

Hence, the asymptotic variance bound for the estimators of λ at (λ, y) is $I^{-1}_{\lambda, y}$. An efficient estimator $\hat{\lambda}_t$ for λ may be obtained as a solution of the estimating equation based on the partially specified likelihood process

$$X_2 \bullet X^{(c)}_{1\lambda t} = 0 . \tag{7.2.62}$$

With $X_1^{(o)}$ the continuous part of X_1, we have the Doob–Meyer decomposition

$$X_1^{(o)} = X^{(c)}_{1\lambda} + \lambda \int_0^{\cdot} X_{2s} ds \tag{7.2.63}$$

and we can obtain the maximum likelihood estimator

$$\hat{\lambda}_t = \left(\int_0^t X_{2s}^2 ds \right)^{-1} X_2 \bullet X_{1t}^{(o)} . \tag{7.2.64}$$

Hence

$$\hat{\lambda}_t - \lambda = \left(\int_0^t X_{2s}^2 ds \right)^{-1} X_2 \bullet X_{1\lambda t}^{(c)} . \tag{7.2.65}$$

Relation (7.2.57) shows that

$$r_t \left(\hat{\lambda}_t - \lambda \right) = r_t^{-1} I_{\lambda,y}^{-1} X_2 \bullet X_{1\lambda t}^{(c)} + o_P(1) , \tag{7.2.66}$$

and we have

$$\mathcal{L}\left(r_t \left(\hat{\lambda}_t - \lambda \right) \middle| P \right) \Rightarrow N\left(0, I_{\lambda,y}^{-1} \right) \text{ as } t \to \infty$$

from (7.2.61). Hence, $\hat{\lambda}_t$ attains the asymptotic variance bound. □

7.3 Partial Likelihood and Asymptotic Efficiency

Consider a statistical model in which the principle objects of interest are an observable stochastic process $\boldsymbol{X} = \{X_t, t \in T\}$ indexed by $T = R$ or $T = N$ where $N = \{1, 2, \ldots\}$ and a parameter $\theta \in \Theta$. Let $\mathcal{F}_t$ be the σ-algebra of all observable events up to and including time t and let $\mathcal{F} = \vee_t \mathcal{F}_t$. Note that X_t is $\mathcal{F}_t$-measurable. In the ideal situation, there is a probability measure P_θ on $(\Omega, \mathcal{F})$ for every $\theta \in \Theta$ and the key role in statistical inference is played by the likelihood process

$$Z_{\theta,\phi}(t) \equiv \left(dP_\theta / dP_\phi \right)\big|_{\mathcal{F}_t} . \tag{7.3.1}$$

There are situations where it is not possible to compute the likelihood ratio process solely in terms of $\boldsymbol{X}$. These can be described in the following manner. **Case (i)** The filtration $\{\mathcal{F}_t\}$ specified above may be generated by the pair $(\boldsymbol{X}, \boldsymbol{Y})$ where $\boldsymbol{Y} = \{Y_t,\ t \in T\}$ is a covariate process. The distribution of $(\boldsymbol{X}, \boldsymbol{Y})$ may depend on θ and another (misance) parameter η with the measure $P_{\theta,\eta}$ indexed by the pair (θ, η) or the distribution of $\boldsymbol{Y}$ may not be specified parametrically but by a "semiparametric" model with a parameter θ and a nonparameteric component.

Let us first consider the case $T = N$. We might know the conditional distribution of X_n given $\mathcal{F}_{n-1}$ and suppose that it is absolutely continuous with density $f_{\theta,n}(\omega, x)$. Following Cox [2], let us consider the process

$$V_{\theta,\phi}(n)(\omega) = \pi_{1 \le i \le n} \frac{f_{\theta,i}\,(\omega, X_i(\omega))}{f_{\phi,i}\,(\omega, X_i(\omega))}\,, \tag{7.3.2}$$

which we call the partial likelihood process. If $\mathcal{F}_n = \sigma(X_i,\ 1 \le i \le n)$, that is $\{\mathcal{F}_n\}$ is generated by $\{X_n\}$, then the true likelihood process $\{Z_{\theta,\phi}(n)\}$ given by (7.2.2). In general $\{V_{\theta,\phi}(n)\}$ is not the likelihood process, but it shares the properties of the true likelihood such as the martingale property.

Let us consider another example when $T = R$. Suppose $\boldsymbol{X}$ is a simple point process with the jump times $T_1 < T_2 < \cdots < T_n < \cdots$ and the intensity function

$$A(t, \theta) = \int_0^t a_s^\theta(\boldsymbol{X}, \boldsymbol{Y})ds\,, \tag{7.3.3}$$

where $\boldsymbol{Y}$ is a covariate process. Note that a_s^θ depends in a non-anticipating way on the point process $\boldsymbol{X}$ and also on the covariate process $\boldsymbol{Y}$. Let

$$V_{\theta,\phi}(t) = \left(\pi_{T_n \le t} \frac{a_{T_n}^\phi(\boldsymbol{X}, \boldsymbol{Y})}{a_{T_n}^\theta(\boldsymbol{X}, \boldsymbol{Y})}\right)$$

$$\exp \int_0^t \left[a_s^\theta(\boldsymbol{X}, \boldsymbol{Y}) - a_s^\phi(\boldsymbol{X}, \boldsymbol{Y})\right] ds\,. \tag{7.3.4}$$

This is not the likelihood process for the model. However, it has the local martingale property [3].

Case (ii) The filtration $\{\mathcal{F}_t\}$ might be generated by the process $\boldsymbol{X}$ but it might not be possible to compute $Z_t^{\phi,\phi}$. For instance this can be seen from the following example.

Suppose $\boldsymbol{X}$ is a solution of the stochastic differential equation

$$dX_t = a_t^{\theta}(X)dt + \sigma_t(X)dW_t, X_0 = x, \ t \geq 0 \tag{7.3.5}$$

where W is the standard Wiener process under P_θ. In other words P_θ is a "weak solution" to (7.3.5). Equation (7.3.5) may admit several weak solutions (say) $P_{\theta,\eta}$ where η can again be considered as a nuisance parameter. The true likelihood depends on θ and η. However, we can ignore η and consider a process which would behave like a likelihood process in case (7.2.5) has a unique weak solution for every θ and it is given by

$$V_{\theta,\phi}^{(t)} = \exp\left[\int_0^t \left(\frac{a_s^{\phi}(X) - a_s^{\theta}(X)}{\sigma_s(X)^2}\right)(dX_s - a_s^{\theta}(X)ds)\right.$$

$$\left. - \frac{1}{2}\int_0^t \left(\frac{(a_s^{\phi}(X) - a_s^{\theta}(X)}{\sigma_s(X)}\right)^2 ds\right]. \tag{7.3.6}$$

Note that $V_{\phi,\theta}$ is not the likelihood process but it is a $P_{\theta,\eta}$ local martingale for all η.

In both the cases, the process $\boldsymbol{X}$ is partially determined by some characteristics such as the conditional distribution or the intensity or the drift and diffusion coefficients, etc., and the partial likelihood process is what one would obtain by writing the likelihood process as if these characteristics were determining the probability law of $\boldsymbol{X}$.

In the light of the discussion earlier, let us again look at the general theory behind partially specified semimartingale experiments.

Suppose we observe two stochastic processes $\boldsymbol{X} = \{X_t, t \geq 0\}$ and $\boldsymbol{Y} = \{Y_t, t \geq 0\}$ defined on a probability space $(\Omega, \mathcal{F}, P_\theta,), \theta \in \Theta$. Let $\zeta_t = \sigma\{(X_s, Y_s) : s \leq t\}$ and $\mathcal{F}_t = \sigma\{X_s; s \leq t\}$. Then $\mathcal{F}_t \subset \zeta_t \subset \mathcal{F}$ for all t. The process of main interest is $\boldsymbol{X}$. Our purpose is to estimate the parameter θ which is possibly infinite dimensional. The process $\boldsymbol{Y}$ is a covariate process which supplies some additional information about θ. If the estimation of θ is to be based only on the observation of $\boldsymbol{X}$ as a function of time, that is, on $\{\mathcal{F}_t\}$, then the likelihood ratio process

$$Z_{\theta,\tau}(t) = \frac{dQ_\theta(t)}{dQ_\tau(t)} \tag{7.3.7}$$

can be used where $Q_\theta(t) = P_\theta|\mathcal{F}_t$, the measure P_θ restricted to $\mathcal{F}_t$ assuming that $Q_\theta(t) << Q_\tau(t)$ for all t and $\theta, \tau \in \Theta$. Suppose we do not know the structure of P_θ for each θ and we would like to use the covariate process $\boldsymbol{Y}$ as well as $\boldsymbol{X}$ for the estimation of θ. Then, probably, one could use

$$V_{\theta,\tau}(t) = \frac{dR_\theta(t)}{dR_\tau(t)} \tag{7.3.8}$$

where $R_\theta(t) = P_\theta|\zeta_t$. This presumes the knowledge of the joint distribution of $(\boldsymbol{X}, \boldsymbol{Y})$ for each θ that is not available in general. We may have partial information on how $\boldsymbol{Y}$ affects $\boldsymbol{X}$ but not the complete structure of $\boldsymbol{Y}$, much less the full structure of $(\boldsymbol{X}, \boldsymbol{Y})$. This leads to the concept of partially specified experiment and partial likelihood.

For $\theta \in \Theta$, let $T_\theta(\boldsymbol{X}|\zeta)$ denote the triplet of predictable characteristics of $\boldsymbol{X}$ with respect to P_θ and the filtration ζ. Recall that $T_\theta(\boldsymbol{X}|\zeta)$ is of the form $(B_\theta, C_\theta, \nu_\theta)$ where $\nu_\theta = \nu_\theta(dt, dx)$ is the (P_θ, ζ)-compensator of the space-time jump measure of the process X,

$$\mu(dt, dx) = \sum_{s:\Delta X_s \neq 0} \varepsilon(s, \Delta X_s)(dt, dx).$$

For an arbitrary truncation function h, B_θ is the (P_θ, ζ)-predictable drift function, that is, B_θ is of finite variation, $B_\theta(0) = 0$ and $X_t - \sum_{s \le t}[\Delta X_s - h(\Delta X_s)] - B_t$ is a (P_θ, ζ)-local martingale; $C_\theta(t) = [\boldsymbol{X}^{(c)}, \boldsymbol{X}^{(c)}]_t$, is the predictable quadratic variation of the continuous compact of $\boldsymbol{X}$. The problem is to estimate θ on the basis of the observed process $\boldsymbol{X}$ and $T_\theta(\boldsymbol{X}|\zeta)$, since T_θ is supposed to be computable from the past of $\boldsymbol{X}$ and $\boldsymbol{Y}$.

7.4 Partially Specified Likelihood and Asymptotic Efficiency (Counting Processes)

Let Θ be an arbitrary parameter space. For $n \ge 1$, let $\{P_{n\theta} : \theta \in \Theta\}$ be a family of probability measures on (Ω_n, ζ_n). Let X_{nj}, $1 \le j \le n$ be counting processes and $Y_{nj}, 1 \le j \le n$ be predictable covariate processes over $[0, 1]$. Suppose that no two counting processes jump simultaneously. Let $\mathcal{F}_n = \{F_{nt}, \le t \le 1\}$ denote the filtration generated by $\boldsymbol{X}_n = (X_{n1}, \ldots, X_{nn})$ and $\zeta_n = \{G_{nt}, 0 \le t \le 1\}$ denote the filtration generated by $\boldsymbol{X}_n$ and

$Y_n = (Y_{n1}, \ldots, Y_{nn})$. Suppose X_{ni} admits a Doob–Meyer decomposition with respect to ζ_n and $P_{n\theta}$ of the form

$$X_{ni}(t) = M_{ni\theta}(t) + \int_0^t a_{ni\theta}(s)ds, \; 0 \le t \le 1 . \tag{7.4.1}$$

Here the intensity process $a_{ni\theta}$ of X_{ni} is predictable and is known up to a possibly infinite-dimensional parameter $\theta \in \Theta$. The problem is to estimate θ or a function of it efficiently in an asymptotic sense.

If $\zeta_n = \mathcal{F}_n$, then there is no additional information about θ in the covariates Y_{ni}, $i \le i \le n$ and it is sufficient to observe the processes X_{ni}, $1 \le i \le n$. Their distribution is determined by the intensity process $a_{ni\theta,\, 1 \le i \le n}$. In other words the model is fully specified. In this case, the following representation of the log-likelihood between τ and θ holds:

$$\log \left(\frac{dP_{n\theta}}{dP_{n\tau}}\right)_{\mathcal{F}_{n1}} = \sum_{i=1}^{n} \left(\int_0^1 \log \left(\frac{a_{ni\tau}(s)}{a_{ni\theta}(s)}\right) dX_{ni}(s) - \int_0^1 (a_{ni\tau}(s) - a_{ni\theta}(s))\, ds \right) . \tag{7.4.2}$$

We will consider the case when ζ_n is larger than $\mathcal{F}_n$. Then the model is partially specified. In this case, the expression (7.4.2) can be considered as a partially specified log-likelihood. Let it be denoted by $\ell_{n\theta\tau}$.

Fix $\theta \in \Theta$. We will introduce a *local model* at θ as follows. Let H be a Hilbert space with the innerproduct $(\cdot, \cdot)$ and the norm $\|\cdot\|$. Let $V \subset H$ be a linear subspace called the *local parameter space*. For $i = 1, \ldots, n$, let

$$D_{ni} : V \times \Omega_n \times [0, 1] \to R$$

be linear in the first variable $v \in V$, there exists a sequence $\{\theta_{nv}\}$ in Θ and $0 < c_n \to \infty$ with the following properties:

(i) the intensity processes $\{a_{ni\theta}\}$ are Hellinger differentiable at θ with derivatives

$\{D_{ni}\}$, that is,

$$\sum_{i=1}^{n} \int_0^1 \left(\left(a_{ni\theta_{nv}}(s)/a_{ni\theta}(s) \right)^{1/2} - 1 - \frac{1}{2} c_n^{-1} D_{ni}(v)(s) \right)^2 a_{ni\theta}(s)ds \to 0\,;$$

$$\text{in } (P_{n\theta}) \ - \text{measure as } n \to \infty\,; \tag{7.4.3}$$

(ii) the derivatives D_{ni} satisfy the Lindeberg condition: for every $\varepsilon > 0$,

$$c_n^{-2} \sum_{i=1}^{n} \int_0^1 D_{ni}(v)(s)^2 I \left\{ |D_{ni}(v)(s)| > \varepsilon c_n \right\} a_{ni\theta}(s)ds \to 0$$

$$\text{in } (P_{n\theta}) - \ \text{measure as } n \to \infty, \text{ and} \tag{7.4.4}$$

(iii)
$$c_n^{-2} \sum_{i=1}^{n} \int_0^1 D_{ni}(v)(s)^2 a_{ni\theta}(s)ds \to \|v\|^2$$

$$\text{in } (P_{n\theta}) \ - \text{measure as } n \to \infty\,. \tag{7.4.5}$$

REMARK 7.1 Observe that $H, (\cdot,\cdot), V, D_{ni}$ and θ_{nv} depend on θ. The family of probability measures

$$\left\{ P_{n\theta_{nv}};\ v \in V \right\}$$

is called a *local model* at θ. Let $\ell_{nv} \equiv \ell_{n\theta\theta_{nv}}$ where $\ell_{n\theta\tau}$ is the partially specified log likelihood defined earlier. ▮

THEOREM 7.4
If the conditions (7.3.3) to (7.3.5) hold, then

$$\ell_{nv} = Z_n(v) - \frac{1}{2}\|v\|^2 + o_{P_{n\theta}}(1) \tag{7.4.6}$$

with

$$Z_n(v) = c_n^{-1} \sum_{i=1}^{n} \int_0^1 D_{ni}(v)(s) dM_{ni\theta}(s) \tag{7.4.7}$$

and $Z_n(v)$ *is asymptotically normal under* $P_{n\theta}$ *with variance* $\|v|^2$.

REMARK 7.2 Proof of Theorem 7.4 is analogous to that of Theorem 6.11. We omit of proof. For details, see [8]. Condition (7.4.3) is a consequence of

$$\sum_{i=1}^{n} \int_0^1 \left(\frac{a_{ni\theta_{nv}}(s)}{a_{ni\theta}(s)} - 1 - c_n^{-1} D_{ni}(v)(s) \right)^2 a_{ni\theta}(s) ds \to 0$$

$$\text{in } (P_{n\theta}) \quad - \text{measure as } n \to \infty . \tag{7.4.8}$$

∎

Example 7.4 **(The additive risk model)**
This model has an intensity process of the form

$$a_{ni\theta}(s) = Y_{ni}(s)\lambda(s), s \in [0, 1]$$

where Y_{ni} is a p-dimensional row vector of predictable covariate processes and λ is a p-dimensional column vector of bounded *hazard functions.* The parameter under consideration here is $\theta = \lambda$.

Fix λ. Let $W_{ni} = (Y_{ni}\lambda)^{-1}$. Define $\boldsymbol{W}_n = diag(W_{n1}, \ldots, W_{nn})$ and $3 = diag(\lambda_1, \ldots, \lambda_p)$. Let $\boldsymbol{Y}_n$ denote the matrix with Y_{ni} is rows and suppose that there exists a matrix function U with the largest and the smallest eigenvalues bounded and bounded away from zero respectively such that

$$\sup_{0 \le s \le 1} \left| c_n^{-2} \boldsymbol{Y}'_n \boldsymbol{W}_n \boldsymbol{Y}_n - U \right| \to 0 \text{ in } P_{n\theta} \ - \text{measure as } n \to \infty \tag{7.4.9}$$

where prime denotes the transpose and $\sup_{0 \le s \le 1} |A(s)|$ of a square matrix $A(s)$ is interpreted in an obvious manner. Let us introduce a local model. Let $H = L_2(\lambda)$. Let V be the set of all bounded functions in H. For each local parameter $v = (v_1, \cdots, v_p) \in V$ and n large, define

$$\lambda_{nvj} = \left(1 + c_n^{-1} v_j\right) \lambda_j \, . \tag{7.4.10}$$

Set

$$a_{niv} = Y_{ni} \lambda_{nv} \text{ where } \lambda_{nv} = \left(\lambda_{nv1}, \ldots, \lambda_{nvp}\right)' \, .$$

We now show that the intensity process $a_{ni\theta}(s)$ satisfies the conditions (7.4.8), (7.4.4), and (7.4.5).

(i) **Hellinger differentiability:** Note that

$$\begin{aligned} \frac{a_{niv}}{a_{ni}} &= Y_{ni} \lambda_{nv} \left(Y_{ni} \lambda\right)^{-1} \\ &= Y_{ni} \lambda_{nv} W_{ni} \\ &= 1 + c_n^{-1} v' 3 Y_{ni} W_{ni} \, . \end{aligned}$$

Hence (7.4.8) holds with $D_{ni}(v) = v' \Lambda Y_{ni} W_{ni}$.

(ii) **Lindeberg condition:** Let $|v_j| \leq c, \ 1 \leq j \leq p$. Then

$$|D_{ni}(v)| = \left|v' \Lambda Y_{ni} W_{ni}\right| \leq c \left|\Lambda Y_{ni} W_{ni}\right| = c \, ,$$

which implies (7.4.4).

(iii) **Nonrandom Hellinger limit:** Relation (7.4.9) implies that

$$\begin{aligned} & c_n^{-2} \sum_{i=1}^{n} \int_0^1 D_{ni}^2(v) a_{ni} ds \\ &= c_n^{-2} \int_0^1 v' \Lambda Y_n' W_n Y_n \Lambda v ds \end{aligned}$$

$$= \int_0^1 v'\Lambda U \Lambda v ds + o_{P_{n\theta}}(1) .$$

In view of (i)–(iii), Theorem 7.4 holds with

$$\|v\|^2 = \int_0^1 v'\Lambda U \Lambda v \, ds . \tag{7.4.11}$$

In other words, the partially specified log-likelihood is locally asymptotically normal and

$$\ell_{nv} = c_n^{-1} \sum_{i=1}^{n} \int_0^1 v'\Lambda Y_{ni} W_{ni} dM_{ni\theta}(s) - \frac{1}{2} \int_0^1 v'\Lambda U \Lambda v ds + o_{P_{n\theta}}(1)$$

where

$$M_{ni\theta}(t) = X_{ni}(t) - \int_0^t Y_{ni}\lambda \, ds . \tag{7.4.12}$$

By (7.4.11), the acuity is

$$(v, w) = \int_0^1 v'\Lambda w \, ds . \tag{7.4.13}$$

We now determine the asymptotic covariance bound for the estimators of the cumulative hazard vector $\int_0^t \lambda ds$.

Let

$$k(\lambda) = \int_0^t \lambda' ds . \tag{7.4.14}$$

From the definition of λ_{nv} given by (7.4.10), it follows that

$$c_n \left(k \left(\lambda_{nv} \right) - k(\lambda) \right)$$

$$= \int_0^t v' \Lambda ds$$

$$= (v, g) \tag{7.4.15}$$

where g is a gradient of the functional $\boldsymbol{k}$. Here g is a matrix of order $p \times p$ such that jth column of g is the gradient of the jth-component k_j of the functional $\boldsymbol{k}$. From (7.4.13), it follows that

$$\int_0^t v' \Lambda ds = \int_0^1 v' \Lambda U \Lambda g ds \,. \tag{7.4.16}$$

Hence

$$g(s) = \Lambda(s)^{-1} U(s)^{-1} I_{[0,t)}(s), \; 0 \le s \le 1 \,. \tag{7.4.17}$$

Note that the gradient is bounded and hence canonical. Assumptions made in the Theorem 6.10 hold, and it follows that the covariance bound is obtained by applying (7.4.16) for $v = g$. Hence

$$(g', g) = \int_0^t g' \Lambda ds$$

$$= \int_0^t U^{-1} \Lambda^{-1} \Lambda ds = \int_0^t U^{-1} ds \,. \tag{7.4.18}$$

This can also be obtained from the relation (7.4.13) with $v = g'$ and $w = g$. Then

$$(g', g) = \int g' \Lambda U \Lambda g \, ds$$

$$= \int_0^t U^{-1}\Lambda^{-1}\Lambda U \Lambda\Lambda^{-1}U^{-1}\,ds = \int_0^t U^{-1}ds\,.$$

McKeague [14] shows that this is the asymptotic covariance of a weighted least-squares estimator introduced by Huffer and McKeague [9, 10]. Hence, this estimator is efficient.

□

Example 7.5 **(Proportional hazards model)**
Here the intensity process is of the form

$$a_{ni\theta}(s) = C_{ni}(s)\lambda(s)\exp(\beta'\boldsymbol{Y}_{ni}(s)),\, s \in [0, 1]\,. \tag{7.4.19}$$

where C_{ni} is a predictable censoring process taking only values 0 and 1, $\boldsymbol{Y}_{ni}$ is a p-dimensional vector of predictable bounded covariate processes, β is a p-dimensional vector of the regression coefficients, and λ is a bounded base line hazard function. Let $\theta = (\beta, \lambda)$. Assume that there exist bounded scalar, vector, and matrix functions S_0, $\boldsymbol{S}_1$, $\boldsymbol{S}_2$, respectively, with S_0 bounded away from zero such that, as $n \to \infty$,

$$\sup_{0\le t\le 1}\left| c_n^{-2}\sum_{i=1} C_{ni}\exp\left(\beta'\boldsymbol{Y}_{ni}\right) - S_0\right| \to 0$$

$$\text{in } P_{n\theta} \;-\text{measure}\,, \tag{7.4.20}$$

$$\sup_{0\le t\le 1}\left| c_n^{-2}\sum_{i=1} C_{ni}\boldsymbol{Y}_{ni}\exp\left(\beta'\boldsymbol{Y}_{ni}\right) - \boldsymbol{S}_1\right| \to 0$$

$$\text{in } P_{n\theta} - \text{measure}\,, \tag{7.4.21}$$

and

$$\sup_{0\le t\le 1}\left| c_n^{-2}\sum_{i=1} C_{ni}\boldsymbol{Y}_{ni}\boldsymbol{Y}'_{ni}\exp\left(\beta'\boldsymbol{Y}_{ni}\right) - \boldsymbol{S}_2\right| \to 0$$

$$\text{in } P_{n\theta} \ - \text{measure} . \tag{7.4.22}$$

Assume that

$$\Sigma = \int_0^1 \left(S_2 - S_0^{-1} S_1 S_1' \right) \lambda ds \tag{7.4.23}$$

is nonsingular.

Let us introduce a local model at $\theta = (\beta, \lambda)$ in the following way. Let $H = R^p \otimes L_2(\lambda)$. Let B be the set of all bounded functions in $L_2(\lambda)$. Set $V = R^p \otimes B$. For each local parameter $(b, v) \in V$ define

$$\beta_{nb} = \beta + c_n^{-1} b , \tag{7.4.24}$$

and

$$\lambda_{nv} = \left(1 + c_n^{-1} v\right) \lambda . \tag{7.4.25}$$

Let

$$a_{nibv} = C_{ni} \lambda_{nv} \exp\left(\beta_{nb}' Y_{ni}\right) . \tag{7.4.26}$$

We now check the conditions (7.4.8), (7.4.4), and (7.4.5) for the intensity process defined by (7.4.19).

(i) **Hellinger differentiability**: Let $r(x) = \exp(x) - 1 - x$. If $C_{ni} = 1$, then, from (7.4.24) and (7.4.25), it follows that

$$\frac{a_{nibv}}{a_{ni}} - 1 = c_n^{-1} \left(b' Y_{ni} + v\right) + c_n^{-2} v b' Y_{ni}$$

$$+ \left(1 + c_n^{-1} v\right) r \left(c_n^{-1} b^{-1} Y_{ni}\right) .$$

Since the covariates are bounded, it follows that

$$\int_0^1 v^2 a_{ni} ds = o_{P_{n\theta}}(1)$$

and the condition (7.4.8) holds with

$$D_{ni}(v) = b' Y_{ni} + v . \tag{7.4.27}$$

(ii) **Lindeberg condition:** Since $D_{ni}(v)$ is bounded, it follows that condition (7.4.4) holds.

(iii) **Nonrandom Hellinger limit:** By (7.4.20) to (7.4.22), we have

$$c_n^{-2}\sum_{i=1}^{n}\int_0^1 D_{ni}^2(v)a_{ni}ds$$

$$= c_n^{-2}\sum_{i=1}^{n}\int_0^1 \left(b'\mathbf{Y}_{ni}+v\right)^2 C_{ni}\lambda\exp\left(\beta'\mathbf{Y}_{ni}\right)ds \tag{7.4.28}$$

$$\to \int_0^1 \left(b'S_2b+2b^1S_1v+S_ov^2\right)\lambda\, ds \text{ in } P_{n\theta} - \text{ measure}$$

as $n \to \infty$. Hence condition (7.4.5) holds with

$$\|(b, v)\|^2 = \int \left(b'S_2b+2b'S_1v+S_0v^2\right)\lambda ds\,. \tag{7.4.29}$$

Hence, by Theorem 7.4, the partially specified loglikelihood is locally asymptotically normal and

$$\ell_{nbv} = n^{-1/2}\sum_{i=1}^{n}\int \left(b'\mathbf{Y}_{ni}+v\right)dM_{ni\theta}(s)$$

$$-\frac{1}{2}\int\left(b'S_2b+2b'S_1v+S_0v^2\right)\lambda ds + o_{P_{n\theta}}(1)\,. \tag{7.4.30}$$

Here

$$M_{ni\theta}(t) = X_{ni}(t) - \int_0^t C_{ni}\lambda\exp\left(\beta'\mathbf{Y}_{ni}\right)ds\,. \tag{7.4.31}$$

By (7.4.29), the acuity is

$$((b, v), (c, w)) = \int \left(b' S_2 c + b' S_1 w + v S_0 w\right) \lambda ds . \tag{7.4.32}$$

We will now determine the asymptotic covariance bound for the estimators of the regression coefficients and the cumulative base line hazard. Let

$$k(\beta, \lambda) = \beta' \tag{7.4.33}$$

where β, the vector of regression coefficients, is considered as a row vector. Note that

$$C_n \left(k\left(\beta_{nb}, \lambda_{nv}\right) - k(\beta, \lambda)\right) = c_n \left(\beta'_{nb} - \beta'\right) = b' .$$

The expression on the right hand side can be considered as a linear functional on V. A gradient (b_β, V_β) of the functional k can be determined by expressing the linear functional in terms of the acuity:

$$b' = \left((b, v), \left(B_\beta, V_\beta\right)\right) \text{ for } (b, v) \in V . \tag{7.4.34}$$

Here (B_β, V_β) is a row of elements $(b_{\beta j}, v_{\beta j})$ in $R^p \times L_2(\lambda)$ which is the gradient of the jth component of the functional k. Using the explicit form (7.4.32) of the activity, we obtain the two equations

$$b' = b' \int \left(S_2 B_\beta + S_1 V_\beta\right) \lambda ds, \; b \in R^p \tag{7.4.35}$$

and

$$0 = \int \left(v S'_1 B_\beta + v S_0 V_\beta\right) \lambda ds, \; v \in B . \tag{7.4.36}$$

The second equation gives $0 = S'_1 B_\beta + S_0 V_\beta$ and hence $V_\beta = -S_0^{-1} S'_1 B_\beta$. This implies that $B_\beta = \sum^{-1}$ where $\sum$ is a given by (7.4.23). Therefore, the gradient is

$$\left(B_\beta, V_\beta\right) = \left(\sum^{-1}, -S_0^{-1} S'_1 \sum^{-1}\right) . \tag{7.4.37}$$

Since $S_0^{-1} S_1$ is bounded, the gradient is canonical. Hence, all the assumptions of the convolution Theorem 6.9 hold. The variance bound for the estimators of

β is now obtained by using the relation (7.4.34) for $(\boldsymbol{b}, v) = (\boldsymbol{B}_\beta, V_\beta)$:

$$\left((\boldsymbol{B}_\beta V_\beta)', (\boldsymbol{B}_\beta V_\beta)\right) = \boldsymbol{B}'_\beta = \sum\nolimits^{-1} . \tag{7.4.38}$$

Andersen and Gill [1, p. 1106] have shown that $\sum^{-1}$ is the asymptotic variance of the Cox estimator. Hence, this estimator is asymptotially efficient.

Let us now consider the asymptotic covariance bound for estimators of the cumulative baseline hazard

$$\int_0^t \lambda ds . \tag{7.4.39}$$

Consider the financial

$$k_t(\beta, \lambda) = \int_0^t \lambda ds . \tag{7.4.40}$$

Define β_{nb} and λ_{nv} as in (7.4.24) and (7.4.25) for the local model. Then, for $(b, v) \in V$,

$$c_n\left(k_t\left(\beta_{nb}, \lambda_{nv}\right) - k_t\left(\beta, \lambda\right)\right) = c_n\left[\int_0^t \lambda_{nv} ds - \int_0^t \lambda ds\right]$$

$$= \int_0^t v\lambda ds . \tag{7.4.41}$$

A gradient, say $(\boldsymbol{B}_t, V_t)$ of the functional k_t is determined by expressing the linear functional on the right-hand side of the above equality in terms of the acuity:

$$\int_0^t v\lambda ds = ((\boldsymbol{B}, v), (\boldsymbol{B}_t, V_t)) \text{ for all } (\boldsymbol{B}, v) \in V . \tag{7.4.42}$$

Using the expression in (7.4.32) which gives the explicit form of the acuity, we obtain the equations

$$0 = \int \left(S_2 B_t + S_1' V_t\right) \lambda ds \,, \tag{7.4.43}$$

and

$$\int_0^t v\lambda ds = \int (vS_1' B_t + vS_0 V_t)\lambda ds, \; v \in B \,. \tag{7.4.44}$$

The second equation shows that

$$I_{[0,t]} = S_1' B_t + S_0 V_t$$

and hence

$$V_t = -S_0^{-1} S_1^1 B_t + S_0^{-1} I_{[0,t]} \,. \tag{7.4.45}$$

Replacing V_t in the first equation by the above expression, we obtain that

$$0 = \sum B_t + G_t$$

with $\sum$ as given by (7.4.23) and

$$G_t = \int_0^t S_0^{-1} S_1 \lambda ds \,.$$

Hence

$$B_t = -\sum\nolimits^{-1} G_t \tag{7.4.46}$$

and the gradient is

$$(B_t, V_t) = \left(-\sum\nolimits^{-1} G_t, S_0^{-1} S_1' G + S_0^{-1} I_{[0,t]}\right) \,, \tag{7.4.47}$$

Since the gradient is bounded, it is canonical. Hence, all the assumptions of the convolution Theorem 6.9 hold and the asymptotic variance bound for the estimators of $k_t(\beta, \lambda) = \int_0^t \lambda ds$ is obtained by applying the analogue of (7.4.34) for $(b, v) = (B_t, V_t)$, namely

$$((B_t, V_t), (B_t, V_t)) = \int_0^t V_t \lambda ds$$

$$= G_t' \Sigma^{-1} G_t + \int_0^t S_0^{-1} \lambda ds . \qquad (7.4.48)$$

Andersen and Gill [1, p. 1108] proved that this is the asymptotic variance of their Breslow-type estimator. Hence, this estimator is asymptotically efficient. □

REMARK 7.3 One can obtain similar results for the model

$$a_{ni\theta}(s) = C_{ni}(s)\lambda(s) r\left(\beta' Y_{ni}(s)\right), s \in [0, 1] \qquad (7.4.49)$$

where $r(\cdot)$ is a known function (cf. [16]). ∎

***Example 7.6* (Counting process regression model)**
Consider the following model. Suppose we observe i.i.d. processes (X_i, C_i, Y_i), $i \geq 1$ over a time interval [0, 1] where X_i is a counting process, Y_i is a covariate process, and C_i is a censoring process taking only values 0 and 1. Let $\{\mathcal{F}_{it}, 0 \leq t \leq 1\}$ be the filtration generated by the process X_i and $\{\zeta_{it}, 0 \leq t \leq 1\}$ be the filtration generated by the process (X_i, C_i, Y_i). Suppose that process X_i admits the Doob–Meyer decomposition with respect to $\{\zeta_{it}, 0 \leq t \leq 1\}$ of the form

$$X_i(t) = M_{\theta i}(t) + \int_0^t C_i(s) a_\theta(s, Y_i(s)) ds . \qquad (7.4.50)$$

The function $a_\theta(s, y)$ is known as the intensity function. The problem is to estimate the parameter θ possibly infinite-dimensional or a function of θ.

Suppose we are interested in estimating the integrated functional

$$A(t, y) = \int_0^t a(s, y) ds , \qquad (7.4.51)$$

which is called the cumulative intensity. Consider the stochastic differential equation

$$dX_i(s) = dM_i(s) + C_i(s)a(s, Y_i(s))ds\,. \qquad (7.4.52)$$

Discretizing the above equation in the space variable, we have

$$\sum_{i=1}^{n} I\left(Y_i(s) \in \mathcal{I}_{nj}\right) dX_i(s)$$

$$= \sum_{i=1}^{n} I(Y_i(s) \in \mathcal{I}_{nj}) dM_{ai}(s)$$

$$+ \sum_{i=1}^{n} C_i(s) I(Y_i(s) \in \mathcal{I}_{nj}) a\,(s, Y_i(s))\, ds \qquad (7.4.53)$$

where $\mathcal{I}_{nj}$ is a small internal. Neglecting the martingale term, we can obtain an estimator for the function $A(t, y)$ for $y \in \mathcal{I}_{nj}$, viz.

$$\bar{A}_n(t, y) = \int_0^t \left(\sum_{i=1}^{n} \left(C_i(s) I\left(Y_i(s) \in \mathcal{I}_{nj}\right)\right)^{-1} \sum_{i=1}^{n} I\left(Y_i(s) \in \mathcal{I}_{nj}\right) dX_i(s)\,. \right.$$

$$(7.4.54)$$

Similarly an estimator for the doubly cumulative intensity functional

$$B(t, z) = \int_0^z \int_0^t a(s, y)\, ds dy \qquad (7.4.55)$$

can be chosen to be

$$\bar{k}_n(t, z) = \int_0^z \bar{A}_n(t, y) dy\,. \qquad (7.4.56)$$

We now obtain an estimator for the doubly cumulative intensity $B(t, z)$ that is asymptotically equivalent to (7.4.55) and asymptotically efficient.

We assume that the function $a(s, y)$ is bounded and identify the parameter θ with the function $a(\cdot, \cdot)$. Let Θ be the parameter space of all bounded functions. We further assume that the covariate process Y takes values in $[0, 1]$ and the following condition holds:

(R) Let δ be the counting measure on $\{0, 1\}$ and let $(Y(s), C(s))$ have $dyd\delta$-density $f(s, y, \delta)$ under P_a-measure such that $f(s, y, 1)$ is bounded away from zero on $[0, 1]$.

Fix $a \in \Theta$. Let us introduce a local model at a as was done in the earlier sections.

Let V be the set of all bounded functions $v(s, y)$. Define

$$a_{nv} = \left(1 + n^{-1/2} v\right) a\,. \tag{7.4.57}$$

Suppose the process $C(s)a_{nv}(s, Y(s))$ is Hellinger differentiable in the sense

$$\int \int_0^t \left(n^{1/2} \left(\left(\frac{c(s)a_{nv}(s, Y(s))}{c(s)a(s, Y(s))} \right)^{1/2} - 1 \right) - \frac{1}{2} D(v)(s) \right)^2$$

$$C(s)a(s, y(s))ds\, dP_a \to 0 \tag{7.4.58}$$

as $n \to \infty$ where

$$D(v)(s) = v(s, Y(s))\,. \tag{7.4.59}$$

Define

$$\begin{aligned} d^2(v)(s) &= \int D^2(v)(s)C(s)a(s, Y(s))dP_a \\ &= \int v^2(s, Y(s))C(s)a(s, (Y(s))dP_a \\ &= \int_0^1 v^2(s, y) f(s, y, 1)a(s, y)dy\,. \end{aligned} \tag{7.4.60}$$

Then

$$\|v\|_a^2 = \int_0^1 d^2(v)(s)ds$$

$$= \int_0^1 \left\{ \int_0^1 v^2(s, y) f(s, y, 1) a(s, y) dy \right\} ds . \qquad (7.4.61)$$

Suppose we observe n i.i.d. copies (X_i, Y_i, C_i) of the process (X, Y, C). Let ℓ_{nv} denote the partially specified log likelihood between a_{nv} and a_v, viz.

$$\ell_{nv} = \sum_{i=1}^{n} \left(\int_0^1 \log\left(\frac{C_i a_{nvi}}{C_i a_i}\right) dX_i(s) - \int_0^1 (a_{nvi} - a_i)\, C_i ds \right) . \qquad (7.4.62)$$

It follows by the arguments similar to those given earlier that

$$\ell_{nv} = Z_{an}(v) - \frac{1}{2}\|v\|_a^2 + o_{P_a}(s) \qquad (7.4.63)$$

where $\|v\|_a^2$ is given by (7.4.61) and

$$Z_{an}(v) = n^{-1/2} \sum_{i=1}^{n} \int_0^1 v\,(s, Y_i(s))\, dM_{ai}(s) \qquad (7.4.64)$$

is asymptotically normal under P_a with Var $\|v\|_a^2$. Note that M_{ai} is a martingale and

$$dM_{ai}(s) = dX_i(s) - C_i(s) a(s, Y_i(s)) ds, 0 \le s \le 1 . \qquad (7.4.65)$$

The acuity for the norm (7.4.61) is

$$(v, w)_a \equiv \int_0^1 \int_0^1 v(s, y) w(s, y) f(s, y, 1) a(s, y) ds\, dy . \qquad (7.4.66)$$

Let us now determine the asymptotic variance bound for the estimators of the doubly cumulative hazard $B(t, z)$ defined by (7.4.55). For fixed $t \in [0, 1]$ and $z \in [0, 1]$, consider

$$k(a) = \int_0^z \int_0^t a(s, y) ds\, dy, \quad a \in \Theta . \tag{7.4.67}$$

Note that

$$n^{1/2}(k(a_{nv}) - k(a)) = \int_0^z \int_0^t n^{1/2}\left(\left[\left(1 + n^{-1/2}v\right)\right) a\right] - a\right) ds\, dy$$

$$= \int_0^z \int_0^t v(s, y) a(s, y) ds dy . \tag{7.4.68}$$

A gradient, say g, of the functional k at a is determined by expressing the linear functional of v on the right-hand side of (7.4.68) in terms of the acuity defined by (7.4.66). Hence,

$$\int_0^z \int_0^t v(s, y) a(s, y) ds\, dy = (v, g)_a \text{ for } v \in V . \tag{7.4.69}$$

From the relation (7.4.66), it follows that

$$\int_0^z \int_0^t v(s, y) a(s, y) ds dy = \int_0^z \int_0^t v(s, y) g(s, y) f(s, y, 1) a(s, y) ds dy . \tag{7.4.70}$$

A solution of this equation is

$$g(s, y) = f(s, y, 1)^{-1} I(s \le t) I(y \le z) . \tag{7.4.71}$$

It follows that the gradient is bounded by the condition (R). Hence, $g \in V$ and the asymptotic variance bound for the regular estimator of k at a is $\|g\|_a^2$ by the convolution Theorem 6.9. Applying the relation (7.4.69) for $v = g$, we have

$$\|g\|_a^2 = \int_0^z \int_0^t f(s, y, 1)^{-1} a(s, y) ds dy . \tag{7.4.72}$$

We now construct an efficient estimator $\hat{k}_n$ for k. A regular and efficient estimator $\hat{k}_n$ for k is characterized by the relation

$$n^{1/2}\left(\hat{k}_n - k(a)\right) = Z_{an}(g) + o_{P_{n\theta}}(1) \tag{7.4.73}$$

where $Z_{an}(g)$ is as given by (7.4.64). It is sufficient to find an estimator $\hat{g}_n$ for g such that

$$Z_{an}\left(\hat{g}_n\right) = n^{1/2}\left(\frac{1}{n}\sum_{i=1}^{n}\int_0^1 \hat{g}_n\left(s, Y_i(s)\right) dX_i(s) - k(a)\right)$$

$$+ o_{P_{n\theta}}(1) . \tag{7.4.74}$$

An estimator for $f(s, Y, 1)$ is

$$\hat{f}_{nj}(s) = d_n^{-1} n^{-1} \sum_{i=1}^{n} C_i(s) I\left(Y_i(s) \in I_{nj}\right), \quad y \in \mathcal{I}_{nj} \tag{7.4.75}$$

where $\{\mathcal{I}_{n1}, \ldots, \mathcal{I}_{nd_n}\}$ is a partition of $[0, 1]$ into interval of length d_n. Define the estimator $\hat{g}_n$ by

$$\hat{g}_n(s, y) = \left[\hat{f}_{nj}(s)\right]^{-1} I(s \le t) I(y \le z), \quad y \in \mathcal{I}_{nj} . \tag{7.4.76}$$

Let

$$Z_{an}\left(\hat{g}_n\right) = n^{1/2}\left(\frac{1}{n}\sum_{i=1}^{n}\int_0^1 \hat{g}_n\left(s, Y_i(s)\right) dX_i(s) - k_n\right) \tag{7.4.77}$$

with

$$k_n = \frac{1}{n}\sum_{i=1}^{n}\int_0^1 \hat{g}_n(s, Y_i(s))C_i(s)a(s, Y_i(s))ds. \tag{7.4.78}$$

Note that the relation (7.4.74) holds if

$$n^{1/2}(k_n - k(a)) = o_{P_{n\theta}}(1). \tag{7.4.79}$$

Observe that

$$k_n - k(a) = \sum_{j=1}^{dn}\int_0^1\int_0^1 \left(\sum_{i=1}^{n}\left\{\frac{C_i(s)I\left(Y_i(s)\in \mathcal{I}_{nj}\right)}{\sum_{\ell=1}^{n} C_\ell(s)I\left(Y_\ell(s)\in \mathcal{I}_{nj}\right)} \times a(s, Y_i(s)) \right.\right.$$

$$\left.\left. I\left(Y_i(s)\le z\right) - a(s, y)I(y\le z)\right\}\right) I(s\le t)I\left(y\in \mathcal{I}_{nj}\right)dsdy. \tag{7.4.80}$$

This is of the order $o_{P_a}(n^{-1/2})$ if $n^{1/2}dn \to 0$ and a satisfies the Lipschitz condition. Hence, the estimator

$$\hat{k}_n = \frac{1}{n}\sum_{i=1}^{n}\int_0^1 \hat{g}_n(s, Y_i(s))\,dX_i(s) \tag{7.4.81}$$

is asymptotically efficient and regular for k at a. This estimator $\hat{k}_n$ can also be rewritten in the form

$$\hat{k}_n = \sum_{j=1}^{dn}\int_0^1 \hat{f}_{nj}(s)^{-1}\frac{1}{n}\sum_{i=1}^{n} I(Y_i(s)\le z)I\left(Y_i(s)\in I_{nj}\right)dX_i(s). \tag{7.4.82}$$

This estimator is stochastically equivalent to the estimator (7.4.56), and hence that estimator is also regular and efficient (cf. [15]).

Improvement of Preliminary Estimators

Let $(\Omega_n, \mathcal{F}_n, \zeta_n, P_{n\theta} : \theta \in \Theta\}$ be a filtered model as before with arbitrary parameter space Θ where Θ might be infinite-dimensional. Suppose we observe counting processes $X_{n1}, \ldots, X_{nn}$ and covariate processes $Y_{n1}, \ldots, Y_{nn}$ on [0, 1]. Assume that the Doob–Meyer decomposition of X_{ni} with respect to ζ_n and $P_{n\theta}$ is of the form

$$X_{ni}(t) = M_{ni\theta} + \int_0^t a_{ni\theta}(s, Y_{ni}(s))\, ds$$

where $M_{ni\theta}$ is a martingale and $a_{ni\theta}(s, Y_{ni}(s))$ is a predictable intensity process.

We assume as before that the covariate process Y_{ni} need not be adapted to the filtration generated by the counting process, and hence the distribution of the counting process is not determined by the intensity process. Consider

$$\ell_{n\theta\tau} = \sum_{i=1}^{n} \left(\int_0^1 \log \left(\frac{a_{ni\tau}(s, Y_{ni}(s))}{a_{ni\theta}(s, Y_{ni}(s))} \right) dX_{ni}(s) \right.$$

$$\left. - \int_0^1 (a_{ni\tau}(s, Y_{ni}(s)) - a_{ni\theta}(s, Y_{ni}(s))\, ds \right) \tag{7.4.83}$$

which is the partially specified log likelihood between τ and θ.

Let us introduce a local parameterization $\xi_{n\theta} : \Theta \to H_\theta$. The map $\xi_{n\theta}$ will rescale $\tau \in \Theta$ around θ as $c_n(\tau - \theta)$ with $c_n \to \infty$ and then we apply an approximate linear isometry from Θ into some Hilbert space, H_θ, with norm $\|\cdot\|_\theta$ determined by the structure of the model.

Suppose that the intensity processes $a_{ni\theta}$ are Hellinger differentiable at θ with random differential operator $D_{ni\theta}$ and that the Hellinger sequence is asymptotically nonrandom. Sufficient conditions for such a property are given in the earlier sections. Then the partially specified log likelihood is locally asymptotically normal for τ near θ, that is,

$$\ell_{n\theta\tau} = c_n^{-1} \sum_{i=1}^{n} \int_0^1 D_{ni\theta}(\xi_{n\theta}(\tau))(s)\, dM_{ni\theta}(s)$$

$$- \frac{1}{2} \|\xi_{n\theta}(\tau)\|_\theta^2 + o_{P_{n\theta}}(1) \tag{7.4.84}$$

where the martingale term is asymptotically normal with Var $\|\xi_{n\theta}(\tau)\|_\theta^2$. Let us call the inner product corresponding to the norm $\|\cdot\|_\theta^2$ as the acuity as defined earlier.

Let $k : \Theta \to R$ and suppose that k is differentiable at θ with gradient $g_\theta \in H_\theta$ in the sense that, for τ near θ,

$$c_n(k(\tau) - k(\theta)) - (\xi_{n\theta}(\tau), g_\theta)_\theta \to 0 \text{ as } n \to \infty . \tag{7.4.85}$$

We assume that g_θ is the canonical gradient.

Recall that an estimator $\hat{k}_n$ for k is said to be regular and asymptotically efficient at θ if

$$c_n\left(\hat{k}_n - k(\theta)\right) = G_n(\theta) + o_{P_{n\theta}}(1) \tag{7.4.86}$$

with

$$G_n(\theta) = c_n^{-1} \sum_{i=1}^{n} \int_0^1 D_{ni\theta}(g_\theta)(s) dM_{ni\theta}(s) . \tag{7.4.87}$$

Suppose $G_n(\theta)$ admits the expansion

$$G_n(\tau) = G_n(\theta) - C_n(k(\tau) - k(\theta)) + o_{P_{n\theta}}(1) \tag{7.4.88}$$

for τ near θ. This expansion follows from the local asymptotic normality given in (7.4.84) and the differentiability of the functional k.

We now study three methods of constructing efficient estimators for $k : \Theta \to R$.

1. Martingale estimator approach: It is possible at times to find an estimator $\tilde{D}_{ni}$ for $D_{ni\theta}(g_\theta)$ such that

$$G_n(\theta) = c_n^{-1} \sum_{i=1}^{n} \int_0^1 \tilde{D}_{ni}(s) dM_{ni\theta}(s) + o_{P_{n\theta}}(1) \tag{7.4.89}$$

and

$$c_n^{-2} \sum_{i=1}^{n} \int_0^1 \tilde{D}_{ni}(s) dM_{ni\theta}(s) = \hat{k}_n - k(\theta) . \tag{7.4.90}$$

Relation (7.4.90) indicates that the difference between the estimator k_n and the value $\hat{k}(\theta)$ of the functional can be written as a stochastic integral with respect to the martingale $M_{ni\theta}$. Such estimators are called *martingale estimators*. In view of (7.4.86) and (7.4.89), it follows that $\hat{k}_n$ is efficient. If the conditions (7.4.89) and (7.4.90) hold for $\theta = \tau$, then the expansion (7.4.88) holds.

2. Newton–Raphson approach: Let $\tilde{k}_n$ be an estimator for $k(\theta)$ and $\tilde{G}_n$ be an "estimator" for G_n such that

$$\tilde{G}_n = G_n(\theta) - c_n\left(\tilde{k}_n - k(\theta)\right) + o_{P_{n\theta}}(1) . \tag{7.4.91}$$

Define

$$\hat{k}_n = \tilde{k}_n + c_n^{-1}\tilde{G}_n .$$

Then $\hat{k}_n$ satisfies (7.4.86) and the estimator is efficient. If the expansion (7.4.88) holds and if $\tilde{\theta}_n$ is a c_n-consistent estimator for θ, then (7.4.91) holds for $\tilde{k}_n = k(\tilde{\theta}_n)$ and $\tilde{G}_n = G_n(\tilde{\theta}_n)$.

3. Estimating equations approach: Let $\hat{\theta}_n$ be a c_n-consistent solution of the estimating equation

$$G_n\left(\hat{\theta}_n\right) = o_{P_{n\theta}}(1) . \tag{7.4.92}$$

Apply the expansion (7.4.88) for $\tau = \hat{\theta}_n$. Then we have the relation (7.4.86) for $\hat{k}_n = k(\hat{\theta}_n)$ and hence $\hat{k}_n$ is efficient. □

Example 7.7

We now illustrate the three methods described earlier for constructing efficient estimators for the problem of estimating the cumulative hazard rate on the basis of n i.i.d. observations with an absolutely continuous distribution function F and the density function f on the interval $[0, 1]$. This model can be written as a multiplicative intensity model. We observe the counting process $X_n = nF_n$ where F_n is the empirical distribution. Its intensity process is of the form $Y_n(s)\lambda(s)$ with $Y_n = n(1 - F_{n-})$ and $\lambda = (1 - F)^{-1}f$. This is a special case of the multivariate intensity model discussed in Section 6.5. Let

$$\Lambda(t) = \int_0^t \frac{f(s)}{1 - F(s)} ds . \tag{7.4.93}$$

The function $\Lambda(\cdot)$ is the cumulative hazard rate. Let Θ be the parameter space of the set of cumulative hazard rates Λ. We are interested in constructing asymptotically efficient estimators for Λ.

Fix Λ. The log likelihood between Λ' and Λ can be written as

$$\log\left(\frac{dP^n_{\Lambda'}}{dP^n_{\Lambda}}\right) = \int_0^1 \log\left(\frac{\lambda'}{\lambda}\right) dX_n(s) - \int_0^1 Y_n\left(\lambda' - \lambda\right) ds\,. \qquad (7.4.94)$$

Let us introduce a local model. For any bounded function v on [0, 1], define a sequence of parameter Λ_{nv} by

$$d\Lambda_{nv} = \left(1 + n^{-1/2}v\right) d\Lambda\,.$$

Let $\Lambda' = \Lambda_{nv}$ in (7.4.94). By the Taylor's expansion,

$$\begin{aligned}\ell_{nv} &= \log\left(dP^n_{\Lambda_{nv}}/dP^n_{\Lambda}\right)\\ &= n^{-1/2}\int_0^1 v\,dM_{n\Lambda}(s) - \frac{1}{2}\|v\|^2_{\Lambda} + o_{P_{\Lambda}}(1)\end{aligned} \qquad (7.4.95)$$

with

$$\mathcal{L}\left(n^{-1/2}\int_0^1 v\,dM_{n\Lambda}(s)\,|P^n_{\Lambda}\right) \Rightarrow M\left(0, \|v\|^2_{\Lambda}\right) \text{ as } n \to \infty \qquad (7.4.96)$$

and $M_{n\Lambda}$ is the martingale defined by

$$\begin{aligned}dM_{n\Lambda(s)} &= dX_n(s) - Y_n\lambda ds\\ &= n dF_n(s) - n\left(1 - F_{n-}\right)\lambda ds\end{aligned} \qquad (7.4.97)$$

Here the $\|\cdot\|_{\Lambda}$ is defined by

$$\|v\|^2_{\Lambda} = \int v^2 f ds\,. \qquad (7.4.98)$$

Relation (7.4.95) indicates the local asymptotic normality. Here $(\cdot,\cdot)_\Lambda$ corresponding to the norm $\|\cdot\|_\Lambda$ is the acuity.

Let us now consider the problem of estimating

$$k_t(\Lambda) = \Lambda(t) = \int_0^t \lambda ds \tag{7.4.99}$$

for fixed $t \in [0, 1]$. The functional k_t is linear in the local parameter v and can be expressed in terms of a gradient $g_{t\Lambda}$ with respect to the acuity $(\cdot,\cdot)_\Lambda$:

$$n^{1/2}(k_t(\Lambda_{nv}) - k_t(\Lambda)) = n^{1/2}(\Lambda_{nv}(t) - \Lambda(t))$$

$$= \int_0^t v\lambda ds = (v, g_{t\Lambda})_\Lambda \tag{7.4.100}$$

with $g_{t\Lambda}(s) = (1-F(s))^{-1}I[0,t](s),\ s \in [0, 1]$. Note that an estimator $\hat{\Lambda}_n(t)$ of $\Lambda(t)$ is regular and efficient in the sense of convolution theorem if

$$n^{1/2}(\Lambda_n(t) - \Lambda(t)) = G_{nt}(\Lambda) + o_{P_\Lambda}(1) \tag{7.4.101}$$

with

$$G_{nt}(\Lambda) = n^{-1/2}\int_0^1 g_{t\Lambda} dM_{n\Lambda}(s)$$

$$= n^{1/2}\int_0^t (1-F)^{-1}(dF_n(s) - (1-F_{n-})\lambda ds)\ . \tag{7.4.102}$$

Let us replace F by a (predictable) estimator F_{n-} to obtain

$$G_{nt}(\Lambda) = \tilde{G}_{nt}(\Lambda(t)) + o_{P_\Lambda}(1) \tag{7.4.103}$$

with

$$\tilde{G}_{nt}(k) = n^{1/2}\left(\int_0^t (1 - F_{n-})^{-1}\, dF_n(s) - k\right), \quad k \text{ real}. \qquad (7.4.104)$$

We will now obtain improved estimators for Λ by the three methods described earlier.

1. Martingale estimator approach: Comparing (7.4.101) and (7.4.103), an efficient estimator $\hat{\Lambda}_n(t)$ for $\Lambda(t)$ is given by

$$\hat{\Lambda}_n(t) = \int_0^t (1 - F_{n-})^{-1}\, dF_n(s). \qquad (7.4.105)$$

This is the *Nelson estimator* for the cumulative hazard rate $\Lambda(t)$. It is a martingale estimator in the sense that

$$\hat{\Lambda}_n(t) - \Lambda(t) = n^{-1}\int_0^t (1 - F_{n-})^{-1} dM_{n\Lambda}(s) \qquad (7.4.106)$$

and the term on the right side is a stochastic integral with respect to $M_{n\Lambda}$.

2. Newton-Raphson approach: Suppose $\tilde{\Lambda}_n(t)$ is a preliminary estimator for $\Lambda(t)$. Let us estimate $G_{nt}(\Lambda)$ by $\tilde{G}_{nt}(\tilde{\Lambda}_n(t))$ with $\tilde{G}_{nt}$ as defined in (7.4.104). Relation (7.4.103) implies that

$$\begin{aligned}\tilde{G}_{nt}\left(\tilde{\Lambda}_n(t)\right) &= \tilde{G}_{nt}(\Lambda(t)) - n^{1/2}\left(\tilde{\Lambda}_n(t) - \Lambda(t)\right) \\ &= G_{nt}(\Lambda) - n^{1/2}\left(\tilde{\Lambda}_n(t) - \Lambda(t)\right) + o_{P_\Lambda}(1). \qquad (7.4.107)\end{aligned}$$

Hence (7.4.101) holds for the improved estimator

$$\check{\Lambda}_n(t) = \tilde{\Lambda}_n(t) - n^{-1/2}\tilde{G}_{nt}\left(\tilde{\Lambda}_n(t)\right). \qquad (7.4.108)$$

This proves that the estimator $\hat{\Lambda}_n(t)$ is efficient. It is easy to check that the improved estimator leads to the Nelson estimator given by (7.4.105) in view of (7.4.104).

3. Estimating equation approach: Let $D[0, 1]$ be the Skorokhod space. Note that k_t and G_{nt} are defined over $\Theta \subset D[0, 1]$. Let us extend k_t to $D[0, 1]$ by the functional $k_t(\Lambda) = \Lambda(t)$, for $\Lambda \in D[0, 1]$ and for $M \notin \Theta$, set

$$G_{nt}(M) = \tilde{G}_{nt}(M(t)) . \tag{7.4.109}$$

Note that the Nelson estimator is a pure jump process and hence it is outside Θ almost surely. In view of the definition (7.4.104) for $\tilde{G}_{nt}$, the Nelson estimator given by (7.4.105) is a solution of the estimating equation

$$G_{nt}\left(\hat{\Lambda}_n\right) = 0, \ 0 \le t \le 1 . \tag{7.4.110}$$

In view of (7.4.103) and (7.4.104), for $M \notin \Theta$,

$$\begin{aligned} G_{nt}(M) - G_{nt}(\Lambda) & \\ &= \tilde{G}_{nt}(M(t)) - G_{nt}(\Lambda) \\ &= \tilde{G}_{nt}(M(t)) - G_{nt}(\Lambda(t)) + o_{P_\Lambda}(1) \\ &= n^{1/2}(M(t) - \Lambda(t)) + o_{P_\Lambda}(1) , \end{aligned} \tag{7.4.111}$$

which proves (7.4.101). Hence $\hat{\Lambda}_n(t)$ is efficient. □

REMARK 7.4 For other applications to Aalen's additive risk model and Cox's proportional hazard model, see [5]. ∎

References

[1] Andersen, P.K. and Gill, R.D. (1982) Cox's regression model for counting processes: a large sample study, *Ann. Statist.*, **10**, 1100–1120.

[2] Cox, D.R. (1975) Partial likelihood, *Biometrika*, **62**, 269–276.

[3] Gill, R.D. (1985) Note on product integration, likelihood and partial likelihood for counting processes, *Tech. Report*, CWI, Amsterdam.

[4] Greenwood, P. (1988) Partially specified semimartingale experiments, *Statistical Inference from Stochastic Processes*, (Ed. N.U. Prabhu), *Contemporary Math.*, **80**, 1–17.

[5] Greenwood, P.E. and Wefelmeyer, W. (1989) Efficient estimating equations for nonparametric filtered models, *Tech. Report*, No. 120, University of Cologne.

[6] Greenwood, P.E. and Wefelmeyer, W. (1989) Efficient estimation in a nonlinear counting process model, *Tech. Report*, No. 121, University of Cologne.

[7] Greenwood, P.E. and Wefelmeyer, W. (1989) Partially and fully specified models and efficiency, *Tech. Report*, No. 124, University of Cologne.

[8] Greenwood, P.E. and Wefelmeyer, W. (1990) Efficiency of estimators for partially specified filtered models, *Stoch. Proc. Appl.*, **36**, 353–370.

[9] Huffer, F.W. and McKeague, I.W. (1987) Survival analysis using additive risk models, *Tech. Report*, No. 396, Stanford University.

[10] Huffer, F.W. and McKeague, I.W. (1991) Weighted least squares estimators for Aalen's additive risk model, *J. Amer. Statist. Assoc.*, **86**, 114–129.

[11] Jacod, J. (1987) Partial likelihood process and asymptotic normalty, *Stoch. Proc. Appl.*, **26**, 47–71.

[12] Jacod, J. (1990) Regularity, partial regularity, partial information process, for a filtered statistical model, *Probab. Theory Relat. Fields*, **86**, 305–335.

[13] Jacod, J. and Shiryayev, A.N. (1987) *Limit Theorems for Stochastic Processes*, Springer, Heidelberg.

[14] McKeague, I.W. (1988) Asymptotic theory for weighted least squares estimators in the Aalen's additive risk model, *Statistical Inference for Stochastic Processes,* (Ed. N.U. Prabhu), *Contemporary Mathematics,* **80**, 139–152.

[15] McKeague, I.W. and Utikal, K.J. (1990) Inference for a nonlinear counting process regression model, *Ann. Statist.*, **18**, 1172–1187.

[16] Prentice, R.L. and Self, S.G. (1983) Asymptotic distribution theory for Cox-type regression models with general relative risk form, *Ann. Statist.*, **11**, 804–812.

Chapter 8

Inference for Counting Processes

8.1 Introduction

Point processes occur widely in applications in modeling of many phenomena in physical, technical, economic, biological sciences, and many other areas (cf. [16, 37]). It is well known that the Poisson process is a satisfactory model for events like times for radioactive decay, arrival times for service, times of rejection of complex electronic equipment, etc. Poisson process of variable intensity has been found useful in modeling the number of calls waiting on telephone at different times of the day, the times of change in cathode temperature, etc. Parametric inference for such processes has been discussed in [32] and more recently in [29]. Statistical inference for counting processes in general is extensively discussed in [6] and [18].

Nonhomogeneous Poisson Processes

Given a probability space $(\Omega, \mathcal{F}, P)$, an integer-valued random process $X = \{X(t), t \geq 0\}$ is called a *nonhomogeneous Poisson process* if $X(0) = 0$, the increments over disjoint intervals are independent and with distributions according to a Poisson law; that is, for every $0 \leq t_1 < t_2,\ k \geq 0$,

$$P\left(X(t_2) - X(t_1) = k\right) = \frac{1}{k!}\left(\int_{t_1}^{t_2} S(t)dt\right)^k \exp\left(-\int_{t_1}^{t_2} S(t)dt\right). \tag{8.1.1}$$

The function $S(t)$ is called the *intensity* function of the process X. It is easy to check that

$$EX(t) = \int_0^t S(\tau)d\tau, \quad \mathrm{Var}X(t) = \int_0^t S(\tau)d\tau . \tag{8.1.2}$$

Let

$$M(t) = X(t) - \int_0^t S(\tau)d\tau \tag{8.1.3}$$

and

$$J(T) = \int_0^T f(t)\, dN(t) \tag{8.1.4}$$

be defined as the Lebesgue–Stieltjes integral of a random process $\{f(t),\ t \geq 0\}$ that is right-continuous and measurable with respect to the σ-algebra $\mathcal{F}_t = \sigma\{X(s), 0 \leq s \leq t\},\ t \geq 0$. The integral $J(T)$ possesses properties similar to those of an Ito stochastic integral with respect to a Wiener process. Suppose that

$$E\left(\int_0^T |f(t)|S(t)dt\right) < \infty . \tag{8.1.5}$$

Then

$$E\int_0^T f(t)dM(t) = 0 . \tag{8.1.6}$$

If, further,

$$E\left(\int_0^T f^2(t)S(t)dt\right) < \infty , \tag{8.1.7}$$

then

$$E\left(\int_0^T f(t)dM(t)\right)^2 = E\left(\int_0^T f^2(t)S(t)dt\right) . \tag{8.1.8}$$

The following result is an analogue of the Ito's lemma.

THEOREM 8.1
Let $\{f(t),\ 0 \le t \le T\}$ be a nonrandom right-continuous function integrable with respect to the weight function $S(\cdot)$ on $[0, T]$. Let $F(t, x)$ be continuously differentiable with respect to t and x. Then

$$F(T, J(T)) = F(0,0) + \int_0^T F_t'(t, J(t-))dt$$

$$+ \int_0^T [F(t, J(t-) + f(t)) - F(t, J(t-))]dM(t)$$

$$+ \int_0^T [F(t, J(t-) + f(t)) - F(t, J(t-)) +$$

$$- F_x'(t, J(t-))f(t)]\, S(t)dt\,. \tag{8.1.9}$$

(Here F_x' and F_t' denote the partial derivatives of F with respect to x and t, respectively.)

For proof, see [32, p. 120–12]. The following result gives a formula for the likelihood ratio corresponding to two nonhomogeneous Poisson processes.

Let $\mathcal{X}$ denote the space of functions $\{x(t), t \ge 0\}$ right-continuous with left limits such that $x(0) = 0,\ x(t) = x(t-) + (1 \text{ or } 0),\ x(t) < \infty$ for all $t \ge 0$. Let $\mathcal{X}_T$ denote the class of all such function on $[0, T]$. Let $\mathcal{B}_T$ be the associated Borel σ-algebra under the Skorokhod topology on $\mathcal{X}_T$. Let $P_S^{(T)}$ be the probability measure induced by a nonhomogeneous Poisson process X on $[0, T]$ with intensity $S(\cdot)$.

THEOREM 8.2
Let $S_i(t) > 0$ for $t \in [0, T],\ i = 1, 2$. Then the measures $P_{S_1}^{(T)}$ and $P_{S_2}^{(T)}$ are equivalent to each other and

$$\frac{dP_{S_1}^{(T)}}{dP_{S_2}^{(T)}} = \exp\left\{\int_0^T \log\frac{S_1(t)}{S_2(t)}dX(t) - \int_0^T (S_1(t) - S_2(t))\,dt\right\} \tag{8.1.10}$$

For proof, see [32].

Processes of Poisson Type

Let $(\Omega, \mathcal{F}, P)$ be a probability space and $\{\mathcal{F}_t, t \geq 0\}$ be a nondecreasing family of right continuous σ-algebras contained in $\mathcal{F}$ completed with respect to the P-null sets. Let $\{\tau_i, i \geq 1\}$ be a sequence of stopping times adapted to $\{\mathcal{F}_t\}$ such that $\tau_1 > 0$ and $\tau_j < \tau_{j+1}$ almost surely $[P]$. Consider the random process

$$N(t) = \sum_{j=1}^{\infty} I\left[\tau_j \leq t\right], \; t \geq 0 .$$

The process N is called a *counting (point) process.*

It is clear that

$$N(0) = 0, \;\; N(t) = N(t-) + (0 \text{ or } 1)$$

and the realizations of $N(\cdot)$ are piecewise constant functions that are right-continuous with left limits and jump size equal to one. We assume that

$$EN(t) < \infty, \; t \geq 0 ,$$

which ensures that the process $\{\tau_i, i \geq 1\}$ does not have an accumulation point in any bounded interval with probability one. The process $\{N(t), \; t \geq 0\}$ is nondecreasing and is a supermartingale. Hence, it admits a decomposition

$$N(t) = \bar{M}(t) + \bar{A}(t), \; t \geq 0 \tag{8.1.11}$$

as the sum of a martingale $\bar{M} = \{\bar{M}(t), \mathcal{F}_t, t \geq 0\}$ and a natural increasing (nondecreasing) process $\bar{A} = \{\bar{A}(t), \mathcal{F}_t, t \geq 0\}$. The process $\bar{A}$ is the *compensator* for the process.

Let $\mathcal{F}_t^N = \sigma\{N(s), \; 0 \leq s \leq t\}, \; t \geq 0$ denote the σ-algebra generated by the process N. Then the process N admits a *unique (minimal) representation*

$$N(t) = M(t) + A(t), \; t \geq 0$$

where $M = \{M(t), \; \mathcal{F}_t^N, \; t \geq 0)$ is a martingale and $A = \{A(t), \mathcal{F}_t^N, \; t \geq 0\}$ is a predictable nondecreasing process. If the compensator $\bar{A}$ is absolutely continuous and

$$\bar{A}(t) = \int_0^t \bar{\lambda}(s)ds$$

for some $\bar{\lambda} = \{\bar{\lambda}(s), \mathcal{F}_t, t \geq 0\}$, then

$$A(t) = \int_0^t \lambda(s)ds \qquad (8.1.12)$$

where

$$\lambda(t) = E\left(\bar{\lambda}(t)|\mathcal{F}_t^N\right), \; t \geq 0 \qquad (8.1.13)$$

(cf. [40, Theorem 18.3]).

A point process N with compensator (8.1.12) is called a *process of Poisson type* with the intensity process

$$\lambda = \left\{\lambda(t), \mathcal{F}_t^N, t \geq 0\right\}.$$

A nonhomogeneous Poisson process with nonrandom intensity function $S(t)$, discussed earlier, is a special case of a process of Poisson type.

Let $f = \{f(t), \mathcal{F}_t^N, t \geq 0\}$ be a predictable process with $P\{|f(t)| < \infty\} = 1$ and

$$E\left\{\int_0^T f(t)^2 \lambda(t)dt\right\} < \infty. \qquad (8.1.14)$$

Let

$$I_T(f) = \int_0^T f(t)dM(t) \qquad (8.1.15)$$

where the integral is defined as a Lebesgue–Stieltjes integral. It can be checked that the process $\{I_T(f), \; T \geq 0\}$ is a square integrable martingale with

$$E\left\{\int_0^T f(t)dM(t)|\mathcal{F}_s^N\right\} = \int_0^s f(t)dM(t), \; s \leq T. \qquad (8.1.16)$$

Furthermore,

$$E\left[I_T(f)\right] = 0\,, \tag{8.1.17}$$

$$E\left[I_T(f)\right]^2 = E\left[\int_0^T f(t)^2\lambda(t)dt\right], \tag{8.1.18}$$

and

$$P\left(|I_T(f)| > \varepsilon\right\} \le \frac{\eta}{\varepsilon^2} + P\left[\int_0^T f(t)^2\lambda(t)dt > \eta\right] \tag{8.1.19}$$

for any $\eta > 0$ and $\varepsilon > 0$. The last inequality is similar to the Lenglart inequality (cf. Chapter 1 or [35]).

Let $P_\lambda^{(T)}$ be the probability measure generated by a process of Poisson type with the intensity process $\{\lambda_i(t), \mathcal{F}_t^N,\ t \ge 0\}$ on the space $(\mathcal{X}_T, \mathcal{B}_T)$ for $i = 1, 2$. Suppose that

$$P\left(\lambda_i(t) > 0 \text{ for } 0 \le t \le T\right) = 1 \text{ for } i = 1, 2\,.$$

The following theorem gives a formula for the likelihood ratio.

THEOREM 8.3
If

$$P_{\lambda_1}^{(T)}\left\{\int_0^T \lambda_2(t)dt < \infty\right\} = 1 \text{ and } P_{\lambda_2}^{(T)}\left\{\int_0^T \lambda_1(t)dt < \infty\right\} = 1\,, \tag{8.1.20}$$

then $P_{\lambda_1}^{(T)}$ and $P_{\lambda_2}^{(T)}$ are equivalent and

$$\frac{dP_{\lambda_1}^{(T)}}{dP_{\lambda_2}^{(T)}}(N_T) = \exp\left\{\int_0^T \log\frac{\lambda_1(t)}{\lambda_2(t)}dN(t) - \int_0^T \left[\lambda_1(t) - \lambda_2(t)\right]dt\right\}. \tag{8.1.21}$$

REMARK 8.1 Note that Theorem 8.2 is a special case of Theorem 8.3. For proofs, see [27, 40]. ▮

8.2 Parametric Inference for Counting Processes

We now study the problem of the estimation of parameters involved in the intensity process. At first, we discuss the case of a nonhomogeneous Poisson process very briefly. An extensive investigation has been carried out in [32].

Estimation for Nonhomogeneous Poisson Process

Suppose Θ is open $\subset R$ and for every $\theta \in \Theta$, suppose that $\boldsymbol{X}_T = \{X(t), 0 \leq t \leq T\}$ is a nonhomogeneous Poisson process with the intensity $S_T(\theta) = \{S(\theta, t), 0 \leq t \leq T\}$. Suppose that $S(\theta, t) > 0$ for all $t \in [0, T]$ and $\theta \in \Theta$. Then $P^T_{\theta_1} \simeq P^T_{\theta_2}$ for all θ_1, θ_2 in Θ and

$$\frac{dP^T_\theta}{dP^T_{\theta_0}}(\boldsymbol{X}_T) = \exp\left\{\int_0^T \log \frac{S(\theta,t)}{S(\theta_0,t)} dX(t) - \int_0^T [S(\theta,t) - S(\theta_0,t)]\, dt\right\}. \tag{8.2.1}$$

Suppose that the process $\{X(t)\}$ is observed over $[0, T]$. The problem is to estimate θ from the observation $\boldsymbol{X}_T = \{X(t),\ 0 \leq t \leq T\}$. For any $f_T \in L_2[0, T]$, define

$$\|f_T\|^2 = \int_0^T f^2(t)\, dt .$$

A function $f_T(\theta) = \{f(\theta, t),\ 0 \leq t \leq T\}$ is said to be *differentiable* in θ at $\theta = \theta_0$ in $L_2[0, T]$ if there exists a function $\dot{f}_T(\theta_0) = \{\dot{f}(\theta_0, t), 0 \leq t \leq T\} \in L_2[0, T]$ such that

$$\lim_{\theta \to \theta_0} \left\| \frac{f_T(\theta) - f_T(\theta_0)}{\theta - \theta_0} - \dot{f}_T(\theta_0) \right\| = 0 .$$

***THEOREM 8.4* (Cramer–Rao inequality)**

Suppose that the following conditions hold:

$$(i)\ \ S(\theta,t) > 0,\ \theta \in \Theta,\ 0 \le t \le T\ ,$$

$$(ii)\ \ f(y,t) = \frac{S(y,t)}{\sqrt{S(\theta,t)}} \text{ is differentiable in } y \text{ in } L_2[0,T] \text{ at } y=\theta$$

for all $\theta \in \Theta$, *and*

$$(iii)\ \ I_T(\theta) \equiv \int_0^T \frac{\dot{S}(\theta,t)^2}{S(\theta,t)} dt > 0,\ \theta \in \Theta\ .$$

Let $\theta_T^* \equiv \theta_T^*(\boldsymbol{X}_T)$ *be any estimator such that*

$$E_\theta\left(\theta_T^* - \theta\right)^2 < \infty,\ \theta \in \Theta\ .$$

Then

(i) $b(\theta) = E_\theta \theta_T^* - \theta$ *is differentiable with respect to* θ, *and*

(ii) *for all* $\theta \in \Theta$,

$$E_\theta\left(\theta_T^* - \theta\right)^2 \ge \frac{\left\{1 + \frac{db(\theta)}{d\theta}\right\}^2}{\int_0^T \frac{\dot{S}(\theta,t)^2}{S(\theta,t)} dt} + b^2(\theta)\ . \tag{8.2.2}$$

PROOF See [32, p. 122]. ∎

Example 8.1

Suppose $\Theta = (\alpha, \beta)$ with $\alpha > 0$ and $S(\theta,t) = \theta f(t)$ and $\int_0^T f(t)dt > 0$. Then it can be checked that the maximum likelihood estimator of θ based on $\boldsymbol{X}_T$ is

$$\theta_T^* = X(T) \Big/ \int_0^T f(t)dt \,,$$

and it attains the Cramer–Rao lower bound. ▯

Let

$$\psi(\theta, t) = 2\sqrt{S(\theta, t)} \tag{8.2.3}$$

and

$$\phi_T(\theta) = \left\{ \int_0^T \frac{\dot{S}(\theta, t)^2}{S(\theta, t)} dt \right\}^{-1/2} = \left\| \dot{\psi}_T(\theta) \right\|^{-1} \equiv I_T(\theta)^{-1/2} \,. \tag{8.2.4}$$

The following theorem gives Hajek–Le Cam lower bound for the local asymptotic minimax mean squared error. Suppose that the following conditions hold:

(D1) $S(\theta, t) > 0$ for all $\theta \in \Theta$ and $0 \le t < \infty$,

(D2) $S(\theta, t)$ is absolutely continuous in θ for almost all $t \ge 0$ and $\dot{\psi}_T(\theta) \in L_2[0, T]$,

(D3) for any $c > 0$ and $\theta \in \Theta$,

$$\lim_{T \to \infty} \sup_{|u| < c} \left\| \dot{\psi}_T(\theta + \phi_T(\theta)u) - \dot{\psi}_T(\theta) \right\| = 0 \,,$$

(D4) for any $c > 0$ and $\theta \in \Theta$,

$$\lim_{T \to \infty} \sup_{|u|+|v| < c} \phi_T^4(\theta) \int_0^T \left\{ \frac{\dot{S}(\theta + \phi_T(\theta)u, t)}{S(\theta + \phi_T(\theta)u, t)} \right\}^4 S(\theta + \phi_T(\theta)v, t)dt = 0 \,.$$

***THEOREM 8.5* (Hajek–Le Cam bound)**
Suppose that the conditions (D1)–(D4) hold and
(D5) $\phi_T(\theta) \to 0$ as $T \to \infty$ for all $\theta \in \Theta$.

Then, for any $\theta_0 \in \Theta$,

$$\liminf_{\delta \to 0} \liminf_{T \to \infty} \inf_{\theta_T^*} \sup_{|\theta - \theta_0| < \delta} \left[E_\theta \left(\theta_T^* - \theta \right)^2 \phi_T \left(\theta_0 \right)^{-2} \right] \geq 1 . \tag{8.2.5}$$

REMARK 8.2 Proof of this theorem consists in establishing the LAN of the family $\{P_\theta^{(T)}, \ \theta \in \Theta\}$ and applying Hajek–Le Cam inequality (cf. [24, 32]). ∎

DEFINITION 8.1 *θ_T^* is said to be asymptotically efficient if for all $\theta_0 \in \Theta$ and some $\delta > 0$,*

$$\lim_{T \to \infty} \sup_{|\theta - \theta_0| < \phi_T(\theta_0)^\delta} \frac{E_\theta \left(\theta_T^* - \theta \right)^2}{\phi_T \left(\theta_0 \right)^2} = 1 . \tag{8.2.6}$$

Asymptotic Properties of an MLE

Let $\boldsymbol{X}_T = \{X(t), \ 0 \leq t \leq T\}$ be a nonhomogeneous Poisson process with the intensity $S_T(\theta) = \{S(\theta, t), 0 \leq t \leq T\}, \ \theta \in \Theta = (\alpha, \beta)$ where $-\infty < \alpha < \beta < \infty$. Let P_θ be the probability measure corresponding to the parameter θ. Suppose the following regularity conditions hold:

(C1) $S(\theta, t) > 0$ for all $\theta \in \Theta$ and $t \geq 0$,

(C2) $S(\theta, t)$ is absolutely continuous in θ for almost all $t \geq 0$ and the derivative $\dot{\psi}_T(\theta)$ of $\psi(\theta, t) = 2\sqrt{S(\theta, t)}$ belongs to $L_2[0, T]$ for all $\theta \in \Theta$ and $T < \infty$,

(C3) $\phi_T(\theta) = \|\dot{\psi}_T(\theta)\|^{-1} \to 0$ as $T \to \infty$ and there exists $c_0 > 0$ such that

$$\phi_T(\theta) \leq c_0 \, \phi_T \left(\theta' \right), \ \theta, \theta' \in \Theta ,$$

(C4) there exists $\delta \in (0, 1)$ such that

(a) $\displaystyle\lim_{T \to \infty} \sup_{|\theta - y| < \phi_T^\delta} \phi_T \|\dot{\psi}_T(y) - \dot{\psi}_T(\theta)\| = 0.$

(b) $\displaystyle\lim_{T \to \infty} \sup_{|\theta - z| + |\theta - y| < \phi_T^\delta} \phi_T^4 \int_0^T \left\{ \frac{\dot{S}(y, t)}{S(y, t)} \right\}^4 S(z, t) dt = 0,$ and for some $\mu > 0$,

(c) $\lim_{T\to\infty} \inf_{|\theta - y| > \phi_T^{\delta}} \phi_T^{\mu} \|\psi_T(y) - \psi_T(\theta)\| > 0,$

where $\phi_T = \phi_T(\theta_0)$ for some θ_0 arbitrary but fixed value in Θ.

REMARK 8.3 The condition (C3) indicates that $\phi_T(\theta)$ tends to zero at a "uniform rate." The conditions (C4) (a) and (b) are the conditions on the smoothness of the derivative of $S(\theta, t)$ and (C4) (c) is a condition on the separability of intensities for adjacent values of θ.

The following result gives the limiting properties of a MLE $\hat{\theta}_T$. ∎

THEOREM 8.6
Suppose the conditions (C1) to (C4) hold. Then the MLE $\hat{\theta}_T$ of θ satisfies the following properties:

$$\begin{aligned}
&(i)\ \hat{\theta}_T \xrightarrow{P} \theta \text{ as } T \to \infty, \\
&(ii)\ \phi_T(\theta)^{-1}\left(\hat{\theta}_T - \theta\right) \xrightarrow{\mathcal{L}} N(0,1) \text{ as } T \to \infty, \\
&(iii)\ E_\theta \left|\hat{\theta}_T - \theta\right|^p \phi_T(\theta)^{-p} \to \frac{2^{p/2}}{\sqrt{\pi}} \Gamma\left(\frac{p+1}{2}\right) \text{ as } T \to \infty, \text{ and} \\
&(iv)\ \hat{\theta}_T \text{ is asymptotically efficient},
\end{aligned} \tag{8.2.7}$$

under P_θ.

Limit Behavior of the Likelihood Ratio Process

Consider the random process

$$Z_T(u) = \frac{dP^{(T)}_{\theta+\phi_T(\theta)u}}{dP^{(T)}_\theta}(\boldsymbol{X}_T), \quad u \in U_{\theta,T} = \{u : \theta + \phi_T(\theta)u \in \Theta\}. \tag{8.2.8}$$

LEMMA 8.1
Under the conditions (C1)–(C4), the family of measures $\{P^{(T)}_\theta, \ \theta \in \Theta\}$ is locally asymptotically normal (LAN) in Θ with the normalizing function $\phi_T(\theta)$ and

$$\Delta_T(\theta, \boldsymbol{X}_T) = \phi_T(\theta) \int_0^T \frac{\dot{S}(\theta,t)}{S(\theta,t)}[dX(t) - S(\theta,t)dt]; \tag{8.2.9}$$

that is, $\Delta_T(\theta, X_T)$ is asymptotically $N(0,1)$ and

$$\log Z_T(u) - u\,\Delta_T(\theta, \boldsymbol{X}_T) + \frac{1}{2}u^2 \xrightarrow{p} 0 \ \textit{under } P_\theta \ \textit{as } T \to \infty\,.$$

This lemma can be proved using the fact that

$$\log Z_T(u) = \int_0^T \log\left\{\frac{S(\theta + \phi_T(\theta)u, t)}{S(\theta, t)}\right\}[dX(t) - S(\theta, t)dt]$$

$$-\int_0^T \left[S(\theta + \phi_T(\theta)u, t) - S(\theta, t) - S(\theta, t)\log\frac{S(\theta + \phi_T(\theta)u, t)}{S(\theta, t)}\right]dt\,.$$

Note that

$$Z_T(u) = \exp\left\{u\Delta_T(\theta, \boldsymbol{X}_T) - \frac{1}{2}u^2 + g_T(\theta, u, \boldsymbol{X}_T)\right\}$$

where $g_T(\theta, u, \boldsymbol{X}_T) \xrightarrow{p} 0$ as $T \to \infty$ under P_θ and under (C4)(b), and it can be shown that

$$\Delta_T(\theta, \boldsymbol{X}_T) \xrightarrow{\mathcal{L}} N(0,1) \text{ under } P_\theta \text{ as } T \to \infty\,.$$

LEMMA 8.2
If the condition (C3) holds, then

$$E_\theta\left|Z_T^{1/2}(u_2) - Z_T^{1/2}(u_1)\right|^2 \le c_1\,|u_2 - u_1|^2$$

for some constant $c_1 > 0$.

LEMMA 8.3
Suppose the conditions (C1)–(C4) hold. Then there exists $T_0 > 0$, $c > 0$ and $\gamma > 0$ such that, for all $T > T_0$,

$$P_\theta^{(T)}\left\{Z_T(u) > e^{-c\,g(u)}\right\} \le e^{-c\,g(u)}$$

where $g(u) = \min(|u^2|,\ |u|^\gamma)$.

The following theorem proves the weak convergence of the process $\{Z_T(u)\}$ as $T \to \infty$.

THEOREM 8.7
Suppose the following conditions hold:

(E1) *the family of measures* $\left\{P_\theta^{(T)}, \theta \in \Theta\right\}$ *is LAN as* $T \to \infty$,

(E2) $E_\theta \left|Z_T^{1/2}(u_2) - Z_T^{1/2}(u_1)\right|^2 \le c\ |u_2 - u_1|^2$,

(E3) *there exists* $c > 0$, $\gamma > 0$ *and* $T_0 > 0$ *such that, for* $T > T_0$, $P_\theta^{(T)}\left\{Z_T(u) > e^{-c\, g(u)}\right\} \le e^{-c\, g(u)}$ *where* $g(u) = |u|^\gamma$.

Then $\nu_\theta^{(T)} \xrightarrow{\mathcal{L}} \nu_\theta$ *as* $T \to \infty$ *where* $\nu_\theta^{(T)}$ *is the measure generated by the process* $Z_\theta^{(T)}$ *and* ν_θ *is the measure generated by the limiting process.*

An application of Theorem 8.6 and the standard methods of weak convergence prove Theorem 8.10 (cf. [32, p. 175–178]). We omit the proofs of Lemmas 8.1 to 8.3. For details, see [32].

REMARK 8.4 One can go through a similar analysis for estimators of parameters for processes of Poisson type (cf. [32, p. 160]). For an extensive discussion, see [32]. We will not pursue the matter here. The analysis depends on the following central limit theorem for stochastic integrals. ∎

Central Limit Theorem

Let

$$I_T(f_T) = \int_0^T f_T(t)dM_T(t) \tag{8.2.10}$$

where $f_T = \{f_T(t), \mathcal{F}_t,\ t \in [0, T]\}$ is $\mathcal{F}_t$-predictable and

$$M_T(t) = X_T(t) - \int_0^t \lambda_T(s)ds,\ 0 \le t \le T$$

is the centered process of Poisson type, that is, $X_T(t)$ is a point process on $[0, T]$ with $\mathcal{F}_t$ stochastic intensity $\lambda_T(\cdot)$.

THEOREM 8.8
Suppose that

$$\begin{aligned}&(i)\ p-\lim_{T\to\infty}\int_0^T f_T(t)^2\lambda_T(t)dt=\sigma^2<\infty,\ \text{and}\\&(ii)\ \text{for every } \varepsilon>0\,,\end{aligned}\tag{8.2.11}$$

$$\lim_{T\to\infty}\int_0^T E\left[f_T(t)^2 I\left\{|f_T(t)|>\varepsilon\right\}\lambda_T(t)\right]dt=0\,.$$

Then

$$\int_0^T f_T dM_T(t)\xrightarrow{\mathcal{L}} N\left(0,\sigma^2\right)\quad \text{as } T\to\infty\,.\tag{8.2.12}$$

PROOF (Sketch) Let

$$X_T^*(t)=\begin{cases}X_T(t), & 0\le t<T\\ Y_T(t), & T\le t\le T+1\end{cases}$$

where $\{Y_T(t), \mathcal{F}_t : T\le t\le T+1\}$ is a Poisson process of intensity T. Then $X_T^*(t)$ is a process of Poisson type with intensity $\lambda_T^*(t)$ (say). Let

$$M_T^*(t)=X_T^*(t)-\int_0^T \lambda_T^*(s)ds\,,$$

$$f_T^*(t)=\begin{cases}f_T(t), & 0\le t\le T\\ \sigma/\sqrt{T}, & T<t\le T+1\,,\end{cases}$$

$$\tau_T=\inf\left\{t:\int_0^t f_T^*(s)^2\lambda_T^*(s)ds\ge\sigma^2\right\}\,,$$

and

$$I_{\tau_T}\left(f_T^*\right) = \int_0^{\tau_T} f_T^*(t)dM_T^*(t)\,.$$

Introduce the process

$$\zeta_T(t) = \exp\left\{i\lambda \int_0^t f_T^*(s)dM_T^*(s) + \frac{1}{2}\lambda^2 \int_0^t f_T^*(s)^2\lambda_T^*(s)ds\right\},$$

$$0 \le t \le T+1$$

for any real λ.

Applying the generalized Ito's lemma for semimartingales, after some algebraic manipulations (cf. [32, p. 167]), it can be shown that

$$\lim_{T\to\infty} E\left(\zeta_T(\tau_T)\right) = 1\,.$$

But $E(\zeta_T(\tau_T)) = \phi_T(\lambda)e^{1/2\lambda^2\sigma^2}$ from the definition of τ_T where $\phi_T(\lambda)$ is the characteristic function of $I_{\tau_T}(f_T^*)$. Hence,

$$\lim_{T\to\infty} \phi_T(\lambda) = \exp\left\{-\frac{1}{2}\lambda^2\sigma^2\right\}.$$

This proves that

$$I_{\tau_T}(f_T^*) \xrightarrow{\mathcal{L}} N\left(0,\sigma^2\right) \text{ as } T\to\infty\,.$$

But, for $\varepsilon > 0$ and $\delta > 0$,

$$P\left\{\left|I_T(f_T) - I_{\tau_T}\left(f_T^*\right)\right| > \varepsilon\right\}$$

$$= P\left\{\left|\int_0^{T+1} f_T^*(t)\left[I(t\le T) - I\left(t \le \tau_T\right)\right]dM_T^*(t)\right| > \varepsilon\right\}$$

$$\leq \frac{\delta}{\varepsilon^2} + P\left\{ \int_0^{T+1} f_T^*(t)^2 \left| I(t \leq T) - I(t \leq \tau_T) \right| \lambda_T^*(t)dt > \delta \right\}$$

$$\leq \frac{\delta}{\varepsilon^2} + P\left\{ \left| \int_0^{T} f_T(t)^2 \lambda(t)dt - \sigma^2 \right| > \delta \right\} .$$

Hence,

$$P\left\{ \left| I_T(f_T) - I_{\tau_T}\left(f_T^*\right) \right| > \varepsilon \right\} \to 0 \text{ as } T \to$$

by (8.2.11). Therefore,

$$I_T(f_T) \xrightarrow{\mathcal{L}} N\left(0, \sigma^2\right) \text{ as } T \to \infty .$$

■

REMARK 8.5 Kutoyants [31] studied asymptotic expansions for the maximum likelihood estimator of the intensity parameter for a nonhomogeneous Poisson process. Bayesian estimation of the intensity parameter is briefly discussed in [32] following the techniques discussed earlier. Multidimensional parameter estimation of the intensity function for a nonhomogeneous Poisson process is discussed in [30] following the same techniques. ■

Bayesian Inference for Counting Processes

Let $N(t) = (N_1(t), \ldots, N_k(t)),\ 0 \leq t \leq 1$ be a multivariate counting process on a complete probability space $(\Omega, \mathcal{F}, P)$. Suppose the component process N_i has the intensity process $\lambda_i(t, \theta) = Y_i(t)\, \alpha_i(t, \theta)$. $0 \leq t \leq 1,\ 1 \leq i \leq k$ with respect to an increasing right continuous family of sub-σ-algebras $\{\mathcal{F}_t, t \geq 0\}$, and the processes Y_i and N_i are observable while θ is an unknown p-dimensional parameter. We assume that $\theta \in \Theta$, an open subset of R^p. Let θ_0 be the true parameter.

Let us first consider the case $k = 1$. We will write $Y(t)$ and $N(t)$ for $Y_1(t)$ and $N_1(t)$, respectively. As we have seen earlier, the likelihood function can be written in the form

$$L_n(\theta) = c \exp\left[\int_0^1 \log \alpha(s,\theta) dN(s) - \int_0^1 Y(s)\alpha(s,\theta) ds\right] \tag{8.2.13}$$

for some constant c independent of $\theta \in \Theta$.

We assume that the following conditions hold in the sequel:

(F1) $\frac{Y(t)}{n} \to y(t)$ uniformly for $t \in [0, 1]$ in probability,

(F2) $\alpha(s, \theta)$ and its derivatives with respect to θ up to the third order exist and are continuous functions of θ for all θ in a neighborhood U_{θ_0} of θ_0 and that they are bounded on $[0, 1] \times U_{\theta_0}$, and

(F3) $\alpha(s, \theta)$ is bounded away from zero on $[0, 1] \times U_{\theta_0}$.

Consider the martingale

$$M(t) = N(t) - \int_0^t Y(s)\alpha(s, \theta_0)\, ds, \quad t \geq 0 \tag{8.2.14}$$

and let

$$\psi_i(s,\theta) = \frac{\partial \log \alpha(s,\theta)}{\partial \theta_i}, \theta = (\theta_1, \ldots, \theta_p) \; . \tag{8.2.15}$$

Then

$$V_{nj}(\theta) \equiv \frac{\partial \log L_n(\theta)}{\partial \theta_j}$$

$$= \int_0^1 \psi_j(s,\theta)[(dN(s) - Y(s)\alpha(s,\theta)ds] \tag{8.2.16}$$

and hence,

$$V_{nj}(\theta_0) = \int_0^1 \psi_j(s, \theta_0)\, dM(s) \; .$$

Let

$$I_{n,j\ell}(\theta) = \frac{\partial^2 \log L_n(\theta)}{\partial\theta_j \partial\theta_l}. \tag{8.2.17}$$

Then

$$I_{n,j\ell}(\theta_0) = \int_0^1 \frac{\partial^2 \log \alpha(s,\theta)}{\partial\theta_j \partial\theta_\ell} |_{\theta=\theta_0} dM(s)$$

$$- \int_0^1 Y(s)\psi_j(s,\theta_0)\,\psi_\ell(s,\theta_0)\,\alpha(s,\theta_0)\,ds$$

for $1 \le j, \ell \le p$. Under some regularity conditions (cf. [12]), it can be checked that

$$-\frac{1}{n} I_{n,j\ell}(\theta_0) \xrightarrow{p} \int_0^1 y(s)\psi_j(s,\theta_0)\,\psi_\ell(s,\theta_0)\,\alpha(s,\theta_0)\,ds$$

$$= \sigma_{j,\ell}(\theta_0) \text{ (say)}, \quad 1 \le j, \ell \le p \tag{8.2.18}$$

as $n \to \infty$ when $y(\cdot)$ is as given by the condition (F1). Furthermore,

$$\frac{1}{n}\boldsymbol{V}_n(\theta_0) = \left(\frac{V_{nj}(\theta_0)}{\sqrt{n}}\right)_{1\le j\le p} \xrightarrow{\mathcal{L}} \boldsymbol{V} \equiv \left(V_j\right)_{1\le j\le p} \text{ as } n \to \infty \tag{8.2.19}$$

where $\boldsymbol{V} \sim N_p(0, \Sigma)$ with $\Sigma = ((\sigma_{j,\ell}(\theta_0))_{p\times p}$. We assume that

(F4) Σ is positive definite.

Borgan [12] showed that

$$-\frac{1}{n} I_n\left(\tilde{\theta}_n\right) \xrightarrow{p} \Sigma \text{ as } n \to \infty \tag{8.2.20}$$

for any sequence $\tilde{\theta}_n \xrightarrow{p} \theta_0$. He also proved that the MLE θ_n^*, or any estimator solving the equations

$$V_{nj}\left(\theta_n^*\right) = 0,\ 1 \le j \le p$$

is consistent and satisfies

$$\sqrt{n}\left(\theta_n^* - \theta_0\right) \xrightarrow{\mathcal{L}} \Sigma^{-1}V \simeq N_p\left(0, \Sigma^{-1}\right) \text{ as } n \to \infty . \tag{8.2.21}$$

We now consider the problem of Bayes estimation of the parameter θ. Suppose that θ has a prior density $p(\theta)$ satisfying the following condition:

(F5) $p(\cdot)$ is continuous and positive in a neighborhood of θ_0 and $\int_{\Theta} \theta p(\theta) d\theta < \infty$.

Here,

$$\int_{\Theta} \theta\ p(\theta) d\theta = \left(\int_{\Theta} \theta_1\ p(\theta) d\theta, \ldots, \int_{\Theta} \theta_p\ p(\theta) d\theta \right) .$$

For convenience, let $p(\theta)$ be defined over R^p with $p(\theta) = 0$ for $\theta \in \Theta^c$.

The posterior density of θ given the data $\{N(t),\ Y(t),\ t \in [0, 1]\}$ is given by

$$\frac{L_n(\theta) p(\theta)}{\int_{\Theta} L_n(\theta) p(\theta) d\theta} . \tag{8.2.22}$$

The Bayes estimator with respect to the quadratic loss function $\ell(\tilde{\theta}, \theta) = (\tilde{\theta} - \theta)' \boldsymbol{W} (\tilde{\theta} - \theta)$, where $\boldsymbol{W}$ is positive definite symmetric matrix, is

$$\hat{\theta}_n = \frac{\int_{R^p} \theta L_n(\theta) p(\theta) d\theta}{\int_{R^p} L_n(\theta) p(\theta) d\theta} . \tag{8.2.23}$$

Define

$$G_n(\boldsymbol{x}) = \frac{L_n(\theta_0 + \boldsymbol{x} n^{-1/2})}{L_n(\theta_0)},\ \boldsymbol{x} \in R^p \tag{8.2.24}$$

and let $\theta = \theta_0 + x n^{-1/2}$. Then

$$\hat{\theta}_n = \theta_0 + n^{-1/2} \frac{\int_{R^p} x\, G_n(x) p\left(\theta_0 + x n^{-1/2}\right) dx}{\int_{R^p} G_n(x) p\left(\theta_0 + x n^{-1/2}\right) dx}$$

or equivalently

$$\sqrt{n}\left(\hat{\theta}_n - \theta_0\right) = \frac{\int_{R^p} x\, G_n(x) p\left(\theta_0 + x n^{-1/2}\right) dx}{\int_{R^p} G_n(x) p\left(\theta_0 + x n^{-1/2}\right) dx}. \tag{8.2.25}$$

Let $C([-A, A]^p)$ be the space of continuous function from $[-A, A]^p \to R$. We assume that A is sufficiently large so that $\Theta \cap [-A, A]^p$ is nonempty. Let

$$G(x) = \exp(x'V - \frac{1}{2} x'\Sigma x), \ x \in R^p \tag{8.2.26}$$

where x' denotes transpose of x and V is as defined by (8.2.19). It can be checked that

$$G_n \xrightarrow{\mathcal{L}} G \text{ as } n \to \infty \tag{8.2.27}$$

on $C([-A, A]^p)$ in the sense of weak convergence (cf. [11, 51]) (cf. [22, Lemma 1]). Let

$$X_n = \begin{pmatrix} \int_{R^p} x G_n(x) p\left(\theta_0 + x n^{-1/2}\right) dx \\ \int_{R^p} G_n(x) p\left(\theta_0 + x n^{-1/2}\right) dx \end{pmatrix} \tag{8.2.28}$$

and

$$X = \begin{pmatrix} \int_{R^p} x G(x) p\left(\theta_0\right) dx \\ \int_{R^p} G(x) p\left(\theta_0\right) dx \end{pmatrix}. \tag{8.2.29}$$

Let $X_{n,A}$ and X_A be as above but with integration only over the set $[-A, A]^p$.

Since $(G_n(\cdot), p(\theta_0 + \frac{\cdot}{\sqrt{n}})) \xrightarrow{\mathcal{L}} (G(\cdot), p(\theta_0))$ in $C([-A, A]^p)^2$ by the continuous mapping theorem (in [11, Theorem 4.4]), it follows that

$$X_{n,A} \xrightarrow{\mathcal{L}} X_A \text{ as } n \to \infty \tag{8.2.30}$$

for each A. Furthermore, $X_A \xrightarrow{\mathcal{L}} X$ by the form of $G(x)$ given by (8.2.26). Hence, $X_n \xrightarrow{\mathcal{L}} X$ if it can be shown that

$$\limsup_{n\to\infty} P\left(\left\|X_n - X_{n,A}\right\|^2 \geq \varepsilon\right) \to 0 \tag{8.2.31}$$

as $A \to \infty$ for each $\varepsilon > 0$ (cf. [11, Theorem 4.2]).

Let

$$B_n(\theta) = \frac{1}{n}\{\log L_n(\theta) - \log L_n(\theta_0)\}, \ \theta \in \Theta . \tag{8.2.32}$$

In addition to the conditions stated earlier, suppose that the following condition holds.

(F6) If $\delta > 0$ is given, then there exists $c_0 > 0$ such that $P(\Omega_n) \to 1$ where $\Omega_n = [\sup_{\|\theta-\theta_0\|>\delta} B_n(\theta) \leq -c_0]$.

Suppose that sufficient conditions for the consistency and asymptotic normality of the MLE θ_n^* hold (cf. [12]). In addition, suppose that the conditions (F1)–(F6) hold. The following theorem is due to Hjort [22]. We omit the proof.

THEOREM 8.9

Under the condition stated above,

$$\sqrt{n}\left(\tilde{\theta}_n - \theta_0\right) \xrightarrow{\mathcal{L}} \Sigma^{-1}V \simeq N_p\left(0, \Sigma^{-1}\right) \quad as\ n \to \infty \tag{8.2.33}$$

and

$$\sqrt{n}\left(\tilde{\theta}_n - \theta_n^*\right) \xrightarrow{p} 0 \quad as\ n \to \infty . \tag{8.2.34}$$

REMARK 8.6 The results obtained above for a single counting process can be extended to k counting processes without any difficulty due to the orthogonal structure of the martingales involved. Consider the k-dimensional counting process $N(t) = (N_1(t), \ldots, N_k(t))$, $t \geq 0$ with intensities $Y_i(t)\alpha_i(t, \theta), t \in [0, T_i]$, $1 \leq i \leq k$. The likelihood function can be written in this case in the form

$$L_n(\theta) = c\ \exp\left\{\sum_{i=1}^{k}\int_0^{T_i}(\log\alpha_i dN_i - Y_i\alpha_i\, ds)\right\} \tag{8.2.35}$$

where c is a constant independent of θ. We omit the details. ∎

Another approach for the study of the asymptotic properties of the Bayes estimators is via the Bernstein–von Mises theorem (cf. [10, 51]). Hjort [22] also discusses this approach and other related results of Bayes estimation for counting processes. For related work, see [9].

M-Estimation for a Nonhomogeneous Poisson Process

Consider a nonhomogeneous Poisson process $\{X(t)\}$ with a parameterized intensity $\lambda(t,\theta), \theta \in \Theta$, a bounded open interval of the real line. We have seen earlier that if $X(t,\theta) > 0$ for $0 \le t \le T$ for all $\theta \in \Theta$, then the log-likelihood function $\ell(T,\theta)$ based on the observation $\{X(t), 0 \le t \le T\}$ is given by

$$\ell(T,\theta) = \frac{dP_\theta^T}{dP_{\theta_0}^T}(X_T) = \int_0^T \log\left\{\frac{\lambda(t,\theta)}{\lambda(t,\theta_0)}\right\} dX(t) - \int_0^T [\lambda(t,\theta) - \lambda(t,\theta_0)]\,dt \tag{8.2.36}$$

for any fixed $\theta_0 \in \Theta$. Hence, the maximum likelihood estimator (MLE) is a solution of the likelihood equation

$$\int_0^T \frac{\lambda'(t,\theta)}{\lambda(t,\theta)} dX(t) - \int_0^T \lambda'(t,\theta)dt = 0 \tag{8.2.37}$$

under some regularity conditions where λ' is the derivative of λ with respect to θ. We have discussed sufficient conditions under which the MLE is consistent, asymptotically normal and asymptotically efficient earlier in this section (cf. [32]).

If the model chosen does not sufficiently reflect the generating mechanism of the data or the data are contaminated by noise, then the true intensity $\mu(t)$ of the process X may not belong to the assumed parametric model $\{\lambda(t,\theta),\ \theta \in \Theta\}$. In such a case the MLE is not necessarily the best choice. For instance, let us consider a nonhomogeneous Poisson process X with the true intensity $\mu(t)$ defined by

$$\mu(t) = (1-\varepsilon)\, f(t) + \varepsilon\, c(t)$$

where f and c are even periodic function with period one and ε denotes the rate of contamination. Suppose we model the process by a nonhomogeneous Poisson process with intensity $\lambda(t-\theta) = f(t-\theta)$ where $\theta \in \Theta = (-\frac{1}{2}, \frac{1}{2})$. If the data are not contaminated, that is, $\varepsilon = 0$, then the MLE is a good estimator. However, the asymptotic efficiency of the MLE decreases as $\varepsilon \to 0$. We now construct a robust estimator of θ with high efficiency even if the data is contaminated.

DEFINITION 8.2 *Let $h(t,\theta)$ and $H(t,\theta)$ be real-valued functions defined on $[0,T]\times\Theta$ such that*

$$C(T,\theta)=\int_0^T h(t,\theta)\,dX(t)-\int_0^T H(t,\theta)dt \tag{8.2.38}$$

is well defined. A solution of the equation

$$C(T,\theta)=0$$

is called an M-estimator for θ.

The MLE corresponds to the M-estimator for

$$h(t,\theta)=\frac{\lambda'(t,\theta)}{\lambda(t,\theta)},\quad H(t,\theta)=\lambda'(t,\theta) \tag{8.2.39}$$

where $\lambda'(t,\theta)$ denotes the derivative of $\lambda(t,\theta)$ with respect to θ.

We assume that the following regularity conditions hold.

(J1) The true intensity $\mu(t)$ of the process X is a bounded measurable function with period τ for some $\tau>0$.

(J2) The functions h and H are periodic with period τ in t for every $\theta\in\Theta$ and are absolutely continuous with respect to θ for any $t\geq 0$. Let h' and H' denote the derivatives of h and H, respectively (which exist almost everywhere with respect to the Lebesgue measure ν) and suppose they are bounded in (t,θ).

(J3) There exists $\theta_0\in\Theta$ such that

$$\int_0^\tau [H(t,\theta_0)-h(t,\theta_0)\,\mu(t)]\,dt=0\,.$$

(J4) Suppose that

$$\Gamma = \frac{1}{\tau} \int_0^\tau \left[H'(t, \theta_0) - h'(t, \theta_0)\, \mu(t) \right] dt > 0 \,,$$

and

$$\Phi = \frac{1}{\tau} \int_0^\tau h(t, \theta_0)^2 \mu(t) dt > 0 \,.$$

(J5) There exist constants C_1 and C_2 that are independent of θ and t such that, for any sufficiently small $\delta > 0$,

$$\nu \left\{ \cup_{|\theta - \theta_0| < \delta} \left[t \in [0, \tau] : \left| h'(t, \theta) - h'(t, \theta_0) \right| \geq C_1 |\theta - \theta_0| \right] \right\} \leq C_2 \delta \,.$$

(J6) $\int_0^\tau |H'(t, \theta) - H'(t, \theta_0)| dt \to 0$ as $\theta \to \theta_0$.

Note that the conditions (J2) and (J5) imply that

$$\int_0^\tau \left| h'(t, \theta) - h'(t, \theta_0) \right| dt \to 0 \text{ as } \theta \to \theta_0 \,. \tag{8.2.40}$$

Consistency

The following result shows that the M-estimator $\hat{\theta}_T$ is consistent under the above conditions.

THEOREM 8.10

Suppose the conditions (J1)–(J6) hold. Then for any $T \geq 0$, there exist $\delta(T) > 0$ such that $\delta(T) \to 0$ as $T \to \infty$, an event $A(T)$ such that $P(A(T)) \to 1$ as $T \to \infty$ and an M-estimator $\hat{\theta}_T$ in $U(\delta(T))$ defined on the event $A(T)$ where $U(\delta) = \{\theta : |\theta - \theta_0| \leq \delta\}$.

PROOF Let

$$m(T, \theta) = \frac{1}{T} \int_0^T h(t, \theta)(dX(t) - \mu(t) dt) \,,$$

$$G(T,\theta) = \frac{1}{T}\int_0^T [H(t,\theta) - h(t,\theta)\mu(t)]dt\,,$$

$$\dot{m}(T,\theta) = \frac{1}{T}\int_0^T h'(t,\theta)(dX(t) - \mu(t)dt)\,,$$

and

$$\dot{G}(T,\theta) = \frac{1}{T}\int_0^T \left[H'(t,\theta) - h'(t,\theta)\mu(t)\right]dt\,.$$

Let $\varepsilon > 0$. Then, for any $T \geq \tau$,

$$\sup_{\theta\in U(\delta)} \left|\dot{G}(T,\theta) - \dot{G}\,(T,\theta_0)\right|$$

$$\leq \sup_{\theta\in U(\delta)} \left\{\frac{1}{T}\int_0^T \left|H'(t,\theta) - H'\,(t,\theta_0)\right| dt\right\}$$

$$+ \|\mu\|_\infty \sup_{\theta\in U(\delta)} \left\{\frac{1}{T}\int_0^T \left|h'(t,\theta) - h'\,(t,\theta_0)\right| dt\right\}$$

$$\leq \sup_{\theta\in U(\delta)} \left\{\frac{2}{\tau}\int_0^\tau \left|H'(t,\theta) - H'\,(t,\theta_0)\right| dt\right\}$$

$$+ \|\mu\|_\infty \sup_{\theta\in U(\delta)} \left\{\frac{2}{\tau}\int_0^\tau \left|h'(t,\theta) - h'\,(t,\theta_0)\right| dt\right\}$$

by the periodicity of H and h with period τ where $\|\mu\|_\infty = \sup_t|\mu(t)|$. By the conditions (J1), (J6), and relation (8.2.40), it follows that the right-hand side of the above inequality tends to zero as $\delta \to 0$. Hence,

$$\sup_{\theta \in U(\delta)} \left|\dot{G}(T,\theta) - \dot{G}(T,\theta_0)\right| \to 0 \text{ as } \delta \to 0 \tag{8.2.41}$$

uniformly for $T \geq \tau$. Since $\dot{G}(T,\theta_0) \to \Gamma$ as $T \to \infty$, it follows that there exists a $T_1 > \tau$ such that, for sufficiently small $\delta > 0$ and $T > T_1$,

$$\begin{aligned}\inf_{\theta \in U(\delta)} \dot{G}(T,\theta) &\geq \inf_{\theta \in U(\delta)} \left[\dot{G}(T,\theta_0) - \left|\dot{G}(T,\theta) - \dot{G}(T,\theta_0)\right|\right] \\ &> \Gamma - 2\varepsilon \end{aligned} \tag{8.2.42}$$

where Γ is a defined by (J4). Furthermore,

$$\begin{aligned}&|\dot{m}(T,\theta) - \dot{m}(T,\theta_0)| \\ &\leq \frac{1}{T}\int_0^T \left|h'(t,\theta) - h'(t,\theta_0)\right| dX(t) \\ &\quad + \frac{1}{T}\int_0^T \left|h'(t,\theta) - h'(t,\theta_0)\right| \mu(t)dt \\ &\leq \frac{1}{T}\int_{D_{\theta,T}} \left|h'(t,\theta) - h'(t,\theta_0)\right| dX(t) \\ &\quad + \frac{1}{T}\int_{[0,T]-D_{\theta,T}} \left|h'(t,\theta) - h'(t,\theta_0)\right| dX(t) \\ &\quad + \frac{2\|\mu\|_\infty}{\tau}\int_0^\tau \left|h'(t,\theta) - h'(t,\theta_0)\right| dt\end{aligned}$$

where

$$D_{\theta,T} = \left\{t \in [0,T] : \left|h'(t,\theta) - h'(t,\theta_0)\right| \geq C_1 \left|\theta - \theta_0\right|\right\}$$

and C_1 is as given by the condition (J5). In view of (8.2.40), there exists $\delta_1 > 0$ such that for any $\theta \in U(\delta_1)$, the last term in the above inequality on the right-hand side is less than ε. Since $X(T)$ follows a Poisson distribution with mean $\int_0^T \mu(t)dt$, for any $\delta > 0$ and $T > 0$,

$$P\left\{\sup_{\theta \in U(\delta)} \frac{1}{T} \int_{[0,T]-D_{\theta,T}} \left|h'(t,\theta) - h'(t,\theta_0)\right| dX(t) \geq \varepsilon\right\}$$

$$\leq P\left\{\frac{C_1\delta}{T} X(T) \geq \varepsilon\right\}$$

$$\leq \frac{C_1\delta}{T\varepsilon} \int_0^T \mu(t)dt$$

$$\leq \frac{C_1\|\mu\|_\infty}{\varepsilon}\delta .$$

For any measurable set B in R, let $X(B) \equiv \int_B dX(t)$. It is clear that $X(B)$ follows a Poisson distribution with mean $\int_B \mu(t)dt$. Then, for any $T \geq \tau$,

$$P\left\{\sup_{\theta \in U(\delta)} \frac{1}{T} \int_{D_{\theta,T}} \left|h'(t,\theta) - h'(t,\theta_0)\right| dX(t) \geq \varepsilon\right\}$$

$$\leq P\left\{\frac{2\left\|h'\right\|_\infty}{T} \sup_{\theta \in U(\delta)} X\left(D_{\theta,T}\right) \geq \varepsilon\right\}$$

$$\leq P\left\{\frac{2\|h'\|_\infty}{T} X\left(\cup_{\theta\in U(\delta)} D_{\theta,T}\right) \geq \varepsilon\right\}$$

$$\leq \frac{2\left\|h'\right\|_\infty}{T\varepsilon} E\left[X\left(\cup_{\theta\in U(\delta)} D_{\theta,T}\right)\right]$$

$$\leq \frac{2\left\|h'\right\|_\infty}{T\varepsilon} \|\mu\|_\infty \frac{T+\tau}{\tau}$$

$$\times \nu\left\{\cup_{|\theta-\theta_0|\leq\delta}\left[t\in[0,\tau] : \left|h'(t,\theta)-h'(t,\theta_0)\right| \geq C_1\left|\theta-\theta_0\right|\right]\right\}$$

where $\nu(\cdot)$ is the Lebesgue measure. From the condition (J5), it follows that, for sufficiently small $\delta > 0$ and any $T \geq \tau$,

$$P\left\{\sup_{\theta\in U(\delta)} \frac{1}{T}\int_{D_{\theta,T}} \left|h'(t,\theta)-h'(t,\theta_0)\right| dX(t) \geq \varepsilon\right\}$$

$$\leq \frac{4\left\|h'\right\|_\infty \|\mu\|_\infty}{\varepsilon T} C_2\delta\,.$$

Hence, for sufficiently small $\delta > 0$ and any $T \geq \tau$,

$$P\left\{\sup_{\theta\in U(\delta)} |\dot{m}(T,\theta)-\dot{m}(T,\theta_0)| \geq 3\varepsilon\right\} \leq \frac{C_3}{\varepsilon}\delta$$

where $C_3 = (C_1 + \{4\|h'\|_\infty C_2/\tau\})\|\mu\|_\infty$. Note that $\dot{m}(T,\theta_0)$ converges to zero in probability as $T \to \infty$. Hence, for sufficiently small $\delta > 0$, there exists a $T_2(\delta) \geq \tau$ such that for $T > T_2(\delta)$,

$$P\left\{\sup_{\theta\in U(\delta)} |\dot{m}(T,\theta)| \geq 4\varepsilon\right\} < \frac{2C_3}{\varepsilon}\delta\,. \tag{8.2.43}$$

In view of (J2), we have

$$C(T,\theta) = C(T,\theta_0) + \int_{\theta_0}^{\theta} \dot{C}(T,u)du \tag{8.2.44}$$

where

$$\dot{C}(T,u) = \int_0^T h'(t,u)dX(t) - \int_0^T \dot{H}(t,u)dt \,.$$

$$= T\,\dot{m}(T,u) - T\,\dot{G}(T,u)$$

from the definitions of $\dot{m}(T,\theta)$ and $\dot{G}(T,\theta)$ given earlier.

Let $A_1(T,\delta) = [\omega : \sup_{\theta\in U(\delta)} |\dot{m}(T,\theta)| < \frac{\Gamma}{3}]$ where Γ is as defined by (J4). Choosing $\varepsilon = \Gamma/12$ in (8.2.43), we obtain that, for sufficiently small $\varepsilon > 0$ and $T > T_2(\delta)$,

$$P\left(A_1(T,\delta)\right) \geq 1 - \frac{24C_3}{\Gamma}\delta$$

$$= 1 - C_4\delta \text{ (say)} \,.$$

It can now be checked that, for sufficiently small $\delta > 0$ and $T > T_1$,

$$\inf_{\theta\in U(\delta)} \left[-\frac{1}{T}\dot{C}(T,u)\right] = \inf_{\theta\in U(\delta)} [-\dot{m}(T,u) + \dot{G}(T,u)]$$

$$> \frac{\Gamma}{2} \tag{8.2.45}$$

on the event $A_1(T,\delta)$ by (8.2.42) for $\varepsilon = \frac{\Gamma}{12}$. Since

$$\frac{1}{T}C(T,\theta_0) = m\,(T,\theta_0) - G\,(T,\theta_0) \xrightarrow{p} 0 \text{ as } T \to \infty\,, \tag{8.2.46}$$

for sufficiently small $\delta > 0$, there exists a $T_3(\delta)(\geq T_2(\delta))$ such that, for any $T > T_2(\delta)$,

$$P\left(A_2(T,\delta)\right) \geq 1 - 2C_4\delta$$

where $A_2(T,\delta) = [\omega : \frac{1}{T}|C(T,\theta_0)| \leq \frac{\Gamma\delta}{4}] \cap A_1(T,\delta)$. Relations (8.2.44)–(8.2.46) imply that for sufficiently small $\delta > 0$ and $T > T_1$,

$$C\left(T, \theta_0 + \delta\right) < 0 \text{ and } C\left(T, \theta_0 - \delta\right) > 0$$

on the event $A_2(T,\delta)$, which implies that there exists $\hat{\theta}_T \in U(\delta)$ such that $C(T,\hat{\theta}_T) = 0$. Hence, for sufficiently small $\delta > 0$, there exists an event $A_2(T,\delta)$ and a $T_0(\delta)$ $(= \max(T_1, T_1(\delta)))$ such that $P(A_2(T,\delta)) \geq 1 - 2C_4\delta$ for any $T > T_0(\delta)$ and an M-estimator $\hat{\theta}_T$ exists in $U(\delta)$ on the event $A_2(T,\delta)$. Now choose an increasing sequence $T_n \to \infty$ such that for $T \geq T_n$, $P(A_2(T, 1/n)) \geq 1 - \frac{2C_4}{n}$. We obtain the theorem by setting $\delta(T) = \frac{1}{n}$ for $T_n \leq T < T_{n+1}$ and $A(T) = A_2(T, \delta(T))$. ▮

Asymptotic Normality

Note that

$$\frac{1}{\sqrt{T}} C\left(T, \theta_0\right) = \frac{1}{\sqrt{T}} \int_0^T h\left(t, \theta_0\right)\left(dX(t) - \mu(t)dt\right) + o_p(1)$$

by (J3). By the central limit theorem for martingales, it follows that

$$\frac{1}{\sqrt{T}} C\left(T, \theta_0\right) \xrightarrow{\mathcal{L}} N(0, \Phi) \text{ as } T \to \infty \tag{8.2.47}$$

where Φ is as defined by (J4). However,

$$\frac{1}{\sqrt{T}} C\left(T, \theta_0\right) - \frac{1}{\sqrt{T}} C(T,\theta) = \sqrt{T}\frac{1}{T} \int_\theta^{\theta_0} \dot{C}(T,u)du$$

$$= \sqrt{T}\left(\int_\theta^{\theta_0} \dot{m}(T,u)du - \int_\theta^{\theta_0} \dot{G}(T,u)du\right)$$

$$= \sqrt{T}\left(\int_{\theta}^{\theta_0} \dot{m}(T,u)du - \int_{\theta}^{\theta_0} \left(\dot{G}(T,u) - \dot{G}(T,\theta_0)\right) du\right.$$

$$\left. - (\theta_0 - \theta)\,\dot{G}(T,\theta_0)\right).$$

Let $\theta = \hat{\theta}_T$ be a consistent M-estimator as obtained earlier. Then $C(T, \hat{\theta}_T) = 0$ and

$$\frac{1}{\sqrt{T}} C(T,\theta_0) = \sqrt{T}\left(\int_{\hat{\theta}_T}^{\theta_0} \dot{m}(T,u)du - \int_{\hat{\theta}_T}^{\theta_0} \left(\dot{G}(T,u) - \dot{G}(T,\theta_0)\right) du\right.$$

$$\left. + \left(\hat{\theta}_T - \theta_0\right) \dot{G}(T,\theta_0)\right).$$

It is clear that

$$\left|\int_{\hat{\theta}_T}^{\theta_1} \dot{m}(T,u)du\right| \le \left|\hat{\theta}_T - \theta_0\right| \sup\left\{|\dot{m}(T,\theta)| : \theta \in U\left(\left|\hat{\theta}_T - \theta_0\right|\right)\right\}$$

and

$$\left|\int_{\hat{\theta}_T}^{\theta_1} \left(\dot{G}(T,u) - \dot{G}(T,\theta_0)\right) du\right|$$

$$\le \left|\hat{\theta}_T - \theta_0\right| \sup\left\{\left|\dot{G}(T,\theta) - \dot{G}(T,\theta_0)\right| : \theta \in U\left(\left|\hat{\theta}_T - \theta_0\right|\right)\right\}.$$

It follows from (8.2.41) and (8.2.43) and the consistency of the M-estimator

$\hat{\theta}_T$ that

$$\frac{1}{\sqrt{T}} C(T, \theta_0) = \sqrt{T}\left(\hat{\theta}_T - \theta_0\right)\left(\Gamma + o_p(1)\right) . \tag{8.2.48}$$

Relations (8.2.47) and (8.2.48) prove the following theorem.

THEOREM 8.11
Under the conditions (J1)–(J6), the M-estimator $\hat{\theta}_T$ is asymptotically normal, that is,

$$\sqrt{T}\left(\hat{\theta}_T - \theta_0\right) \xrightarrow{\mathcal{L}} N\left(0, \Phi\Gamma^{-2}\right) \quad \text{as } T \to \infty \tag{8.2.49}$$

where Φ and Γ are positive constants as given by (J4).

Special Case: Yoshida and Hayashi [58] consider the problem of the estimation of the phase parameter θ in the periodic intensity $\lambda(t, \theta) = f(t - \theta)$ where f is the C^2-class strictly positive functions with period unity and $\theta \in \Theta = \left(-\frac{1}{2}, \frac{1}{2}\right)$. Suppose further that

$$\int_0^1 f(t)dt = 1$$

and the true intensity

$$\mu(t) = (1 - \varepsilon)\, f(t - \theta_1) + \varepsilon\, C(t - \theta_1) \tag{8.2.50}$$

where $\theta_1 \in \Theta$, $\varepsilon \in [0, M_\varepsilon]$, $0 < M_\varepsilon < 1$ and $C(\cdot)$ is a periodic even bounded function. Without loss of generality, suppose that $\theta_1 = 0$. Let $h(t, \theta) = \psi(t - \theta)$ and $H(t, \theta) = \phi(t - \theta)$ where ϕ is odd and both ψ and ϕ are periodic functions satisfying the conditions (J1)–(J6) with $\theta_0 = 0 (= \theta_1)$. It is easy to see that Γ of (J4) is given by

$$\Gamma = \int_0^1 \psi'(t)\mu(t)dt . \tag{8.2.51}$$

Yoshida and Hayashi [58] construct an M-estimator that has the minimax asymptotic variance over a suitable class M of intensity functions μ. Let

$$V(\mu, \psi) = \frac{\int_0^1 \psi(t)^2 \mu(t) dt}{\left(\int_0^1 \psi'(t) \mu(t) dt \right)^2} . \tag{8.2.52}$$

Note that $V(\mu, \psi)$ is the information corresponding to ψ when μ is the intensity function. Let Q_1 be the class of all periodic continuously differentiable functions ψ with

$$\int_0^1 \psi(t)^2 \mu(t) dt > 0$$

and μ_0 be the intensity function minimizing $I(\mu)$ given by

$$I(\mu) = \sup_{\psi \in Q_1} \frac{\left(\int_0^1 \psi'(t) \mu(t) dt \right)^2}{\int_0^1 \psi(t)^2 \mu(t) dt} . \tag{8.2.53}$$

In addition, let $\mathcal{M}$ be a class of all periodic even functions $\mu(t)$ satisfying

$$\int_0^1 \mu(t) dt \leq \int_0^1 \mu_0(t) dt . \tag{8.2.54}$$

Under some technical conditions, Yoshida and Hayashi [58] prove that the M-estimator corresponding to $h(t, \theta) = \psi(t - \theta)$ has the minimax asymptotic variance, that is,

$$\inf_{\psi \in Q_1} \sup_{\mu \in \mathcal{M}} V(\mu, \psi) = V(\mu_0, \psi_0) \tag{8.2.55}$$

where $\psi_0(t) = -\frac{\mu_0'(t)}{\mu_0(t)}$ and μ_0 is as specified earlier. Yoshida and Hayashi [58] further study the minimax problem when the true intensity $\mu(t)$ is given by (8.2.50). They consider the function

$$h(t,\theta) = \beta\,\psi\left(\frac{S(t-\theta)}{\beta}\right) \tag{8.2.56}$$

where $\beta > 0$ and $S(t-\theta) = \frac{\partial}{\partial\theta}\log f(t-\theta)$ and ψ is a piecewise continuously diferentiable, continuous, nondecreasing odd function concave on $[0,\infty)$. We omit the details. Further details are given in [58].

8.3 Semiparametric Inference

Let us consider an r-dimensional counting process

$$\boldsymbol{N}(t) = (N_1(t),\ldots,N_r(t)),\ t \geq 0 \tag{8.3.1}$$

defined on a probability space $(\Omega, \mathcal{F}, P)$ adapted to the filtration $\{\mathcal{F}_t, t \geq 0\}$. As before the components N_i are integer-valued, right-continuous with jumps of size unity only. We assume that any two components of $\boldsymbol{N}$ cannot have jumps at the same time and $N_i(0) = 0,\ 1 \leq i \leq r$. Suppose the process $\boldsymbol{N}$ has the stochastic intensity

$$\lambda(t) = (\lambda_1(t),\ldots,\lambda_r(t)),\ t \geq 0 \tag{8.3.2}$$

of the form

$$\lambda_i(t) = \exp(\beta_i)\ \alpha(t)Y_i(t),\ 1 \leq i \leq r \tag{8.3.3}$$

where $\{Y_i\}$ $1 \leq i \leq r$, are nonnegative predictable processes relative to the filtration $\{\mathcal{F}_t\}$. Here, the function $\alpha(\cdot)$ is an unknown nuisance *parameter* and the problem of interest is the estimation of the parameters $\beta_i,\ 1 \leq i \leq r$ which describe the proportionalities between the intensities of the component processes. We use the normalization $\beta_r = 0$ and let

$$\boldsymbol{\beta} = (\beta_1,\ldots\beta_{r-1})\ .$$

We assume that the process $\boldsymbol{N}$ is observed during a long period of time and the problem is to estimate $\boldsymbol{\beta}$ based on such an observation.

Suppose the process $\{(N_i(t), Y_i(t)),\ 1 \le i \le r\}$ is observed up to time $t = u$. Consider the partial maximum likelihood estimator (cf. [7, 15]) for estimating β; that is, we will estimate β by maximizing the function

$$H(\beta, u) = \sum_{i=1}^{r} \beta_i N_i(u) - \int_0^u \log\left(\sum_{j=1}^{r} \exp(\beta_j)\, Y_j(t)\right) d\bar{N}(t) \qquad (8.3.4)$$

where $\beta_r = 0$ and $\bar{N}(t) = \sum_{i=1}^{r} N_i(t)$. Let $\hat{\beta}(u)$ denote this estimator. The estimator $\hat{\beta}(u)$ will be a solution of the equation

$$G_i(\beta, u) = \frac{\partial H(\beta, u)}{\partial \beta_i} = N_i(u) - \int_0^u p_i(\beta, t) d\bar{N}(t) = 0,\ 1 \le i \le r - 1 \qquad (8.3.5)$$

where

$$p_i(\beta, t) = \exp(\beta_i) Y_i(t) \Big/ \left\{\sum_{j=1}^{r} \exp(\beta_j)\, Y_j(t)\right\}. \qquad (8.3.6)$$

Note that

$$G(\beta, u) = (G_1(\beta, u), \ldots, G_{r'}(\beta, u)),\ r' = r - 1 \qquad (8.3.7)$$

is a martingale with the predictable quadratic variation

$$\boldsymbol{W}(\beta, u) = \int_0^u \boldsymbol{P}(\beta, t) \bar{\lambda}(t) du \qquad (8.3.8)$$

where $\bar{\lambda}(t) = \sum_{i=1}^{r} \lambda_i(t)$ and $\boldsymbol{P}(\beta, t)$ is the $r' \times r'$ matrix with the elements

$$P_{ij}(\beta, t) = \frac{\partial p_i(\beta, t)}{\partial \beta_j}$$

$$= \delta_{ij}\, p_i(\beta, t) - p_i(\beta, t) p_j(\beta, t)$$

with $\delta_{ij} = 1$ if $i = j$ and $\delta_{ij} = 0$ if $i \neq j$. Let

$$S(\beta, u) = \int_0^u P(\beta, t) d\bar{N}(t),$$

and

$$U_{ij}(\beta, u) = \sum_{k=1}^{r} \int_0^u (\delta_{ik} - p_i(\beta, t)) \left(\delta_{jk} - p_j(\beta, t)\right) dN_k(t).$$

Then $U(\beta, u) = ((U_{ij}(\beta, u))_{r' \times r'}$ is the optional quadratic variation matrix. Note that $\{S(\beta, u) - W(\beta, u), u \geq 0\}$ is a zero mean martingale.

Let $F \in \mathcal{F}$ be a set on which

(i) $\int_0^u \bar{\lambda}(t)dt \to \infty$ as $u \to \infty$ and

(ii) the smallest eigenvalue ν of $\frac{W(\beta,u)}{\int_0^u \bar{\lambda}(t)dt}$ is bounded away from zero for large values of u.

LEMMA 8.4
Under the conditions stated above, almost surely on F,

(i) $S(\beta, u)\, W^{-1}(\beta, u) \to I$ *as* $u \to \infty$,

(ii) $S(\beta, u)\, U^{-1}(\beta, u) \to I$ *as* $u \to \infty$, *and*

(iii) $W(\beta, u)\, U^{-1}(\beta, u) \to I$ *as* $u \to \infty$.

PROOF Let the norm of a matrix A be denoted by $\|A\| = \sup_{|x|=1} |Ax'|$. Then

$$\left\| S(\beta, u)\, W^{-1}(\beta, u) - I \right\|$$

$$\leq \nu^{-1} \| \boldsymbol{S}(\boldsymbol{\beta}, u) - \boldsymbol{W}(\boldsymbol{\beta}, u) \| \left\{ \int_0^u \bar{\lambda}(t) dt \right\}^{-1} .$$

The (i, j)-th element of $\boldsymbol{S}(\boldsymbol{\beta}, u) - \boldsymbol{W}(\boldsymbol{\beta}, u)$ is

$$R_{ij}(u) = \int_0^t P_{ij}(\boldsymbol{\beta}, t)(d\bar{N}(t) - \bar{\lambda}(t)dt)$$

and is a martingale with the predictable quadratic variation

$$\int_0^t P_{ij}^2(\boldsymbol{\beta}, t)\bar{\lambda}(t)dt \leq \int_0^u \bar{\lambda}(t)dt .$$

Hence, by the Lepingle's law of large numbers [36]

$$\frac{R_{ij}(u)}{\int_0^u \bar{\lambda}(t)dt} \to 0 \text{ a.s. on } F \text{ as } u \to \infty .$$

Therefore,

$$\frac{\| \boldsymbol{S}(\boldsymbol{\beta}, u) - \boldsymbol{W}(\boldsymbol{\beta}, u) \|}{\int_0^u \bar{\lambda}(t)dt} \to 0 \text{ a.s. on } F \text{ as } u \to \infty$$

and

$$\boldsymbol{S}(\boldsymbol{\beta}, u)\, \boldsymbol{W}^{-1}(\boldsymbol{\beta}, u) \to \boldsymbol{I} \text{ a.s. on } F \text{ as } u \to \infty . \tag{8.3.9}$$

Since $\{U_{ij}(\boldsymbol{\beta}, u) - S_{ij}(\boldsymbol{\beta}, u)\}$, $1 \leq i, j \leq r'$ are martingales with the predictable quadratic variations that are dominated by $\int_0^u \bar{\lambda}(t)dt$, it follows, again by the Lepingle's law of large numbers, that

$$\frac{\| \boldsymbol{U}(\boldsymbol{\beta}, u) - \boldsymbol{S}(\boldsymbol{\beta}, u) \|}{\int_0^u \bar{\lambda}(t)dt} \to 0 \text{ a.s. on } F \text{ as } u \to \infty .$$

Hence,

$$S(\beta, u)\, U^{-1}(\beta, u) \rightarrow I \text{ a.s. on } F \text{ as } u \rightarrow \infty\,. \tag{8.3.10}$$

Relations (8.3.9) and (8.3.10) together prove that

$$W(\beta, u)\, U^{-1}(\beta, u) \rightarrow I \text{ a.s. on } F \text{ as } u \rightarrow \infty\,. \tag{8.3.11}$$

▌

LEMMA 8.5
For any $\varepsilon > 0$ and almost all sample paths in F, there exists a $\rho > 0$ and a $u_0 < \infty$ such that

$$\left\| S(\gamma, u)\, S^{-1}(\beta, u) - I \right\| \leq \varepsilon \;\; \textit{if}\; \|\gamma - \beta\| \leq \rho$$

and $u > u_0$.

PROOF If $Y_i(t)Y_j(t) \neq 0$ and $\max_j |\gamma_j - \beta_j| = \delta$, then one can check that

$$\exp(-4\delta) \leq \frac{p_i(\gamma, t)p_j(\gamma, t)}{p_i(\beta, t)p_j(\beta, t)}$$

$$\leq \exp\left(\gamma_i - \beta_i + \gamma_j - \beta_j\right) \frac{\left(\sum_1^r \exp(\beta_k)\, Y_k(t)\right)^2}{\left(\sum_1^r \exp(\gamma_k)\, Y_k(t)\right)^2}$$

$$\leq \exp(4\delta)\,.$$

If $Y_i(t)Y_j(t) = 0$, then either $p_i(\gamma, t) = 0$ or $p_j(\gamma, t) = 0$ for all values of γ. This implies that for any $\eta > 0$ there exists a $\rho > 0$ such that

$$\left| P_{ij}(\gamma, t) - P_{ij}(\beta, t) \leq \eta | P_{ij}(\beta, t) \right| \tag{8.3.12}$$

for $i, = 1, \ldots, r'$ if $\|\gamma - \beta\| \leq \rho$ and $i \neq j$. The inequality (8.3.12) holds for

$i = j$ from the fact that

$$P_{ii}(\gamma, t) = p_i(\gamma, t) \sum_{j \neq i} p_j(\gamma, t) .$$

Inequality (8.3.12) implies that

$$\begin{aligned} \left|S_{ij}(\gamma, u) - S_{ij}(\beta, u)\right| &\leq \int_0^u \left|P_{ij}(\gamma, t) - P_{ij}(\beta, t)\right| d\bar{N}(t) \\ &\leq \eta \int_0^u \left|P_{ij}(\beta, t)\right| d\bar{N}(t) \\ &= \eta \left|S_{ij}(\beta, u)\right| , \end{aligned}$$

since $P_{ij}(\beta, t)$ has the same sign for all t. Then

$$\begin{aligned} \|S(\gamma, t) - S(\beta, t)\|^2 &\leq \sum_{ij} \left(S_{ij}(\gamma, t) - S_{ij}(\beta, t)\right)^2 \\ &\leq \eta^2 \sum_{ij} S_{ij}^2(\beta, t) \\ &\leq r\eta^2 \max_i \sum_k S_{ij}^2(\beta, t) \\ &= r\eta^2 \|S(\beta, t)\|^2 . \end{aligned}$$

From Lemma 8.4, it follows that, for almost all sample paths in F, there exists a u_0 such that

$$\|S(\beta, u)\| \leq (1 + \eta)\|W(\beta, u)\|$$

$$\leq (1+\eta) \int_0^u \bar{\lambda}(t)dt$$

and

$$\left\| S^{-1}(\beta, u) \right\| \leq \frac{1}{1-\eta} \left\| W^{-1}(\beta, u) \right\|$$

$$\leq (1+\eta) \left(\nu \int_0^u \bar{\lambda}(t)dt \right)^{-1}$$

for $u \leq u_0$. For these sample paths,

$$\|S(\gamma, u)\, S^{-1}(\beta, u) - I\|$$

$$\leq \|S(\gamma, u) - S(\beta, u)\| \; \|S^{-1}(\beta, u)\|$$

$$\leq \sqrt{r}\eta \|S(\beta, u)\| \; \left\| S^{-1}(\beta, u) \right\|$$

$$\leq \frac{\sqrt{r}\eta(1+\eta)}{\nu(1-\eta)}$$

if $\|\gamma - \beta| \leq \rho$ and $u \geq u_0$. We can choose η such that $\varepsilon = \sqrt{r}\eta(1+\eta)/\{\nu(1-\eta)\}$. This completes the proof of the lemma. ∎

We now study the asymptotic properties of the estimator $\hat{\beta}(u)$.

Consistency

THEOREM 8.12
Under the conditions stated above,

$$\hat{\beta}(u) - \beta \rightarrow 0 \;\; \textit{a.s. on } F \textit{ as } \; u \rightarrow \infty\,.$$

PROOF The martingale $G_i(\beta, u)$ defined by (8.3.5) has the predictable quadratic variation

$$W_{ii}(\beta, u) = \int_0^u P_{ii}(\beta, u)\bar{\lambda}(t)dt$$

$$\leq \int_0^u \bar{\lambda}(t)dt \, .$$

It follows from the Lepingle's law of large numbers that

$$\frac{G_i(\beta, u)}{\int_0^u \bar{\lambda}(t)dt} \to 0 \text{ a.s. on } F \text{ as } u \to \infty \, .$$

Hence,

$$G(\beta, u)\, \boldsymbol{W}^{-1}(\beta, u) \to 0 \text{ a.s. on } F \text{ as } u \to \infty \, .$$

Lemma 8.4 implies that

$$G(\beta, u)\, \boldsymbol{S}^{-1}(\beta, u) \to 0 \text{ a.s. on } F \text{ as } u \to \infty \, .$$

Applying the Taylor's expansion, we have

$$G(\gamma, u)\, \boldsymbol{S}^{-1}(\beta, u)\, (\gamma - \beta)^t = G(\beta, u)\, \boldsymbol{S}^{-1}(\beta, u)\, (\gamma - \beta)^t$$

$$- (\gamma - \beta)\, \boldsymbol{S}\left(\gamma^*, u\right)\, \boldsymbol{S}^{-1}(\beta, u)\, (\gamma - \beta)^t$$

for some γ^* such that $\|\gamma^* - \beta\| \leq \|\gamma - \beta\|$. Here, α^t denotes the transpose of α. Combining this with Lemma 8.5, we find that on F, if ρ is small and u is large, then

$$G(\gamma, u)\, \boldsymbol{S}^{-1}(\beta, u)\, (\gamma - \beta)^t < 0$$

for all $\|\gamma - \beta\| = \rho$. An application of the Lemma 8.6 at the end of the proof shows that the equation

$$G(\gamma, u)\, \boldsymbol{S}^{-1}(\beta, u) = 0$$

or equivalently

$$G(\gamma, u) = 0$$

has a solution in the set $\|\gamma - \beta\| < \rho$. This proves the theorem. ∎

LEMMA 8.6
Let $f(\gamma)$ be a continuous function from R^r to R^r such that $f(\gamma)(\gamma - \beta)' < 0$ for all γ such that $\|\gamma - \beta\| = \rho$. Then there exists a $\hat{\gamma}$ such that $\|\hat{\gamma} - \beta\| < \rho$ and $f(\hat{\gamma}) = 0$.

For proof, see [5].

REMARK 8.7 Lemma 8.5 and Theorem 8.12 together prove the following result. ∎

LEMMA 8.7
Under the conditions stated above,

$$S\left(\hat{\beta}(u), u\right) S^{-1}(\beta, u) \to I \text{ a.s. on } F \text{ as } u \to \infty$$

where $\hat{\beta}(u)$ is as given by the Theorem 8.12.

Asymptotic Distribution of the Estimator $\widehat{\beta}(u)$

We have the following result proving the asymptotic normality of the estimator $\hat{\beta}(u)$.

THEOREM 8.13
Suppose there exists an $\mathcal{F}_0$-measurable scalar valued function $b(u)$, a positive definite matrix D, and an $\mathcal{F}$-measurable random matrix ψ such that

(i) $b(u) \to \infty$,

(ii) $\frac{U(\beta,u)}{b(u)} \xrightarrow{p} \psi$, *and*

(iii) $E\left\{\frac{U(\beta,u)}{b(u)}\right\} \to D$

as $u \to \infty$, then

$$\left(\hat{\beta}(u) - \beta\right) S^{1/2}\left(\hat{\beta}(u), u\right) \xrightarrow{\mathcal{L}} N(0, I) \text{ conditionally on } \{\psi > 0\} \cap F .$$

PROOF [(Sketch)] We now apply a central limit theorem for martingales (cf. [23]). Observe that $G(\beta, \cdot)$ is an r'-dimensional martingale. Since $|\Delta G(\beta, u)| \leq 1$ for all u, it can be checked that the conditions for the central limit theorem of Hutton and Nelson [23] hold. Then

$$G(\beta, u)\, U^{-1/2}(\beta, u) \xrightarrow{\mathcal{L}} N(0, I) \text{ as } u \to \infty$$

conditionally on the event $F \cap [\psi > 0]$. Using Lemma 8.4, it follows that

$$G(\beta, u)\, W^{-1/2}(\beta, u) \xrightarrow{\mathcal{L}} N(0, I)$$

and

$$G(\beta, u)\, S^{-1/2}(\beta, u) \xrightarrow{\mathcal{L}} N(0, I)$$

conditionally on the event $F \cap [\psi > 0]$. Applying the Taylor's expansion, it follows that, for any $\alpha \in R^{r'}$,

$$0 = G\left(\hat{\beta}(u), u\right) S^{-1/2}(\beta, u)\, \alpha'$$

$$= G(\beta, u)\, S^{-1/2}(\beta, u)\, \alpha'$$

$$+ \left(\hat{\beta}(u) - \beta\right) S^{1/2}(\beta, u)\, (S(\gamma_\alpha^*, u)\, S^{-1}(\beta, u))\, \alpha'$$

where $\|\gamma_\alpha^* - \beta\| \leq \|\hat{\beta}(u) - \beta\|$. Since $\hat{\beta}(u) \to \beta$ a.s. on F as $u \to \infty$ by Theorem 8.12, it follows following Lemma 8.5 that

$$S\left(\gamma_\alpha^*, u\right) S^{-1}(\beta, u) \to I \text{ a.s. on } F \text{ as } u \to \infty .$$

Hence, $G(\beta, u)\, S^{-1/2}(\beta, u)\, \alpha'$ and $(\hat{\beta}(u) - \beta)\, S^{1/2}(\beta, u)\, \alpha'$ are asymptotically equivalent on F for all $\alpha \in R^{r'}$. This proves the result concerning the asymptotic normality of $\hat{\beta}(u)$. In view of Lemma 8.7, it follows that

$$\left(\hat{\beta}(u) - \beta\right) S^{1/2}\left(\hat{\beta}(u), u\right) \xrightarrow{\mathcal{L}} N(0, I) \text{ as } u \to \infty$$

conditionally on $F \cap [\psi > 0]$, proving Theorem 8.13. ∎

Example 8.2

Svensson [55] considers an example of a system of n particles that move between two states (say) A and B independently of each other. The intensity for a jump from state A to state B (and B to A) is assumed to be of the form $\mu_1 \theta(t)$ (and $\mu_2 \theta(t)$) at the time t. The unknown nuisance parameter is $\theta(\cdot)$. Let $N_1(t)$ (and $N_2(t)$) denote the number of jumps from A to B (and B to A) up to time t. If the system starts at time $t = 0$ with n_A particles in the state A and n_B in the state B, then, just before time t there are $F(t) = n_A + N_2(t-) - N_1(t-)$ particles in the state A. Now the process $N(t) = (N_1(t), N_2(t))$ is a two-dimensional continuous process with intensities $\lambda_1(t) = \exp(\beta)\, \alpha(t)F(t)$ and $\lambda_2(t) = \alpha(t)\,(n - F(t))$ where $\beta = \log(\mu_1/\mu_2)$ and $\alpha(t) = \mu_2\, \theta(t)$. Svensson [55] discusses the estimation of β by the techniques describe above. □

8.4 Nonparametric Inference

Estimation by the Kernel Method (Nonhomogeneous Poisson Process)

Let $\boldsymbol{X}_T = \{X(t), 0 \le t \le T\}$ be a nonhomogeneous Poisson process with intensity function $\boldsymbol{S}_T = \{S(t), 0 \le t \le T\}$. Suppose the function $S(t)$ is periodic with a known period τ. We now consider nonparametric estimation of the intensity function $S(\cdot)$.

Let Θ_0 be a set of periodic functions on $[0, \infty)$ with period τ that are uniformly bounded and uniformly continuous on $[0, 2\tau]$. Suppose $S(\cdot) \in \Theta_0$.

Let $K(u)$ be a bounded function defined on the real line with $K(u) = 0$ for $u < A < 0$ or $u > B > 0$, and

$$\int_{-\infty}^{\infty} K(u)du = 1$$

where $-\infty < A < 0 < B < \infty$.

Consider the estimator

$$\hat{S}_T(t) = \frac{\tau}{Th_T}\sum_{\ell=1}^{N}\int_{\tau(\ell-1)}^{\tau\ell} K\left(\frac{v-(\ell-1)\tau-t}{h_T}\right)dX(v)$$

$$+\frac{\tau}{Th_T}\int_{\tau N}^{T} K\left(\frac{v-N\tau-t}{h_T}\right)dX(v) \qquad (8.4.1)$$

where N is the number of periods of length τ in $[0,T]$ and $0<h_T\to 0$ as $T\to\infty$. This is a kernel-type estimator (cf. [50]) for the intensity function $S(\cdot)$ based on the observation $\boldsymbol{X}_T=\{X(t),\ 0\le t\le T\}$. We now study some properties of this estimator.

Uniform Consistency

THEOREM 8.14
Let $[a,b]\subset(0,\tau)$ and $S(\cdot)\in\Theta_0$. If $h_T\to 0$ and $Th_T\to\infty$, then $\hat{S}_T(t)$ is uniformly consistent in quadratic mean as $T\to\infty$, that is,

$$\lim_{T\to\infty}\ \sup_{S(\cdot)\in\Theta_0}\ \sup_{a\le t\le b} E\left|\hat{S}_T(t)-S(t)\right|^2=0\,. \qquad (8.4.2)$$

PROOF Let

$$M(t)=X(t)-\int_0^t S(v)dv$$

and

$$I_T(f)=\int_0^T f(t)dM(t)$$

where f is any function continuous from the left. It is known that

$$E\left(I_T(f)\right)=0 \text{ and } EI_T(f)^2=\int_0^T f(t)^2S(t)dt\,.$$

The estimator $\hat{S}_T(t)$ can be written in the form

$$\hat{S}_T(t) = \frac{\tau}{Th_T}\sum_{\ell=1}^{N}\int_{\tau(\ell-1)}^{\tau\ell} K\left(\frac{v_{\ell,t}}{h_T}\right) S(v)dv$$

$$+\frac{\tau}{Th_T}\sum_{\ell=1}^{N}\int_{\tau(\ell-1)}^{\tau\ell} K\left(\frac{v_{\ell,t}}{h_T}\right) dM(v) + o\left(T^{-2}\right) \qquad (8.4.3)$$

where $v_{\ell,t} = v - (\ell-1)\tau - t$. Let $u = v_{\ell,t}/h_T$ and define

$$\tilde{M}_t(u) = h_T^{-1}\left[M\left(t+\tau(\ell-1)+uh_T\right) - M(t+\tau(\ell-1))\right]. \qquad (8.4.4)$$

Let $T_0 = \min(T', T'')$ where T' and T'' are the solutions of the equations

$$\frac{-a}{h_{T'}} = A, \quad \frac{\tau-b}{h_{T''}} = B .$$

Then, for $T > T_0$,

$$\sup_{a\le t\le b} E\left|\hat{S}_T(t) - S(t)\right|^2 = \sup_{a\le t\le b} E\left\{\int_{-\infty}^{\infty} K(u)\left[S\left(t+uh_T\right) - S(t)\right]du\right.$$

$$\left. + \frac{\tau}{Th_T}\int_{-\infty}^{\infty} K(u)d\tilde{M}_t(u)\right\}^2 + o\left(T^{-1}\right)$$

$$\le 2\sup_{a\le t\le b}\left(\int_{-\infty}^{\infty} K(u)\left[S\left(t+uh_T\right) - S(t)\right]du\right)^2$$

$$+\frac{2\tau^2}{Th_T}\sup_{a\le t\le b}\int_{-\infty}^{\infty} K^2(u)S\left(t+uh_T\right)du$$

$$+\,o\left(T^{-1}\right). \tag{8.4.5}$$

Since $S(\cdot) \in \Theta_0$, it follows that $S(t + uh_T) \to S(t)$ uniformly over $S(\cdot) \in \Theta_0$ and hence,

$$\lim_{T\to\infty} \sup_{S(\cdot)\in\Theta_0} \sup_{a\le t\le b} E\left|\hat{S}_T(t) - S(t)\right|^2 = 0\,. \tag{8.4.6}$$

▮

Asymptotic Efficiency

Let Θ_β be a set of periodic functions $m(t)$, $t \geq 0$ with period τ which are j-times differentiable and for which

$$\left|m^{(j)}(t) - m^{(j)}(s)\right| \leq L|t - s|^\alpha \tag{8.4.7}$$

where L is a constant, $\beta = j + \alpha$ with $0 < \alpha \leq 1$, $j \geq 0$. Here, $m^{(j)}(\cdot)$ denotes the jth derivative of $m(\cdot)$.

In addition to the conditions assumed earlier on the kernel $K(\cdot)$, suppose that

$$\int_{-\infty}^{\infty} K(u)u^\ell du = 0,\ 1 \leq \ell \leq j\,. \tag{8.4.8}$$

THEOREM 8.15

Let $S(\cdot) \in \Theta_\beta$. Then

$$\limsup_{T\to\infty} \sup_{S(\cdot)\in\Theta_\beta} \sup_{a\le t\le b} E\left|\hat{S}_T(t) - S(t)\right|^2 T^{\frac{2\beta}{2\beta+1}} < \infty\,. \tag{8.4.9}$$

PROOF Under the condition $S(\cdot) \in \Theta_\beta$, it follows, by the Taylor's expansion, that

$$S(t+\delta) - S(t) = \sum_{r=1}^{j} \frac{S^{(r)}(t)}{r!}\delta^r + \frac{S^{(j)}(t+\gamma\delta) - S^{(j)}(t)}{j!}\delta^j$$

for some $\gamma \in (0, 1)$. Hence, for $T > T_0$,

$$\left| \int_{-\infty}^{\infty} K(u) [S(t + uh_T) - S(t)]\, du \right|$$

$$= \left| \frac{h_T^j}{j!} \int_{-\infty}^{\infty} u^j K(u) \left[S^{(j)}(t + u\gamma h_T) - S^{(j)}(t) \right] du \right|$$

$$\leq \frac{L h_T^{j+\alpha}}{j!} \int_{-\infty}^{\infty} \left| u^{j+\alpha} K(u) \right| du .$$

Let

$$\sigma^2 = \int_{-\infty}^{\infty} K^2(u) du .$$

Then

$$\sup_{a \leq t \leq b} E \left| \hat{S}_T(t) - S(t) \right|^2 \leq \frac{2\tau^2 \sigma^2}{T h_T} + 2 \frac{L^2 h_T^{2\beta}}{(j!)^2} \left(\int_{-\infty}^{\infty} \left| u^{j+\alpha} K(u) \right| du \right)^2 .$$

Let us choose h_T such that $h_T^{2\beta} = (T h_T)^{-1}$. Then $h_T = T^{\frac{1}{2\beta+1}}$, which leads to the result stated in the theorem. ∎

REMARK 8.8 The above proof leads to the representation

$$\left[\hat{S}_T(t) - S(t) \right] T^{\frac{\beta}{2\beta+1}} = \frac{h_T^j}{j!} \int_{-\infty}^{\infty} K(u) u^j \left[S^{(j)}(t + \gamma u h_T) - S^{(j)}(t) \right] du$$

$$+ \sqrt{\frac{\tau}{T h_T}} \zeta_T + o\left(T^{-1} \right) \qquad (8.4.10)$$

where ζ_T is asymptotically normal with mean 0 and variance $S(t)\sigma^2$ (cf. [32]). Hence, the following result holds. ∎

THEOREM 8.16

Let $S(\cdot) \in \Theta_\beta$ and suppose that

$$\lim_{h_T \to 0} \left[\left\{ \int_{-\infty}^{\infty} K(u) \left[S(t + uh_T) - S(t) \right] du \right\} h_T^{-j-\alpha} \right] = q(t) . \qquad (8.4.11)$$

Then $\hat{S}_T(t)$ is asymptotically normal as $T \to \infty$, that is, uniformly on $S(\cdot) \in \Theta_\beta$ and $t \in [a, b]$,

$$\left[\hat{S}_T(t) - S(t) \right] T^{\frac{\beta}{2\beta+1}} \xrightarrow{\mathcal{L}} N(q(t), \tau S(t)\sigma^2) \text{ as } T \to \infty . \qquad (8.4.12)$$

The following theorem proves that it is not possible to construct estimators that have better rate of convergence then $\hat{S}_T(t)$.

THEOREM 8.17

Let $S(\cdot) \in \Theta_\beta$. Then, for every $t \in (0, \tau)$,

$$\liminf_{T \to \infty} \inf_{S_T^*(t)} \sup_{S(\cdot) \in \Theta_\beta} E \left[S_T^*(t) - S(t) \right]^2 T^{\frac{\beta}{2\beta+1}} > 0 . \qquad (8.4.13)$$

PROOF (Sketch of the Proof) Let

$$S_T(t) = S_0(t) + \theta T^{-\frac{\beta}{2\beta+1}} g\left((t - t_0) T^{\frac{1}{2\beta+1}} x \right) \qquad (8.4.14)$$

where $g(\cdot), S_0(\cdot) \in \Theta_\beta$ with $\frac{1}{2}L$ as the constant in (8.4.7) and suppose that $S_0(t) > 0$ for $t \geq 0$ and $g(\cdot) \neq 0$ with $g(u) = 0$ for $u < -\frac{1}{2}$ and $u > \frac{1}{2}$. Consider the problem of estimation of θ (for $|\theta| < x$) by observations from a Poisson process with intensity given by (8.4.14). The family of probability measures $\{P_\theta^T, |\theta| < x\}$ induced by the process on the space of sample paths of the process is locally asymptotically normal (cf. [32]) and the result follows from the techniques of Ibragimov and Khasminskii [24]. ∎

Estimation by the Kernel Method (Counting Processes)

We first introduce the notation as in the last chapter. Let $(\Omega, \mathcal{F}, P)$ be a probability space and $\{\mathcal{F}_t, t \in [0, 1]\}$ be an increasing family of sub-σ-algebras of $\mathcal{F}$. Let $\{N(t), \mathcal{F}_t, \ 0 \le t \le 1\}$ be a counting process with stochastic intensity process $\{\lambda(t), \mathcal{F}_t, \ t \ge 0\}$ where $\lambda(t)$ is a nonnegative left-continuous process with right limits. We assume that $N(0) = 0$ and that there are only a finite number of jumps of size unity in each interval $[0, t]$. Further we assume that $EN(1) < \infty$. It is known that

$$M(t) = N(t) - \int_0^t \lambda(s)ds, \ 0 \le t \le 1 \tag{8.4.15}$$

is a zero mean square integrable martingale with the quadratic variation process

$$\langle M \rangle_t = \int_0^t \lambda(s)ds, \ 0 \le t \le 1 \, . \tag{8.4.16}$$

Suppose that

$$\lambda(t) = \alpha(t) \ Y(t), \ t \in [0, 1] \tag{8.4.17}$$

where α is an unknown nonrandom function and Y is an observable process. The function $\alpha(\cdot)$ is called the *intensity function* and the model is the *multiplicative intensity model* of Aalen [1, 2]. The function $\alpha(\cdot)$ and the sample paths of Y are nonnegative and left continuous with right-hand limits. We assume that $Y(t)$ is adapted to $\{\mathcal{F}_t\}$. The intensity function $\alpha(\cdot)$ can be interpreted as the transition intensity on the individual level and $Y(t)$ measures the size of the risk population just before time t.

It is easy to see that (8.4.1) can be written in the form

$$dN(t) = \alpha(t) \ Y(t)dt + dM(t), \ 0 \le t \le 1 \tag{8.4.18}$$

and a natural estimator for the cumulative intensity function

$$\beta(t) = \int_0^t \alpha(s)ds, \ 0 \le t \le 1 \tag{8.4.19}$$

is

$$\tilde{\beta}(t) = \int_0^t \frac{\alpha(s)}{Y(s)} dN(s), \ 0 \le t \le 1 . \tag{8.4.20}$$

Since $Y(s)$ may be zero for some value of s, one can consider an estimator of

$$\beta^*(t) = \int_0^t \alpha(s) J(s) ds, \ 0 \le t \le 1 \tag{8.4.21}$$

where $J(s) = I(Y(s) > 0)$. If $Y(s) = 0$, we define $\frac{J(s)}{Y(s)} = 0$. Assume that

$$E\left[\int_0^1 \left\{\frac{J(s)}{Y(s)}\right\} \alpha(s) ds\right] < \infty . \tag{8.4.22}$$

This will ensure the existence of the stochastic integral defined below and a natural estimator for $\beta^*(t)$ is

$$\hat{\beta}(t) = \int_0^t \left\{\frac{J(s)}{Y(s)}\right\} dN(s) . \tag{8.4.23}$$

It follows, from the general theory of stochastic integrals, that $\hat{\beta} - \beta^*$ is a square integrable martingale with the quadratic variation process

$$\langle \hat{\beta} - \beta^* \rangle_t = \int_0^t \frac{\alpha(s) J(s)}{Y(s)} ds \tag{8.4.24}$$

(cf. [2]). Furthermore, the mean square error of $\hat{\beta}(t)$ is

$$\eta(t) = E\left\{\hat{\beta}(t) - \beta^*(t)\right\}^2$$

$$= E \int_0^t \left\{\frac{\alpha(s) J(s)}{Y(s)}\right\} ds . \tag{8.4.25}$$

Let

$$\hat{\eta}(t) = \int_0^t \left\{ \frac{J(s)}{Y^2(s)} \right\} dN(s) .$$

Note that

$$\hat{\eta}(t) = \int_0^t \left\{ \frac{\alpha(s) J(s)}{Y(s)} \right\} ds + \int_0^t \left\{ \frac{J(s)}{Y^2(s)} \right\} dM(s)$$

and the second term on the right-hand side in a zero-mean martingale. It is easy to see that

$$E\left(\hat{\eta}(t)\right) = \eta(t)$$

and $\hat{\eta}(t)$ gives an unbiased estimator for $\eta(t)$.

We now consider the problem of estimation of the intensity function $\alpha(t)$ instead of the cumulative intensity $\beta(\cdot)$ by the kernel method (cf. [50]). Let $K(\cdot)$ be a suitable kernel and define

$$\hat{\alpha}(t) = \frac{1}{h} \int_0^t K\left(\frac{t-s}{h}\right) d\hat{\beta}(s) \tag{8.4.26}$$

where $\hat{\beta}(\cdot)$ is as defined by (8.4.23). Note that

$$\hat{\alpha}(t) = \frac{1}{h} \sum_{T_i} \left\{ K\left(\frac{t - T_i}{h}\right) \Big/ Y(T_i) \right\} \tag{8.4.27}$$

where the jump times of the process $N(\cdot)$ are $T_1, T_2, \ldots$. If $K(\cdot)$ is continuous, then $\hat{\alpha}(\cdot)$ will have continuous sample paths. The estimator $\hat{\alpha}(t)$ can be considered as a moving average of $\frac{1}{Y(T_i)}, i = 1, 2, \ldots$ with weights integrating to unity.

Suppose the kernel $K(\cdot)$ has its support contained in the interval $[-1, 1]$ and $0 < h < 1/2$. Define

$$\alpha^*(t) = \frac{1}{h} \int_0^t K\left(\frac{t-s}{h}\right) d\beta^*(s)$$

$$= \int_{-1}^{1} K(u)\,\alpha(t-hu)\,J(t-hu)du \tag{8.4.28}$$

for $t \in [h, 1-h]$. Then

$$\hat{\alpha}(t) - \alpha^*(t) = \frac{1}{h}\int_0^t K\left(\frac{t-s}{h}\right) d\left(\hat{\beta} - \beta^*\right)(s) \tag{8.4.29}$$

and hence,

$$E\left[\hat{\alpha}(t) - \alpha^*(t)\right] = 0\,; \tag{8.4.30}$$

since $\hat{\beta} - \beta^*$ is a martingale. In other words $E[\hat{\alpha}(t)] = E[\alpha^*(t)]$. Furthermore, it follows that

$$\sigma^2(t) \equiv E\left[\hat{\alpha}(t) - \alpha^*(t)\right]^2 = \frac{1}{h^2}E\left[\int_0^1 K^2\left(\frac{t-s}{h}\right)\left\{\frac{\alpha(s)J(s)}{Y(s)}\right\}ds\right] \tag{8.4.31}$$

from (8.4.24). Interchanging the order of integration, we have

$$E\left[\hat{\alpha}(t) - \alpha^*(t)\right]^2 = \frac{1}{h}\int_{-1}^{1} K^2(u)\alpha(t-hu)E\left(\frac{J(t-hu)}{Y(t-hu)}\right)du\,. \tag{8.4.32}$$

An unbiased estimator of $\sigma^2(t)$ is

$$\hat{\sigma}^2(t) = \frac{1}{h^2}\int_0^1 K^2\left(\frac{t-s}{h}\right)\left\{\frac{J(s)}{Y^2(s)}\right\}dN(s)\,. \tag{8.4.33}$$

This can be seen from the observation that

$$\hat{\sigma}^2(t) = \frac{1}{h}\int_{-1}^{1} K^2(u)\left\{\frac{J(t-hu)}{Y(t-hu)}\right\}\alpha(t-hu)du$$

$$+\frac{1}{h^2}\int_0^1 K^2\left(\frac{t-s}{h}\right) J(s)dM(s)\,. \tag{8.4.34}$$

Note that the expectation of the second term on the right side of (8.4.34) is zero since M is a martingale defined by (8.4.15). Let $\sigma^{*^2}(t)$ denote the first term on the right-hand side of (8.4.34). Then

$$E\left[\hat{\sigma}^2(t)\right] = E\left[\sigma^{*^2}(t)\right]. \tag{8.4.35}$$

Furthermore,

$$E\left[\hat{\sigma}^2(t) - \sigma^{*^2}(t)\right]^2 = \frac{1}{h^2}\int K^4(u)\alpha(t-hu)E\left(\frac{J(t-hu)}{Y(t-hu)}\right)du\,. \tag{8.4.36}$$

Let $\{N_n\}$ be a sequence of independent one-dimensional counting processes each with the stochastic intensity process of the form $\Lambda_n(t) = \alpha(t)Y_n(t)$. The function $\alpha(t)$ is sometimes known as the *base line intensity function.* Keeping the kernel $K(\cdot)$ fixed and letting the "window" or "bandwidth" h depending on n, we construct the estimator $\hat{\alpha}_n(t)$ for $\alpha(t)$ following (8.4.26). Let $J_n = I([Y_n(s) > 0]$.

The following theorem is an easy consequence.

THEOREM 8.18

(i) *If the intensity function $\alpha(\cdot)$ is continuous at a point t and if $E[J_n(s)] \to 1$ uniformly in a neighborhood of t, then $E[\hat{\alpha}_n(t)] \to E[\alpha(t)]$ as $n \to \infty$.*

(ii) *If $nE\{\frac{J_n(s)}{Y_n(s)}\} \to \frac{1}{\tau(s)}$ uniformly in a neighborhood of t as $h \to \infty$ and $\alpha(\cdot)$ and $\tau(\cdot)$ are continuous at the point t, then*

$$\sigma_n^2(t) = E\left(\hat{\alpha}_n(t) - \alpha_n^*(t)\right)^2$$

$$= (nh_n)^{-1}\left\{\frac{\alpha(t)}{\tau(t)}\right\}\int_{-1}^{1} K^2(u)du + o\left((nh_n)^{-1}\right) .$$

(8.4.37)

Note that

$$E\left(\hat{\alpha}(t)\right) = E\left(\alpha^*(t)\right) = K_h \star (\alpha j)(t) \qquad (8.4.38)$$

where $K_h(s) = \frac{1}{h}K(\frac{s}{h})$, $j(s) = E[J(s)]$ and $\star$ denotes the convolution operation. The sequence $\{K_{h_n}\}$ is a Dirac delta sequence as $h_n \to 0$. Theorem 8.18 is now an easy consequence of these observations. We omit the details.

The above theorem proves that the estimator $\hat{\alpha}_n(t)$ is asymptotically unbiased. Furthermore, if $nh_n \to \infty$, then

$$E\left(\hat{\alpha}_n(t) - \alpha_n^*(t)\right)^2 \to 0 \text{ as } n \to \infty , \qquad (8.4.39)$$

which in turn implies that

$$\hat{\alpha}_n(t) - \alpha_n^*(t) \xrightarrow{p} 0 \text{ as } n \to \infty . \qquad (8.4.40)$$

Under the conditions stated in Theorem 8.18, it follows also that the estimator $\hat{\alpha}_n(t)$ is mean square consistent, that is,

$$E\left[\hat{\alpha}_n(t) - \alpha(t)\right]^2 \to 0 \text{ as } n \to \infty , \qquad (8.4.41)$$

whenever $n \to \infty$, $h_n \to 0$ such that $nh_n \to \infty$. In view of (8.4.39) it is sufficient to check that

$$E\left[\alpha_n^*(t) - \alpha(t)\right]^2 \to 0 \text{ as } n \to \infty \qquad (8.4.42)$$

whenever $n \to \infty$, $h_n \to 0$ such that $nh_n \to \infty$. Note that

$$\alpha_n^*(t) - \alpha(t) = \int_{-1}^{1} K(u)\left\{\alpha\left(t - h_n u\right) J_n\left(t - h_n u\right) - \alpha(t)\right\} du . \qquad (8.4.43)$$

Since $J_n \xrightarrow{p} 1$ *uniformly* in a neighborhood of t, it follows that

$$\alpha_n^*(t) - \alpha(t) \xrightarrow{p} 0 \text{ as } n \to \infty \tag{8.4.44}$$

and there exists a constant c such that $|\alpha_n^*(t) - \alpha(t)| \leq c$. Hence,

$$E\left[\alpha_n^*(t) - \alpha(t)\right]^2 \to 0 \text{ as } n \to \infty\,. \tag{8.4.45}$$

The following theorem shows the mean square uniform consistency of $\hat{\alpha}_n(t)$ to $\alpha(t)$ on an interval (z_0, z_1) with $0 < z_0 < z_1 < 1$.

THEOREM 8.19
Suppose that

(i) $J_n \xrightarrow{p} 1$ *uniformly on* $[0, 1]$ *as* $n \to \infty$,

(ii) $\alpha(\cdot)$ *is continuous on* $[0, 1]$,

(iii) $n\eta_n(1) = n\int_0^1 E\left(\frac{J_n(s)}{Y_n(s)}\right)\alpha(s)ds$ *is bounded as* $n \to \infty$, *and*

(iv) *the kernel* K *is of bounded variation.*

Then

$$E\left[\sup_{t\in(z_0,z_1)} |\hat{\alpha}_n(t) - \alpha(t)|^2\right] \to 0 \tag{8.4.46}$$

as $n \to \infty$, $h_n \to 0$ *such that* $nh_n^2 \to \infty$.

PROOF It is sufficient to prove that

$$E\left[\sup_{t\in(z_0,z_1)} |\hat{\alpha}_n(t) - \alpha_n^*(t)|^2\right] \to 0 \tag{8.4.47}$$

and

$$E\left[\sup_{t\in(z_0,z_1)} |\alpha_n^*(t) - \alpha(t)|^2\right] \to 0\,. \tag{8.4.48}$$

Note that

$$\hat{\alpha}_n(t) - \alpha_n^*(t) = \frac{1}{h_n}\int_0^1 K\left(\frac{t-s}{h_n}\right) d\left(\hat{\beta}_n - \beta_n^*\right)(s). \tag{8.4.49}$$

Since $K(\cdot)$ is of bounded variation,

$$\left|\hat{\alpha}_n(t) - \alpha_n^*(t)\right| \leq \frac{2}{h_n} V(K) \sup_{t\in[0,1]} \left|\hat{\beta}_n(s) - \beta_n^*(s)\right| \tag{8.4.50}$$

where $V(K)$ is the total variation of $K(\cdot)$ over $[-1, 1]$. Hence, the relation (8.4.47) holds provided

$$\frac{1}{h_n^2} E\left[\sup_{t\in[0,1]} \left|\hat{\beta}_n(s) - \beta_n^*(s)\right|^2\right] \to 0 \text{ as } n \to \infty .$$

Applying the Doob's inequality for the submartingale $(\hat{\beta}_n - \beta_n^*)^2$, it follows that

$$\frac{1}{h_n^2} E\left\{\sup_{t\in[0,1]} \left|\hat{\beta}_n(s) - \beta_n^*(s)\right|^2\right\} \leq \left(nh_n^2\right)^{-1} 4n\eta_n(1) \tag{8.4.51}$$

from (8.4.25). Since $n\eta_n(1)$ is bounded and $nh_n^2 \to \infty$ by the hypothesis, it follows that the relation (8.4.47) holds. Proof of (8.4.48) is similar to that of (8.4.45). ∎

In order to study the asymptotic normality of the estimator $\hat{\alpha}_n(t)$, we first state a limit theorem for stochastic integrals.

Let $\{N_n\}$ be a sequence of counting processes on [0, 1] with the corresponding sequence of martingales

$$M_n(t) = N_n(t) - \int_0^t \lambda_n(s)ds \tag{8.4.52}$$

as discussed earlier. Let $\{H_n\}$ be a sequence of predictable processes where $E\{\int_0^1 H_n^2(s)\lambda_n(s)ds\} < \infty$. Consider

$$\tilde{M}_n(t) = \int_0^t H_n(s)dM_n(s)\,. \tag{8.4.53}$$

THEOREM 8.20

Suppose that

$$\text{(i)} \quad \textit{for all } \varepsilon > 0, \int_0^1 H_n^2(s)I\left(|H_n(s)| > \varepsilon\right)\lambda_n(s)ds \xrightarrow{p} 0 \;\; \textit{as } n \to \infty \tag{8.4.54}$$

and

$$\text{(ii)} \qquad \int_0^1 H_n^2(s)\lambda_n(s)ds \xrightarrow{p} 1 \;\; \textit{as } n \to \infty\,. \tag{8.4.55}$$

Then

$$\tilde{M}_n(1) = \int_0^1 H_n(B)dM_n(s) \xrightarrow{\mathcal{L}} N(0,1) \;\; \textit{as } n \to \infty\,. \tag{8.4.56}$$

This result is a consequence of the results in [41] on the functional central limit theorem for semimartingales. The following theorem gives sufficient conditions for the asymptotic normality of the estimator $\hat{\alpha}_n(t)$.

THEOREM 8.21

Suppose that (i) $\frac{nJ_n(s)}{Y_n(s)} \xrightarrow{p} \frac{1}{\tau(s)}$ uniformly in a neighborhood of t as $n \to \infty$ and (ii) the function $\alpha(\cdot)$ and $\tau(\cdot)$ are continuous at the point t. Then

$$(nh_n)^{1/2}\left(\hat{\alpha}_n(t) - \alpha_n^*(t)\right) \xrightarrow{\mathcal{L}} N(0, \gamma(t)) \tag{8.4.57}$$

as $n \to \infty$, $h_n \to 0$ such that $nh_n \to \infty$ where

$$\gamma(t) = \frac{\alpha(t)}{\tau(t)} \int_{-1}^{1} K^2(u)du \, . \tag{8.4.58}$$

It is sufficient to check that the conditions (i) and (ii) hold for any fixed t for the sequence $\{H_n(s)\}$ defined by

$$H_n(s) = \left(\frac{n}{h_n}\right)^{1/2} K\left(\frac{t-s}{h_n}\right) \frac{J_n(s)}{Y_n(s)} \, . \tag{8.4.59}$$

Observe that

$$(nh_n)^{1/2}\left(\hat{\alpha}_n(t) - \alpha_n^*(t)\right) = \int_0^1 H_n(s)dM_n(s) \, . \tag{8.4.60}$$

We omit the details.

REMARK 8.9 (i) If $\alpha(\cdot)$ has a bounded derivative in a neighborhood of t and $nh_n^3 \to 0$ as $n \to \infty$, then it can be shown that $(nh_n)^{1/2}(\alpha_n^*(t) - \alpha(t)) \xrightarrow{p} 0$, and hence,

$$(nh_n)^{1/2}\left(\hat{\alpha}_n(t) - \alpha(t)\right) \xrightarrow{\mathcal{L}} N(0, \gamma(t)) \text{ as } n \to \infty \, . \tag{8.4.61}$$

It can be further proved that $\hat{\alpha}_n(t)$ and $\hat{\alpha}_n(s)$ are asymptotically independent for $s \neq t$ as $n \to \infty$, $h_n \to 0$ with $nh_n \to \infty$.

(ii) Ramlau–Hansen [52] discussed the estimation of the hazard rate for random censored data using the above approach. He has also extended the results to the estimation of derivatives of intensity function $\alpha(\cdot)$. Thavaneswaran and Samanta [56] considered a recursive version of the estimator $\hat{\alpha}_n(t)$ defined by

$$\tilde{\alpha}_n(t) = \frac{1}{c_n} \sum_{j=1}^{n} \frac{1}{\sqrt{h_j}} \int_0^1 K\left(\frac{t-s}{h_j}\right) d\hat{\beta}_j(t) \tag{8.4.62}$$

for $\alpha(t)$ where $c_n = (knh_n \log n)^{1/2}$ with $k > 1,\ h_n \to c$ and $nh_n \to \infty$ as $n \to \infty$. They proved the asymptotic normality of the estimator and showed that the estimator $\tilde{\alpha}_n(t)$ has a smaller asymptotic variance that $\hat{\alpha}_n(t)$. We omit the details. For a study of recursive versions of density estimators, see [50].

(iii) McKeague [44] studies the asymptotic theory for weighted least squares estimators in the additive risk model due to Aalen [3]. In the case of a single covariate, the smoothed weighted least squares estimator suggested by McKeague [43] reduces to the kernel type estimator of Ramlau–Hansen [52] discussed above. McKeague and Utikal [45] study the problem of inference for conditional hazard function in the case of time dependent covariates. The conditional hazard function is given by

$$\lambda\,(t|Z_i) = \lim_{\varepsilon \downarrow 0} \frac{1}{\varepsilon} P\,(T_i \leq t + \varepsilon \mid T_i > t;\ Z_i(s), s \leq t) \tag{8.4.63}$$

for the survival time T_i of an individual with the covariate process $Z_i = (Z_i(t))$ and suppose it is of the form

$$\lambda\,(t|Z_i) = \alpha\,(t, Z_i(t))\,,\ 1 \leq i \leq n \tag{8.4.64}$$

where α is a function of the time t and the state of the covariate process at the time t. In the counting process frame work, we assume that $\boldsymbol{N}(t) = (N_i(t), \ldots, N_n(t))$ is a multivariate counting process under the usual conditions with the stochastic intensity

$$\lambda_i(t) = Y_i(t)\,\alpha\,(t, Z_i(t)) \tag{8.4.65}$$

for the process $N_i,\ 1 \leq i \leq n$. The problem studied by McKeague and Utikal [45] is to estimate $\alpha(.,.)$ over a region in the (t, z)-plane given the observations on $(N_i, Y_i, Z_i),\ 1 \leq i \leq n$ over $[0, 1]$. Examples of such processes include right censored survival data, semi-Markov processes, an illness–death process with duration-dependence and age-dependent birth and death processes. Mckeague and Utikal [45] construct a kernel type estimator for $\alpha(\cdot, \cdot)$ and study its asymptotic properties. We omit the details. We come back to the discussion of more general models later in this book. Nielsen and Linton [46] construct a kernel-type estimator for $\alpha(\cdot, \cdot)$ using nonparametric regression estimator of the type of developed by Nadaraya and Watson (cf. [50]). They establish pointwise and global convergence of these estimators. Borgan, Goldstein and Langholz [13] study methods for the analysis of sampled cohort data in the Cox proportional hazard model for which the intensity process is of the form

$$\lambda_i(t) = Y_i(t)\, \alpha_0(t)\, \exp\left(\beta_0^T Z_i(t)\right)$$

for the subject i. ∎

Estimation by the Method of Sieves

With respect to a probability measure $P_\alpha, \alpha \in I$, let $(N_i, \lambda_i), i \geq 1$ be i.i.d. counting processes with stochastic intensity process $\alpha(t)\, Y(t)$. Suppose that the index set I consists of all nonnegative left-continuous functions with right limits such that $\alpha(\cdot) \in L^2[0, 1]$. Let

$$N^{(n)} = \sum_{i=1}^{n} N_i \text{ and } \lambda^{(n)} = \sum_{i=1}^{n} \lambda_i . \tag{8.4.66}$$

The log-likelihood function $\ell_n(\alpha)$ is given by

$$\ell_n(\alpha) = \int_0^1 \lambda^{(n)}(s)(1 - \alpha(s))ds + \int_0^1 (\log \alpha(s))dN^{(n)}(s) . \tag{8.4.67}$$

The method of sieves consists in choosing a sequence of subsets $\{I_n\} \subset I$ such that for each n, there does exist a maximum likelihood estimator $\hat{\alpha} = \hat{\alpha}_n$ of α in I_n and such that I_n increases with $\cup_n I_n$ dense in I.

The problem is to find suitable conditions so that $\|\hat{\alpha}_n - \alpha\|_1 \to 0$, that is, $\hat{\alpha}_n$ is consistent for α.

For each $a > 0$, let $I(a)$ denote the family of absolutely continuous functions $\alpha \in I$ satisfying

$$a \leq \alpha(\cdot) \leq \frac{1}{a} ,$$

and

$$\frac{|\alpha'(\cdot)|}{\alpha(\cdot)} \leq \frac{1}{a}$$

everywhere on $[0, 1]$. As the sieve mesh $a \downarrow 0$, $I(a)$ increases to a dense subset of I. In the following, we prove that for each n and a there exists a maximum likelihood estimator $\hat{\alpha}(n, a)$ in $I(a)$ and we choose $a = a_n$ converging to zero so that $\hat{\alpha}(n, a_n) \to \alpha$ in $L^1[0, 1]$.

THEOREM 8.22
Suppose the following conditions hold:

(i) *the function* $m_s(\alpha) = E_\alpha[\lambda(s)]$ *is bounded and bounded away from zero on the interval* $[0, 1]$,

(ii) *the entropy*

$$H(\alpha) = -\int_0^1 [(1 - \alpha(s) + \alpha(s)\log\alpha(s)]m_s(\alpha)ds$$

is finite ,

(iii) $\int_0^1 var_\alpha[\lambda(s)]ds < \infty$, *and*

(iv) $E_\alpha[N_1^2] < \infty$.

Then

(a) *for each* n *and each* $\alpha > 0$, *there exists a (not necessarily unique) maximum likelihood estimator* $\alpha(n, a) \in I(a)$ *such that*

$$\ell_n\left(\hat{\alpha}(n, a)\right) \geq \ell_n(\alpha)$$

for all $\alpha \in I(u)$ *and*

(b) *for* $a_n = n^{-\frac{1}{4}+\eta}$ *with* $0 < \eta < \frac{1}{4}$ *and* $\hat{\alpha} = \hat{\alpha}(n, a_n)$,

$$\|\hat{\alpha} - \alpha\|_1 \to 0 \ \ a.s. \ \ [P_\alpha] \ \ as\ n \to \infty\,. \tag{8.4.68}$$

This result is from [28] following the techniques of Grenander [21, Section 8.2, Theorem 1]). We omit the proof.

REMARK 8.10 Suppose $\{c_n\}$ is a sequence such that $c_n \to 0$ and $c_n n^{-1/2} \to 0$. Under conditions (i)–(iii) of Theorem 8.22, it can be shown that, for $a_n = \left(\frac{c_n}{n^{1/2}}\right)^{1/2}$ and $\hat{\alpha} = \hat{\alpha}(n, a_n)$,

$$\lim_{d\to\infty} \limsup_{n\to\infty} \sup_{\|\alpha_0-\alpha\|_1<\varepsilon} \left(c_n \left\|\hat{\alpha} - \alpha_0\right\| > d\right) = 0 \tag{8.4.69}$$

for every $\alpha \in I$ and $\varepsilon > 0$. This result gives an improved rate of convergence for the estimator $\hat{\alpha}$ in the sense of weak consistency. However, if $\alpha \in I(u)$ for some a and if the conditions (i)–(iii) of Theorem 8.22 hold, then the maximum likelihood estimator $\hat{\alpha} = \hat{\alpha}(n, a)$ is $n^{\frac{1}{2}}$-consistent; that is, for every $\varepsilon > 0$,

$$\lim_{d\to\infty} \limsup_{n\to\infty} \sup_{\|\alpha_0-\alpha\|_1<\varepsilon,\ \alpha_0\in I(a)} P_{\alpha_0}\left(n^{1/2}\|\hat{\alpha}-\alpha_0\| > d\right) = 0\,. \tag{8.4.70}$$

For details, see [28]. ∎

Let us now consider the log-likelihood function as a random function of t and study the limit behavior of the difference

$$\ell_n\left(\alpha + n^{-1/2}\alpha^*, t\right) - \ell_n(\alpha, t) \tag{8.4.71}$$

for fixed elements α and $\alpha^* \in I$ where

$$\ell_n(\alpha, t) = \int_0^t \lambda^{(n)}(s)(1-\alpha(s))ds + \int_0^t (\log\alpha(s))dN^{(n)}(s)\,. \tag{8.4.72}$$

THEOREM 8.23
Suppose that

$$\begin{aligned} &(i)\ \int_0^1 \frac{\alpha^*(s)^2}{\alpha(s)} m_s(\alpha)ds < \infty\,,\\ &\textit{and}\\ &(ii)\ \int_0^1 \frac{\alpha^*(s)^3}{\alpha(s)^2} m_s(\alpha)ds < \infty\,. \end{aligned} \tag{8.4.73}$$

Then, under P_α, the stochastic process

$$\left(\left\{\ell_n\left(\alpha + n^{-1/2}\alpha^*, t\right) - \ell_n(\alpha, t) + \frac{1}{2}\int_0^t \frac{\alpha^*(s)^2}{\alpha(s)} m_s(\alpha)ds\right\}, 0 \le t \le 1\right) \tag{8.4.74}$$

converges in distribution to a Gaussian martingale with the quadratic variation

$$V_t\left(\alpha,\alpha^*\right)=\int_0^t \frac{\alpha^*(s)^2}{\alpha(s)} m_s(\alpha)ds\ . \tag{8.4.75}$$

PROOF In view of the central limit theorem due to Rebolledo [53, 54], it follows that the process

$$\left\{n^{-1/2}\left[N^{(n)}(t)-\int_0^t \alpha(s)\lambda^{(n)}(s)\right],\ 0\le t\le 1\right\} \tag{8.4.76}$$

converges in distribution on $D[0, 1]$ to

$$\{M_t(\alpha),\ 0\le t\le 1\} \tag{8.4.77}$$

where $M_t(\alpha)$ is a Gaussian martingale with the quadratic variation

$$V_t(\alpha)=\lim \frac{1}{n}\langle M^{(n)}\rangle_t=\int_0^t m_s(\alpha)\,\alpha(s)ds\ . \tag{8.4.78}$$

Recall that

$$M^{(n)}(t)=N^{(n)}(t)-\int_0^t \alpha(s)\lambda^{(n)}(s)ds,\ 0\le t\le 1\ . \tag{8.4.79}$$

In order to prove the theorem, it is sufficient to prove that the family of measures generated by the processes $\{Z_n(t),\ 0\le t\le 1\}$, given by (8.4.74), form a tight family on $D[0, 1]$ and that there is convergence of the finite dimensional distributions (cf. [11, 49, 51]) to the appropriate distributions of the limiting process. Tightness of the sequence of processes follows from the facts stated in (8.4.76) and (8.4.77). This can be seen from the following discussion. In view of the independent increments, it is sufficient to prove the convergence of the one-dimensional distributions. For fixed n and t,

$$\ell_n\left(\alpha+n^{-1/2}\alpha^*,t\right)-\ell_n(\alpha,t)$$

$$= -n^{-1/2}\int_0^t \lambda^{(n)}(s)\alpha^*(s)ds + \int_0^t \log\left(1+n^{-1/2}\frac{\alpha^*}{\alpha}\right)dN^{(n)}(s)$$

$$= -n^{-1/2}\int_0^t \lambda^{(n)}(s)\alpha^*(s)ds$$

$$+\int_0^t \left[n^{-1/2}\left(\frac{\alpha^*}{\alpha}\right) - (2n)^{-1}\left(\frac{\alpha^*}{\alpha}\right)^2\right.$$

$$\left.+\,O\left(n^{-3/2}\left(\frac{\alpha^*}{\alpha}\right)^3\right)\right]dN^{(n)}(s)$$

$$= n^{-1/2}\int_0^t \left(\frac{\alpha^*(s)}{\alpha(s)}\right)\left(dN^{(n)}(s) - \alpha(s)\lambda^{(n)}(s)ds\right)$$

$$-\frac{1}{2n}\int_0^t \left[\frac{\alpha^*(s)}{\alpha(s)}\right]^2 dN^{(n)}(s) + O\left(n^{-1/2}\int_0^1 \left(\frac{\alpha^*}{\alpha}\right)^3\frac{dN^{(n)}}{n}\right)$$

(8.4.80)

where the $O(.)$ term is uniform in t. This term can be shown to converge to zero in probability by the law of large numbers. The fact that the error bound is uniform in t implies the tightness of the stochastic processes defined by (8.4.74). Furthermore, the second term converges in probability to

$$-\frac{1}{2}\int_0^t \left(\frac{\alpha^*(s)}{\alpha(s)}\right)^2 m_s(\alpha)\,\alpha(s)ds\,. \tag{8.4.81}$$

The first term converges in distribution to

$$\int_0^t \frac{\alpha^*(s)}{\alpha(s)}dM_s(\alpha) \tag{8.4.82}$$

by (8.4.76) and (8.4.77) and the continuous mapping theorem (cf. [11]). Let the limit be denoted by $M(\alpha^*, \alpha)$. Then $M(\alpha^*, \alpha)$ is the integral of a predictable process with respect to a Gaussian martingale and is, hence, a Gaussian martingale with the quadratic variation $V_t(\alpha, \alpha^*)$ defined by (8.4.75). ∎

The following result proves the asymptotic normality of an estimator for the integral of the $\alpha(\cdot)$.

THEOREM 8.24
Suppose that $\alpha \in I(a)$ for some $a > 0$ and that the conditions (i)–(iii) of Theorem 8.22 and the conditions (i), (ii) of Theorem 8.23 hold. Let $\hat{\alpha}(n, a)$ be a maximum likelihood estimator of α. Then the stochastic processes

$$W_n(t) = n^{1/2} \int_0^t \left(\hat{\alpha}(s) - \alpha(s)\right) ds \tag{8.4.83}$$

converge in distribution to a gaussian martingale with variance function

$$H_t(\alpha) = \int_0^t \left[\frac{\alpha(s)}{m_s(\alpha)}\right] ds \, . \tag{8.4.84}$$

PROOF For fixed n, define a mapping $T(\cdot)$ of $L^1[0, 1]$ into itself by the relation

$$T(\alpha) = \int_0^{\cdot} \lambda^{(n)}(s)(1 - \alpha(s))ds + \int_0^{\cdot} \log \alpha(s)\, dN^{(n)}(s) \tag{8.4.85}$$

Following Luenberger [42], the first, second, and third Frechet derivatives of T are given by

$$T^{(1)}(\alpha)[\beta] = -\int_0^{\cdot} \beta\lambda^{(n)} ds + \int_0^{\cdot} \frac{\beta}{\alpha} dN^{(n)}(s) \, , \tag{8.4.86}$$

$$T^{(2)}(\alpha)[\beta][\gamma] = -\int_0^{\cdot} \frac{\beta\gamma}{\alpha^2} dN^{(n)}(s) \, , \tag{8.4.87}$$

and

$$T^{(3)}[\beta][\gamma][\delta] = 2\int_0^{\cdot} \frac{\beta\gamma\delta}{\alpha^3} dN^{(n)}(s)\,, \tag{8.4.88}$$

respectively. Then, by Leunberger [42, Section 7.3], treating $n^{-1/2}$ as a constant function, it follows that

$$\begin{aligned}
&\ell_n\left(\alpha + n^{-1/2}\right) - \ell_n(\alpha) \\
&= T^{(1)}(\alpha)\left[n^{-1/2}\right] + \frac{1}{2}T^{(2)}(\alpha)\left[n^{-1/2}\right]\left[n^{-1/2}\right] \\
&\quad + O\left(T^{(3)}(\alpha)\left[n^{-1/2}\right]\left[n^{-1/2}\right]\left[n^{-1/2}\right]\right) \\
&= T^{(1)}(\alpha)\left[n^{-1/2}\right] + \frac{1}{2}T^{(2)}(\alpha)\left[n^{-1/2}\right]\left[n^{-1/2}\right] \\
&\quad + O\left(n^{-1/2}\int_0^1 \alpha^{-3}\left[\frac{dN^{(n)}}{n}\right]\right) \\
&= T^{(1)}(\alpha)\left[n^{-1/2}\right] + \frac{1}{2}T^{(2)}(\alpha)\left[n^{-1/2}\right]\left[n^{-1/2}\right] + O_p(1)
\end{aligned} \tag{8.4.89}$$

by the law of large numbers. Convergence here is with respect to the norm in $L^1[0,1]$. Furthermore,

$$\begin{aligned}
\frac{1}{2}T^{(2)}(\alpha)\left[n^{-1/2}\right]\left[n^{-1/2}\right] &= -\frac{1}{2}\int_0^{\cdot} \alpha^{-2}\frac{dN^{(n)}}{N} \\
&\to -\frac{1}{2}\int_0^{\cdot} \frac{m_s(\alpha)}{\alpha(s)} ds
\end{aligned} \tag{8.4.90}$$

as $n \to \infty$ in the sense of convergence in probability. Theorem 8.23 and the relations (8.4.89) and (8.4.90) imply that

$$T^{(1)}(\alpha)\left[n^{-1/2}\right] \xrightarrow{\mathcal{L}} M \text{ as } n \to \infty \tag{8.4.91}$$

where M is a Gaussian martingale with the quadratic variation

$$V_t(\alpha, 1) = \int_0^1 \frac{m_s(\alpha)}{\alpha(s)} ds\,. \tag{8.4.92}$$

By the mean value theorem for functionals (cf. [20]), for some $r \in (0, 1)$,

$$T^{(1)}(\alpha)\left[n^{-1/2}\right] = T^{(2)}\left(\hat{\alpha} + r\left(\alpha - \hat{\alpha}\right)\right)\left[\alpha - \hat{\alpha}\right]\left[n^{-1/2}\right]$$

$$= -\int_0^{\cdot} \left[\left(\alpha - \hat{\alpha}\right) n^{-1/2} / \left(\hat{\alpha} + r\left(\alpha - \hat{\alpha}\right)\right)^2\right] dN^{(n)}(s)$$

$$= n^{1/2} \int_0^{\cdot} \left[\left(\hat{\alpha} - \alpha\right)/\alpha^2\right]\left[\frac{dN^{(n)}}{n}\right]$$

$$+ O_p\left(n^{-1/2} \int_0^1 [\hat{\alpha} - \alpha)^2/\alpha^3] dN^{(n)}\right)$$

$$= n^{1/2} \int_0^{\cdot} \left[\left(\hat{\alpha} - \alpha\right)/\alpha^2\right]\left[\frac{dN^{(n)}}{N}\right] + O_p(1)$$

(by (8.4.70) and the law of large numbers)

$$= n^{1/2} \int_0^{\cdot} \left(\hat{\alpha}(s) - \alpha(s)\right)\left[\frac{m_s(\alpha)}{\alpha(s)}\right] ds\,. \tag{8.4.93}$$

This, together with the earlier remark on the convergence in distribution of $T^{(1)}(\alpha)[n^{-1/2}]$, it follows that

$$n^{1/2} \int_0^{\cdot} \left(\hat{\alpha} - \alpha\right) \left[\frac{m(\alpha)}{\alpha}\right] ds \xrightarrow{\mathcal{L}} M \text{ as } n \to \infty . \qquad (8.4.94)$$

Hence

$$n^{1/2} \int_0^{\cdot} \left(\hat{\alpha} - \alpha\right) ds \xrightarrow{\mathcal{L}} \int_0^{\cdot} \left[\frac{\alpha}{m(\alpha)}\right]^2 dM \text{ as } n \to \infty \qquad (8.4.95)$$

with the limit a Gaussian martingale with the quadratic variation $H_t(\alpha)$ given by (8.4.84). ∎

Estimation by the Method of Penalty Functions

As discussed earlier, let us consider a sequence of n-component multivariate counting processes

$$N^{(n)} = \left(N_1^{(n)}, \ldots, N_n^{(n)}\right) \qquad (8.4.96)$$

where each component of $\{N^{(n)}(t), t \in [0, T]\}$ is a counting process over $[0, T]$. The sample paths of $N^{(n)}$ are step functions, zero at time zero, with jumps of size unity only. We assume that no two components jumps simultaneously. Assume that $N^{(n)}$ is defined on a probability space $(\Omega^{(n)}, \mathcal{F}^{(n)}, P^{(n)})$ and $\{\mathcal{F}_t^{(n)}, t \in [0, T]\}$ be the filtration on nth probability space. We assume that

$$E_{P^{(n)}}\left[N_i^{(n)}(T)\right] < \infty, \; n \geq 1, 1 \leq i \leq n . \qquad (8.4.97)$$

Suppose $N^{(n)}$ has a stochastic intensity function

$$\Lambda^{(n)} = \left(\lambda_1^{(n)}, \ldots, \lambda_n^{(n)}\right) \qquad (8.4.98)$$

such that

$$\lambda_i^{(n)} = \alpha(t)\, Y_i^{(n)}(t) , \qquad (8.4.99)$$

where $\alpha(t)$ is a nonrandom nonnegative function and $Y_i^{(n)}$ are predictable observable nonnegative processes adapted to $\{\mathcal{F}_t^{(n)}\}$.

Suppose that $\alpha(\cdot)$ is a continuous function and we assume that the sample paths of $Y_i^{(n)}$ are left-continuous with right limits. Let

$$Y^{(n)}(t) = \sum_{i=1}^{n} Y_i^{(n)}(t) \tag{8.4.100}$$

and

$$N^{(n)}(t) = \sum_{i=1}^{n} N_i^{(n)}(t) . \tag{8.4.101}$$

For each n, let $P_0^{(n)}$ be a probability measure on $(\Omega^{(n)}, \mathcal{F}^{(n)})$ that makes the components $N_i^{(n)}$ of $N^{(n)}$ independent homogeneous Poisson processes each with parameter one. Then, the distribution of $N^{(n)}$ under $P^{(n)}$ is absolutely continuous with respect to the distribution of $N^{(n)}$ under $P_0^{(n)}$ and up to a multiplication random variable constant in $\alpha(\cdot)$, the likelihood function is given by

$$L_T^{(n)}(\alpha) = \exp\left(-\int_0^T \alpha(s)Y^{(n)}(s)ds + \int_0^T \log[\alpha(s)]dN^{(n)}(s)\right) \tag{8.4.102}$$

(cf. [1, 25, 53]).

Suppose $\theta = \log\alpha$ belongs to a separable Hilbert space Θ with norm $\|\cdot\|$. A *penalized estimator* of θ is obtained by minimizing a functional of the form

$$A_{n,\lambda}^*(\theta) = \ell_{n,T}(\theta) + \lambda\, J(\theta) \tag{8.4.103}$$

over θ where $\ell_{n,T}(\theta) = -\log L_T^{(n)}(\theta)$ and $J(\theta) : \Theta \to R_+$ is a penalty or a roughness functional. Note that

$$A_{n,\lambda}^*(\theta) = -\int_0^T \theta(s)dN^{(n)}(s) + \int_0^T e^{\theta(s)}Y^{(n)}(s)ds + \lambda\, J(\theta) . \tag{8.4.104}$$

Since the process $Y^{(n)}(s)$ may take zero value on $[0, T]$ for some s, consider the modified log-likelihood functional

$$\ell^*_{n,T}(\theta) = -\int_0^T \theta(s)\frac{J_n(s)}{\bar{Y}_n(s)}dN^{(n)}(s)$$

$$+\int_0^T e^{\theta(s)} J_n(s)ds \tag{8.4.105}$$

where $J_n(s) = I[Y^{(n)}(s) > 0]$ with $J_n(s)[Y^{(n)}(s)]^{-1} = 0$ whenever $Y^{(n)}(s) = 0$. The maximizers of $\ell_{n,T}(\theta)$ and $\ell^*_{n,T}(\theta)$ need not be the same.

We assume that the processes $\{[J_n(s)][Y^{(n)}(s)]^{-1},\ 0 \leq t \leq T\}$ are $P^{(n)}$ almost surely uniformly bounded, that is, there exists $c_n > 0$ for every $n \geq 1$ such that

$$\sup\left\{J_n(s)\left[Y^{(n)}(s)\right]^{-1},\ 0 \leq s \leq T\right\} \leq c_n \text{ a.s. } \left[P^{(n)}\right].$$

This assumption ensures that the stochastic integrals involved in $L^*_{n,T}(\theta)$ exist.

Let $\hat{\theta}_n$ be the *maximum penalized likelihood estimator* (MPLE); that is, an estimator that is a solution of the unconstrained optimization problem: for fixed $\lambda > 0$ and $n \geq 1$, minimize the functional

$$A_{n,\lambda}(\theta) = -\int_0^T \theta(s)\frac{J_n(s)}{Y^{(n)}(s)}dN^{(n)}(s)$$

$$+\int_0^T e^{\theta(s)} J_n(s)ds + \frac{\lambda}{2}J(\theta) \tag{8.4.106}$$

where $J(\theta)$ is a suitable penalty functional on Θ as defined below.

Suppose the following regularity conditions hold:

(A1) Θ is a real separable Hilbert space of real-valued functions $\theta : [0, T] \to R$ with the inner product $\langle\cdot,\cdot\rangle$ and the norm $\|\cdot\|$,

(A2)   for some real $m > 1$, $\Theta = H_2^m([0, T])$ and they have equivalent norms (here H_2^m is the Sobolev space of order m),

(A3)   the penalty functional $J(\cdot)$ is defined by $J(\theta) = \|\pi\theta\|^2$ where π is a projection operator on Θ with the finite dimensional null space Θ_0 containing the constant functions on $[0, T]$,

(A4)   there exist positive constants M_1 and M_2 such that

$$M_1\|\theta\|^2 \leq J(\theta) + \|\theta\|^2_{L^2([0,T])} \leq M_2\|\theta\|^2 ,$$

and

(A5)   the true function θ_0 lies in a Sobolev space on $[0, T]$ of order p where $p \geq m$.

Recall that if m is an integer, then $H_2^m([0, T])$ is the Hilbert space of functions on $[0, T]$ whose derivatives $f^{(i)}$ up to the order $(m-1)$ exist and are absolutely continuous and such that $f^{(m)}$ is in $L^2([0, T])$. Then norm of $H_2^m([0, T])$ is $\|f\|_m = \{\sum_{i=1}^m \|f^{(i)}\|^2_{L^2([0,T])}\}^{1/2}$. If m is a positive number not necessarily an integer, one can define $H_2^m([0, T])$ using interpolation theory (cf. [4]).

The set of assumptions (A) are about the space Θ and the true but unknown function $\theta_0 \in \Theta$. Suppose the penalty function $J(\theta)$ is chosen to be

$$J(\theta) = \int_0^T (D\theta)^2(s)ds \tag{8.4.107}$$

where $D : H_2^m \to L^2([0, T])$ is a differential operator of order m with $\Theta_0 = \ker(D)$ (cf. [14]).

We assume that the following conditions hold in addition to conditions (A1) to (A5).

(B1) $j_n(s) = E_{P(n)}(J_n(s))$ converges uniformly to 1 in $[0, T]$ at a rate $O(n^{-1})$,

(B2) $nE_{P(n)}(J_n(s)/Y_\theta^{(n)}(s))$ is uniformly bounded below and away from 0 and ∞,

(B3) $\{nJ_n(s)/Y^{(n)}(s); 0 \leq s \leq T\}$ converges uniformly in probability to a continuous function ζ on $[0, T]$.

Antoniadis [8] proved the following results regarding the existence and the asymptotic properties of a MPLE. We omit the proofs.

Existence of an MPLE

THEOREM 8.25

Under the conditions (A) and (B) stated above the functional $A_{n,\lambda}$ defined by (8.4.106) has a unique minimizer in Θ whenever there exists a minimizer θ_0 of $\ell^*_{n,T}$ defined by (8.4.105) in the space of infinitely smooth log intensity functions, which is the null family of the projection operator π defined by (A3).

Consistency of an MPLE

THEOREM 8.26

Suppose the conditions (A) and (B) hold and the smoothing parameter λ satisfies $\lambda \to 0$ and $n^{m-\delta}\lambda \to \infty$ as $n \to \infty$ for some $\delta > 0$. Then $\hat{\theta}$ is uniformly consistent as an estimator of θ_0 and in addition, for all $\varepsilon > 0$ sufficiently small, as $n \to \infty$,

$$\left\|\hat{\theta} - \theta_0\right\|_{\infty}^{2} = O_p\left[\lambda^{-\varepsilon/2m}\left(n^{-1}\lambda^{-1/m} + \lambda^{(2p-1)/2m}\right)\right].$$

Asymptotic Normality of MPLE

Suppose $J(\theta) = \|\pi\theta\|^2 = \int_0^T (\theta^{(m)}(s))^2 ds$.

THEOREM 8.27

Suppose the conditions (A) and (B) hold and, as $n \to \infty$, $\lambda \to 0$ and $n^{(\frac{2}{3})m-\delta}\lambda \to \infty$ for some $\delta > 0$. Then, for each t in $[0, T]$,

$$n^{1/2}\lambda^{1/4m}\left(\hat{\theta}(t) - \theta_0(t)\right) \xrightarrow{\mathcal{L}} N\left(0, \sigma^2(t)\right)$$

where $\sigma^2(t)$ depends on $\theta_0(t)$.

REMARK 8.11 Antoniadis [8] has obtained a closed form for the variance of the limiting normal distribution. Proof of this theorem relies on a central limit theorem for stochastic integrals with respect to point processes due to

Rebolledo [53] and an approximation $\tilde{\theta}(t)$ for the MPLE $\hat{\theta}(t)$ that behaves like a kernel smoothing estimator with a suitable kernel. He has also pointed out that one can prove that the finite dimensional distributions of the process $\{\hat{\theta}(s),\ s \in [0, T]\}$ can be shown to be asymptotically multivariate Gaussian by using a multivariate extension of the Rebolledo's theorem as given in Andersen and Gill [7] and the standard Cramer–Wold technique. For applications to the estimation of hazard rate for a random censored data and for estimation in the competing risk model, see [8]. ∎

Estimation by the Method of Martingale Estimators

Let $\boldsymbol{N}_t = (N_1(t), \ldots, N_K(t))$ be a K-variate counting process with stochastic intensity $\boldsymbol{\lambda}_t = (\lambda_1(t), \ldots, \lambda_K(t))$ defined on $(\Omega, \mathcal{F}, P)$ with filtration $\{\mathcal{F}_t\}$ and $E[N_j(t)] < \infty,\ 1 \le j \le K$. Further assume that λ is bounded and predictable with respect to $\{\mathcal{F}_t\}$ ($\{\mathcal{F}_t\}$ right-continuous and complete as usual). It is known that

(i) $M_j(t) = N_j(t) - \int_0^t \lambda_j(s)ds$ is a square integrable $\mathcal{F}_t$-martingale on $(\Omega, \mathcal{F}, P)$,

(ii) $\langle M_j \rangle_t = \int_0^t \lambda_j(s)ds$, and

(iii) for $\ell \ne j$, the martingales M_ℓ and M_j are orthogonal.

Suppose the intensity $\lambda_j(t)$ depends on a parameter α. Let P_α be the probability measure corresponding to α. Suppose that

$$\lambda_j(t, \alpha) = H_j(t, \alpha)\, \lambda_j^*(t),\ 1 \le j \le K,\ \alpha \in I$$

where $H_j(t, \alpha)$ and $\lambda_j^*(t)$ are nonnegative predictable processes such that

$$E_\alpha \left[\sum_{j=1}^{K} \int_0^1 H_j(t, \alpha)\, \lambda_j^*(t)dt \right] < \infty .$$

Further assume that

$$\mathcal{F}_t = \mathcal{F}_t^N \vee \sigma\left(\lambda_u^*,\ 0 \le u \le t\right)$$

where $\boldsymbol{\lambda}_u^* = (\lambda_1^*(u), \ldots, \lambda_k^*(u))$. The process $\boldsymbol{\lambda}^*$ is called the *baseline intensity*. Here $\mathcal{F}_t^N$ is the filter corresponding to the process N and $\sigma(\boldsymbol{\lambda}_u^*,\ 0 \le u \le t)$ is the σ-algebra generated by $\boldsymbol{\lambda}_u^*,\ 0 \le u \le t$. In practice $H_j(t, \alpha)$ is nonrandom. If $H_j(t, \alpha) = \alpha_j(t)$, for all $\alpha \in I$, then one has a multiplicative intensity model discussed later in this section.

The general principle of estimation via martingale methods is to choose an estimator for which the error is a martingale. If $\mathcal{F}_t = \mathcal{F}_t^N$, then it follows by the general representation of $\mathcal{F}_t^N$-martingales M_t^* over $(\Omega, \mathcal{F}, P_\alpha)$, that

$$M_t^* = M_0^* + \sum_{j=1}^{K} \int_0^t Y_j(s, \alpha) dM_j(s, \alpha) \tag{8.4.108}$$

where $Y_j(s, \alpha),\ 1 \le j \le K$ are $\mathcal{F}_t^N$-predictable. Hence, the relation (8.4.108) defines the class of allowable error processes. Note that

$$\begin{aligned}\int_0^t Y_j(s, \alpha) dM_j(s, \alpha) &= \int_0^t Y_j(s, \alpha) dN_j(s) \\ &\quad - \int_0^t Y_j(s, \alpha) \lambda_j(s, \alpha) ds \\ &= \int_0^t Y_j(s, \alpha) dN_j(s) \\ &\quad - \int_0^t Y_j(s, \alpha) H_j(s, \alpha) \lambda_j^*(s) ds\,, \end{aligned} \tag{8.4.109}$$

and the expectation of the expression on the left-hand side of (8.4.109) is zero.

DEFINITION 8.3 *Let Y be a predictable process with*

$$\sum_{j=1}^{K}\int_0^1 \left|Y_j(s)\right| H_j(s,\alpha)\lambda_j^*(s)ds < \infty \ a.s. \ [P_\alpha]\ .$$

The martingale estimator of the process

$$B_j(t,\alpha) = \int_0^t Y_j(s)H_j(s,\alpha)I\left(\lambda_j^*(s) > 0\right)ds \tag{8.4.110}$$

is defined to be the process

$$\hat{B}_j(t) = \int_0^t Y_j(s)I\left(\lambda_j^*(s) > 0\right)\lambda_j^*(s)^{-1}dN_j(s)\ . \tag{8.4.111}$$

We assume that
(C1) $\lambda_j^*(s) \geq \delta > 0$ for $1 \leq j \leq K$ and for $0 \leq s \leq 1$ and that $\{Y_j\}$ are bounded.

Note that

$$\begin{aligned}\hat{B}_j(t) - B_j(t,\alpha) &= \int_0^t Y_j(s)\lambda_j^*(s)^{-1}dN_j(s)\\ &\quad - \int_0^t Y_j(s)H_j(s,\alpha)ds\\ &= \int_0^t Y_j(s)\lambda_j^*(s)^{-1}dM_j(s,\alpha)\ .\end{aligned} \tag{8.4.112}$$

Since $Y_j(s)\lambda_j^*(s)^{-1}$ is bounded and predictable under (C1), it follows that the process $\{\hat{B}_j(t) - B_j(t,\alpha),\ 0 \leq t \leq 1\}$ is a mean zero square integrable martingale with

$$\left\langle \hat{B}_j - B_j(\alpha) \right\rangle_t = \int_0^t Y_j^2(s) \lambda_j^*(s)^{-2} d\left\langle M_j \right\rangle_s$$

$$= \int_0^t Y_j^2(s) \lambda_j^*(s)^{-2} H_j(s,\alpha) \lambda_j^*(s) ds$$

$$= \int_0^t Y_j^2(s) \lambda_j^*(s)^{-1} H_j(s,\alpha) ds \,. \tag{8.4.113}$$

Hence,

$$E_\alpha\left[\hat{B}_j(t)\right] = E_\alpha\left[B_j(t,\alpha)\right] \tag{8.4.114}$$

and

$$E_\alpha\left[\hat{B}_j(t) - B_j(t,\alpha)\right]^2$$

$$= E_\alpha\left[\left\langle \hat{B}_j - B_j(\alpha) \right\rangle_t\right]$$

$$= \int_0^t E_\alpha\left[Y_j^2(s) \lambda_j^*(s)^{-1} H_j(s,\alpha)\right] ds \,.$$

Note that the mean square error given above can be estimated in turn by the following approach. Let

$$C_j(t,\alpha) = \int_0^t Y_j^2(s) \lambda_j^*(s)^{-1} H_j(s,\alpha) ds \,. \tag{8.4.115}$$

A martingale estimator of $C_j(t,\alpha)$ is

$$\hat{C}_j(t) = \int_0^t (Y_j^2(s)/\lambda_j^{*2}(s))dN_j(s)\,. \tag{8.4.116}$$

Note that $E_\alpha[\hat{C}_j(t)] = E_\alpha[C_j(t,\alpha)]$.

As a special case of the model described above, we have the following model known as the multiplicative intensity model.

(Multiplicative intensity model) Suppose that $\lambda_j(t,\alpha) = \alpha_j(t)\,\lambda_j^*(t)$ where $\alpha_j(t)$ is nonrandom, right-continuous with left limits and nonnegative such that

$$\int_0^1 |\alpha_j(s)|\,ds < \infty, \quad 1 \le j \le K\,.$$

Assume that

$$\text{(C1)}' \qquad \lambda_j^*(t) \ge \delta > 0 \ \text{ for all } 1 \le j \le K \text{ and } 0 \le t \le 1\,.$$

The problem is to estimate $\alpha(\cdot) = (\alpha_1(\cdot), \cdots, \alpha_K(\cdot))$ or integrals of $\alpha_j(\cdot)$. We now discuss the estimation of the integrals of $\alpha_j(\cdot)$ using the approach of the martingale estimators. Note that the cumulative intensity defined by

$$B_j(t,\alpha) = \int_0^t \alpha_j(s)ds \tag{8.4.117}$$

has a martingale estimator

$$\hat{B}_j(t) = \int_0^t \lambda_j^*(s)^{-1}dN_j(s)\,. \tag{8.4.118}$$

Example 8.3

Let X_i, $1 \le i \le n$ be i.i.d. random variables on [0, 1] with a distribution function $F(\cdot)$ and the failure rate $r(\cdot) = f(\cdot)/1 - F(\cdot)$, f being the density of X_1. Consider the point process $\{N_t\}$ with $\mathcal{F}^N$-stochastic intensity $\lambda_t(r) = r(t-)(n - N_{t-})$ and the base line intensity $\lambda_t^* = (n - N_{t-})$. Note that $N_t = \#X_i \leq t$ in $1 \leq i \leq n$. An estimator of the integrated failure rate

$$B(r,t) = \int_0^t r(s) I\,(N_s < n)\, ds$$

is

$$\hat{B}(t) = \int_0^t \frac{1}{n - N_{s-}} dN_s = \sum_{j=1}^{N_t} \frac{1}{n - j + 1}\,.$$

Since $F(t) = 1 - \exp(-\int_0^t r(s)ds)$, an estimator of $F(\cdot)$, for large n, is

$$\hat{F}(t) = 1 - \exp(-\hat{B}(t))$$

$$= 1 - \exp\left(-\int_0^t \frac{1}{n - N_{s-}} dN_s\right)$$

and formally

$$d\hat{F}(t) = \left(1 - \hat{F}(t)\right) \lambda_t^{*-1} dN_t\,.$$

Now we have the product limit estimator

$$\hat{F}(t) = 1 - \Pi_{s \leq t}\left(1 - \frac{\Delta N_s}{n - N_{s-}}\right)$$

as the solution of the above stochastic differential equation which is the empirical distribution function. □

REMARK 8.12 In the censored data case, one can obtain Kaplan–Meier estimator as a special case of the above discussion (cf. [29, p. 176]). For related discussion see [51]. ∎

Asymptotic Properties of the Martingale Estimators

Consider the model discussed at the beginning in this section. For each $n \geq 1$, let $N^{(n)}$ be a K-variate point process as discussed above. Define

$$B_j^{(n)}(t,\alpha) = \int_0^t Y_j^{(n)}(s) H_j^{(n)}(s,\alpha) ds \tag{8.4.119}$$

and

$$\hat{B}_j(t) = \int_0^t Y_j^{(n)}(s) \lambda_j^{*(n)}(s)^{-1} dN_k^{(n)}(s) \,. \tag{8.4.120}$$

We assume that $\lambda_j^{*(n)}(\cdot) > 0$ for $1 \leq j \leq K$ and $n \geq 1$.

Uniform Consistency in Quadratic Mean

THEOREM 8.28
Suppose that

$$E_\alpha \left[\int_0^1 \left\{ \frac{Y_j^{(n)}(s)}{\lambda_j^{*(n)}(s)} \right\}^2 dN_j^{(n)}(s) \right] \to 0$$

$$\textit{as } n \to \infty \textit{ for each } \alpha \textit{ and } j \,. \tag{8.4.121}$$

Then $\hat{B}_j^{(n)}$ are uniformly consistent in quadratic mean, that is,

$$E_\alpha \left[\sup_{0 \leq t \leq 1} \left| \hat{B}_j^{(n)}(t) - B_j^{(n)}(t,\alpha) \right|^2 \right] \to 0 \textit{ as } n \to \infty \tag{8.4.122}$$

for each α and j.

PROOF Since $\{\hat{B}_j^{(n)}(t) - B_j^{(n)}(t,\alpha),\ 0 \leq t \leq 1\}$ is a zero mean square integrable martingale,

$$E_\alpha\left[\sup_{0\le t\le 1}\left|\hat{B}_j^{(n)}(t)-B_j^{(n)}(t,\alpha)\right|^2\right]$$

$$\le 4\;E_\alpha\left[\left|\hat{B}_j^{(n)}(1)-B_j^{(n)}(1,\alpha)\right|^2\right]$$

$$=4\;E_\alpha\left[\left\langle\hat{B}_j^{(n)}(1)-B_j^{(n)}(1,\alpha)\right\rangle_1\right]$$

$$=4\;E_\alpha\left[\int_0^1 Y_j^{(n)^2}(s)\lambda_j^{*(n)}(s)^{-1}H_j^{(n)}(s,\alpha)ds\right]$$

$$=4\;E_\alpha\left[\int_0^1\left\{\frac{Y_j^{(n)}(s)}{\lambda_j^{*(n)}(s)}\right\}^2 dN_j^{(n)}(s)\right] \qquad (8.4.123)$$

and the last term tends to zero by (8.4.121). ▮

THEOREM 8.29 **(Asymptotic normality)**
Suppose that, for each α,

$$\sum_{j=1}^{k}\int_0^1 Y_j^{(n)^2}(s)\lambda_j^{*(n)}(s)H_j^{(n)}(s,\alpha)ds<\infty \;\; a.s. \;\; [P_\alpha]$$

and there exists $0<b_n\uparrow\infty$ *such that, for each* α *and* j,

$$b_n^{1/2}\left[\hat{B}_j^{(n)}-B_j^{(n)}(\alpha)\right]$$

satisfies the condition

$$P\left\{\sup_{0\le t\le 1}\left|\Delta D_j^{(n)}(t)\right|\le c_n\right\}\to 1 \;\; as\; n\to\infty$$

for some $c_n \downarrow 0$ as $n \to \infty$ where

$$D_j^{(n)}(t) = b_n^{1/2}\left[\hat{B}_j^{(n)}(t) - B_j^{(n)}(t,\alpha)\right] .$$

Further suppose that there are functions $V_k(t,\alpha)$ such that

$$b_n \int_0^t \left[\frac{Y_j^{(n)}(s)}{\lambda_j^{*(n)}(s)}\right]^2 H_j^{(n)}(s,\alpha)\lambda_j^{*(n)}(s)ds \xrightarrow{p} V_j(t,\alpha) \ \ as\ n \to \infty \qquad (8.4.124)$$

for each t and j where $V_j(0,\alpha) = 0$, $V_j(t,\alpha)$ is continuous and nondecreasing in t for each j and α. Then there exists a continuous Gaussian martingale $M(\alpha)$ such that

$$b_n^{1/2}\left[\hat{B}^{(n)} - B^{(n)}(\alpha)\right] \xrightarrow{w} M(\alpha) \ \ on\ D[0,1]^K \qquad (8.4.125)$$

and

$$\langle M_\ell(\alpha), M_j(\alpha)\rangle = I((\ell = j)V_\ell(\alpha), \ 1 \le \ell \le K . \qquad (8.4.126)$$

This result follows from the central limit theorem for martingales stated in Chapter 1.

Results in this section are due to Karr [29].

Maximum Likelihood Estimation

Let $\{N_t\}$ be a counting process with a continuous compensator $\{A_t\}$. Suppose that $\alpha_0 = \frac{dA}{d\mu} \in \mathcal{A}$ where μ is a given dominating measure and $\mathcal{A}$ is a given set of intensities with respect to μ. The log-likelihood ratio for the process $\{N_t\}$ observed up to time T is equal to

$$\ell_T(\alpha,\alpha_0) = \int_0^T \log\left(\frac{\alpha}{\alpha_0}\right) dN - \int_0^T (\alpha - \alpha_0)\, d\mu, \ \alpha \in \mathcal{A} . \qquad (8.4.127)$$

A *maximum likelihood estimator* $\hat{\alpha}$ is defined by

$$\ell_T\left(\hat{\alpha},\alpha_0\right) = \sup_{\alpha\in\mathcal{A}} \ell_T(\alpha,\alpha_0) . \qquad (8.4.128)$$

We assume that such an estimator $\hat{\alpha}$ exists. It need not be unique. Let

$$h_T^2(\alpha, \tilde{\alpha}) = \frac{1}{2} \int_0^T \left(\sqrt{\alpha} - \sqrt{\tilde{\alpha}}\right)^2 d\mu \,. \tag{8.4.129}$$

Note that $h_T^2(\alpha, \tilde{\alpha})$ is the Hellinger process. Let

$$g_{\alpha_i} = \frac{1}{2} \log\left(\frac{\alpha_i}{\alpha_0}\right), \; i = 1, 2 \tag{8.4.130}$$

and

$$d_T^2\left(g_{\alpha_1}, g_{\alpha_2}\right) = \left(\frac{1}{2} \int_0^T \left(e^{g_{\alpha_1}} - e^{g_{\alpha_2}}\right)^2 dA\right)$$

$$= h_T^2\left(\alpha_1, \alpha_2\right) \,. \tag{8.4.131}$$

We now prove some results on the Hellinger process.

LEMMA 8.8
For $\bar{g}_{\hat{\alpha}} = \frac{1}{2} \log\left(\frac{1}{2}(\hat{\alpha} + \alpha_0)\right)$,

$$\int_0^T \bar{g}_{\hat{\alpha}} \, d(N - A) - h_T^2\left(\frac{1}{2}\left(\hat{\alpha} + \alpha_0\right), \alpha_0\right) \geq 0 \,. \tag{8.4.132}$$

PROOF Note that $\ell_T(\hat{\alpha}, \alpha_0) \geq 0$. Furthermore,

$$\ell_T\left(\frac{1}{2}\left(\hat{\alpha} + \alpha_0\right), \alpha_0\right) = \int_0^T \log\left(\frac{\frac{1}{2}\left(\hat{\alpha} + \alpha_0\right)}{\alpha_0}\right) dN$$

$$- \int_0^T \left(\frac{1}{2}\left(\hat{\alpha} + \alpha_0\right) - \alpha_0\right) d\mu$$

$$\geq \frac{1}{2}\int_0^T \log\left(\frac{\hat{\alpha}}{\alpha_0}\right) dN - \frac{1}{2}\int_0^T (\hat{\alpha} - \alpha_0)\, d\mu$$

$$= \frac{1}{2}\ell_T(\hat{\alpha}, \alpha_0) \geq 0 \tag{8.4.133}$$

by the concavity of the log function. For any $\alpha \geq 0$ and $g_\alpha = \frac{1}{2}\log\left(\frac{\alpha}{\alpha_0}\right)$,

$$\frac{1}{2}\ell_T(\alpha, \alpha_0) = \int_0^T g_\alpha\, d(N - A) + \int_0^T g_\alpha\, dA - \frac{1}{2}\int_0^T (\alpha - \alpha_0)\, d\mu$$

and

$$\int_0^T g_\alpha dA - \frac{1}{2}\int_0^T (\alpha - \alpha_0)\, d\mu$$

$$= \int_0^T \log\left(\sqrt{\frac{\alpha}{\alpha_0}}\right) dA - \frac{1}{2}\int_0^T (\alpha - \alpha_0)\, d\mu$$

$$\leq \int_0^T \left(\sqrt{\frac{\alpha}{\alpha_0}} - 1\right) dA - \frac{1}{2}\int_0^T (\alpha - \alpha_0)\, d\mu$$

$$= \int_0^T \sqrt{\alpha\alpha_0}\, d\mu - \frac{1}{2}\int_0^T \alpha d\mu - \frac{1}{2}\int_0^T \alpha_0 d\mu$$

$$= -h_T^2(\alpha, \alpha_0)\ .$$

Hence,

$$\frac{1}{2}\ell_T(\alpha,\alpha_0) \le \int_0^T g_\alpha\, d(N-A) - h_T^2(\alpha.\alpha_0)\ . \tag{8.4.134}$$

Choosing $\alpha = \frac{\hat{\alpha}+\alpha_0}{2}$ in (8.4.133) and combining with (8.4.118) proves the inequality stated in the lemma. ∎

LEMMA 8.9
Let α_0 and α be in $\mathcal{A}$ and define

$$\alpha_u = u\alpha + (1-u)\alpha_0 \tag{8.4.135}$$

for $0 < u < 1$. Then

$$\frac{1}{4(1-u)}\left(\sqrt{\alpha}-\sqrt{\alpha_u}\right)^2 \le \left(\sqrt{\alpha}-\sqrt{\alpha_0}\right)^2$$

$$\le \frac{4}{(1-u)^2}\left(\sqrt{\alpha}-\sqrt{\alpha_u}\right)^2\ . \tag{8.4.136}$$

PROOF Note that $\alpha_u(x) = 0$ if and only if $\alpha(x) = 0$ and $\alpha_0(x) = 0$. On the set $D = \{x : \alpha_u(x) = 0\}$, the result is true. On D^c,

$$\left(\sqrt{\alpha}-\sqrt{\alpha_u}\right)^2 = (1-u)^2\left(\sqrt{\alpha}-\sqrt{\alpha_0}\right)^2\left(\frac{\sqrt{\alpha}+\sqrt{\alpha_0}}{\sqrt{\alpha}+\sqrt{\alpha_u}}\right)^2$$

$$= (1-u)^2\left(\sqrt{\alpha}-\sqrt{\alpha_0}\right)^2\left\{\left(\frac{1+\sqrt{\alpha_0/\alpha}}{1+\sqrt{u+(1-u)\frac{\alpha_0}{\alpha}}}\right)^2\right.$$

$$I(\alpha_0 \le \alpha)$$

$$\left.+\left(\frac{\sqrt{\alpha/\alpha_0}+1}{\sqrt{\frac{u\alpha}{\alpha_0}+(1-u)}+\sqrt{\frac{\alpha}{\alpha_0}}}\right)^2 I(\alpha_0 > \alpha)\right\}$$

$$\leq (1-u)^2 \left(\sqrt{\alpha} - \sqrt{\alpha_0}\right)^2 \left(\frac{4}{1-u}\right)$$

and

$$\left(\sqrt{\alpha} - \sqrt{\alpha_0}\right)^2 = \frac{1}{(1-u)^2} \left(\sqrt{\alpha} - \sqrt{\alpha_u}\right)^2 \left(\frac{\sqrt{\alpha} + \sqrt{\alpha_u}}{\sqrt{\alpha} + \sqrt{\alpha_0}}\right)^2$$

$$= \frac{\left(\sqrt{\alpha} - \sqrt{\alpha_u}\right)^2}{(1-u)^2} \left\{ \left(\frac{1 + \sqrt{u + (1-u)\alpha_0/\alpha}}{1 + \sqrt{\alpha_0/\alpha}}\right)^2 I\left(\alpha_0 \leq \alpha\right) \right.$$

$$\left. + \left(\frac{\sqrt{\alpha/\alpha_0} + \sqrt{u\alpha/\alpha_0 + (1-u)}}{\sqrt{\alpha/\alpha_0} + 1}\right)^2 I\left(\alpha_0 > \alpha\right) \right\}$$

$$\leq \frac{4}{(1-u)^2} \left(\sqrt{\alpha} - \sqrt{\alpha_0}\right)^2.$$

∎

Choosing $u = \frac{1}{2}$, in Lemma 8.9, the following lemma follows, which proves that the Hellinger process evaluated at the convex combination $\frac{1}{2}(\hat{\alpha}+\alpha_0)$ instead of at $\hat{\alpha}$ behaves in an equivalent way as the Hellinger process evaluated at $\hat{\alpha}$.

LEMMA 8.10
For any nonnegative α,

$$2\, h_T^2 \left(\frac{1}{2}\left(\alpha + \alpha_0\right), \alpha_0\right) \leq h_T^2 \left(\alpha, \alpha_0\right)$$

$$\leq 16\, h_T^2 \left(\frac{1}{2}\left(\alpha + \alpha_0\right), \alpha_0\right). \tag{8.4.137}$$

Let ζ be the class of functions $\frac{1}{2}\log(\frac{1}{2}\frac{(\alpha+\alpha_0)}{\alpha_0})$ for $\alpha \in \mathcal{A}$ and $H(\delta, b, B)$ be its entropy as defined in [57, Section 1.11]. Note that $H(\delta, b, B)$ can be seen as the entropy for the Hellinger metric from (8.4.131). We now state the main theorem giving the rate of convergence of the MLE $\hat{\alpha}$.

THEOREM 8.30

Let $B \subset \{A_T \leq \sigma_T^2\}$. Then there exist universal constants C_i, $1 \leq i \leq 4$ such that, under the condition $\frac{\phi(b)}{b}$ is nonincreasing with

$$\phi(b) = \left\{ \int_{b^2/(c_2\sigma_T)\wedge b/8}^{b} \sqrt{H(x, b, T)}dx \ \vee b \right\}, \tag{8.4.138}$$

if

$$\frac{b_*^2}{C_1} \geq \phi(b_*) , \tag{8.4.139}$$

then

$$P\left(h_T\left(\hat{\alpha}, \alpha_0\right) > b_*\right) < C_4 \exp\left(\frac{-b_*^2}{C_3}\right) + P\left(B^c\right) . \tag{8.4.140}$$

PROOF Lemmas 8.8 and 8.10 show that it is sufficient to prove that

$$P\left(\int_0^T \bar{g}_{\hat{\alpha}}\, d(N - A) \geq h_T^2\left(\frac{1}{2}\left(\hat{\alpha} + \alpha_0\right), \alpha_0 v\right)\right.$$

$$\left.\text{and } h_T\left(\frac{1}{2}\left(\hat{\alpha} + \alpha_0\right), \alpha_0\right) > \frac{b_*}{4}\right)$$

$$\leq C_4 \exp\left(-\frac{b_*^2}{C_3}\right) + P\left(B^c\right) .$$

Let $\bar{\alpha} = \frac{1}{2}(\alpha + \alpha_0)$ and $\bar{g}_\alpha = \frac{1}{2}\log(\frac{\bar{\alpha}}{\alpha_0}) = g_{\bar{\alpha}}$, $\alpha \in \mathcal{A}$. It can be checked that

$$P\left(\int_0^T \bar{g}_\alpha\, d(N - A) \geq h_T^2\left(\bar{\alpha}, \alpha_0\right) \text{ and } h_T\left(\bar{\alpha}, \alpha_0\right) > \frac{b_*}{4} \text{ for some } \alpha \in \mathcal{A}\right)$$

$$\leq \sum_{j=1}^{\infty} P\left(\left\{\int_0^T \bar{g}_\alpha \, d(N-A) \geq \left(2^{j-1}\frac{b_*}{4}\right)^2 \text{ and}\right.\right.$$

$$\left.\left. h_T(\bar{\alpha}, \alpha_0) \leq 2^j \frac{b_*}{4} \text{ for some } \alpha \in \mathcal{A}\right\} \cap B\right) + P\left(B^c\right)$$

$$= \sum_{j=1}^{\infty} p_j + P\left(B^c\right) .$$

Since $\frac{\phi(b)}{b}$ is nonincreasing and the inequality (8.4.139) holds, it follows that

$$\frac{\left(2^j b_*\right)^2}{C_1} \geq \phi\left(2^j b_*\right) .$$

Hence, we can apply Theorem 3.1 from [57] (cf. Theorem 1.43, Chapter 1) to get a bound on each p_j. These bounds combined will prove the theorem for suitable constants C_4 and C_3. ∎

Example 8.4

Let X_i^*, $1 \leq i \leq n$ be i.i.d. failure times and U_i, $1 \leq i \leq n$ be the corresponding independent censoring times. It is possible to observe (X_i, Δ_i), $1 \leq i \leq n$ where $X_i = \min(X_i^*, U_i)$ and $\Delta_i = I(X_i \leq U_i)$ for $1 \leq i \leq n$. Consider the counting process

$$N_{it} = I\left(X_i \leq t, \Delta_i = 1\right), \; 1 \leq i \leq n . \tag{8.4.141}$$

The number of observed failures at the time t is

$$N_t = \sum_{i=1}^{n} N_{it} , \tag{8.4.142}$$

and the number of individuals at risk immediately before the time t is

$$R_n(t-) = \sum_{i=1}^{n} I(X_i \geq t) \ . \tag{8.4.143}$$

Let A_{it} be the compensator of N_{it} for $1 \leq i \leq n$. We assume a multiplicative intensity model with

$$\frac{dA_{it}}{d\mu} = \beta_{ot} I(X_i \geq t),\ t \geq 0,\ 1 \leq i \leq n\ , \tag{8.4.144}$$

where $\beta_0 \in \mathcal{B}$ and $\mathcal{B}$ is a given family of hazard rates. The log-likelihood ratio $\ell_\infty(\beta, \beta_0)$ is given by

$$\ell_\infty(\beta, \beta_0) = \int_0^\infty \log\left(\frac{\beta}{\beta_0}\right) dN - \int_0^\infty (\beta_t - \beta_{ot}) R_n(t-)dt\ . \tag{8.4.145}$$

(cf. [19, 26]).

Let $\hat{\beta}$ be a maximum likelihood estimator, that is, a maximizer of $\ell_\infty(\beta, \beta_0)$ over all $\beta \in \mathcal{B}$. We now obtain the rate of convergence of the Hellinger process $h_\infty(\hat{\beta}, \beta_0)$ following Theorem 8.30. Note that

$$h_\infty^2(\beta, \beta_0) = \frac{1}{2}\int_0^\infty \left(\sqrt{\beta_t} - \sqrt{\beta_{ot}}\right)^2 R_n(t-)dt\ .$$

Let

$$\bar{h}^2(\beta, \beta_0) = \frac{1}{n} h_\infty^2(\beta, \beta_0)\ .$$

We assume that β_0 is bounded (say) $\beta_0 \leq 1$ and that

$$\rho_{0n} = \frac{1}{n}\int_0^\infty R_n(t-)dt \xrightarrow{p} \rho_0,\ 0 < \rho_0 < \infty\ .$$

Since $A_t = \sum_{i=1}^{n} A_{it}$ is the compensator for $\{N_t\}$, it follows that

$$A_\infty = \int_0^\infty \beta_0 \, R_n(t-)dt \le n\rho_{0n}$$

and hence,

$$P\left(A_\infty > 2n\rho_0\right) \to 0 \text{ as } n \to \infty \,.$$

Let us first consider the case when $\mathcal{B} = \{\beta \equiv constant\}$. Let

$$B = \left\{\frac{1}{2}\rho_0 \le \rho_{0n} \le 2\rho_0\right\} .$$

Then, for $\sigma_\infty^2 = 2n\rho_0, \; B \subset [A_\infty \le 2n\rho_0]$. In this case the maximum likelihood estimator $\hat{\beta}$ can be explicitly computed. In fact $\hat{\beta} = N_\infty/\rho_{0n}$. On the set $\{h_\infty(\beta, \beta_0) \le b\} \cap B$,

$$\begin{aligned}\frac{1}{4}n\rho_0 \left(\sqrt{\beta_1} - \sqrt{\beta_2}\right)^2 &\le \frac{1}{2}n\rho_{0n} \left(\sqrt{\beta_1} - \sqrt{\beta_2}\right)^2 \\ &= h_\infty^2 \left(\beta_1, \beta_2\right) \\ &\le n\rho_0 \left(\sqrt{\beta_1} - \sqrt{\beta_2}\right)^2 .\end{aligned}$$

Using the fact that an interval of length b can be covered by constant (b/δ) intervals of length δ, we have

$$H(\delta, b, B) \le \text{constant } \log(b/\delta) \,.$$

Hence, $\phi(b) \le$ constant b so that one can take $b_* =$ constant which shows that $h_\infty(\hat{\beta}, \beta_0) = O_p(1)$. This proves that

$$\bar{h}\left(\hat{\beta}, \beta_0\right) = O_p\left(n^{-1/2}\right) .$$

□

REMARK 8.13 In the above example, if the class $\mathcal{B}$ is the class of increasing functions or $\mathcal{B}$ is the class of unimodal functions, then Van de Geer [57] proved that $\bar{h}(\hat{\beta}, \beta_0) = O_p(n^{-1/3})$. This can be compared with the results in [47, 48] on the estimation of a unimodal density and the distribution with the monotone failure rate (cf. [50]) where it was proved that the appropriate norming factor for the existence of a limiting distribution for the maximum likelihood estimator is $n^{1/3}$. ∎

8.5 Inference for Additive–Multiplicative Hazard Models

Semiparametric regression models based on the hazard function or intensity process are generally used for studying the influence of the covariate history on the failure time or the counting process. Proportional hazard model [15] or the multiplicative hazard model assumes that the hazard function associated with a multidimensional covariate process $\boldsymbol{X}(\cdot)$ is of the form

$$\lambda(t|\boldsymbol{X}) = \lambda_0(t)\exp\left[\gamma_0^T \boldsymbol{X}(t)\right] \tag{8.5.1}$$

where $\lambda_0(\cdot)$ is an unspecified baseline hazard function and γ_0 is an unknown parameter (here γ_0^T denotes the transpose of the vector γ_0). Another alternate model is the additive hazard model

$$\lambda(t|\boldsymbol{X}) = \lambda_0(t) + \beta_0^T \boldsymbol{X}(t) \tag{8.5.2}$$

where β_0 is an unknown parameter vector (cf. [17, 38]). A general class of additive-multiplicative hazard models is

$$\lambda(t|\boldsymbol{Z}) = g(\beta_0^T \boldsymbol{W}(t)) + \lambda_0(t)h(\gamma_0^T \boldsymbol{X}(t)) \tag{8.5.3}$$

where $\boldsymbol{Z} = (\boldsymbol{W}^T, \boldsymbol{X}^T)^T$ is a p-vector of covariates, $\theta_0 = (\beta_0^T, \gamma_0^T)^T$ is a p-vector of unknown regression parameters, g and h are known link functions and λ_0 is an unspecified "base line hazard function." Under $g = 0$ and $h = 1$, this model includes the models given by (8.5.1) and (8.5.2) as special cases. Some examples of link functions h are $h(x) = e^x$ or $h(x) = 1 + x$ and those of g are $g(x) = x$ and $g(x) = e^x$. For $g(x) = e^x$, one can choose the first component of $\boldsymbol{W}$ to be unity.

Let $N_1(t), N_2(t), \ldots, N_k(t)$ be n independent counting processes adapted to the filtration $\{\mathcal{F}_t, t \in [0, \tau]\}$ satisfying the usual conditions (cf. [6, II.4]). However, we allow here that each N_i takes multiple jumps. Associated with each $N_i(t)$, there is a p-dimensional covariate process $\boldsymbol{Z}_i(t) = (\boldsymbol{W}_i^T(t), \boldsymbol{X}_i^T(t))$ that is $\mathcal{F}_t$-predictable. Suppose the compensator for $N_i(t)$ is of the form

$$\int_0^t Y_i(s)\lambda(s|\boldsymbol{Z}_i)\, ds$$

where $\lambda(t|\boldsymbol{Z}_i)$ is given by (8.5.3) and $Y_i(t)$ is a $\{0, 1\}$-valued left continuous $\mathcal{F}_t$-adapted process. Note that

$$M_i(t) = N_i(t) - \int_0^t Y_i(s)\left[g\left(\beta_0^T \boldsymbol{W}_i(s)\right) + h\left(\gamma_0^T \boldsymbol{X}_i(s)\right)\right]\lambda_0(s)ds \quad (8.5.4)$$

is a zero-mean $\mathcal{F}_t$-martingale. We assume that there exists $\delta_\tau > 0$ such that

$$\left[n^{-1} \inf_{t\le\tau:\|\gamma-\gamma_0\|\le\delta_\tau} \sum_{i=1}^n Y_i(t)h\left(\gamma^T \boldsymbol{X}_i(t)\right)\right]^{-1} = O_p(1)\,.$$

Estimation of the Parameter θ_0 $\mathcal{D}$ $\triangleright\beta_0^T, G_0^T\triangleleft$

If the baseline hazard function λ_0 is known, then the likelihood for $\theta_0 = (\beta_0^T, \gamma_0^T)$ is proportional to

$$\Pi_{i=1}^n \left[\left\{\Pi_{t\le\tau}\lambda(t|\boldsymbol{Z}_i)^{dN_i(t)}\right\}\exp\left\{-\int_0^\tau Y_i(t)\lambda(t|\boldsymbol{Z}_i)\, dt\right\}\right] \quad (8.5.5)$$

(cf. [6, pp. 58–59]). The corresponding p-dimensional score function is

$$\left(\sum_{i=1}^{n}\int_{0}^{\tau}\frac{g'\left(t^T W_i(t)\right) W_i(t)}{g\left(t^T W_i(t)\right)+h\left(\gamma^T X_i(t)\right)\lambda_0(t)}dM_i(\theta,t)\,,\right.$$

$$\left.\sum_{i=1}^{n}\int_{0}^{\tau}\frac{h'\left(\gamma^T X_i(t)\right) X_i(t)\lambda_0(t)}{g\left(t^T W_i(t)\right)+h\left(\gamma^T X_i(t)\right)\lambda_0(t)}dM_i(\theta,t)\right) \tag{8.5.6}$$

where

$$M_i(\theta,t)=N_i(t)-\int_0^t Y_i(s)\left[g\left(\beta^T W_i(s)\right)+h\left(\gamma^T X_i(s)\right)\lambda_0(s)\right]ds\,. \tag{8.5.7}$$

For $\theta=\theta_0$, the score function (8.5.6) is a zero-mean martingale.

If λ_0 is unknown, then we modify the score function. We replace the integrands in the components of (8.5.6) by a p-dimensional predictable process $D_i(\theta,t)$ which is a smooth function of Z_i and θ not involving λ_0.

Following the martingale equation

$$\sum dM_i(t)=\sum dN_i(t) - \sum Y_i(t)\left[g\left(\beta_0^T W_i(t)\right)dt+h\left(\gamma_0^T X_i(t)\right)d\Lambda_0(t)\right] \tag{8.5.8}$$

with $\Lambda_0(t)=\int_0^t \lambda_0(s)ds$, we estimate $\Lambda_0(t)$, if θ_0 is known, by

$$\hat{\Lambda}_0(\theta_0,t)=\int_0^t\frac{\sum_{i=1}^n\left[dN_i(s)-Y_i(s)g\left(\beta_0^T W_i(s)\right)ds\right]}{\sum_{i=1}^n Y_i(s)h\left(\gamma_0^T X_i(s)\right)}\,. \tag{8.5.9}$$

We first obtain an ad hoc (p-dimensional) estimating function $S(\theta,t)$ free of

λ_0 where

$$S(\theta, t) = \sum_{i=1}^{n} \int_0^t D_i(\theta, s) \Big[dN_i(s) - Y_i(s) g\left(\beta^T \boldsymbol{W}_i(s)\right) ds$$

$$- Y_i(s) h\left(\gamma^T \boldsymbol{X}_i(s)\right) d\hat{\Lambda}_0(\theta, s) \Big]$$

$$= \sum_{i=1}^{n} \int_0^t \left[D_i(\theta, s) - \bar{D}(\theta, s) \right] \left[dN_i(s) - Y_i(s) g\left(\beta^T \boldsymbol{W}_i(s)\right) ds \right] \tag{8.5.10}$$

where

$$\bar{D}(\theta, t) = \left(\sum_{i=1}^{n} Y_i(t) h\left(\gamma^T \boldsymbol{X}_i(t)\right) D_i(\theta, t) \right) \Big/ \left(\sum_{i=1}^{n} Y_i(t) h\left(\gamma^T \boldsymbol{X}_i(t)\right) \right).$$

Since

$$\sum_{i=1}^{n} (D_i(\theta, t) - \bar{D}(\theta, s) Y_i(s) h\left(\gamma^T \boldsymbol{X}_i(s)\right) = 0$$

by the definition of $\bar{D}(\theta, s)$, it follows that

$$S(\theta, t) = \sum_{i=1}^{n} \int_0^t \left(D_i(\theta, s) - \bar{D}(\theta, s) \right) dM_i(\theta, s),$$

which shows that $S(\theta_0, t)$ is a martingale.

Let $\hat{\theta}$ be a solution of the equation $S(\theta, \tau) = 0$. Under some technical conditions, Lin and Ying [39] proved that there exists a unique consistent and asymptotically normal estimator $\hat{\theta}$ for θ as $n \to \infty$. One choice for $D_i(\theta, t)$ is

$$\begin{pmatrix} \tilde{W}_i(\theta, t) \\ \tilde{X}_i(\theta, t) \end{pmatrix} = \begin{pmatrix} g'\left(\beta^T \boldsymbol{W}_i(t)\right) \boldsymbol{W}_i(t) / h\left(\gamma^T \boldsymbol{X}_i(t)\right) \\ h'\left(\gamma^T \boldsymbol{X}_i(t)\right) \boldsymbol{X}_i(t) / h\left(\gamma^T \boldsymbol{X}_i(t)\right) \end{pmatrix}.$$

In this case

$$S(\theta, \tau) = \begin{bmatrix} U_\beta(\theta) \\ U_\gamma(\theta) \end{bmatrix}$$

where

$$U_{\beta}(\theta) = \sum_{i=1}^{n} \int_{0}^{\tau} \left\{ \tilde{W}_i(\theta, t) - \bar{W}(\theta, t) \right\} \left[dN_i(t) - Y_i(t) g\left(\beta^T \boldsymbol{W}_i(t) \right) dt \right]$$

and

$$U_{\gamma}(\theta) = \sum_{i=1}^{n} \int_{0}^{\tau} \left\{ \tilde{X}_i(\theta, t) - \bar{X}(\theta, t) \right\} \left[dN_i(t) - Y_i(t) g\left(\beta^T \boldsymbol{W}_i(t) \right) dt \right]$$

with

$$\bar{W}(\theta, t) = \frac{\sum Y_i(t) h\left(\gamma^T \boldsymbol{X}_i(t)\right) \tilde{W}_i(\theta, t)}{\sum Y_i(t) h\left(\gamma^T \boldsymbol{X}_i(t)\right)}$$

and

$$\bar{X}(\theta, t) = \frac{\sum Y_i(t) h\left(\gamma^T \boldsymbol{X}_i(t)\right) \tilde{X}_i(\theta, t)}{\sum Y_i(t) h\left(\gamma^T \boldsymbol{X}_i(t)\right)} .$$

It $\boldsymbol{Z} = \boldsymbol{X}$, then U_γ is the partial likelihood score function under the multiplicative hazard model. If $\boldsymbol{Z} = \boldsymbol{W}$ and $g(x) = x$, then U_β reduces to the estimating function of Lin and Ying [38] for the additive hazard model (8.5.2).

Estimation of $\Lambda_0(\cdot)$

Given $\hat{\theta}$, we may estimate $\Lambda_0(t)$ by

$$\hat{\Lambda}_0\left(\hat{\theta}, t\right) = \int_{0}^{t} \frac{\sum_{i=1}^{n} \left[dN_i(s) - Y_i(s) g\left(\hat{\beta}^T \boldsymbol{W}_i(s) \right) ds \right]}{\sum_{i=1}^{n} Y_i(s) h\left(\hat{\gamma}^T \boldsymbol{X}_i(s) \right)} .$$

Ling and Ying [39] investigated the asymptotic properties of the estimator $\hat{\Lambda}_0(\hat{\theta}, t)$. The estimator $\hat{\Lambda}_0(\hat{\theta}, t)$ may not be monotonic increasing in t. One can choose

$$\hat{\Lambda}_0^*(\theta, t) = \sup_{0 \le s \le t} \hat{\Lambda}_0(\theta, s)$$

and the asymptotic properties of $\hat{\Lambda}_0^*(\hat{\theta}, t)$ and $\hat{\Lambda}_0(\hat{\theta}, t)$ are the same. Lin and Ying [39] also construct estimators that are asymptotically efficient. We omit the details.

References

[1] Aalen, O.O. (1975) *Statistical Inference for a Family of Counting Processes,* Ph.D. Thesis, Univ. of California, Berkeley.

[2] Aalen, O.O. (1978) Nonparametric inference for a family of counting processes, *Ann. Statist.,* **6**, 701–726.

[3] Aalen, O.O. (1980) A model for nonparametric regression analysis of counting processes, *Lecture Notes in Statistics,* **2**, Springer, New York, 1–25.

[4] Adams, R. (1975) *Sobolev Spaces,* Academic Press, New York.

[5] Aitchinson, J. and Silvey, S.D. (1958) Maximum likelihood estimation of parameters subject to restraints, *Ann. Math. Statist.,* **29**, 813–828.

[6] Andersen, P.K., Borgan, Φ., Gill, R.D., and Keiding, N. (1993) *Statistical Methods for Counting Processes,* Springer, New York.

[7] Andersen, P.K. and Gill, R.D. (1982) Cox's regression model for counting processes: a large sample study, *Ann. Statist.,* **10**, 1100–1125.

[8] Antoniadis, A. (1989) A penalty method for nonparametric estimation of the intensity function of a counting process, *Ann. Inst. Statist. Math.,* **41**, 781–807.

[9] Aven, T. (1986) Bayesian inference in a parametric counting process model, *Scand. J. Statist.,* **13**, 87–97.

[10] Basawa, I.V. and Prakasa Rao, B.L.S. (1980) *Statistical Inference for Stochastic Processes,* Academic Press, London.

[11] Billingsley, P. (1968) *Convergence of Probability Measures,* Wiley, New York.

[12] Borgan, Φ. (1984) Maximum likelihood estimation in parametric counting process models, with applications to censored failure time data, *Scand. J. Statist.*, **11**, 1–16, (Corrigendum, *ibid*, p. 275).

[13] Borgan, Φ., Goldstein, L., and Langholz, B. (1995) Methods for analysis of sample cohort data in the Cox proportional hazard model, *Ann. Statist.*, **23**, 1749–1778.

[14] Cox, D.D. (1984) Multivariate smoothing spline functions, *SIAM J. Numer. Anal.*, **21**, 789–813.

[15] Cox, D.R. (1972) Regression models and life tables, *J. Roy. Statist. Soc.*, **34**, 187–220.

[16] Cox, D.R. and Lewis, P.A.W. (1966) *The Statistical Analysis of Series of Events*, Chapman and Hall, London.

[17] Cox, D.R. and Oakes, D. (1984) *Analysis of Survival Data*, Chapman and Hall, London.

[18] Fleming, T.R. and Harrington, D.P. (1991) *Counting Processes and Survival Analysis*, Wiley, New York.

[19] Gill, R.D. (1980) *Censoring and Stochastic Integrals*, Mathematical Centre Tracts, *Math. Centrum*, Amsterdam.

[20] Graves, L.M. (1929) Riemann integration and Taylor's theorem in general analysis, *Trans. Amer. Math. Soc.*, **27**, 163–177.

[21] Grenander, U. (1981) *Abstract Inference*, Wiley, New York.

[22] Hjort, N.L. (1986) Bayes estimators and asymptotic efficiency in parametric counting process models, *Scand. J. Statist.*, **13**, 63–85.

[23] Hutton, J.E. and Nelson, P.I. (1986) Quasi-likelihood estimation for semimartingales, *Stoch. Proc. Appl.*, **22**, 245–257.

[24] Ibragimov, I.I. and Khasminskii, R.Z. (1981) *Statistical Estimation: Asymptotic Theory*, Springer, Berlin.

[25] Jacobsen, M. (1982) *Statistical Analysis of Counting Processes*, Lecture Notes in Statistics, **12**, Springer, New York.

[26] Jacobsen, M. (1989) Right censoring and martingale methods for failure time data, *Ann. Statist.*, **17**, 1133–1156.

[27] Kabanov, Yu.M., Liptser. R.S., and Shiryayev, A.N. (1975) Martingale methods in the theory of point processes, *Workshop Seminar on Theory of Random Processes*, **2**, 269–354 (in Russian).

[28] Karr, A.F. (1987) Maximum likelihood estimation in the multiplicative intensity model via sieves, *Ann. Statist.*, **15**, 473–490.

[29] Karr, A.F. (1991) *Point Processes and Their Statistical Inference*, Marcel Dekker, New York.

[30] Kutoyants, Y.A. (1982) Multidimensional parameter estimation of intensity function of inhomogeneous Poisson process, *Problems of Control and Information Theory*, **11**, 325–334.

[31] Kutoyants, Y.A. (1983) Asymptotic expansion of the maximum likelihood estimate of the intensity parameter for inhomogeneous Poisson observations, *Trans. Ninth Prague Conf.*, Vol. B, D. Reidel, Holland, 35–41.

[32] Kutoyants, Y.A. (1984) *Parameter Estimation for Stochastic Processes*, (Trans. and Ed. B.L.S. Prakasa Rao), Heldermann, Berlin.

[33] Kutoyants, Y.A. (1985) On nonparametric estimation of trend coefficients in a diffusion process, *Statistics and Control of Stochastic Processes*, (Ed. N.V. Krylov et al.), Optimization Software, New York, 230–250.

[34] Kutoyants, Y.A. (1985) Efficient nonparametric estimation of trend coefficients (preprint).

[35] Lenglart, E. (1977) Relation de domination entre deux processus, *Ann. Inst. Henri Poincaré*, **13**, 171–179.

[36] Lepingle, D. (1978) Sur les comportement asymptotique des martingales locales, *Seminaire de Probabilites XII*, *Lecture Notes in Math.*, **649**, Springer, Berlin, 148–161.

[37] Lewis, P.A.W. (1972) *Stochastic Point Processes: Statistical Analysis, Theory and Applications*, Wiley, New York.

[38] Lin, D.Y. and Ying, Z. (1994) Semiparametric analysis of the additive risk model, *Biometrika*, **81**, 61–71.

[39] Lin, D.Y. and Ying, Z. (1995) Semiparametric analysis of general additive-multiplicative hazard models for counting processes, *Ann. Statist.*, **23**, 1712–1734.

[40] Liptser, R.S. and Shiryayev, A.N. (1978) *Statistics of Random Processes: Applications*, Springer, New York.

[41] Liptser, R.S. and Shiryayev, A.N. (1980) A functional central limit theorem for semimartingales, *Theor. Probab. Appl.*, **25**, 667–688.

[42] Luenberger, D.G. (1969) *Optimization by Vector Space Methods*, Wiley, New York.

[43] McKeague, I.W. (1988) A counting process approach to the regression analysis of grouped survival data, *Stoch. Proc. Appl.*, **28**, 221–239.

[44] McKeague, I.W. (1988) Asymptotic theory for weighted least squares estimators in Aalen's additive risk model, *Statistical Inference from Stochastic Processes*, (Ed. N.U. Prabhu), *Contemporary Mathematics*, **80**, 139–152.

[45] McKeague, I.W. and Utikal, K. (1990) Identifying nonlinear covariate effects in semimartingale regression models, *Probab. Th. Rel. Fields*, **87**, 1–25.

[46] Nielsen, J.P. and Linton, O.B. (1995) Kernel estimation in a nonparametric marker dependent hazard model, *Ann. Statist.*, **23**, 1735–1748.

[47] Prakasa Rao, B.L.S. (1969) Estimation of a unimodal density, *Ann. Math. Statist.*, **31**, 23–36.

[48] Prakasa Rao, B.L.S. (1970) Estimation for distributions with monotone failure rate, *Ann. Math, Statist.*, **41**, 507–519.

[49] Prakasa Rao, B.L.S. (1975) Tightness of probability measures generated by stochastic processes on metric spaces, *Bull. Inst. Math. Acad. Sinica*, **3**, 353–367.

[50] Prakasa Rao, B.L.S. (1983) *Nonparametric Functional Estimation*, Academic Press, Orlando, Fl.

[51] Prakasa Rao, B.L.S. (1987) *Asymptotic Theory of Statistical Inference*, Wiley, New York.

[52] Ramlau–Hansen, H. (1983) Smoothing counting process intensities by means of kernel functions, *Ann. Statist.*, **11**, 453–466.

[53] Rebolledo, R. (1978) Sur les applications de la theorie des martingales a l'etude statistique d'une famille de processus ponctuels, *Lecture Notes in Math.*, **636**, Springer, New York, 27–70.

[54] Rebolledo, R. (1980) Central limit theorem for local martingales, *Z. Warsch. verw Gebiete*, **51**, 269–286.

[55] Svensson, A. (1989) Estimation in some counting process models with multiplicative structure, *Ann. Statist.*, **17**, 1501–1509.

[56] Thavaneswaran, A. and Samanta, M. (1992) A note on smoothed estimate of counting process intensities, *Sankhya Ser. A*, **54**, 449–454.

[57] Van de Geer, S. (1995) Exponential inequalities for martingales with application to maximum likelihood estimation for counting processes, *Ann. Statist.*, **23**, 1779–1801.

[58] Yoshida, N. and Hayashi, T. (1990) On the robust estimation in Poisson processes with periodic intensities, *Ann. Inst. Statist. Math.*, **42**, 489–507.

Chapter 9

Inference for Semimartingale Regression Models

9.1 Estimation by the Quasi-Least-Squares Method

Consider the following class of linear regression models

$$\boldsymbol{Y}_t = \boldsymbol{Y}_0 + \boldsymbol{B} \int_0^t \boldsymbol{X}_s d\theta_s + \boldsymbol{M}_t, t \geq 0, \boldsymbol{Y}_0 = 0 = \boldsymbol{X}_0 . \tag{9.1.1}$$

The m-dimensional process $\{\boldsymbol{Y}_t\}$ is a semimartingale generated as the integral of a k-dimensional input process $\{\boldsymbol{X}_t\}$ with respect to an increasing process $\{\theta_t\}$ and weighted with a $(m \times k)$-parameter matrix $\boldsymbol{B}$ plus an unobservable disturbance (noise) $\{\boldsymbol{M}_t\}$ which is an m-dimensional martingale. The model (9.1.1) includes the discrete time model with stochastic regressor

$$Y_t = BX_t + \varepsilon_t,\ t = 1, 2, \ldots \tag{9.1.2}$$

and the continuous time diffusion model

$$dY_t = BX_t dt + CdW_t,\ t \geq 0 \tag{9.1.3}$$

where $\{W_t, t \geq 0\}$ is the standard Wiener process.

Let $0 = t_0 < t_1 < \cdots < t_n = T$ be a partition of $[0, T]$ and consider the weighted sum of squares

$$\sum_{i=1}^{n} \left\| Y_{t_i} - Y_{t_{i-1}} - B \int_{t_{i-1}}^{t_i} X_s d\theta_s \right\|^2 \delta_i \tag{9.1.4}$$

where $\delta_i = (\theta_{t_i} - \theta_{t_{i-1}})^{-1}$ if $\theta_{t_i} - \theta_{t_{i-1}} > 0$ and $\delta_i = 0$ otherwise.

An approximation for (9.1.4) is

$$\sum_{i=1}^{n} \|Y_{t_i} - Y_{t_{i-1}}\|^2 \delta_i - 2\sum_{i=1}^{n} X'_{t_{i-1}} B' (Y_{t_i} - Y_{t_{i-1}})$$

$$+ \left\{ \sum_{i=1}^{n} X'_{t_{i-1}} B' B X_{t_{i-1}} (\theta_{t_i} - \theta_{t_{i-1}}) \right\} . \tag{9.1.5}$$

Suppose the process (X_t, Y_t) is observed over $[0, T]$. Then the expression obtained by minimizing (9.1.5) with respect to B converges to

$$\hat{B}'_T = \left(\int_0^T X_s X'_s d\theta_s \right)^{-1} \int_0^T X_s dY'_s \tag{9.1.6}$$

under some conditions as the partition becomes finer and finer provided the matrix

$$\int_0^t X_s X'_s d\theta_s$$

is nonsingular. We call the estimator (9.1.6) as the *quasi-least-squares* estimator of B.

We will now study the asymptotic properties of the estimation $\hat{B}_T$ as an estimator of B. Let $(\Omega, \mathcal{F}, P)$ be a probability space with a filtration $\{\mathcal{F}_t\}$ satisfying the usual conditions. Recall that a process $\{Z_t, t \geq 0\}$ is said to be regular if it is adapted to $\{\mathcal{F}_t\}$ with right and left limits and a martingale $\{Z_t, t \geq 0\}$ is an L^2-martingale if $E|Z_t|^2 < \infty$ for every $t \geq 0$. We assume that the following regularity conditions hold in the sequel.

(A1) The components $\{M_t^{(i)}\}$, $1 \leq i \leq m$ of M_t belong to the class M^2, the space of regular right continuous L^2-martingales with $M_0^{(i)} = 0, 1 \leq i \leq m$.

(A2) The process $\{X_t\}$ is regular left-continuous process.

(A3) The increasing process $\{\theta_t\}$ is predictable and belongs to $\mathcal{A}$, the set of regular right-continuous increasing processes.
Let

$$M_t^{(i)} = M_t^{ic} + M_t^{id} \tag{9.1.7}$$

where M_t^{ic} is the continuous component and $M_t^{id} = M_t^{(i)} - M_t^{ic}$. Let $\langle M^i \rangle_t$ and $[M^i]_t$ be defined as before. In addition to (A1)-(A3), suppose the following conditions hold.

(A4) For all $\delta > 0$
(i) $\int_0^t \phi_s d\langle M^{ic}\rangle_s = O[(\int_0^t \phi_s d\theta_s^c)^{1+\delta}]$
and
(ii) $\int_0^t \phi_s d\langle M^{id}\rangle_s = O[(\int_0^t \phi_s d\theta_s)^{1+\delta}]$
holds with probability one for $1 \leq i \leq m$ and for every nonnegative predictable process $\{\phi_t\}$ where θ_s^c denotes the continuous component of θ_s.
Let

$$Z_t = \int_0^t X_s X_s' d\theta_s \tag{9.1.8}$$

and $\lambda_{\max}(t)(\lambda_{\min}(t))$ be the maximal (minimal) eigenvalue of the matrix Z_t.
In addition to the conditions (A1)-(A4), suppose the following condition holds.

(A5) $\lambda_{\min}(T) \to \infty$ a.s. and

$$(\log \lambda_{\max}(T))^{1+\delta} = O\left(\lambda_{\min}(T)\right) \text{ a.s.}$$

for some $\delta > 0$.

Consistency

The following result is due to Christopeit [5] on the strong consistency of the quasi-least-squares estimator.

THEOREM 9.1
Suppose that conditions (A1) to (A5) hold. Then the quasi-least-squares estimator $\hat{\boldsymbol{B}}_T$ defined by (9.1.6) converges almost surely to the true parameter matrix $\boldsymbol{B}$.

Following Lai and Wei [10], the proof is based on the study of the asymptotic behaviour of the quadratic form

$$Q_T = \left\| \left(\int_0^T \boldsymbol{X}_t \boldsymbol{X}_t' d\theta_t \right)^{-1/2} \int_0^T \boldsymbol{X}_t d\boldsymbol{M}_t' \right\|^2$$

$$= \sum_{i=1}^{m} Q_T^{(i)} \tag{9.1.9}$$

where $\|\cdot\|$ is the ordinary Euclidean matrix norm and

$$Q_T^{(i)} = \left\| \left(\int_0^T \boldsymbol{X}_t \boldsymbol{X}_t' d\theta_t \right)^{-1/2} \int_0^T \boldsymbol{X}_t d M_t^i \right\|^2$$

$$= \left(\int_0^T \boldsymbol{X}_t' d M_t^i \right) \left(\int_0^T \boldsymbol{X}_t \boldsymbol{X}_t' d\theta_t \right)^{-1} \left(\int_0^T \boldsymbol{X}_t d M_t^i \right). \tag{9.1.10}$$

LEMMA 9.1
Suppose that, for some $c > 0$, $\tau = \inf\{t : \lambda_{\min}(t) \geq c\} < \infty$ a.s. Then, under the assumption (A1) to (A4),

$$Q_T = o\left((\log \lambda_{\max}(T))^{1+\delta} \right) + O(1) \ \ a.s. \tag{9.1.11}$$

for every $\delta > 0$.

We first state and prove some results which lead to a proof of the Lemma 9.1 which in turn will lead to a proof of the Theorem 9.1.

LEMMA 9.2
Let τ be defined as in Lemma 9.1. Define

$$\tilde{Z}_t = c\boldsymbol{I} + \chi([t \geq \tau]) \int_{\tau}^{t} \boldsymbol{X}_s \boldsymbol{X}_s' \, d\theta_s \,. \tag{9.1.12}$$

Then $\{\tilde{Z}_t, \mathcal{F}_t, t \geq 0\}$ is a right-continuous semimartingale on $(\Omega, \mathcal{F}, P)$ and

$$\tilde{Z}_t^{-1} = c^{-1}\boldsymbol{I} - \chi([t \geq \tau]) \int_{\tau}^{t} \tilde{Z}_{s-}^{-1} \boldsymbol{X}_s \boldsymbol{X}_s' \tilde{Z}_{s-}^{-1} d\theta_s^c + \sum_{\tau < s \leq t} \Delta \tilde{Z}_s^{-1} \tag{9.1.13}$$

where $\boldsymbol{I}$ is the identity matrix and $\chi(A)$ denotes the indicator function of a set A.

PROOF Let $\tilde{Z}^{ij}$ be the (i, j)th element of $\tilde{Z}^{-1}$. Note that

$$\frac{\partial \tilde{Z}^{ij}}{\partial \tilde{Z}_{k\ell}} = -\tilde{Z}^{ik} \tilde{Z}^{\ell j} \,. \tag{9.1.14}$$

Consider the transformation $\phi_{ij}(\tilde{Z}) = \tilde{Z}^{ij}$. Applying the semimartingale transformation theorem (cf. [17, Theorem 27.1]), we have

$$\tilde{Z}_t^{ij} = \tilde{Z}_0^{ij} - \sum_{k,\ell} \int_0^t \tilde{Z}_{s-}^{ik} \tilde{Z}_{s-}^{\ell j} d\tilde{Z}_{k\ell}(s)$$

$$+ \sum_{s \leq t} \left[\Delta \tilde{Z}_s^{ij} + \sum_{k,\ell} \tilde{Z}_{s-}^{ik} \tilde{Z}_{s-}^{\ell j} \Delta \tilde{Z}_{k\ell}(s) \right] \tag{9.1.15}$$

where we have used the fact that $[U, V]_t = \sum_{s \leq t} \Delta U_s \, \Delta V_s$ for any two processes

U and V of bounded variation. From the definition of $\tilde{Z}$, we have, from (9.1.15), that

$$\tilde{Z}_t^{ij} = c^{-1}\delta_{ij} - \chi([t \geq \tau]) \sum_{k,\ell} \int_\tau^t \tilde{Z}_{s-}^{ik} \tilde{Z}_{s-}^{\ell j} X_k(s) X_\ell(s) d\theta_s$$

$$+ \sum_{\tau < s \leq t} \left[\Delta \tilde{Z}_s^{ij} + \sum_{k,\ell} \tilde{Z}_{s-}^{ik} \tilde{Z}_{s-}^{\ell j} X_k(s) X_\ell(s) \Delta\theta_s \right]$$

$$= c^{-1}\delta_{ij} - \chi([t \geq \tau]) \sum_{k,\ell} \int_t^\tau \tilde{Z}_{s-}^{ik} \tilde{Z}_{s-}^{\ell j} X_k(s) X_\ell(s) d\theta_s^c$$

$$+ \sum_{\tau < s \leq t} \Delta \tilde{Z}_s^{ij}, \tag{9.1.16}$$

which proves the lemma. ∎

PROOF of Lemma 9.1: Note that $\boldsymbol{Y}_0 = 0$. In view of (9.1.9) and (9.1.10), it is sufficient to prove that

$$Q_T = \gamma_T' \boldsymbol{Z}_T^{-1} \gamma_T = o\left(\lambda_{\max}(T)\right) + O(1) \text{ a.s.} \tag{9.1.17}$$

where $\boldsymbol{Z}_t$ is as defined by (9.1.8) and

$$\gamma_t = \int_0^t \boldsymbol{X}_s dM_s \tag{9.1.18}$$

with $\{M_t\}$ a real-valued L^2-martingale. Suppose we show that

$$\tilde{Q}_T = \gamma_T' \tilde{\boldsymbol{Z}}_T^{-1} \gamma_T = o\left(\tilde{\lambda}_{\max}(T)\right) + O(1) \text{ a.s.} \tag{9.1.19}$$

where $\tilde{\boldsymbol{Z}}_T$ is as defined by (9.1.12) in Lemma 9.2 and $\tilde{\lambda}_{\max}(T)$ is the largest eigenvalue of $\tilde{\boldsymbol{Z}}_T$. Note that $\tilde{\lambda}_{\max}(T)$ increases with T. Since $\boldsymbol{Z}_t - \tilde{\boldsymbol{Z}}_t = \boldsymbol{P}_t$ is a positive semidefinite matrix for $t \geq \tau$, the relation (9.1.19) will imply (9.1.17). Hence, it is sufficient to prove (9.1.19).

Let us now compute $\tilde{Z}^{ij}\gamma_i\gamma_j$ using the product formula for semimartingales:

$$\tilde{Z}_t^{ij}(\gamma_i\gamma_j)_t = \int_0^t \tilde{Z}_{s-}^{ij} d(\gamma_i\gamma_j)_s + \int_0^t (\gamma_i\gamma_j)_s d\tilde{Z}_s^{ij}$$

$$= \int_0^t \tilde{Z}_{s-}^{ij}\gamma_i(s-)X_j(s)dM_s + \int_0^t \tilde{Z}_{s-}^{ij}\gamma_j(s-)X_i(s)dM_s$$

$$+ \int_0^t \tilde{Z}_{s-}^{ij}X_i(s)X_j(s)d[M]_s$$

$$- \int_\tau^t \gamma_i(s)\left[\tilde{Z}_{s-}^{-1}X_sX_s'\tilde{Z}_{s-}^{-1}\right]_{ij}\gamma_j(s)d\theta_s^c$$

$$+ \sum_{\tau<s\leq t}\gamma_i(s)\,\Delta\tilde{Z}_s^{ij}\gamma_j(s) \qquad (9.1.20)$$

following Lemma 9.2 and the product formula for semimartingales. Let us sum over i, j. Let $V_t = \tilde{Z}_t^{-1}$. Then

$$\tilde{Q}_t = 2\int_0^t \gamma_{s-}'V_{s-}X_sdM_s + \int_0^t X_s'V_{s-}X_sd[M]_s$$

$$- \int_t^\tau \gamma_s'V_{s-}X_sX_s'V_{s-}\gamma_s\, d\theta_s^c$$

$$+ \sum_{\tau<s\leq t}\gamma_s'\Delta V_s\gamma_s \qquad (9.1.21)$$

where

$$\Delta V_s = \left(\tilde{Z}_{s-}^{-1} + X_s X_s' \Delta\theta_s\right)^{-1} - \tilde{Z}_{s-}^{-1}$$

$$= -\tilde{Z}_{s-}^{-1} \frac{X_s X_s' \Delta\theta_s}{1 + X_s' \tilde{Z}_{s-}^{-1} X_s \Delta\theta_s} \tilde{Z}_{s-}^{-1} . \tag{9.1.22}$$

Since $\Delta\gamma_s = X_s \Delta M_s$, it follows that

$$\sum_{\tau<s\leq t} \gamma_s' \Delta V_s \gamma_s = - \sum_{\tau<s\leq t} \frac{|\gamma_{s-}' V_{s-} X_s|^2}{1 + X_s' V_{s-} X_s \Delta\theta_s} \Delta\theta_s$$

$$-2 \sum_{\tau<s\leq t} \gamma_{s-}' V_{s-} X_s \frac{X_s' V_{s-} X_s \Delta\theta_s}{1 + X_s' V_{s-} X_s \Delta\theta_s} \Delta M_s$$

$$+ \sum_{\tau<s\leq t} X_s' \Delta V_s X_s \, \|\Delta M_s\|^2 . \tag{9.1.23}$$

Let $M_t = M_t^c + M_t^d$ be the unique decomposition of M into its continuous and purely discontinuous parts. Then

$$\tilde{Q}_T = 2 \int_0^t \gamma_{s-}' V_{s-} X_s dM_s^c - \int_0^t |\gamma_s' V_{s-} X_s|^2 \, d\theta_s^c$$

$$+ 2 \int_0^t \gamma_{s-}' V_{s-} X_s dM_s^d - \sum_{\tau<s\leq t} \frac{|\gamma_s' V_{s-} X_s|^2}{1 + X_s' V_{s-} X_s \Delta\theta_s} \Delta\theta_s$$

$$+ \int_0^t X_s' V_s X_s d[M]_s - \int_0^t X_s' \Delta V_s X_s d[M]_s$$

$$+ \sum_{\tau<s\leq t} X_s' \Delta V_s X_s \, \|\Delta M_s\|^2$$

$$-2\sum_{\tau<s\le t}\gamma'_{s-}V_{s-}X_s\frac{X'_sV_{s-}X_s\Delta\theta_s}{1+X'_sV_{s-}X_s\Delta\theta_s}\Delta M_s\,. \qquad (9.1.24)$$

The stochastic integrals with respect to $\{M_t^c\}$ and $\{M_t^d\}$ are well defined since the integrands are locally bounded. Note that

$$\text{(i)}\quad \int_\tau^t \gamma'_s\, V_{s-}X_s|^2\,d\theta_s^c = \int_\tau^t |\gamma'_{s-}V_{s-}X_s|^2\,d\theta_s^c\,,$$

$$\text{(ii)}\quad \int_\tau^t |\gamma'_sV_{s-}X_s|^2\,d\theta_s^c + \sum_{\tau<s\le t}\frac{|\gamma'_sV_{s-}X_s|^2}{1+X'_sV_{s-}X_s\Delta\theta_s}\Delta\theta_s$$

$$\ge \int_\tau^t \frac{|\gamma'_sV_{s-}X_s|^2}{1+X'_sV_{s-}X_s\Delta\theta_s}d\theta_s\,,$$

and

(iii) the sixth and seventh terms in (9.1.24) cancel, since the process $([M]_t - \sum_{s\le t}|\Delta M_s|^2)$ is continuous.

Hence

$$\tilde{Q}_t \le 2\int_0^t \gamma'_{s-}V_{s-}X_s dM_s^c - \frac{1}{2}\int_\tau^t |\gamma'_{s-}V_{s-}X_s|^2\,d\theta_s^c$$

$$+2\int_0^t \gamma'_{s-}V_{s-}X_s dM_s^d - \frac{1}{2}\int_\tau^t \frac{|\gamma'_{s-}V_{s-}X_s|^2}{1+X'_sV_{s-}X_s\Delta\theta_s}d\theta_s$$

$$+\int_0^t X'_sV_sX_s d[M]_s - \frac{1}{2}\sum_{\tau<s\le t}\frac{|\gamma'_{s-}V_{s-}X_s|^2}{1+X'_sV_{s-}X_s\Delta\theta_s}\Delta\theta_s$$

$$-2\sum_{\tau<s\le t}\gamma'_{s-}V_{s-}X_s\frac{X'_sV_{s-}X_s\Delta\theta_s}{1+X'_sV_{s-}X_s\Delta\theta_s}\Delta M_s\,. \qquad (9.1.25)$$

The first term on the right side of the above inequality satisfies

$$\int_0^t \gamma'_{s-} V_{s-} X_s dM_s^c = o\left(\int_\tau^t \left\| \gamma'_{s-} V_{s-} X_s \right\|^2 d\theta_s^c \right) + O(1) \qquad (9.1.26)$$

by Theorem 1.51, Chapter 1 with $\frac{1}{2} < \alpha < 1$ and the assumption (A4) (i) with $\delta = \frac{1-\alpha}{\alpha}$. The third term can be written in the form

$$2 \int_0^\tau \gamma'_{s-} V_{s-} X_s dM_s^d + 2 \int_\tau^t \frac{\gamma'_{s-} V_{s-} X_s}{1 + X'_s V_{s-} X_s \Delta\theta_s} dM_s^d$$

$$+ 2 \int_\tau^t \gamma'_{s-} V_{s-} X_s \frac{X'_s V_{s-} X_s \Delta\theta_s}{1 + X'_s V_{s-} X_s \Delta\theta_s} dM_s^d \,. \qquad (9.1.27)$$

Applying again Theorem 1.51 of Chapter 1, under the assumption (A4) (ii), it follows that

$$\int_0^t \gamma'_{s-} V_{s-} X_s dM_s^d + \int_\tau^t \frac{\gamma'_{s-} V_{s-} X_s}{1 + X'_s V_{s-} X_s \Delta\theta_s} dM_s^d$$

$$= o[\text{4th term of (9.1.25)}] + O(1) \,.$$

By Theorem 1.52 of Chapter 1, the third term in (9.1.27) and the seventh term in (9.1.25) cancel. Hence,

$$\tilde{Q}_T + \left(\frac{1}{2} + o(1) \right) \int_\tau^t \left| \gamma'_{s-} V_{s-} X_s \right|^2 d\theta_s^c$$

$$+ \left(\frac{1}{2} + o(1) \right) \int_\tau^t \frac{\left\| \gamma'_{s-} V_{s-} X_s \right\|^2}{1 + X'_s V_{s-} X_s \Delta\theta_s} d\theta_s + O(1)$$

$$\leq \int_0^t X_s' V_s X_s d[M]_s \,. \tag{9.1.28}$$

Applying the Corollary E.1 of Appendix E for $\tilde{Z}_t$ which is of the form $M(t)$ in the corollary (with $\xi_s = 0$ for $0 \leq s \leq \tau$), it follows that

$$\int_0^t X_s' \tilde{Z}_{s-}^{-1} X_s d\theta_s = o\left(\log \tilde{\lambda}_{\max}(t)\right) \,. \tag{9.1.29}$$

Since $[M]_t = \langle M^c \rangle_t + [M^d]_t$, we obtain, from Theorem 1.51 of Chapter 1, that (apply for $\xi_t = \int_0^t (X_s' V_{s-} X_s)^{1/2} dM_s^{id}$), under (A4),

$$\frac{\int_0^t X_s' \tilde{Z}_s^{-1} X_s d[M]_s}{\left(\log \tilde{\lambda}_{\max}(t)\right)^{1+\delta}} \leq \left(\frac{\int_0^t X_s' \tilde{Z}_{s-}^{-1} X_s d[M]_s}{\left(1 + \int_0^t X_s' \tilde{Z}_{s-}^{-1} X_s d\theta_s\right)^{1+\delta}} \right) \left(\frac{1 + \int_0^t X_s' \tilde{Z}_{s-}^{-1} X_s d\theta_s}{\log \tilde{\lambda}_{\max}(t)} \right)^{1+\delta}$$

$$= O(1) \,. \tag{9.1.30}$$

Since this is true for every $\delta > 0$, the right-hand side of (9.1.28), and hence, each term on the left hand side behave as $o((\log \tilde{\lambda}_{\max}(t))^{1+\delta}) + O(1)$. This completes the proof of Lemma 9.1. ∎

PROOF of Theorem 9.1: Since $\lambda_{\min}(T) \to \infty$ a.s., the assumption about

τ in Lemma 9.1 holds and Z_T is nonsingular for T large. Hence

$$\begin{aligned}\left\|\hat{B}_T - B\right\|^2 &= \left\|Z_T^{-1}\int_0^T X_s dM_s'\right\|^2 \\ &\le \left\|Z_T^{-1/2}\right\|^2 \left\|Z_T^{-1/2}\int_0^T X_s dM_s'\right\|^2 \\ &\le \sqrt{k}\,\lambda_{\min}^{-1}(T)\, Q_T \\ &= \sqrt{k}\, o(1)\left(\log \lambda_{\max}(T)^{1+\delta}/\lambda_{\min}(T)\right) \\ &\to 0 \text{ a.s.}\end{aligned}$$

by (A5). This proves the almost sure consistency of $\hat{B}_T$ in norm. ∎

Asymptotic Normality

We shall now prove that suitably normalized rows of the quasi-least-squares estimator $\hat{B}_T$ are asymptotically normal. For simplicity of exposition, we assume that Y and M are one-dimensional. Denote B and $\hat{B}_T$ by β' and $\hat{\beta}_T'$, respectively. We assume that the following conditions hold.

(D) There exist nonrandom positive definite symmetric matrices $P_T, T \ge 0$ such that

(i) $P_T^{-1}\left(\int_0^T X_t X_t' d\theta_t\right)^{1/2} \xrightarrow{p} Q,$

and

(ii) $P_T^{-1}\left(\int_0^T X_t X_t' d\langle M\rangle_t\right)^{1/2} \xrightarrow{p} I$

where Q is a nonrandom positive definite matrix, and

(iii) for every nonrandom k-dimensional vector c and each $T \ge 0$, let $X^{T,C}$

denote the local L^2-martingale with respect to $\{\mathcal{F}_{t,T}\}$ given by

$$X^{T,c}(t) = c' P_T^{-1} \int_0^{tT} X_s dM_s, 0 \leq t \leq 1$$

and let $\mu^{T,c}$ denote the random measure corresponding to the jumps of $X^{T,c}$ and $\nu^{T,c}$ its compensator. Then

$$\int_{(0,1]} \int_{|x|>\varepsilon} x^2 \nu^{T,c}(ds, dx) \xrightarrow{p} 0 \text{ as } T \to \infty \text{ for all } \varepsilon > 0 .$$

THEOREM 9.2
Suppose the conditions (A1) to (A3) and (B) hold. Then the quasi-least-squares estimator $\hat{\beta}_T$ has an asymptotically normal distribution; in fact,

$$Z_T^{1/2} \left(\hat{\beta}_T - \beta \right) \xrightarrow{\mathcal{L}} N\left(0, Q^{-1}\right) \tag{9.1.31}$$

where $Z_T = \int_0^T X_s X_s' d\theta_s$.

PROOF Let us suppress the index c in the sequel. For an arbitrary sequence $T_n \uparrow \infty$, let

$$Y^T(t) = c' P_T^{-1} X_t, \; Y_n(t) = Y^{T_n}(t) , \tag{9.1.32}$$

and

$$X_n(t) = X^{T_n}(t) = \int_0^{tT_n} Y_n(s) dM_s . \tag{9.1.33}$$

By the Cramer–Wold technique and the condition D(i), it is sufficient to prove that

$$X_n(1) \xrightarrow{\mathcal{L}} N\left(0, \; \|c\|^2\right) . \tag{9.1.34}$$

Since $\{X_n(t)\}$ are local L^2-martingales with

$$\langle X_n \rangle (1) = \int_0^{T_n} Y_n^2(s) d\langle M \rangle_s \xrightarrow{p} \|c\|^2 \text{ as } n \to \infty ,$$

the relation (9.1.34) follows from the central limit theorem for stochastic integrals with respect to martingales. (See Chapter 1, cf. [12, 13]). ∎

REMARK 9.1 Christopeit [5] gives sufficient conditions for (D2) to hold and gives some special cases where the results can be obtained. Typical applications include models of the following form.

(i) **(Discrete time case)** $Y_n = BX_n + \varepsilon_n$ where $\{\varepsilon_n\}$ is a martingale difference sequence with respect to a filtration $\{\mathcal{F}_n\}$ such that $\sup_n E(\varepsilon_n^2|\mathcal{F}_{n-1}) < \infty$ a.s. and the X_n are $\mathcal{F}_{n-1}$-measurable random vectors.

(ii) **(Diffusion model)**

$$dY_t = BX_t dt + \sigma_t \, dW_t$$

where $\{W_t\}$ is a standard Wiener process. Here $d\theta_t = dt$ and

$$d[M]_t = d\langle M\rangle_t = \sigma_t^2 dt \, .$$

(iii) Both (i) and (ii) are special cases of a general class of models in which $\theta = \langle|M|\rangle$.

(iv) **(Skorokhod equation model)** Here $p(\omega, ds, dz)$ is a point process, $\alpha(dt)$ its Lévy measure and let q be the martingale measure

$$q(\omega, ds, dz) = p(\omega, ds, dz) - ds \otimes \alpha(dz) \, .$$

Consider the Skorokhod equation of the type

$$Y_t = Y_0 + B\int_0^t X_s ds + \int_0^t \sigma_s dW_s + \int_0^t \int_Z c_s(\omega, z)\, q\,(\omega, ds, dz)$$

for some predictable process $c_s(\omega, z)$ such that the last integral is well defined. In this case M_t^c coincides with the Ito integral and M_t^d coincides with the point process integral. Furthermore,

$$\langle M_t^c \rangle = \int_0^t \sigma_s^2 ds \text{ and } \langle M_t^d \rangle = \int_0^t \lambda_s ds ,$$

with

$$\lambda_s = \int |c_s(\omega, z)|^2 \alpha(dz) .$$

∎

9.2 Estimation by the Maximum Likelihood Method

We now consider the problem of parameter estimation in a system modeled by a scalar linear stochastic differential equation excited by the Gaussian and Poisson disturbances or noises.

Let D be the space of all real-valued right-continuous functions with left limits and D be endowed with the Skorokhod topology on every compact subset of R. Let $\mathcal{D}_T$ denote the σ-algebra generated by the coordinates π_s, $0 \leq s \leq T$ and $\mathcal{D}$ be the σ-algebra generated by $\{\mathcal{D}_T, T \geq 0\}$.

Suppose a process $\{X(t), t \geq 0\}$, governed by the stochastic dynamical system (cf. [6]) defined by

$$X_t = \int_0^t \phi_1 X_s ds + \int_0^t \phi_2 u_s ds + \int_0^t C\, dW_s + \int_R v\, r([0, t], dv), t \geq 0, \quad (9.2.1)$$

is observed over $[0, T]$. Here X_t is the state of the system at the time t, u_t is a known control function, W_t is the standard Wiener process, and r is a centered Poisson random measure independent of $\{W_t\}$ with the associated measure μ. Note that μ is a positive measure on R finite on every compact set not containing the origin and possessing the first and second moments. Suppose measure μ and the coefficients ϕ_1, ϕ_2 and C are unknown. We consider the problem of estimation of ϕ_1, ϕ_2, C and μ when the process $\{X_t\}$ is observed $[0, T]$. We will also consider the problem of choosing an optimal control (u_s) to optimize the precision of the estimator. For simplicity, we consider the scalar case.

Let $\phi = \binom{\phi_1}{\phi_2}$ and let $\binom{\phi^0}{C^0}$ be the true value of $\binom{\phi}{C}$ and μ^o be the true value of μ. Suppose $\phi \in R^2$, $C \in R_+$ and $\mu \in M$ where M is a set of positive measures on R finite on every compact set not containing the origin and possessing moments of first and second order. Let $P^T_{(\phi,C,\mu)}$ the probability measure generated on $(D, \mathcal{D}_T)$ by (9.2.1) and $Q^T_{(C,\mu)}$ be the probability measure generated on $(D, \mathcal{D}_T)$ by a process that is a strong solution of the equation

$$X_t = \int_0^t C\,dW_s + \int v\,r([0,t], dv), \; t \geq 0 . \tag{9.2.2}$$

We denote by $P_{(\phi,C,\mu)}$ and $Q_{(C,\mu)}$ the measures generated by (9.2.1) and (9.2.2), respectively, on $(D, \mathcal{D})$. A process $\{\pi_t\}$ satisfies the equation (9.2.2) relative to $Q_{(C,\mu)}$ for the Poisson measure q defined by the jumps of π_t given by

$$P_t^{\mu} = \sum_{s \leq t} (\pi_s - \pi_{s-}) - t \int v\,d\mu(v) = \int v q([0,t], dv), \; t \geq 0 \tag{9.2.3}$$

and the standard Wiener process $M_t^{C,\mu}$ defined by

$$M_t^{C,\mu} = C^{-1}\left(\pi_t - P_t^{\mu}\right); \; t \geq 0 . \tag{9.2.4}$$

LEMMA 9.3

For every $T > 0$ and for every $(\phi, C, \mu) \in R^2 \times R^+ \times M$, the measures $P^T_{(\phi,C,\mu)}$ and $Q^{(T)}_{(C,\mu)}$ are equivalent and the Radon–Nikodym derivative is given by

$$\frac{dP^T_{(\phi,C,\mu)}}{dQ^T_{(C,\mu)}} = \exp\left\{ C^{-1} \int_0^T [\phi_1 \pi_s + \phi_2 u_s]\,dM_s^{C,\mu} \right.$$

$$\left. - \frac{C^{-2}}{2} \int_0^T [\phi_1 \pi_s + \phi_2 u_s]^2\,ds \right\} \tag{9.2.5}$$

where $M_s^{C,\mu}$ is as defined by (9.2.3) and (9.2.4), and the stochastic integral is with respect to the measure $Q^T_{(C,\mu)}$.

The above lemma follows from an extension of the Girsanov's theorem, which shows also that, relative to the measure $P_{(\phi,C,\mu)}$, $\{\pi_t\}$ is a solution of the equation (9.2.1) for the Poisson measure q and the Wiener process $\tilde{W}_t$ defined by

$$\tilde{W}_t = C^{-1}\left[\pi_t - \int_0^t (\phi_1\pi_s + \phi_2 u_s)\,ds - \int v\ q([0,t],dv)\right], t \geq 0 \quad (9.2.6)$$

(cf. [21]). Hence

$$\pi_t = m_t + Y_t + Z_t,\ t \geq 0 \quad (9.2.7)$$

where $\{m_t\}$, $\{Y_t\}$ and $\{Z_t\}$ satisfy the differential equations

$$\dot{m}_t = \phi_1 m_t + \phi_2 u_t,\ t \geq 0\,, \quad (9.2.8)$$

$$dY_t = \phi_1 Y_t dt + C d\tilde{W}_t,\ t \geq 0\,, \quad (9.2.9)$$

and

$$dZ_t = \phi_1 Z_t dt + \int v\ q(dt,dv),\ t \geq 0 \quad (9.2.10)$$

with the initial conditions $m_0 = Y_0 = Z_0 = 0$.

Estimation of Parameters when the Characteristics of the Noise are Known

Suppose the parameter vector ϕ is unknown but the system is stable, that is $\phi_1^0 < 0$. Further assume that

$$\int |v|^k d\mu(v) < \infty,\ 1 \leq k \leq 5 \quad (9.2.11)$$

and the control $\{u_t\}$ is ergodic in the sense that, for every $s \geq 0$,

$$\lim_{T\to\infty} \frac{1}{T}\int_0^T u_t u_{t+s} dt = \rho_u(s) \quad (9.2.12)$$

where ρ_u is a continuous function.

The statistical structure for the problem of estimation can be written in the form

$$\left(D, \mathcal{D}_T, \left\{P_\phi^T, \phi \in R^2,\ \phi_1 < 0\right\}\right) \tag{9.2.13}$$

omitting the indices (C, μ) in $P^T_{(\phi, C, \mu)}$ and $\{M_t^{C,\mu}\}$. The family of measures is dominated by Q^T and the log-likelihood function $\ell_T(\phi)$ is given by

$$\begin{aligned}\ell_T(\phi) \doteq C^{-1}&\left[\phi_1 \int_0^T \pi_s dM_s + \phi_2 \int_0^T u_s\, dM_s\right]\\ &-\frac{C^{-2}}{2}\left[\phi_1^2 \int_0^T \pi_s^2 ds + \phi_2^2 \int_0^T u_s ds\right.\\ &\left.+2\phi_1\phi_2 \int_0^T \pi_s u_s ds\right]\end{aligned} \tag{9.2.14}$$

(here we are writing $\ell_T(\phi)$ or $\ell_T(\phi, x)$ where x denotes the sample path on $[0, T]$).

Let $\ell_T^{(i)}(\phi)$ denote the ith derivative of $\ell_T(\phi)$ for $i = 1, 2$. It can be checked that

$$\begin{aligned}\ell_T^{(1)}(\phi) &= \begin{bmatrix} \frac{\partial}{\partial\phi_1}\ell_T(\phi) \\ \frac{\partial}{\partial\phi_2}\ell_T(\phi)\end{bmatrix}\\ &= C^{-1}\begin{bmatrix} \int_0^T \pi_s\, d\tilde{W}_s \\ \int_0^T u_s\, d\tilde{W}_s\end{bmatrix}\end{aligned} \tag{9.2.15}$$

and

$$\ell_T^{(2)}(\phi) = \left(\left(\frac{\partial^2 \ell_T(\phi)}{\partial\phi_i \partial\phi_j}\right)\right)_{2\times 2}$$

$$= -C^{-2} \begin{bmatrix} \int_0^T \pi_s^2 ds & \int_0^T \pi_s u_s ds \\ \int_0^T \pi_s u_s ds & \int_0^T u_s^2 ds \end{bmatrix} \tag{9.2.16}$$

where $\tilde{W}_s$ is defined by (9.2.6) and π_s is defined by (9.2.7)–(9.2.10).

We first prove few lemmas needed in the sequel for studying the asymptotic properties of an estimator $\hat{\phi}$ solving the likelihood equation

$$\ell_T^{(1)}(\phi) = 0\,. \tag{9.2.17}$$

LEMMA 9.4
Suppose the following conditions hold:

(G1) Y_0^* *is* $N\left(0, \frac{-C^2}{2\phi_1}\right)$,

(G2) Z_0^* *has the characteristic function*

$$\exp\left[\frac{1}{\phi_1}\int \Phi(uv)d\mu(v) + \frac{iu}{\phi_2}\int v\, d\mu(v)\right]$$

where $\Phi(x) = \int_0^x \phi(z)dz$ *with* $\phi(z) = \frac{1-e^{-iz}}{z}$, *and*

(G3) Y_0^*, Z_0^*, $\left\{\tilde{W}_t\right\}$ *and* q *are independent.*

Then Equations (9.2.9) and (9.2.10) admit solutions $\{Y_t^*\}$ *and* $\{Z_t^*\}$ *that are stationary in the strict sense with the initial conditions* Y_0^* *and* Z_0^*, *respectively. In particular* $Y_t^* + Z_t^*$ *is also strict sense stationary process.*

PROOF Since $\{Y_t^*\}$ is Gaussian, it is well known that $\{Y_t^*\}$ is stationary in the strict sense. It is sufficient to prove the lemma for the process $\{Z_t^*\}$ since $\{Y_t^*\}$ and $\{Z_t^*\}$ are independent. Note that

$$Z_t^* = e^{t\phi_1} Z_0^* + e^{t\phi_1} \int_0^t \int e^{-\phi_1 s} v\, q(ds, dv)$$

$$= e^{t\phi_1} Z_0^* + e^{t\phi_1} V_t \text{ (say)}. \tag{9.2.18}$$

Let us first calculate the characteristic function of V_t. Let

$$\tilde{V}_t = \int_0^t \int e^{-\phi_1 s} v \, p(ds, dv) \tag{9.2.19}$$

where p is a random measure defined such that for t, t' in R_+ with $t' < t$ and Borel set A in R not containing the origin,

$$p\left([t,t'), A\right) = q\left([t,t'), A\right) + \left(t' - t\right) \mu(A). \tag{9.2.20}$$

The random variable $\tilde{V}_t$ can be obtained as the limit in quadratic mean of sums of the form

$$\Delta = \sum_{k,j} e^{-\phi_1 t_k} v_j p\left([t_k, t_{k+1}), A_j\right) \tag{9.2.21}$$

where the sets A_j are disjoint (cf. [7]).

The random variables $p([t_k, t_{k+1}), A_j)$ are independent Poisson random variables with parameter $(t_{k+1} - t_k)\mu(A_j)$. Hence, the characteristic function of Δ is given by

$$\phi_\Delta(u) = \exp\left[\sum_{k,j} (t_{k+1} - t_k) \mu\left(A_j\right) e^{i \exp(-t_k\phi_1) v_j u}\right]. \tag{9.2.22}$$

Passing to the limit, we obtain the characteristic function of $\tilde{V}_t$ given by

$$\phi_{\tilde{V}_t}(u) = \exp\left[\int_0^t \int e^{i \exp(-\phi_1 s)} ds d\mu(v)\right]. \tag{9.2.23}$$

Define

$$\phi(z) = \frac{1 - e^{iz}}{z} \text{ and } \Phi(x) = \int_0^x \phi(z) dz. \tag{9.2.24}$$

Then the characteristic function of $e^{t\phi_1} V_t$ is given by

$$\phi_{e^{t\phi_1} V_t}(u) = \exp\left[\frac{1}{\phi_1}\int \Phi(uv)d\mu(v) - \frac{1}{\phi_1}\int \Phi(uv)e^{t\phi_1}d\mu(v)\right.$$

$$\left. + \frac{iu}{\phi_1}\left(1 - e^{t\phi_1}\right)\int v\, d\mu(v)\right]. \qquad (9.2.25)$$

Under the condition (G2) and from the definition of Φ, it follows that

$$\lim_{t\to+\infty} \phi_{e^{t\phi_1} V_t}(u) = \exp\left[\frac{1}{\phi_1}\int \Phi(uv)d\mu(v) + \frac{iu}{\phi_1}\int v d\mu(v)\right].$$

Since the limit is a continuous function, it follows that it is a characteristic function which is the characteristic function of Z_0^* as given in (ii). On the other hand, following (9.2.18), the characteristic function of Z_t^* is

$$\phi_{Z_t^*}(u) = \exp\left[\frac{1}{\phi_1}\left(\int \Phi(uv)d\mu(v) + iu\int v d\mu(v)\right)\right]$$

$$= \phi_{Z_0^*}(u). \qquad (9.2.26)$$

For $0 \le t_1 \le t_2$ and $0 \le t$, the characteristic function of $(Z_{t_1+t}^*, Z_{t_2+t}^*)$ is given by

$$\phi_{Z_{t_1+t}^*, Z_{t_2+t}^*}(u_1, u_2) = \exp\left[\frac{1}{\phi_1}\int \Phi\left[\left(u_1 + u_2 e^{(t_2-t_1)\phi_1}\right)v\right]d\mu(v)\right.$$

$$+ \frac{1}{\phi_1}\int \Phi\left(u_2 v\right)d\mu(v) - \frac{1}{\phi_1}$$

$$\int \Phi\left(e^{(t_2-t_1)\phi_1}u_2 v\right)d\mu(v)$$

$$\left. + \frac{i}{\phi_1}(u_1 + u_2)\int v d\mu(v)\right]$$

$$= \phi_{Z_{t_1}^*, Z_{t_2}^*}(u_1, u_2). \qquad (9.2.27)$$

This result can be extended to higher dimensions proving the stationarity in the strict sense of the process $\{Z_t^*\}$. ▌

LEMMA 9.5
Under the conditions stated earlier,

$$\lim_{T\to\infty}\frac{1}{T}\int_0^T \left(Y_s^* + Z_s^*\right)^2 ds = \frac{-K}{2\phi_1} \tag{9.2.28}$$

in the sense of convergence in the quadratic mean with respect to P_ϕ *where*

$$K = C^2 + \int v^2 d\mu(v)\,. \tag{9.2.29}$$

PROOF Let

$$\eta_t = \left(Z_t^* + Y_t^*\right)^2\,. \tag{9.2.30}$$

Note that

$$Y_t^* = Y_0^* e^{t\phi_1} + Ce^{t\phi_1}\int_0^t e^{-\phi_1 s} d\tilde{W}_s\,. \tag{9.2.31}$$

It is easy to check that

$$E_\phi\left(Y_t^{*^2} Y_0^{*^2}\right) = \frac{C^4}{4\phi_1^2} + c^{2t\phi_1}\frac{C^4}{2\phi_1^2}\,. \tag{9.2.32}$$

From the definition of Z_t^* given by (9.2.18), it follows that

$$E_\phi\left(Z_t^{*^2} Z_0^{*^2}\right) =$$

$$e^{2t\phi_1} E_\phi\left(Z_0^{*^4}\right) - \tfrac{e^{2t\phi_1}}{2\phi_1}\left(e^{-2t\phi_1} - 1\right)\left(\textstyle\int v^2 d\mu(v)\right) E\left(Z_0^{*^2}\right)\,. \tag{9.2.33}$$

Condition (9.2.11) implies that the function

$$F(u) = \int \Phi(uv)d\mu(v) \tag{9.2.34}$$

is differentiable up to order 4 and

$$F^{(k)}(u) = \int v^k \phi^{(k-1)}(uv)d\mu(v) . \tag{9.2.35}$$

This result allows us to show that the characteristic function of Z_0^* is differentiable four times and it can be checked that

$$E_\phi\left(Z_0^{*^2}\right) = -\frac{1}{2\phi_1}\int v^2 d\mu(v) , \tag{9.2.36}$$

and

$$E_\phi\left(Z_0^{*^4}\right) = -\frac{1}{4\phi_1}\int v^4 d\mu(v) + \frac{3}{4\phi_1^2}\left(\int v^2 d\mu(v)\right)^2 . \tag{9.2.37}$$

Following (9.2.18) for Z_t^* and (9.2.31) for Y_t^*, it can be checked that

$$E_\phi\left(Z_t^{*^2} Z_0^{*^2}\right) = \frac{1}{4\phi_1^2}\left(\int v^2 d\mu(v)\right)^2$$

$$+ e^{2t\phi_1}\left[-\frac{\int v^4 d\mu(v)}{4\phi_1} + \frac{\left(\int v^2 d\mu(v)\right)^2}{2\phi_1^2}\right] \tag{9.2.38}$$

and

$$E_\phi\left(Y_t^* Z_t^* Y_0^* Z_0^*\right) = e^{2t\phi_1}\frac{C^2\int v^2 d\mu(v)}{4\phi_1^2} . \tag{9.2.39}$$

Relations (9.2.32), (9.2.38), and (9.2.39) show that there exists a constant k such that

$$E_\phi\left(\eta_0\eta_t\right) - \left[E_\phi\left(\eta_t\right)\right]^2 = ke^{2t\phi_1}$$

$$= -k\frac{2\phi_1}{\pi}\int \frac{e^{i\lambda t}d\lambda}{\lambda^2 + 4\phi_1^2} .$$

An application of the classical ergodic theorem proves that

$$\lim_{T\to\infty} \frac{1}{T} \int_0^T \left(\eta_t - E_\phi(\eta_t)\right) dt = 0$$

in the sense of convergence in quadratic mean with respect to the probability measure P_ϕ. The lemma is proved by noting that

$$E_\phi(\eta_t) = \frac{-K}{2\phi_1} .$$

This completes the proof of lemma. ∎

The condition (9.2.12) on the ergodicity of $\{u_t\}$ implies that (cf. [9]) the family $\{u_t, t \geq 0\}$ admits a spectral measure M^u in the following sense.

Let h be a bounded measurable function on R_+ with values in R^k such that for every $s \geq 0$,

$$\lim_{T\to\infty} \frac{1}{T} \int_0^T h_t h'_{t+s} dt = \rho_h(s)$$

where ρ_h is a continuous function associated to a bounded measure on R with values in the set of all positive Hermitian matrices of order $k \times k$ such that

$$\rho_h(s) = \int_R e^{is\lambda} dM^h(\lambda)$$

and setting $\hat{h}_T(\lambda) = \int_0^T e^{-it\lambda} h_t dt$, one has

$$M^h = \lim_{T\to\infty} \left[\frac{1}{2\pi T} \hat{h}_T(\cdot)\overline{\hat{h}'_T(\cdot)}\beta\right]$$

where β is the Lebesgue measure on R and the limit is in the sense of weak convergence.

LEMMA 9.6
The function $\left\{\begin{bmatrix} m_t \\ u_t \end{bmatrix}, t \geq 0\right\}$ *where* $\{m_t, t \geq 0\}$ *is a solution of (9.2.8) with the initial condition* $m_0 = 0$ *admits a spectral measure* $M^{\phi,u}$ *defined by*

$$dM^{\phi,u}(\lambda) = \begin{bmatrix} \frac{\phi_2^2}{\phi_1^2+\lambda^2} & \frac{-\phi_2}{\phi_1-i\lambda} \\ \frac{-\phi_2}{\phi_1+i\lambda} & 1 \end{bmatrix} dM^u(\lambda) .$$

PROOF Since $\{m_t, t \geq 0\}$ is bounded and measurable, the family $\left(\begin{bmatrix} m_t \\ u_t \end{bmatrix}, t \geq 0\right)$ has a spectral measure if and only if the weak limit of

$$(2\pi T)^{-1} \begin{bmatrix} m_T(\cdot) \\ u_T(\cdot) \end{bmatrix} \begin{bmatrix} \bar{m}_T(\cdot) \\ \bar{u}_T(\cdot) \end{bmatrix}' \beta$$

exists, in which case the limit is that given by $M^{\phi,u}$ (cf. [9]).

It is easy to check that

$$m_T(\lambda) = \int_0^\infty e^{-i\lambda t} \tilde{m}_t^T dt - \int_T^\infty e^{-i\lambda t} e^{(t-T)\phi_1} m_T dt$$

where, if

$$\tilde{u}_t^T = \begin{cases} u_t & \text{if } t \leq T \\ 0 & \text{if } t > T \, , \end{cases}$$

then $\{\tilde{m}_t^T, t \geq 0\}$ denotes the solution of (9.2.8) for the control $\{\tilde{u}_t^T, t \geq 0\}$ with the initial condition $\tilde{m}_0^T = 0$.

Let $\dot{m}_t^T$ denote the derivative of $\tilde{m}_t^T$ with respect to t. Note that

$$\int_0^\infty e^{-i\lambda t} \tilde{m}_t^T dt = \left[-\frac{e^{-i\lambda t}}{i\lambda} \tilde{m}_t^T\right]_0^\infty + \int_0^\infty \frac{e^{-i\lambda t}}{i\lambda} \dot{\tilde{m}}_t^T dt$$

$$= \frac{1}{i\lambda} \int_0^\infty e^{-i\lambda t} \left[\phi_1 \tilde{m}_t^T + \phi_2 \tilde{u}_t^T\right] dt$$

and hence,

$$(\phi_1 - i\lambda) \int_0^\infty e^{-it\lambda} \tilde{m}_t^T \, dt = -\phi_2 \int_0^T e^{-i\lambda T} u_t \, dt \,,$$

which implies that

$$\int_0^\infty e^{-it\lambda} \tilde{m}_t^T \, dt = -\phi_2 \, (\phi_1 - i\lambda)^{-1} \, u_T(\lambda) \, .$$

On the other hand,

$$\int_T^\infty e^{-i\lambda t} e^{(t-T)\phi_1} m_T dt = e^{-i\lambda T} \, (\phi_1 - i\lambda)^{-1} \, m_T \, .$$

Let

$$C_T(\lambda) = e^{-i\lambda T} \, (\phi_1 - i\lambda)^{-1} \, m_T \, .$$

Then

$$m_T(\lambda) = -\,(\phi_1 - i\lambda)^{-1} \, \phi_2 u_T(\lambda) + C_T(\lambda)$$

where

$$|C_T(\lambda)| = \left| e^{-i\lambda T} \, (\phi_1 - i\lambda)^{-1} \, m_T \right|$$

$$\leq \nu \left(\phi_1^2 + \lambda^2 \right)^{-1/2}$$

with $\nu = \sup_{t \geq 0} |m_t|$. Furthermore,

$$\tfrac{1}{T} \int_R |C_T(\lambda) u_T(\lambda)| \, d\lambda$$

$$\leq \left[T^{-1} \int_R |C_T(\lambda)|^2 \, d\lambda \right]^{1/2} \left[T^{-1} \int_R |u_T(\lambda)|^2 \, d\lambda \right]^{1/2}$$

$$\leq T^{-1/2}\left[\int_R \nu^2\left(\phi_1^2+\lambda^2\right)^{-1} d\lambda\right]^{1/2}\left[T^{-1}\int_R |u_T(\lambda)|^2\,d\lambda\right]^{1/2}$$

$$\leq \rho T^{-1/2}$$

for some constant $\rho > 0$ since

$$\int_R \left[\phi_1^2+\lambda^2\right]^{-1} d\lambda < \infty$$

and

$$\lim_{T\to\infty}\frac{1}{2\pi T}\int_R |u_T(\lambda)|^2\,d\lambda = M^u(R) < +\infty\,.$$

Hence

$$\lim_{T\to\infty}\frac{1}{T}\int_R |C_T(\lambda)u_T(\lambda)|\,d\lambda = 0\,.$$

Similarly, it can be shown that for any bounded continuous function g,

$$\lim_{T\to\infty}\frac{1}{T}\int_R |g(\lambda)C_T(\lambda)u_T(\lambda)|\,d\lambda = 0\,.$$

From the bound on $C_T(\lambda)$, one also has the relation

$$\lim_{T\to\infty}\frac{1}{T}\int \left|g(\lambda)C_T^2(\lambda)\right| d\lambda = 0\,.$$

All the above remarks imply that the weak limit of

$$(2\pi T)^{-1}\begin{bmatrix} |m_T(\cdot)|^2 & m_T(\cdot)\bar{u}_T(\cdot) \\ u_T(\cdot)\bar{m}_T(\cdot) & |u_T(\cdot)|^2 \end{bmatrix}\beta$$

exists, which is the same as that of

$$\frac{d\lambda}{2\pi T}\begin{bmatrix} \phi_2^2\left(\phi_1^2+\lambda^2\right)^{-1}|u_T(\lambda)|^2 & -\phi_2\left(\phi_1-i\lambda\right)^{-1}|u_T(\lambda)|^2 \\ -\phi_2\left(\phi_1+i\lambda\right)^{-1}|u_T(\lambda)|^2 & |u_T(\lambda)|^2 \end{bmatrix}.$$

This proves the required result. ∎

LEMMA 9.7

Under the conditions stated earlier, with respect to the probability measure P_ϕ,

$$\lim_{T\to\infty}\frac{1}{T}\int_0^T\begin{bmatrix} Y_s+Z_s\\ m_s\\ u_s\end{bmatrix}^{2\otimes} ds$$

$$=\begin{bmatrix}\frac{-K}{2\phi_1} & 0\\ 0 & M^{\phi,u}(R)\end{bmatrix}=\Gamma_u(\phi) \qquad (9.2.40)$$

in the sense of convergence in L^1 and

$$\lim_{T\to\infty}\frac{1}{\sqrt{T}}\int_0^T\begin{bmatrix} Y_s+Z_s\\ m_s\\ u_s\end{bmatrix} d\tilde{W}_s=\Delta_u(\phi) \qquad (9.2.41)$$

in the sense of convergence in law where $\Delta_u(\phi)=(\Delta_u^i(\phi),\ i=1,2,3)$ is a Gaussian random vector with the covariance matrix $\Gamma_u(\phi)$.

PROOF We show at first that

$$\lim_{T\to\infty}\frac{1}{T}\int_0^T\begin{bmatrix} Y_s^*+Z_s^*\\ m_s\\ u_s\end{bmatrix}^{2\otimes} ds=\Gamma_u(u) \qquad (9.2.42)$$

in quadratic mean. The relation (9.2.42) is an immediate consequence of Lemmas 9.5 and 9.6, since

$$\lim_{T\to\infty}\frac{1}{T}\int_0^T\begin{bmatrix} m_s\\ u_s\end{bmatrix}^{2\otimes} ds=M^{\phi,u}(R) \qquad (9.2.43)$$

if we show that

$$\lim_{T\to\infty} \frac{1}{T}\int_0^T \left(Y_s^* + Z_s^*\right)\begin{bmatrix} m_s \\ u_s \end{bmatrix}' ds = 0 \tag{9.2.44}$$

in quadratic mean. Following the calculations given in Lemma 9.5, one can show that

$$E_\phi\left[\left(Y_0^* + Z_0^*\right)\left(Y_t^* + Z_t^*\right)\right] = e^{t\phi_1} E_\phi\left(Y_0^{*^2} + Z_0^{*^2}\right)$$

$$= \frac{-Ke^{t\phi_1}}{2\phi_1} = \frac{K}{2\pi}\int \frac{e^{i\lambda t}}{\lambda^2+\phi_1^2} d\lambda$$

and

$$E_\phi \left\| \frac{1}{T}\int_0^T \left(Y_s^* + Z_s^*\right)\begin{bmatrix} m_s \\ u_s \end{bmatrix}' ds \right\|^2$$

$$= \frac{1}{T^2}\int_0^T\int_0^T \frac{K}{2\pi}\left(\int_R \frac{e^{i\lambda(s-s')}}{\lambda^2+\phi_1^2} d\lambda\right)\left(m_s m_{s'} + u_s u_{s'}\right) ds ds'$$

$$= \frac{1}{T^2}\int_R d\lambda \int_0^T\int_0^T \frac{K}{2\pi}\frac{e^{i\lambda(s-s')}}{\lambda^2+\phi_1^2}\left(m_s m_{s'} + u_s u_{s'}\right) ds ds'$$

$$= \frac{1}{T}\int_R \frac{K}{2\pi}\frac{1}{\lambda^2+\phi_1^2}\left(\frac{1}{T}\left|\int_0^T e^{i\lambda s} m_s ds\right|^2 + \frac{1}{T}\left|\int_0^T e^{i\lambda s} u_s ds\right|^2\right) d\lambda .$$

Note that the function

$$\frac{1}{\lambda^2+\phi_1^2}$$

is a continuous and bounded function in λ and

$$\lim_{T\to\infty} T E_\phi \left\| \frac{1}{T}\int_0^T (Y_s^* + Z_s^*)\begin{bmatrix} m_s \\ u_s \end{bmatrix}' ds \right\|^2$$

$$= K \int_K \frac{1}{\lambda^2+\phi_1^2}\left(dM_{1,1}^{\phi,u} + dM_{2,2}^{\phi,u}\right) < +\infty$$

where $M^{\phi,u} = ((M_{i,j}^{\phi,u}))_{2\times 2}$. This implies (9.2.44), and hence, (9.2.42).

The relation (9.2.40) is now a consequence of (9.2.42) by observing that $\{u_t\}$ and $\{m_t\}$ are bounded and the inequality

$$E_\phi \left| \frac{1}{T}\int_0^T (Y_s + Z_s)^2 - \frac{1}{T}\int_0^T (Y_s^* + Z_s^*)^2\, ds \right| \le \frac{K\sqrt{2}}{T\phi_1^2} ,$$

which follows from the fact that

$$Y_t^* + Z_t^* = e^{t\phi_1}\left(Y_0^* + Z_0^*\right) + Y_t + Z_t \,.$$

The relation (9.2.41) now follows from the central limit theorem for stochastic integrals (cf. [1, 20]) and the relation

$$\lim_{T\to\infty} \frac{1}{T}\int_0^T E_\phi \begin{bmatrix} Y_s^* + Z_s^* \\ m_s \\ u_s \end{bmatrix}^{2\otimes} ds = \Gamma_u(\phi) ,$$

which is easy to check. This completes the proof of this lemma. ∎

We now study the asymptotic behavior of the log-likelihood ratio $\ell_T(\phi)$ defined by (9.2.14).

THEOREM 9.3

The following results hold with respect to the probability measure P_ϕ:

$$(i)\ \lim_{T\to\infty} \frac{1}{T}\ell_T^{(1)}(\phi) = 0 \ \textit{in quadratic mean} ,$$

(ii) $\lim_{T\to\infty} \frac{1}{T}\ell_T^{(2)}(\phi) = -\boldsymbol{J}_u(\phi)$ *in first mean, and*

(iii) $\lim_{T\to\infty} \frac{1}{\sqrt{T}}\ell_T^{(1)}(\phi) = \nabla_u(\phi)$ *in law*

where the matrix $\boldsymbol{J}_u(\phi)$ *is given by*

$$J_u^{i,j}(\phi) = \boldsymbol{H}_i\Gamma_u(\phi)\boldsymbol{H}_j',\ 1 \le i, j \le 2 \tag{9.2.45}$$

with

$$\boldsymbol{H}_i = C^{-1}\left(\frac{\partial\phi_1}{\partial\phi_i}, \frac{\partial\phi_1}{\partial\phi_i}, \frac{\partial\phi_2}{\partial\phi_i}\right), i = 1, 2$$

and $\Gamma_u(\phi)$ *is as defined by (9.2.40) and* $\nabla_u(\phi)$ *is a Gaussian random vector with mean zero and covariance matrix* $J_u(\phi)$*.*

PROOF As a consequence of Lemma 9.3, we have

$$E_\phi\left\|\frac{\ell_T^{(1)}(\phi)}{T}\right\|^2 = \frac{1}{T^2}\int_0^T C^{-2}E_\phi\,|\pi_s|^2\,ds + \frac{1}{T^2}\int_0^T C^{-2}u_s^2 ds\,. \tag{9.2.46}$$

From the decomposition (9.2.7), it follows that

$$\int_0^T E_\phi\,|\pi_s|^2\,ds \le \alpha e^{2t\phi_1} + \beta e^{t\phi_1} + \gamma t + \delta$$

where α, β, γ, and δ are constants. This bound together with the fact that $\{u_s\}$ is bounded implies (i) in view of (9.2.46). Part (ii) of the theorem is a consequence of (9.2.40) of Lemma 9.7 by noting that $\ell_T^{(2)}(\phi)$ given by (9.2.16) can be written in the form

$$\ell_{T,i,j}^{(2)}(\phi) = -\int_0^T \boldsymbol{H}_i\begin{bmatrix} Y_s + Z_s \\ m_s \\ u_s\end{bmatrix}^{2\otimes} \boldsymbol{H}_j' d\tilde{W}_s,\ i, j = 1, 2$$

and the relation (9.2.41). Simple calculations prove that the covariance matrix of the limiting Gaussian distribution is as stated in the theorem. ∎

We now state and prove the main result describing the asymptotic behavior of a maximum likelihood estimator of the parameter ϕ.

THEOREM 9.4
Let ϕ° denote the true parameter and suppose that the matrix $J_u(\phi^\circ)$ is nonsingular. Then the system is identifiable. Let

$$D_T = \left\{x \in D|\ell_T^{(2)}(x) \text{ is nonsingular}\right\} .$$

Then

$$\lim_{T\to\infty} P_{\phi^0}(D_T) = 1$$

Furthermore, define

$$\hat{\phi}_T(x) = \begin{cases} -\left[\ell_T^{(2)}(x)\right]^{-1} \ell_T^{(1)}(0, x) & \text{if } x \in D_T \\ \hat{\phi} & \text{if } x \notin D_T . \end{cases}$$

Then the family $\{\hat{\phi}_T, T \geq 0\}$ satisfies

$$(i) \quad \lim_{T\to\infty} P_{\phi^o}\left(\ell_T^{(1)}\left(\hat{\phi}_T\right) = 0\right) = 1 ,$$

$$(ii) \quad \lim_{T\to\infty} \hat{\phi}_T = \phi^o \quad \text{in } P_{\phi^o} - \text{probability, and}$$

$$(iii) \quad T^{1/2}\left(\hat{\phi}_T - \phi^o\right) \xrightarrow{\mathcal{L}} N\left(0, \left[J_u\left(\phi^o\right)\right]^{-1}\right) \quad \text{as } T \to \infty$$

under P_{ϕ^o} where $J_u(\phi^o)$ is as defined by (9.2.45) .

PROOF Since $J_u(\phi^o)$ is a covariance matrix, it is nonnegative definite. Since it is nonsingular by hypothesis, the matrix is positive definite. Hence

$$\inf_{v\in R^2, \|v\|=1} \left\{v' J_u\left(\phi^o\right) v\right\} = \gamma^o > 0$$

and

$$v' J_u(\phi^o)\, v \geq \gamma^o \|v\|^2 \text{ for all } v\ .$$

Note that

$$\left\langle \frac{1}{T}\ell_T^{(2)} v, v \right\rangle = \langle -J_u(\phi^o)\, v, v\rangle + \left\langle \left[\frac{1}{T}\ell_T^{(2)} + J_u(\phi^o)\right] v, v \right\rangle ,$$

which implies that

$$\left\langle \frac{1}{T}\ell_T^{(2)} v, v \right\rangle < 0 \text{ for } v \neq 0$$

if

$$\left\| \frac{1}{T}\ell_T^{(2)} + J_u(\phi^o) \right\| < \frac{\gamma^o}{2}\ .$$

Therefore,

$$D_T \supset \left[\left\| \frac{1}{T} L_T^{(2)} + J_u(\phi^o) \right\| < \frac{\gamma^o}{2} \right] .$$

Part (ii) of Theorem 9.3 implies that

$$\lim_{T\to\infty} P_{\phi^o}(D_T) = 1\ .$$

By construction,

$$\left[\ell_T^{(1)}\left(\hat{\phi}_T\right) = 0 \right] \supset D_T\ .$$

Hence, $\lim_{T\to\infty} P_{\phi^o}[\ell_T^{(1)}(\hat{\phi}_T) = 0] = 1$ proving part (i) of the theorem. Observe that

$$\phi^o - \hat{\phi}_T(x) = \begin{cases} \left[\frac{1}{T}\ell_T^{(2)}(x)\right]^{-1} \frac{1}{T}\ell_T^{(1)}(\phi^o, x) & \text{for } x \in D_T \\ \phi^o - \hat{\phi} & \text{for } x \notin D_T\ , \end{cases}$$

which implies part (ii) of the theorem following (i) and (ii) of Theorem 9.3 and the fact that

$$\lim_{T\to\infty} P_{\phi^o}(D_T) = 1\ .$$

Finally part (iii) of the theorem is a consequence of (ii) and (iii) of Theorem 9.3 and the fact

$$T^{1/2}\left(\phi^o - \hat{\phi}_T(x)\right) = \left[\frac{1}{T}\ell_T^{(2)}(x)\right]^{-1} \left[\frac{1}{T^{1/2}}\ell_T^{(1)}(\phi^o, x)\right]$$

for $x \in D_T$. ▮

Estimation of the Characteristics of the Noise

Let p be the Poisson measure related to q by

$$p(\Delta, A) = q(\Delta, A) + |\Delta|\mu(A)$$

where $|\Delta|$ denotes the Lebesgue measure of a Borel subset Δ of R_+.

With the aid of p and with respect to the measure Q, the process $\{\pi_t\}$ can be written in the form

$$\pi_t = A_t + B_t$$

where

$$A_t = CM_t - t \int v \, d\mu(v)$$

and

$$B_t = \int v \, p([0, t], dv) .$$

Estimation of C

Consider the interval $[0, T]$. For every integer N, partition $[0, T]$ into K_N intervals by an increasing sequence t_i^N, $i = 0, 1, \ldots, K_N$ and let

$$\delta_N = \max_{1 \le i \le K_N} \left| t_i^N - t_{i-1}^N \right| .$$

Let

$$G_N = \frac{1}{T} \sum_{i=1}^{K_N} \left| A_{t_i^N} - A_{t_{i-1}^N} \right|^2 .$$

THEOREM 9.5

If $\sum_{N=1}^{\infty} \delta_N < \infty$, *then* $G_N \xrightarrow{a.s.} C^2$ *as* $N \to \infty$ *with respect to the probability measure* P_ϕ^T.

PROOF Let

$$\int v \, d\mu(v) = \lambda .$$

Then it can be checked that

$$E_{Q^T}\left|G_N - C^2\right|^2 = \frac{2C^4}{K_N} + 4\frac{\lambda^2 C^2 T}{K_N^2} + \frac{\lambda^4 T^2}{K_N^2}.$$

This equality implies the convergence in quadratic mean of G_N to C^2 relative to the measure Q^T. Since $\sum_N \delta_N < \infty$ by hypothesis, it follows that

$$\sum_N E_{Q^T}\left|G_N - C^2\right| < \infty,$$

which implies that $G_N \overset{a.s.}{\to} C^2$ as $N \to \infty$ with respect to the measure Q^T, and hence, with respect to the measure P_ϕ^T-a.s., since the two measures are equivalent. ∎

Estimation of μ

Let I be a bounded interval of R not containing the origin. For every $t > 0$, $p([0, t], I)$ represents the number of jumps of the process $\{B_t\}$ with height belonging to I. This random variable has a Poisson distribution with parameter $t\,\mu(I)$ relative to the measure Q and P_ϕ (cf. [6]). The following result holds. We omit the proof.

THEOREM 9.6

With respect to the probability measure P_ϕ,

$$\lim_{t\to\infty} \frac{1}{t} p([0, t], I) = \mu(I).$$

REMARK 9.2 Theorems 9.5 and 9.6 allow us theoretically to estimate C and μ, since, for every $t > 0$,

$$B_s = \sum_{v \le s} (\pi_v - \pi_{v-}) \text{ and } A_s = \pi_s - B_s, s \le t$$

a.s. $[Q^t]$ for every trajectory $x(\cdot)$ observed on $[0, t]$. If μ is finite, it can be shown that, Q^t-a.s., the number of discontinuities of B_s in $[0, t]$ is finite and B_s and

A_s for $s \leq t$ can be computed. Brodeau and Le Breton [4] discuss the choice of the optimal control to improve the precision of the estimators. Le Breton [11] discusses parameter estimation and input design for linear dynamical systems. ∎

9.3 Estimation by the Method of Sieves

Suppose n subjects and p covariates for each subject are observed over [0, 1]. Let $X_i(t)$ denote the state of the ith subject at the time t and suppose that $\boldsymbol{X} = (X_1, \ldots, X_n)$ satisfies

$$\boldsymbol{X}(t) = \boldsymbol{X}(0) + \int_0^t \boldsymbol{Y}(s)\alpha(s)\, ds + \boldsymbol{M}(t), \ 0 \leq t \leq 1\,; \tag{9.3.1}$$

that is,

$$d\boldsymbol{X}(t) = \boldsymbol{Y}(t)\alpha(t)dt + d\boldsymbol{M}(t), \ 0 \leq t \leq 1 \tag{9.3.2}$$

where $\alpha = (\alpha_1, \ldots, \alpha_p)'$ is a vector of unknown nonrandom functions and $\boldsymbol{Y} = ((Y_{ij}))_{n\times p}$ is a matrix of covariate processes, Y_{ij} is the jth covariate for the ith subject, and $\boldsymbol{M} = (M_1, \ldots, M_n)'$ where M_i is a square integrable martingale.

The model (9.3.2) includes several special models.

Example 9.1 **(Diffusion process)**
Suppose

$$dX_t = \alpha(t)b\,(X_t)\ dt + \sigma\,(t, X_t)\,dW_t, \ 0 \leq t \leq 1, \ X_0 = \eta$$

where η is $\mathcal{F}_0$ -measurable. Here $Y(t) = b(X_t)$ and

$$M_t = \int_0^t \sigma\,(s, X_s)\,dW_s\,.$$

□

Example 9.2 **(Point process) (Aalen model)**
Suppose $N = (N(t), \mathcal{F}_t)$ is a point process with $\mathcal{F}_t$-intensity

$$\lambda(t) = \sum_{j=1}^{p} \alpha_j(t) Y_j(t) .$$

Here α_j is unknown, nonnegative, continuous and $Y_j(t)$ is observable, nonnegative, and $\mathcal{F}_t$-predictable for $1 \leq j \leq p$. If $EN(1) < \infty$, then

$$M_t = N_t - \int_0^t \lambda(s) ds$$

is an $\mathcal{F}_t$-martingale and is square integrable. An example for Aalen model is obtained when $\lambda(t)$ is the hazard rate for the incidence of cancer in a subject at age t, $Y_j(t)$ is the cumulative exposure at age t for each of $j = 1, \ldots, p$ carcinogens and N is the point process with a jump at the time of initial detection of cancer. Here $\lambda(\cdot)$ is set to be equal to zero after the cancer is detected and α_i, $1 \leq i \leq p$ represent the changes in the hazard rates for the p carcinogens with age. □

Example 9.3 **(Processes with both diffusion process and point process components)**
Suppose

$$dX(t) = \beta(t) b(X_t)\, dt + \sigma(t, X_t)\, dW_t + \varepsilon dN_t, \quad X_0 = \eta, \quad N_0 = 0, t \geq 0 ,$$

or equivalently,

$$X(t) = \eta + \int_0^t \beta(s) b(X_s)\, ds + \int_0^t \sigma(s, x_s)\, dW_s + \varepsilon N_t ,$$

$$X_0 = \eta, N_0 = 0, \ t \geq 0 .$$

The process $\{X(t)\}$ is a diffusion process between the jump times. The size of the jumps is ε whenever there is a jump. □

For simplicity, let us consider the case $p = 1$ in the model defined by (9.3.2).

Let $(\Omega, \mathcal{F}, P)$ be a complete probability space and $\{\mathcal{F}_{it},\ t \in [0,1]\}$ be a filtration (right-continuous) for $1 \le i \le n$ such that $\mathcal{F}_{i0}$ continuous all P-null sets of $\mathcal{F}$. Further suppose that $M_i \equiv \{M_i(t),\ \mathcal{F}_{it}, t \in [0,1]\}$ is a square integrable martingale with sample paths right continuous and having left limits for $1 \le i \le n$. Let $\langle M_i \rangle$ be the predictable quadratic variation of M_i, that is $\langle M_i \rangle_0 = M_i^2(0)$ and $M_i^2 - \langle M_i \rangle$ is a martingale for $1 \le i \le n$. We further assume that the covariate process Y_i is $\mathcal{F}_{it}$-predictable for $1 \le i \le n$. Note that $\mathcal{F}_{it}$ represents the state and covariate history up to the time t. For simplicity, we assume that the subjects are i.i.d.

The method of sieves consists in obtaining an estimator of $\alpha(\cdot)$ from an increasing sequence of sets of functions indexed by the sample size n.

Let $\{\phi_r, t \ge 1\}$ be a complete orthonormal basis in $L^2[0,1]$. Define

$$\hat{\alpha}^{(n)}(t) = \sum_{r=1}^{d_n} \hat{\alpha}_n^{(n)} \phi_r(t) \tag{9.3.3}$$

where d_n is an increasing sequence of positive integers. Let

$$B_r^{(n)} = \sum_{i=1}^{n} \int_0^1 \phi_r(t) Y_i(t)\, dX_i(t),\ r \ge 1\,, \tag{9.3.4}$$

$$A_{r\ell}^{(n)} = \sum_{i=1}^{n} \int_0^1 \phi_r(t)\phi_\ell(t) Y_i^2(t) dt,\ r \ge 1, \ell \ge 1\,, \tag{9.3.5}$$

$$B^{(n)} = \left(B_1^{(n)}, \ldots, B_{d_n}^{(n)}\right)',\ A^{(n)} = \left(A_{r\ell}^{(n)}\right)_{d_n \times d_n} \text{ and}$$

$$\hat{\alpha}^{(n)} = \left(\hat{\alpha}_1^{(n)}, \ldots, \hat{\alpha}_{d_n}^{(n)}\right)' \tag{9.3.6}$$

where

$$\hat{\alpha}^{(n)} = A^{(n)^-} B^{(n)} \tag{9.3.7}$$

and $A^{(n)^-}$ is a generalized inverse of $A^{(n)}$.

Suppose the following conditions hold:

(C1) $\int_0^1 \alpha^2(t)dt < \infty$,

(C2) $\sup_t EY_1^2(t) < \infty$,

(C3) $\inf_t EY_1^2(t) > 0$,

and

(C4) $\nu(t) = E\left[\int_0^t Y_1^2(s)d\langle M_1\rangle_s\right]$, $t \in [0, 1]$ is absolutely continuous

with bounded derivative (a.e., Lebesgue).

Note that

$$E\left[\int_0^1 \phi_r(t)Y_1(t)dM_1(t)\right]^2 = E\left[\int_0^1 \phi_r^2(t)\, Y_1^2(t)\, d\, \langle M_1\rangle_t\right]$$

$$= \int_0^1 \phi_r^2(t)\, d\nu(t) < \infty. \tag{9.3.8}$$

Hence,

$$\int_0^1 \phi_r(t)Y_1(t)\, dM_1(t)$$

is well defined. Define a measure μ by

$$d\mu(t) = EY_1^2(t)\, dt.$$

Then $L^2([0,1],dt) = L^2([0,1], d\mu(t))$ and the norms are equivalent and there exists a complete orthonormal sequence $\{\psi_r, r \geq 1\}$ in $L^2([0,1]), d\mu(t))$ such that

$$\text{span } \{\psi_r, 1 \leq r \leq d_n\} = \text{ span } \{\phi_r, 1 \leq r \leq d_n\}, \; n \geq 1 .$$

Denote the coordinates of α and $\hat{\alpha}^{(n)}$ in the basis $\{\psi_r, \; r \geq 1\}$ by $\xi_r, r \geq 1$ and $\hat{\xi}_r^{(n)}, 1 \leq r \leq d_n$. Let

$$\boldsymbol{\xi}^{(n)} = (\xi_1, \ldots, \xi_{d_n}, 0, 0, \ldots), \boldsymbol{\xi} = (\xi_1, \xi_2, \ldots) , \tag{9.3.9}$$

and

$$\hat{\boldsymbol{\xi}}^{(n)} = \left(\hat{\xi}_1^{(n)}, \ldots, \hat{\xi}_{d_n}^{(n)}, 0, 0, \ldots\right) . \tag{9.3.10}$$

Note that

$$\begin{aligned} \left\|\hat{\alpha}^{(n)} - \alpha\right\|_2 &= \left\|\hat{\boldsymbol{\xi}}^{(n)} - \boldsymbol{\xi}\right\|_2 \\ &\leq \left\|\hat{\boldsymbol{\xi}}^{(n)} - \boldsymbol{\xi}^{(n)}\right\|_2 + \left\|\boldsymbol{\xi}^{(n)} - \boldsymbol{\xi}\right\|_2 . \end{aligned} \tag{9.3.11}$$

We shall denote $(\hat{\xi}_1^{(n)}, \ldots, \hat{\xi}_{d_n}^{(n)})'$ and $(\xi_1, \ldots, \xi_{d_n})'$ also by $\hat{\boldsymbol{\xi}}^{(n)}$ and $\boldsymbol{\xi}^{(n)}$, respectively, whenever they are considered as vectors in R^{d_n}.

THEOREM 9.7
Under the conditions (C1)–(C4), for any $d_n \uparrow \infty$ such that $d_n = o(n)$,

$$\left\|\hat{\alpha}^{(n)} - \alpha\right\|_2^2 = \int_0^1 \left|\hat{\alpha}^{(n)}(t) - \alpha(t)\right|^2 dt \xrightarrow{p} 0 \; as \; n \to \infty . \tag{9.3.12}$$

Before we give a proof of this theorem, we find further relations between $\boldsymbol{\xi}$ and α. Note that

$$\hat{\boldsymbol{\xi}}^{(n)} = \boldsymbol{a}^{(n)^-} \boldsymbol{b}^{(n)} \tag{9.3.13}$$

where $\boldsymbol{a}^{(n)^-}$ is a generalized inverse of $\boldsymbol{a}^{(n)} = ((a_{rl}^{(n)}))_{d_n \times d_n}$,

$$a_{rl}^{(n)} = n^{-1} \sum_{i=1}^{n} \int_0^1 \psi_r(t)\psi_\ell(t) Y_i^2(t) dt , \tag{9.3.14}$$

$$b_r^{(n)} = n^{-1} \sum_{i=1}^{n} \int_0^1 \psi_r(t) Y_i(t) dX_i(t) , \tag{9.3.15}$$

$$b^{(n)} = \left(b_1^{(n)}, \ldots, b_{d_n}^{(n)}\right)' , \tag{9.3.16}$$

and

$$\hat{\xi}^{(n)} = \left(\hat{\xi}_1^{(n)}, \ldots, \hat{\xi}_{d_n}^{(n)}\right)' . \tag{9.3.17}$$

Note that $\boldsymbol{\xi}^{(n)} = (\xi_1, \ldots, \xi_{d_n})'$ and

$$\alpha^{(n)} = \sum_{r=1}^{d_n} \xi_r \psi_r . \tag{9.3.18}$$

Applying the definition of $\boldsymbol{X}(t)$, it can be checked that

$$\hat{\boldsymbol{\xi}}^{(n)} - \boldsymbol{\xi}^{(n)} = \boldsymbol{a}^{(n)^-} \mathbf{c}^{(n)} \tag{9.3.19}$$

where

$$\mathbf{c}^{(n)} = \left(c_1^{(n)}, \ldots, c_{d_n}^{(n)}\right)' , \tag{9.3.20}$$

and

$$c_r^{(n)} = n^{-1} \sum_{i=1}^{n} \int_0^1 \psi_r(t) Y_i^2(t) \left(\alpha(t) - \alpha^{(n)}(t)\right) dt$$

$$+ n^{-1} \sum_{i=1}^{n} \int_0^1 \psi_r(t) Y_i(t) dM_i(t) . \tag{9.3.21}$$

LEMMA 9.8
Under the conditions (C1) to (C4),

$$\left\|\mathbf{c}^{(n)}\right\| \xrightarrow{p} 0 \quad as\ n \to \infty .$$

PROOF Let

$$c_r^{(n)} = \gamma_r^{(n)} + \eta_r^{(n)} + \rho_r^{(n)} , \tag{9.3.22}$$

where

$$\gamma_r^{(n)} = \frac{1}{n}\sum_{i=1}^{n}\left(\tilde{\gamma}_{ri}^{(n)} - E\tilde{\gamma}_{ri}^{(n)}\right) , \tag{9.3.23}$$

$$\tilde{\gamma}_{ri}^{(n)} = \int_0^1 \psi_r(t)Y_i^2(t)\left(\alpha(t) - \alpha^{(n)}(t)\right) dt . \tag{9.3.24}$$

$$\eta_r^{(n)} = E\tilde{\gamma}_{r1}^{(n)} , \tag{9.3.25}$$

and

$$\rho_r^{(n)} = n^{-1}\sum_{i=1}^{n}\int_0^1 \psi(t)Y_i(t)dM_i(t) . \tag{9.3.26}$$

Let

$$\eta^{(n)} = \left(\eta_1^{(n)}, \ldots, \eta_{d_n}^{(n)}\right)' . \tag{9.3.27}$$

Now

$$\left\|\eta^{(n)}\right\|^2 = \sum_{r=1}^{d_n}\left[E\tilde{\gamma}_{r1}^{(n)}\right]^2 = \sum_{r=1}^{d_n}\left(\int_0^1 \psi_r(t)\left(\alpha(t) - \alpha^{(n)}(t)\right) EY_1^2(t)dt\right)^2$$

$$= \sum_{r=1}^{d_n}\left(\int_0^1 \psi_r(t)\left(\alpha(t) - \alpha^{(n)}(t)\right) d\mu(t)\right)^2$$

$$\le \int_0^1 \left(\alpha(t) - \alpha^{(n)}(t)\right)^2 d\mu(t)$$

$$= \int_0^1 \left(\alpha(t) - \alpha^{(n)}(t)\right)^2 EY_1^2(t)dt \,. \tag{9.3.28}$$

Since $\alpha^{(n)} \to \alpha$ in $L^2([0, 1]), dt)$ and $\sup_t EY_1^2(t) < \infty$, it follows that

$$\left\|\eta^{(n)}\right\| \to 0 \text{ as } n \to \infty \,. \tag{9.3.29}$$

Note that $E(\gamma_r^{(n)}) = 0$ and

$$\begin{aligned} \operatorname{Var}\left(\gamma_r^{(n)}\right) &= \frac{1}{n}\operatorname{Var}\left[\tilde{\gamma}_{r1}^{(n)}\right] \quad \text{(by the i.i.d. nature of } X_i) \\ &\le \frac{1}{n}E\left[\tilde{\gamma}_{r1}^{(n)}\right]^2 \end{aligned}$$

and hence, for $\gamma_n = (\gamma_1^{(n)}, \ldots, \gamma_{d_n}^{(n)})'$,

$$\begin{aligned} E\left\|\gamma^{(n)}\right\|^2 &= \sum_{r=1}^{d_n} E\left(\gamma_r^{(n)}\right)^2 = \sum_{r=1}^{d_n} \operatorname{Var}\left(\gamma_r^{(n)}\right) \\ &\le \frac{1}{n}\sum_{r=1}^{d_n} E\left[\tilde{\gamma}_{r1}^{(n)}\right]^2 \\ &\le \frac{1}{n}\int_0^1 \left(\alpha(t) - \alpha^{(n)}(t)\right)^2 EY_1^2(t)dt \quad \to 0 \,. \end{aligned} \tag{9.3.30}$$

Note that

$$E\left[\int_0^1 \psi_r(t)Y_i(t)dM_i(t)\right]^2 = E\left[\int_0^1 \psi_r^2(t)Y_i^2(t)d\left\langle M_i\right\rangle_t\right]$$

$$= \int_0^1 \psi_r^2(t)dv(t) ,$$

which is uniformly bounded in r. It can be checked that

$$E \left\| \rho^{(n)} \right\|^2 = o(1) \text{ for } \rho^{(n)} = \left(\rho_1^{(n)}, \ldots, \rho_{d_n}^{(n)} \right)' . \tag{9.3.31}$$

Combining (9.3.29) to (9.3.31), we have

$$E \left\| c^{(n)} \right\|^2 \leq 3 \left(E \left\| \gamma^{(n)} \right\|^2 + E \left\| \rho^{(n)} \right\|^2 + \left\| \eta^{(n)} \right\|^2 \right)$$

$$= o(1) . \tag{9.3.32}$$

Therefore, $\|c^{(n)}\| \xrightarrow{p} 0$ as $n \to \infty$. ∎

LEMMA 9.9
The sequence $\{\|a^{(n)^-}\|, n \geq 1\}$ is a tight sequence of random variables and $P\{a^{(n)}$ is invertible$\} \to 1$ as $n \to \infty$ under the conditions (C1) to (C4).

PROOF (Sketch) Let

$$a^{(n)} = \beta^{(n)} + \zeta^{(n)} , \tag{9.3.33}$$

where

$$\beta_{r\ell}^{(n)} = n^{-1} \sum_{i=1}^{n} \left(\tilde{\beta}_{r\ell}^{(i)} - E\tilde{\beta}_{r\ell}^{(i)} \right) , \tag{9.3.34}$$

$$\tilde{\beta}_{r\ell}^{(i)} = \int_0^1 \psi_r(t)\psi_\ell(t)Y_i^2(t)dt , \tag{9.3.35}$$

and

$$\zeta_{r\ell}^{(n)} = E\tilde{\beta}_{r\ell}^{(1)} . \tag{9.3.36}$$

It can be shown as in Lemma 9.8 that

$$E\left\|\beta^{(n)}\right\|^2 = O\left(d_n n^{-1}\right) = o(1)\,. \tag{9.3.37}$$

Let $I^{(n)}$ denote identity matrix of order $d_n \times d_n$. Note that

$$\zeta_{r\ell}^{(n)} = E\tilde{\beta}_{r\ell}^{(1)} = \int_0^1 \psi_r(t)\psi_\ell(t)\, d\mu(t) = \delta_{r\ell}\,.$$

Hence, $\zeta^{(n)} = I^{(n)}$ where $I^{(n)}$ is the identity matrix of order $d_n \times d_n$. ∎

Therefore, $\|a^{(n)} - I^{(n)}\| \xrightarrow{p} 0$ as $n \to \infty$ by (9.3.37), that is, $P(a^{(n)}$ is invertible) $\to 1$ as $n \to \infty$, and for any $0, < c < 1$,

$$P\left(\left\|a^{(n)^-}\right\| < (1-c)^{-1}\right) \to \quad \text{as } n \to \infty\,.$$

PROOF of Theorem 9.7: Lemmas 9.8 and 9.9 prove the Theorem 9.7 on the L_2-consistency in view of (9.3.19). ∎

REMARK 9.3

(i) One can obtain the asymptotic distribution for the integrated estimator

$$\hat{A}^{(n)}(t) = \int_0^t \hat{\alpha}^{(n)}(s)ds$$

(cf. [15]).

(ii) McKeague [15] considered smoothened sieve estimators of the type

$$\tilde{\alpha}(t) = \frac{1}{h_n}\int_0^1 K\left(\frac{t-s}{h_n}\right)\hat{\alpha}^{(n)}(s)ds$$

and obtained their rates of convergence.

(iii) Results in Theorem 9.7 can be extended to the case when the subjects are not i.i.d. but mixing under suitable conditions on the mixing coefficient (cf. [14]).

∎

9.4 Nonlinear Semimartingale Regression Models

Let $(\Omega, \mathcal{F}, P)$ be a complete probability space and $\{\mathcal{F}_t, 0 \leq t \leq 1\}$ be a right-continuous filtration with $\mathcal{F}_0$ containing all P-null sets in $\mathcal{F}$. Suppose the process $\{M_t, \mathcal{F}_t\}$ is a zero-mean L^4-martingale with sample paths that are right-continuous with left limits on $[0, 1]$. Let $\langle M \rangle$ be the quadratic characteristic and $[M]$ be the quadratic variation of M. Suppose that Y and Z are predictable processes with Y an indicator process. For simplicity, we assume that Z is a process taking values in the interval $[0, 1)$.

Consider a nonlinear semimartingale regression model in which a process X is related to the indicator process Y and the covariate process Z by

$$X_t = X_0 + \int_0^t \lambda_s ds + M_s, \; 0 \leq t \leq 1 , \tag{9.4.1}$$

$$\lambda_t = Y_t \, \alpha(t, Z_t) \tag{9.4.2}$$

where α is an unknown function. Here Y is an indicator process taking the value 1 when X and Z are observed and 0 otherwise.

Equation (9.4.1) can also be written in the form

$$dX_t = Y_t \; \alpha(t, Z_t) \, dt + dM_t, \; 0 \leq t \leq 1 . \tag{9.4.3}$$

We assume that α is Lipschitzian and

$$\langle M \rangle_t = \int_0^t \gamma(t, Z_s) \; Y_s ds \tag{9.4.4}$$

where γ is a continuous function. Let

$$A(t, z) = \int_0^t \alpha(s, z)ds \,. \tag{9.4.5}$$

The function $A(t, z)$ is called the *conditional cumulative hazard function for a fixed level* z. Let

$$\mathcal{A}(t, z) = \int_0^z \int_0^t \alpha(s, x)dsdx = \int_0^z A(t, x)dx \,. \tag{9.4.6}$$

The function $\mathcal{A}(t, z)$ is called the *doubly cumulative hazard function.*

The problem considered here is the estimation of $\mathcal{A}(t, z)$ based on the process (X_i, Y_i, Z_i, M_i), $1 \leq i \leq n$ and a filtration $\{\mathcal{F}_t^{(n)}\}$. Suppose the process M_i, $1 \leq i \leq n$ are orthogonal $\{\mathcal{F}_t^{(n)}\}$-martingales. This is the case if $\mathcal{F}_t^{(n)} = \mathcal{F}_{1t} \vee \mathcal{F}_{2t} \vee \ldots \vee \mathcal{F}_{nt}$ where $\mathcal{F}_{it}$, $1 \leq i \leq n$ are independent filtrations and each M_i is an $\mathcal{F}_{it}$-martingale.

We first estimate $A(\cdot, x)$ by stratifying over the covariate and then integrate with respect to x. Let

$$\mathcal{I}_r = [x_{r-1}, x_r) \,, \ 1 \leq r \leq d_n \tag{9.4.7}$$

be the strata where $x_r = \frac{r}{d_n}$ and d_n is an increasing sequence of positive integrals. Note that the width of each stratum is $\omega_n = \frac{1}{d_n}$. Let us denote $\mathcal{I}_z = \mathcal{I}_r$ if $z \in \mathcal{I}_r$. Note that $\mathcal{I}_z$ depends on n. Let $X_i(t, z)$ denote the contribution of $X_i(t)$ from the stratum $\mathcal{I}_z$:

$$X_i(t, z) = \int_0^t \chi\left(Z_i(s) \in \mathcal{I}_z\right) \, dX_i(s) \tag{9.4.8}$$

and set

$$X^{(n)}(s, z) = \sum_{i=1}^n X_i(s, z) \,. \tag{9.4.9}$$

Here $\chi(A)$ denotes the indicator function of the set A. The number of covariate processes observed to be in stratum $\mathcal{I}_z$ at the time s is given by

$$Y^{(n)}(s, z) = \sum_{i=1}^{n} \chi\left(Z_i(s) \in \mathcal{I}_z\right) Y_i(s). \tag{9.4.10}$$

An estimator for $A(t, z)$ is

$$\hat{A}(t, z) = \int_0^t \frac{X^{(n)}(ds, z)}{Y^{(n)}(s, z)}$$

(cf. [3]) (here we interpret $1/0 \equiv 0$). We assume that the stratum width ω_n tends to zero as $n \to \infty$ at a suitable rate which we will specify later. An estimator for the doubly cumulative hazard function $\mathcal{A}(t, z)$ is defined by

$$\hat{\mathcal{A}}(t, z) = \int_0^z \hat{A}(t, x)dx .$$

Suppose the following conditions hold:

(A1) there exists a nonnegative continuous function $\phi(\cdot, \cdot)$ such that $\int\int_{[0,1]^2} \left|\frac{n\omega_n}{Y^{(n)}(s,x)} - \phi(s, x)\right| ds\, dx \xrightarrow{p} 0$,

(A2) the two dimensional Lebesgue measure of the set $\{(s, x) \in [0, 1]^2 : Y^{(n)}(s, x) = 0\}$ is $o_p\left(\frac{1}{\sqrt{n}}\right)$, and

(A3) $\sup_{s,x,n} E\left[\frac{n\omega_n}{Y^{(n)}(s, x)}\right]^3 < \infty.$

Let $W = \{W(t, z) : (t, z) \in [0, 1]^2\}$ be a two-parameter Wiener process, that is, a Gaussian process with zero mean and $EW(t, z)W(t', z') = \min(t, t')\min(z, z')$. For $\psi \in L^2([0, 1]^2, ds\, dx)$ define

$$\int_0^t \int_0^z \psi(s, x)dW(s, x)$$

as a continuous version of the Wiener integral (cf. [2, 8, 22]). Let C_2 be the space of continuous functions on $[0, 1]^2$ with the supremum norm and D_2 denote the extension of $D[0, 1]$ to $[0, 1]^2$ (cf. [18, 19]).

The following theorem is due to McKeague and Utikal [16]. We omit the proof of this theorem.

THEOREM 9.8

Suppose that the conditions (A1) to (A3) hold. Further suppose that $n\omega_n^2 \to 0$ and $d_n = o(n)$. In addition assume that either X is a counting process or X has continuous sample paths. Let $h = \gamma\phi$. Then

$$\sqrt{n}\left(\hat{\mathcal{A}} - \mathcal{A}\right) \xrightarrow{\mathcal{D}} m$$

in D_2 as $n \to \infty$ where

$$m(t, z) = \int_0^t \int_0^z \sqrt{h(s, x)}\, dW(s, x)\,.$$

REMARK 9.4

(i) Note that m is a continuous Gaussian random field with mean zero and covariance function

$$\operatorname{Cov}\left(m\left(t_1, z_1\right),\ m\left(t_2, z_2\right)\right) = \int_0^{z_1 \wedge z_2} \int_0^{t_1 \wedge t_2} h(s, x)\, ds dx\,.$$

(ii) McKeague and Utikal [16] develop tests for independence of X from the covariate Z, a test for homogenity of α, and a goodness of fit test for the general proportional hazards model $\alpha(t, z) = \alpha_1(t)\alpha_2(z)$ from the above result.

(iii) Important examples of the model (9.4.3) are the counting processes (for which $\gamma = \alpha$) and the diffusion processes (for which α is the drift and γ is the infinitesimal variance) and $Y \equiv 1$.

∎

References

[1] Basawa, I.V. and Prakasa Rao, B.L.S. (1980) *Statistical Inference for Stochastic Processes,* Academic Press, London.

[2] Bass, R.F. (1988) Probability estimates for multiparameter Brownian motion, *Ann. Probab.,* **16**, 251–264.

[3] Beran, R. (1981) Nonparametric regression with randomly censored survival data, *Tech. Report,* University of California, Berkeley.

[4] Brodeau, F. and Le Breton, A. (1979) Identification de parametres pour un systeme excite par des bruits Gaussien and Poissonien, *Ann. Inst. Henri Poincaré,* **15**, 1–23.

[5] Christopeit, N. (1986) Quasi-least-squares estimation in semimartingale regression models, *Stochastics,* **16**, 255–278.

[6] Gikhman, I.I. and Skorokhod, A.V. (1969) *Introduction to the Theory of Random Processes,* Saunders, New York.

[7] Gikhman, I.I. and Skorokhod, A.V. (1972) *Stochastic Differential Equations,* Springer, Berlin.

[8] Ito, K. (1951) Multiple Wiener integral, *J. Math. Soc. Japan,* **3**, 157–169.

[9] Kholevo, A.S. (1969) On estimates of regression coefficients, *Theor. Probab. Appl.,* **14**, 79–104.

[10] Lai, T.L. and Wei, C.Z. (1982) Least squares estimation in stochastic regression models with applications to identification and control of dynamical systems, *Ann. Statist.,* **10**, 154–166.

[11] Le Breton, A. (1977) Parameter estimation and input design in a linear dynamical system, *Report de Recherches,* No. 86, Inst. Rech. Math. Avan., Grenoble, France.

[12] Liptser, R.S. and Shiryayev, A.N. (1977) *Statistics of Random Processes: General Theory,* Springer, Berlin.

[13] Liptser, R.S. and Shiryayev, A.N. (1980) A functional central limit theorem for semimartingales, *Theor. Probab. Appl.*, **25**, 667–688.

[14] McKeague, I.W. (1986) Estimation for a semimartingale regression model using the method of sieves, *Ann. Statist.*, **14**, 579–589.

[15] McKeague, I.W. (1987) Asymptotic theory for sieve estimators in semimartingale regression models, *Tech. Report,* Florida State University.

[16] McKeague, I.W. and Utikal, K.J. (1990) Identifying nonlinear covariate effects in semimartingale regression models, *Probab. Th. Rel. Fields,* **87**, 1–25.

[17] Métivier, M. (1982) *Semimartingales,* Walter de Gryuter, Berlin.

[18] Neuhaus, G. (1971) On weak convergence of stochastic processes with multidimensional time parameter, *Ann. Math. Statist.*, **42**, 1285–1295.

[19] Prakasa Rao, B.L.S. (1975) Tightness of probability measures generated by stochastic processes on metric spaces, *Bull. Inst. Math. Acad. Sinica,* **3**, 353–367.

[20] Taraskin, A.F. (1974) Some limit theorems for stochastic integrals, *Theory of Stochastic Processes,* **1**, 136–151.

[21] Van Schuppen , J.H. (1973) Estimation theory for continuous time processes, a martingale approach, *Tech. Report,* No. ERL-M405, University of California.

[22] Wong, E. and Zakai, M. (1974) Martingale and stochastic integrals for processes with a multidimensional parameter, *Z. Warsch. verw Gebiete,* **51**, 109–122.

Chapter 10

Applications to Stochastic Modeling

10.1 Introduction

Traditional applications of diffusion processes for stochastic modeling can be found in [10]. Some recent applications to mathematical finance, forest management, and market policy involving modeling by diffusion-type processes are discussed in [14] with main emphasis on the statistical inference for such models. Numerical approximation methods for simulation of stochastic differential equations are also discussed in [14]. We are not giving the applications of techniques of counting processes to survival analysis and reliability, as there are books covering these areas extensively (cf. [1, 4, 11]). We will now discuss a couple of applications involving stochastic modeling through semimartingales.

10.2 Applications to Engineering and Economic Systems

For modeling the random evolution of different economic and engineering systems, the theory of continuous time stochastic processes is indispensable, as we have seen in the earlier sections. Modeling through stochastic differential equations driven by a Wiener process has been the subject of discussion until now. However, there are situations where the discrete changes in the state of the system cannot be modeled by diffusion alone, and the inclusion of jump component in the model might be appropriate (cf. [12]).

Suppose the process $\{X(t), t \geq 0\}$ satisfies the equation

$$dX(t) = f(X,t;\alpha)dt + g(X,t;\beta)dW(t) + h(X,t;\gamma)dN(t), t \geq 0 \quad (10.2.1)$$

where $\{W(t)\}$ is the standard Wiener process and $\{N(t)\}$ is a Poisson process (jump size =1) independent of $\{W(t)\}$ with intensity λ.

In addition to the condition which ensure the existence and uniqueness of the solution to (10.2.1) (cf. [2, Ch. 6]), suppose the following conditions hold.

(A1) (i) The coefficient functions f, g and h are known up to parameter vectors α, β, γ, and λ. The true but unknown parameters $\alpha_0, \beta_0, \gamma_0$, and λ_0 lie in the interior of a compact parameter space $\Theta = A \otimes B \otimes \Gamma \otimes \Lambda$. Let $\theta_0 = (\alpha_0, \beta_0, \gamma_0, \lambda_0)$.

(ii) The functions f, g, and h are twice continuously differentiable in (x, t) and three times continuously differentiable in θ.

Suppose the process $\{X(t)\}$ is observed at the times $t_1, t_2, \ldots, t_n$ *not* necessarily equally spaced where $0 < t_1 < \cdots < t_n$. Let $X_i = X(t_i)$, $1 \leq i \leq n$ and $X(t_0) = x_0$. Suppose x_0 is known.

Since the processes $\{W_t\}$ and $\{N_t\}$ are independent, the probability distribution of $(X(t_1), \ldots, X(t_n))$ is absolutely continuous with respect to the Lebesgue measure on R^n. Let $\rho(x_1, \ldots x_n)$ be its joint density. Given f, g and h, the joint density function $\rho(x)$ of the sample $(X(t_1), \ldots, X(t_n))$ can be derived from the relation

$$\rho(x_1, \ldots, x_n) = \rho_1(x_1)\rho_2(x_2|x_1)\ldots\rho_n(x_n|x_{n-1})$$

due to the Markov property of the process $\{X(t)\}$ where $\rho_k(x_k|x_{k-1}) \equiv \rho_k(x_k, x_{k-1})$ denotes the transition density. Observe that the transition density ρ_k can be obtained by solving the Fokker–Planck backward equation or forward equation. For the process (10.2.1), this equation can be derived in the following way heuristically.

Let ψ be any infinitely differentiable function. Applying Ito's lemma, we have

$$d\Psi = \left(\Psi_X f + \frac{1}{2}\Psi_{XX} g^2\right)dt + \Psi_X g dW$$

$$+ [\Psi(X+h) - \Psi(X)]dN \quad (10.2.2)$$

where

$$\Psi_x \equiv \frac{\partial \Psi}{\partial x}, \Psi_{xx} = \frac{\partial^2 \Psi}{\partial x^2} . \tag{10.2.3}$$

Let $D_{P,k}$ be the Dynkin operator at the time t_k, that is,

$$D_{P,k} = \frac{d}{dt} E_{t_k}[\cdot] \tag{10.2.4}$$

where $E_{t_k}[\cdot]$ is the conditional expectation with respect to the probability measure P given X_{t_k}. Applying the same to Ψ, we have

$$D_{P,k}(\Psi) = E_{t_k}\left[\Psi_X f + \frac{1}{2}\Psi_{XX} g^2\right] + \lambda E_{t_k}[\Psi(X+h) - \Psi(X)]$$

$$= \int_\Omega \left\{\Psi_x f + \frac{1}{2}\Psi_{xx} g^2 + \lambda(\Psi(x+h) - \Psi(x))\right\} \rho_{k+1}\left(x, X_{t_k}\right) dx$$

$$= \int_\Omega \left\{-\Psi \frac{\partial}{\partial x}(f\tilde{\rho}_k) + \frac{1}{2}\Psi \frac{\partial^2}{\partial x^2}\left(g^2 \tilde{\rho}_k\right) - \Psi\lambda\tilde{\rho}_k\right\} dx$$

$$+ \lambda \int_\Omega \Psi(x+h)\tilde{\rho}_k dx$$

where $\tilde{\rho}_k \equiv \tilde{\rho}_k(x, t_k) = \rho_{k+1}(x, X(t_k))$. Let

$$y = \tilde{h}(x, t, \gamma) = h(x, t, \gamma) + x . \tag{10.2.5}$$

Suppose $\tilde{h}$ is an *onto* map of Ω to Ω for all (t, γ) and further suppose that

$$\left|\frac{\partial}{\partial x}(\tilde{h}) + 1\right| \neq 0 \quad \text{for all} \quad (t, \gamma) \quad \text{and} \quad x \in \Omega . \tag{10.2.6}$$

Applying the implicit function theorem guaranteeing the inverse map $x = \tilde{h}^{-1}(y, t, \gamma)$ and applying the change of variable techniques, we have

$$\int_{\Omega} \Psi(x+h)\tilde{\rho}_k(x, t_k)\,dx \tag{10.2.7}$$

$$= \int_{\Omega} \Psi(y)\tilde{\rho}_k(\tilde{h}^{-1}(y, t_k, \gamma), t_k) \left| \frac{\partial}{\partial y}\left(\tilde{h}^{-1}(y, t_k, \gamma)\right) \right| dy\,.$$

Let

$$\rho_k^* \equiv \tilde{\rho}_k\left(\tilde{h}^{-1}\left(\tilde{h}^{-1}(y, t_k, \gamma), t_k\right)\right)\,.$$

Therefore

$$D_{P,k}(\Psi) = \int_{\Omega} \left\{ -\frac{\partial}{\partial x}(f\tilde{\rho}_k) + \frac{1}{2}\frac{\partial^2}{\partial x^2}\left(g^2\tilde{\rho}_k\right) - \lambda\tilde{\rho}_k \right.$$

$$\left. + \lambda\rho_k^* \left| \frac{\partial}{\partial x}\tilde{h}^{-1} \right| \right\} \Psi(x)dx\,. \tag{10.2.8}$$

Assuming that $\Psi(x)\tilde{\rho}_k(x,t)$ is continuous on $\Omega \times [0,\infty)$, the function $D_{P,k}(\Psi)$ may also be calculated from the relation

$$D_{P,k}(\Psi) = \frac{d}{dt}E_{t_k}(\Psi)$$

$$= \int_{\Omega} \Psi(x)\frac{\partial}{\partial t}\left[\tilde{\rho}_k(x, t_k)\right] dx\,. \tag{10.2.9}$$

Equating (10.2.8) and (10.2.9) and noting that these relations hold for all smooth function Ψ, it follows that

$$\frac{\partial}{\partial t}\left[\tilde{\rho}_k\right] = -\frac{\partial}{\partial x}\left[f\tilde{\rho}_k\right] + \frac{1}{2}\frac{\partial^2}{\partial x^2}\left[g^2\tilde{\rho}_k\right]$$

$$- \lambda\tilde{\rho}_k + \lambda\rho_k^* \left| \frac{\partial}{\partial x}\left(\tilde{h}^{-1}\right) \right| \tag{10.2.10}$$

with the boundary condition

$$\rho_k\left(x, t_{k-1} \mid x_{k-1}, t_{k-1}\right) = \delta\left(x - x_{k-1}\right) \qquad (10.2.11)$$

where $\delta(x - x_{k-1})$ is the Dirac delta function centered at x_{k-1}.

In general, it is difficult to solve (10.2.11) for ρ_k for general f, g, and h. If $h = 0$, then the process X reduces to a diffusion process and the estimation of parameters for such processes is discussed in [3].

Example 10.1 **(Combined lognormal and jump process)**
Let $\{X(t)\}$ be the solution of the stochastic differential equation

$$dX(t) = \alpha_o X(t)dt + \beta_o X(t)dW(t) + \gamma_0 X(t)dN(t)\,. \qquad (10.2.12)$$

By using the log-transformation $Y = log X$ and the generalized Ito's lemma, it can be seen that $\{Y(t)\}$ is a solution of

$$dY(t) = \left(\alpha_0 - \frac{1}{2}\beta_0^2\right)dt + \beta_0 dW(t) + \log\left(1 + \gamma_0\right)dN(t),\ t \geq 0\,. \quad (10.2.13)$$

The likelihood function can be derived by computing the convolution of the densities of the diffusion and the Poisson components. Let V and Z be the corresponding diffusion and the jump components. Then

$$dY = dV + dZ\,.$$

The conditional likelihood of Y can be obtained by calculating the convolution of $\rho_V(v, t)$ and $\rho_Z(z, t)$ where

$$\rho_V(v, t) = \frac{1}{\sqrt{2\pi\sigma^2 t}}\exp\left[-\frac{(v - v_0 - \mu t)^2}{2\sigma^2 t}\right], \sigma = \beta_0, \mu = \alpha_0 - \frac{1}{2}\beta_o^2$$

and

$$\rho_Z(z, t) = c\sum_{k=o}^{\infty} e^{-\lambda t}\frac{(\lambda t)^k}{k!}\delta(cz - k), c = \frac{1}{\log\left(1 + \gamma_0\right)}$$

with $\delta(\cdot)$ as the Dirac delta function. After some computations, the convolution reduces to

$$p\left(y_k, t_k \mid y_{k-1}, t_{k-1}\right) = \sum_{j=0}^{\infty} e^{-\lambda} \frac{\lambda^j}{j!} \phi$$

$$\left(\frac{y_k - y_{k-1} - \log\left(1 - \gamma_0\right)^j - \left(\alpha_0 - \frac{1}{2}\beta_0^2\right)\Delta t_k}{\beta_0 \sqrt{\Delta t_k}} \right)$$

where ϕ is the standard normal density function.

Approximate MLE of the parameters can be completed by using a finite number of terms from the above series. □

10.3 Application to Modeling of Neuron Movement in a Nervous System

Information Processing in the Nervous System [15]

The central nervous system (CNS) can be viewed as a communication system that receives, processes, and transmits a large amount of information. Some degree of uncertainty is inherent in the behavior of the neural communication system due to its anatomical and functional complexity and the nondeterministic nature of the responses of its components to identical stimuli or experience.

The base unit of the nervous system that receives and transmits information is the nerve cell or *neuron.* A neuron has three morphological regions: the cell body (or *soma*), the *dendrites,* and the *axon.* The soma contains the nucleus and many of the organelles involved in the metabolic processes. The dendrites form a series of branched protoplasmic tree-like outgrowths from the cell body. The dendrites and the soma are the sites of the junctions where the signals are received from other neurons. The axon is a typically a protoplasmic extension that exists in the soma at the initial segment. Near its end, the axon branches into numerous axonal terminals that are responsible for transmitting signals from the neuron to other parts of the system. The junction between two neurons is called the *synapse,* which is an anatomically specialized junction between two neurons where the electrical activity in one neuron influences the activity of the other.

The electric activity in the nervous system is due to the presence of organic as well as inorganic electrically charged ions. These ions are present outside as well as inside the cell. The cell's membrane is selectively permeable to different ions. This leads to a difference in concentration of ions on both sides of the membrane, which in turn leads to a difference in potential across the membrane. The ions are transmitted across the membrane through structures which are called channels. These channels exist in active or inactive states (cf. [13]). Ionic channels open and close in a stochastic manner. The state of the neuron is assumed to be characterized by the difference in potential across its somal membrane (membrane potential) near a spatially restricted area of the soma. This area is thought to be the action potential (or spike) generating area and is called the *trigger zone.*

The membrane potential at any time t is modeled by a stochastic process $V(t)$. The process $V(t)$ is assumed to be subject to instantaneous changes due to the occurrence of (idealized) post-synaptic potentials (PSP) of two different types: excitatory post-synaptic potential (EPSP) and inhibitory post-synaptic potential (IPSP). In the absence of post-synaptic activity, the membrance potential decays exponentially with rate (say) ρ. Hence, the incremental decay $\Delta V(t)$ during a short time interval Δt is $\Delta Vt = -\rho V(t)\,\Delta t$. It is believed that there may be thousands of synapses on the surface of certain types of neurons. In response to a stimulus, a small number of synapses may be activated. We model two types of synaptic activity.

In the first post-synaptic input received due to stimulus presentation, suppose there are n_1 excitatory synapses and n_2 inhibitory synapses. The EPSPs are assumed to arrive according to stochastic point processes $N_k^{(c)}(t)$ with stochastic intensities $\lambda_k^{(c)}(t)$ and stochastic amplitude (or potential displacements) $\alpha_k^{(c)}(t)$, $1 \leq k \leq n_1$. Similarly, the IPSPs are assumed to arrive according to stochastic point processes $N_k^{(i)}(t)$ with intensities $\lambda_k^{(i)}(t)$ and amplitudes $\alpha_k^{(i)}(t)$, $1 \leq k \leq n_2$. It is also assumed that this synaptic input is summed linearly at the trigger zone. The rest of the synaptic input is lumped together, and it is assumed that their potential displacements are small in magnitude and occur frequently and independently. This input may be modeled by a diffusion process with jump components (cf. [15]).

Based on the above assumptions and experimentally established neurophysiological observations, the process $V(t)$ may be modeled as the solution of the following stochastic differential equation:

$$dV(t) = (\mu - \rho V(t))dt + \sigma dW(t) + \sum_{k=1}^{n_1} \alpha_k^{(c)}(t) dN_k^{(c)}(t)$$

$$- \sum_{k=1}^{n_2} \alpha_k^{(i)}(t) dN_k^{(i)}(t) . \tag{10.3.1}$$

An extended model can be obtained by replacing the term $\mu - \rho V(t)$ by a continuous function $f(V(t), t)$ that need not be linear in V. The process W may be replaced by a martingale M^c with conti uous paths leading to the model

$$dV(t) = f(V(t), t)dt + \sigma(t)dM^c(t) + \int_u G(V(t), u)N(du, dt) \tag{10.3.2}$$

where N is a linear combination of compensated point processes representing the stimulus evoked synaptic potentials (cf. [15]). In the case when $M = W$ and N is a compensated Poisson process, that is, $N(du, dt) = \tilde{N}(du, dt) - \mu(du)dt$ with $E[N(du, dt)] = \mu(du)dt$, Kallianpur and Wolpert [9] discussed the existence and uniqueness of a strong solution of (10.3.2).

Another characteristic of the central nervous system information processing is the dependence of both the magnitude and the time course of the post-synaptic potential evoked by a given synapses on the spatial location of the active junction. This leads to modeling the subthreshold behavior of the membrane potential $V(t, x)$ by a stochastic partial differential equation (cf. [15]).

Suppose the process $\{V(t)\}$ is modeled by the equation

$$dV_t = (-\rho V_t + \mu)\, dt + \sigma\, dM_t, \; t \geq 0 \tag{10.3.3}$$

where $\langle M_t \rangle$ is a square integrable martingale. One can discuss the problem of the estimation of ρ and μ using the quasi-likelihood method discussed in the earlier sections following the approach of estimating functions due to Godambe and Heyde [5]. Habib and Thavaneswaran [8] derive the asymptotic properties of optimal estimates. For related work on inference for neuron models, see [6, 7].

References

[1] Andersen, P.K., Borgan, Φ., Gill, R.D., and Keiding, N. (1993) *Statistical Methods for Counting Processes,* Springer, New York.

[2] Arnold, L. (1974) *Stochastic Differential Equations: Theory and Applications,* Wiley, New York.

[3] Basawa, I.V. and Prakasa Rao, B.L.S. (1980) *Statistical Inference for Stochastic Processes,* Academic Press, London.

[4] Fleming, T.R. and Harrington, D.P. (1991) *Counting Processes and Survival Analysis,* Wiley, New York.

[5] Godambe, V. and Heyde, C.C. (1987) Quasi-likelihood and optimal estimation, *Internat. Statist. Rev.*, **55**, 231–244.

[6] Habib, M.K. (1985) Parameter estimation for randomly stopped processes and neuronal modeling, *UNC Institute of Statistics, Mimeo Series,* No.1492.

[7] Habib, M.K. and Thavaneswaran, A. (1990) Inference for stochastic neuronal models, *Appl. Math. Comput.*, **38**, 51–73.

[8] Habib, M.K. and Thavaneswaran, A. (1992) Optimal estimation for semimartingale neuronal models, *J. Stat. Plan. Inf.*, **33**, 143–156.

[9] Kallianpur, G. and Wolpert, R. (1984) Infinite dimensional stochastic differential equation models for spatially distributed neurons, *Appl. Math. Optim.*, **12**, 125–172.

[10] Karlin, S. and Taylor, H. (1975) *A First Course in Stochastic Processes,* Academic Press, New York.

[11] Karr, A.F. (1991) *Point Processes and Their Statistical Inference,* Marcel Dekker, New York.

[12] Lo, A. (1988) Maximum likelihood estimation of generalised Ito processes with discretely sampled data, *Econometric Theory,* **4**, 231–247.

[13] Neher, E. and Stevens, C.F. (1977) Conductance fluctuations and ionic pores in membranes, *Ann. Rev. Biophys. Bioeng.*, **6**, 345–381.

[14] Prakasa Rao, B.L.S. (1999) *Statistical Inference for Diffusion Type Processes,* Arnold, London.

[15] Wegman, E.J. and Habib, M.K. (1992) Stochastic models for neural systems, *J. Stat. Plan. Inf.,* **33**, 5–25.

Appendix A

Doléans Measure for Semimartingales and Burkholder's Inequality for Martingales

A.1 Doléans Measure

Let $(\Omega, \mathcal{F}, P)$ be a probability space and $\{\mathcal{F}_t\}$ be a filtration defined on $(\Omega, \mathcal{F}, P)$ satisfying the "usual" conditions. Suppose $\{X_t, t \in I\}$ is an $\mathcal{F}_t$-adapted process such that X_t is integrable. Define

$$\lambda_X((0,t] \times F) = E\left[\chi_F\left(X_t - X_s\right)\right], \ (s,t) \in I \times I, \ F \in \mathcal{F}_{s-}$$

and

$$\lambda_X([0, \inf I] \times F) = 0 \text{ for } F \in \mathcal{F}_0 .$$

Here $\chi_F(\cdot)$ denotes the indicator function of a set F.

The process $\{X_t, \ t \in I\}$ is called a *quasi-martingale* if λ_X has a bounded variation on every set $(0,t] \times \Omega, \ t \in I$.

Let $I = R^+$ and $\{X_t, \ t \geq 0\}$ be a quasi martingale. The process X is said to belong to the class $[D]$ on I if the family $\{X_\tau : \tau \in \mathcal{F}_I^f\}$ of random variables is uniformly integrable (here $\mathcal{F}_I^f$ is the set of stopping times that take values in a finite subset of I). The process X is said to be in the class $[LD]$ if it is in class $[D]$ on every bounded interval $[0,\alpha] \subset R$.

THEOREM A.1

Suppose $\{X_t, t \geq 0\}$ is a right-continuous quasi martingale. Let $\alpha \in R^+$. The function λ_X defined above can be extended to a σ-additive measure with *bounded variation if and only if X belongs to the class $[LD]$ and $|\lambda_X(R^+ \times \Omega)| < \infty$. If X belongs to the class $[D]$ on the interval I, then for every two I-valued stopping times σ and τ with $\sigma \leq \tau$,*

$$\lambda_X((\sigma, \tau]) = E\,(X_\tau - X_\sigma)\ .$$

REMARK A.1 The σ-additive measure λ_X corresponding to X is called the *Doléans measure* of the quasi martingale X. ▮

For details, see [10, p. 86].

A.2 Burkholder's Inequality for Martingales

THEOREM A.2
For any $p \geq 2$, there exist positive constants $C_i = C_i(p), i = 1, 2$ such that

$$C_1 E\left[\langle M\rangle_t^{p/2}\right] \leq E\left[|M_t|^p\right] \leq C_2 E\left[\langle M\rangle_t^{p/2}\right]$$

for all $0 \leq t \leq T$ for any continuous martingale M_t such that $M_0 = 0$ and $E|M_T|^p < \infty$.

For proof (as an application of the Ito formula), see [6, p. 66].

Appendix B

Interchanging Stochastic Integration and Ordinary Differentiation and Fubini-Type Theorem for Stochastic Integrals

B.1 Interchanging Stochastic Integration and Ordinary Differentiation

Let $(\Omega, \mathcal{F}, P)$ be a complete probability space and $\{\mathcal{F}_t, t \geq 0\}$ be an increasing family of sub-σ-algebras of $\mathcal{F}$ such that $\mathcal{F}_0$ contains all P-null sets in $\mathcal{F}$. Let $\{W_t, t \geq 0\}$ be an $\mathcal{F}_t$-adapted Wiener process. The following theorem generalizes Kolomogrov's theorem for Banach space valued random variables.

THEOREM B.1

Let B be a separable Banach space endowed with the norm $\|\cdot\|$ and let $\{Z(\theta) : \theta \in R^d\}$ be a family of B-valued random variables. Suppose there exists $0 < c < \infty$, $0 < \alpha \leq 1$ and $p \geq d + \alpha$ such that

$$E\,\|Z(\theta_1) - Z(\theta_2)\|^p \leq c\,\|\theta_1 - \theta_2\|^{d+\alpha}, \theta_1, \theta_2 \in R^d .$$

Then there exists a continuus version $\{Z_1(\theta)\}$ of the process $\{Z(\theta)\}$, that is, $P(Z(\theta = Z_1(\theta)) = 1$ for all θ such that for all $\omega \in \Omega$ and $\theta \to Z_1(\theta, \omega)$ is a continuous map from R^d into B.

For proof, see [13].

Let $\mathcal{L}_2$ be the family of all progressively measurable processes f on $[0, 1]$

such that $E\{\int_0^1 f^2(t,\omega)dt\} < \infty$.

THEOREM B.2

Let $p \geq 2$. Then there exists a universal constant c_p such that

$$E \sup_{0\leq t\leq 1} \left| \int_0^t f(u,\cdot)dW(u) \right|^p \leq c_P E \int_0^1 |f(u,\cdot)|^p du \, .$$

For proof of Theorem B.2, see [13] or [14].

THEOREM B.3

Suppose there exists constants $c > 0$ and $0 < \beta \leq 1$ such that

$$(C1)|g(\theta_1,t,\omega) - g(\theta_2,t,\omega)| \leq c\|\theta_1 - \theta_2\|^\beta$$

for all t, ω and θ_1 and θ_2 in R^d. Further suppose that $\{g(\theta,t,\cdot)\} \in \mathcal{L}_2$ for all $\theta \in R^d$. Then there exists a continuous version $Z(\theta, t.\omega)$ of the stochastic integral $\int_0^1 g(\theta,u,\cdot)dW(u)$, that is, for all $\omega \in \Omega$, the map $\theta \to Z(\theta,\cdot,\omega)$ is a continuous map from R^d into $C[0,1]$ equipped with the sup norm.

PROOF Let $p \geq 2$ such that $p\beta > d$. Let

$$Z(\theta,t) = \int_0^t g(\theta,u)\, dW(u)\, .$$

Then, by Theorem B.2, it follows that

$$E \sup_{0\leq t\leq 1} |Z(\theta_1,t) - Z(\theta_2,t)|^p$$

$$\leq c_p E\left[\int_0^1 |g(\theta_1,u) - g(\theta_2,u)|^p\, du\right]$$

$$\leq c_p c \left\|\theta_1 - \theta_2\right\|^{p\beta} .$$

The result now follows from Theorem B.1. ∎

The following theorem gives sufficient conditions for the differentiability under the stochastic integral.

THEOREM B.4

Let $\{f(\theta,\cdot,\cdot), \theta \in R\} \subset \mathcal{L}_2$ *be such that for all* t, ω, $\frac{d}{d\theta} f(\theta,t,\omega) = f'(\theta,t,\omega)$ *exists and* $\{f'(\theta,\cdot,\cdot), \theta \in R\} \subset \mathcal{L}_2$. *Further assume that there exists constants* $0 < c < \infty$, $0 < \beta_i \leq 1$, $i = 1, 2$ *such that*

$$(C2) \quad \begin{cases} |f(\theta,t,\omega) - f(\theta_2,t,\omega)| \leq c\,|\theta_1 - \theta_2|^{\beta_1} \\ \left|f'(\theta,t,\omega) - f'(\theta_2,t,\omega)\right| \leq c\,|\theta_1 - \theta_2|^{\beta_2} . \end{cases}$$

Then there exists a version $X(\theta,t,\omega)$ *of* $\int_0^t f(\theta,u)dW(u)$ *such that for all* $\omega, t, \theta \to X(\theta,t,\omega)$ *is differentiable in* θ *and*

$$\frac{d}{d\theta} X(\theta,t,\omega) = \int_0^t f'(\theta,t,\omega)dW(u) .$$

PROOF Let us first choose a continuous version $X_1(\theta,t)$ of $\int_0^t f(\theta,u)dW(u)$. This is possible under the condition (C2) since (C2) implies (C1) of Theorem B.3. Let

$$Y(\theta_1,\theta_2,t,\omega) = \frac{X_1(\theta_1,t,\omega) - X_1(\theta_2,t,\omega)}{\theta_1 - \theta_2} \text{ if } \theta_1 \neq \theta_2$$

$$= \int_0^t f'(\theta_1,u)\,dW(u) \text{ if } \theta_1 = \theta_2 .$$

Note that Y is continuous on the set $[(\theta_1,\theta_2) : \theta_1 \neq \theta_2]$ and any two continuous versions agree outside a null set N. Hence, $X_1(\theta,t,\omega)$ is a differentiable function of θ for all t outside a null set N. Define $X_1(\theta, t, \omega) = 0$ on the null set N and equal to $X_1(\theta, t, \omega)$ elsewhere.

For any θ_1 and θ_2,

$$Y(\theta_1, \theta_2, t, \cdot) = \int_0^t \left[\int_0^1 f'(\lambda\theta_1 + (1-\lambda)\theta_2, u, \cdot)\, d\lambda \right] dW(u) .$$

Let

$$g(\theta_1, \theta_2, u, \omega) = \int_0^1 f'(\lambda\theta_1 + (1-\lambda)\theta_2, u, \omega)\, d\lambda .$$

Then

$$\begin{aligned}
&|g(\theta_1, \theta_2, u, \omega) - g(\theta_3, \theta_4, u, \omega)| \\
&\leq \int_0^1 \left| f'(\lambda\theta_1 + (1-\lambda)\theta_2, u, \omega) - f'(\lambda\theta_3 + (1-\lambda)\theta_4, u, \omega) \right| d\lambda \\
&\leq c \int_0^1 |\lambda\theta_1 + (1-\lambda)\theta_2 - \lambda\theta_3 - (1-\lambda)\theta_4|^{\beta_2}\, d\lambda \\
&\leq c_1 \left(|\theta_1 - \theta_3|^{\beta_2} + |\theta_2 - \theta_4|^{\beta_2} \right) \\
&\leq c_2 \left(|\theta_1 - \theta_3|^2 + |\theta_2 - \theta_4|^2 \right)^{\beta_2/2}
\end{aligned}$$

for some constants c_1 and c_2. An application of Theorem B.3 proves that Y has a continuous modification. ∎

REMARK B.1 The above result can be extended to include process on $[0, \infty)$ instead of $[0, 1]$. A more general result valid for semimartingales is as follows.

Suppose $\{g(\theta, t, \omega),\ \theta \in R^d\}$ satisfies the following conditions.

(C3) For all $n \geq 1$, there exist constants c_n, α_n, $0 < c_n < \infty, 0 < \alpha_n \leq 1$ and stopping times T_n such that

(i) $T_n \uparrow \infty$ a.s.,

(ii) $|g(0, t, \omega)| \leq c_n$ on $[t < T_n(\omega)]$, and

(iii) $|g(\theta_1, t, \omega) - g(\theta_2, t, \omega) \leq c_n |\theta_1 - \theta_2|^{\alpha_n}$ if $|\theta_1| \leq n, |\theta_2| \leq n$ and on $[t < T_n(\omega)]$.

Then the following result holds. ∎

THEOREM B.5

Let S be a continuous semimartingale and $\{f(\theta, \cdot, \cdot), \theta \in R^d\}$ be a family of progressively measurable processes. The following properties hold:

(i) if $f(\theta, \cdot, \cdot) \in R^d$ satisfies (C3), then there exists a continuous version $X(\theta, t, \cdot)$ of $\int_0^t f(\theta, u), \cdot)dS(u)$;

(ii) if $g(\theta, t, \omega) = \frac{d}{d\lambda} f(\theta + \lambda e, t, \omega)|_{\lambda=0}$ exists for all θ where $e \in R^d$ and $\{g(\theta, \cdot, \cdot) \in R^d\}$ satisfies (C3), then there exists a version of $X(\theta, t, \cdot)$ of $\int_0^t f(\theta, u)dS(u)$ such that

$$Y(\theta, t, \omega) = \frac{d}{d\lambda} X(\theta + \lambda e, t, \omega)|_{\lambda=0}$$

exists and

$$Y(\theta, t, \omega) = \int_0^t g(\theta, u, \cdot)dS(u) .$$

Results given above are due to Karandikar [5].

B.2 Fubini-Type Theorem for Stochastic Integrals

Let $(\Omega, \mathcal{F}, P)$ and $(\tilde{\Omega}, \tilde{f}, \tilde{P})$ be two probability spaces. Define

$$\left(\bar{\Omega}, \bar{\mathcal{F}}, \bar{P}\right) = \left(\Omega \times \tilde{\Omega}, \mathcal{F} \times \tilde{\mathcal{F}}, P \times \tilde{P}\right) .$$

Let $\{\mathcal{F}_t, \ 0 \leq t \leq 1\}$ and $\{\tilde{\mathcal{F}}_t, \ 0 \leq t \leq 1\}$ be filtrations in $\mathcal{F}$ and $\tilde{\mathcal{F}}$, respectively. Let $W = \{W_t, 0 \leq t \leq 1\}$ be a Wiener process on $(\Omega, \mathcal{F}, P)$ such that W_t is $\mathcal{F}_t$ measurable. Let $\mathcal{F}_t^W = \sigma\{W_s, s \leq t\}$.

THEOREM B.6
Let $\{g_t(\omega, \tilde{\omega}), \mathcal{F}_t^W \times \tilde{\mathcal{F}}_t, \ 0 \leq t \leq 1\}$ be a random process such that

$$E\left[\int_0^1 g_t^2 (\omega, \tilde{\omega})\, dt\right] < \infty$$

where E denotes the expentation with respect to $P \times \tilde{P}$. Then, for each t, $0 \leq t \leq 1$,

$$\int_{\tilde{\Omega}} \left[\int_0^t g_s (\omega, \tilde{\omega})\, dW_s(\omega)\right] d\tilde{P} (\tilde{\omega}) = \int_0^t \left[\int_{\tilde{\Omega}} g_s (\omega, \tilde{\omega})\, d\tilde{P} (\tilde{\omega})\right] dW_s(\omega) .$$

For proof, see [8, p. 187].

B.3 Sufficient Conditions for the Differentiability of an Ito Stochastic Integral

THEOREM B.7
Let a random process $\{f(u, t), \mathcal{F}_t, 0 \leq t \leq T\}$ be, with probability one, $k + 1$

times continuously differentiable with respect to u and

$$E\left[\int_0^T \left[\frac{\partial^j f(u,t)}{\partial u^j}\right]^2 dt\right] \le D_j < \infty, \ j = 0, 1, \ldots, k+1 .$$

Suppose the stochastic integral

$$\zeta(u) = \int_0^T f(u,t)dW(t) \tag{B.3.1}$$

exists with respect to a Wiener process $\{W(t),\ 0 \le t \le T\}$. *Then the stochastic integral has, with probability one, k continuous derivatives and*

$$\frac{\partial^j \zeta(u)}{\partial u^j} = \int_0^T \frac{\partial^j f(u,t)}{\partial u^j} dW(t), \ 1 \le j \le k . \tag{B.3.2}$$

PROOF For simplicity, we consider the case $k = 1$. The general case can be proved analogously. Define the process

$$\dot{\zeta}(u) = \int_0^T \dot{f}(u,t)dw(t)$$

where $\dot{f}(u,t) = \frac{\partial}{\partial u} f(u,t)$. It can be seen, from the continuous differentiability of $\dot{f}(u,t) = \frac{\partial}{\partial u} f(u,t)$ with respect to u and the properties of a stochastic integral, that

$$\begin{aligned} E\left|\dot{\zeta}(u+h) - \dot{\zeta}(u)\right|^2 &= E\left\{\int_0^T \left[\dot{f}(u+h) - \dot{f}(u)\right]^2 dt\right\} \\ &\le h \int_0^h \left(E\int_0^T \ddot{f}(u+v,t)^2 dt\right) dv \\ &\le D_2 h^2 \end{aligned} \tag{B.3.3}$$

where $\ddot{f}(u,t) = \frac{\partial^2}{\partial u^2} f(u,t)$. Hence, the process $\{\zeta(u)\}$ is continuous with probability one by Kolmogorov's theorem. Taking limit as $h \to 0$ in the expression, we have

$$\frac{\zeta(u+h) - \zeta(u)}{h} = \frac{1}{h}\int_0^h \left(\int_0^T \dot{f}(u+v,t)dW(t) \right) dv$$

$$\overset{a.s}{\to} \int_0^T \dot{f}(u,t)dW(t)\,. \tag{B.3.4}$$

This proves (B.3.2). ∎

Appendix C

The Fundamental Identity of Sequential Analysis

Consider a filtered space $(\Omega, \mathcal{F}, \{\mathcal{F}_t\}.P_1, P_2)$ where P_1 and P_2 are two probability measures that are not typically equivalent or singular with respect to each other. Let P_i^t be the restriction of P_i to $\mathcal{F}_t$. Suppose $P_2 \overset{loc}{<<} P_1$. Let

$$L_t = \frac{dP_2^t}{dP_1^t} . \tag{C.1.1}$$

For a stopping time τ with respect to $\{\mathcal{F}_t\}$, the σ-algebra of events happening before τ is given by

$$A \in \mathcal{F}_\tau \text{ if and only if } A \cap [\tau \le t] \in \mathcal{F}_t \text{ for all } t \ge 0 . \tag{C.1.2}$$

We denote the restriction of P_i to $\mathcal{F}_\tau$ by P_i^τ.

THEOREM C.1
Let τ be a stopping time. Then

$$P_2[A \cap \{\tau < \infty\}] = \int_{A\cap[\tau<\infty]} L_\tau dP_1, \ A \in \mathcal{F}_\tau . \tag{C.1.3}$$

PROOF Since $\mathcal{F}_{\tau\wedge n} \subset \mathcal{F}_n$, we obtain that $P_2^{\tau\wedge n} << P_1^{\tau\wedge n}$, and since $\tau \wedge n$ is a bounded stopping time, it follows by the optional stopping theorem that

$$\frac{dP_2^{\tau\wedge n}}{dP_1^{\tau\wedge n}} = E_i(L_n | \mathcal{F}_{\tau\wedge n}) = L_{\tau\wedge n} \tag{C.1.4}$$

where E_i denotes the expectation with respect to P_i. To prove (C.1.3), it is sufficient to prove the result for every $A \in \mathcal{F}_{\tau\wedge k}$ for all $k \geq 1$ since

$$(\tau < \infty) \cap \mathcal{F}_\tau = (\tau < \infty) \cup \sigma\left(\cup_{k=1}^{\infty} \mathcal{F}_{\tau\wedge k}\right) . \tag{C.1.5}$$

Therefore fix k and choose $A \in \mathcal{F}_{\tau\wedge k}$. Since $A \cap [\tau \leq m] \in \mathcal{F}_{\tau\wedge m}$ for all $m \geq k$, we have that

$$P_2\left(A \cap [\tau \leq m]\right) = \int_{A\cap[\tau\leq m]} L_{\tau\wedge m} dP_1 = \int_{A\cap[\tau\leq m]} L_\tau dP_1 \tag{C.1.6}$$

by (C.1.4). Let $m \to \infty$ and apply the monotone convergence theorem. Then it follows that

$$P_2\left(A \cap [\tau \leq m]\right) = \int_{A\cap[\tau\leq m]} L_\tau dP . \tag{C.1.7}$$

This completes the proof. ∎

As a corollary, we have the following important result.

THEOREM C.2
Let τ be a stopping time such that

$$P_1(\tau < \infty) = P_2(\tau < \infty) = 1 .$$

Then

$$P_2^\tau << P_1^\tau \tag{C.1.8}$$

and

$$\frac{dP_2^\tau}{dP_1^\tau} = L_\tau . \tag{C.1.9}$$

The above remarks follow, for instance, from [12].

Appendix D

Stieltjes–Lebesgue Calculus

DEFINITION D.1 *Let $f(t)$ be a function such that, for all $t \geq 0$,*

$$V_f(t) = \sup_{\mathcal{D}} \sum_{i=1}^{N} |f(t_i) - f(t_{i-1}| < \infty$$

where $\mathcal{D}$ ranges over all the subdivisions of $[0, t] : 0 = t_0 < t_1 < \cdots < t_N = t$. The function $f(t)$ is said to be of bounded variation over finite intervals. The function $V_f(t)$ is called the variation of f over $(0, t]$.

Example D.1

Let $f(t) = \int_0^t f(s)ds$ where $g(\cdot)$ is locally integrable. Then $f(\cdot)$ is of bounded variation over finite intervals. □

Example D.2

If $f(t) = a(t)$, where $a(\cdot) \uparrow$, then f is of bounded variation over finite intervals and $V_f(t) = a(t) - a(0)$. □

REMARK D.1 If f is right-continuous from $[0, \infty)$ to R and is of bounded variation, then $f(t) = f(0) + a(t) - b(t)$ where $a(\cdot)$ and $b(\cdot)$ are right-continuous increasing functions. Such a decomposition is called *canonical* if $a(t) = V_f(t)$ and $b(t) = -f(t) + f(0) + V_f(t)$. ▮

REMARK D.2 Hereafter, we consider $f(\cdot)$ which (i) are right-continuous,

(ii) have left limits, and (iii) are of bounded variation over finite intervals. Further, we assume that $f(0) = 0$ later in the sequel. Such a function is called a b.v. function in the following discussion. ∎

REMARK D.3 Let f be a b.v. function with $f(t) = f(0) + a(t) - b(t)$ in the canonical decomposition. Let μ_a and μ_b be σ-finite measures on $(0, \infty)$ defined by

$$\mu_a((0, t]) = a(t) = V_f(t), \ \ \mu_b((0, t]) = b(t) \ .$$

Note that $\mu_b \leq 2\mu_a$.
We write $d\mu_a(t) = \mu_a(df) = |df(t)|$.

Suppose $u(t)$ is a measurable function such that

$$\int_{(0,\infty)} |u(s)| d\mu_a(s) = \int_{(0,\infty)} |u(s)| \, |df(s)| < \infty \, .$$

Then

$$\int_a^b |u(s) \, d\mu_b(s) < \infty \text{ since } \mu_b \leq 2\mu_a \, . \quad ∎$$

DEFINITION D.2 A function $u(t)$ is said to be Stieltjes–Lebesgue integrable with respect to $f(t)$ if

$$\int_{(0,\infty)} |u(s)| d\mu_a(s) = \int_{(0,\infty)} |u(s)| \, |df(s)| < \infty$$

and the Stieltjes–Lebesgue integral is defined by

$$\int_{(0,\infty)} u(s) df(s) = \int_{(0,\infty)} u(s) d\mu_a(s) - \int_{(0,\infty)} u(s) d\mu_b(s)$$

$$\triangleq \int_{(0,\infty)} u(s) da(s) - \int_{(0,\infty)} u(s) db(s).$$

Notation: We define

$$\int_A u(s)df(s) \triangleq \int_{(0,\infty)} u(s)I_A(s)df(s),\ A \in \mathcal{B}(0,\infty)\,,$$

$$\int_a^b u(s)df(s) \triangleq \int_{(a,b)} u(s)df(s)\,,$$

$$\int_{(0,t]} df(s) = f(t) - f(0), \text{ and}$$

$$\int_{(0,t)} df(s) = f(t-) - f(0)\,.$$

DEFINITION D.3 *A function $u(\cdot)$ is said to be locally Stieltjes–Lebesgue integrable with respect to the b.v. function $f(t)$ if*

$$\int_0^t |u(s)|\,|df(s)| < \infty,\ t \geq 0.$$

Then $\psi(t) = \int_0^t u(s)df(s)$ is a b.v. function and

$$V_\psi(t) = \int_0^t |u(s)|\,|df(s)|\,.$$

DEFINITION D.4 **(Stieltjes–Lebesgue integral on R^2)**
Let f_1 and f_2 be b.v. functions with canonical decompositions $a_1 - b_1$ and $a_2 - b_2$, respectively. Then

$$\int\int_{(0,\infty]\times(0,\infty)} u(x,y)df_1(x)df_2(y)$$

is defined by the relation

$$df_1(x)df_2(y) = d\mu_{a1}(x)d\mu_{a_2}(y) + d\mu_{b_2}(x)d\mu_{b_2}(y)$$

$$- d\mu_{a_1}(x)d\mu_{b_2}(y) - d\mu_{a_2}(y)d\mu_{b_1}(x) .$$

REMARK D.4 It is easy to see that Fubini's theorem holds under this definition. ▮

D.1 Product Formula

THEOREM D.1

Let $f(t)$ and $g(t)$ be two functions of bounded variation over finite intervals, right-continuous with left limits. Then

$$f(t)g(t) = f(0)g(0) + \int_0^t f(s)dg(s) + \int_0^t g(s-)df(s) . \tag{D.1.1}$$

PROOF Note that

$$(f(t) - f(0))(g(t) - g(0))$$

$$= \int_0^t df(x) \int_0^t dg(y)$$

$$= \int_{(0,t]\times(0,t]}\int df(x)\, dg(y) \quad \text{(by Fubini's theorem)}$$

$$= \int_{D_1}\int df(x)dg(y) + \int_{D_2}\int df(x)dg(y) \tag{D.1.2}$$

where

$$D_1 = \{(x, y) : x \leq y,\ 0 < x \leq t,\ 0 < y \leq t\}, \text{ and}$$

$$D_2 = \{(x, y) : x > y,\ 0 < x \leq t,\ 0 < y \leq t\}.$$

But

$$\int\int_{D_1} df(x)dg(y) = \int_{(0,t]} \left(\int_{(0,y]} df(x) \right) dg(y) \text{ (by Fubini's theorem)}$$

$$= \int_0^t [f(y) - f(0)]dg(y)$$

$$= \int_0^t f(y)dg(y) - f(0)(g(t) - g(0)). \qquad \text{(D.1.3)}$$

Similarly,

$$\int\int_{D_2} df(x)dg(y) = \int_{(0,t]} \left(\int_{(0,x)} dg(y) \right) df(x)$$

$$= \int_{(0,t]} (g(x-) - g(0))df(x)$$

$$= \int_0^t g(x-)df(x) - g(0)[f(t) - f(0)]. \qquad \text{(D.1.4)}$$

Relations (D.1.2), (D.1.3), and (D.1.4) prove (D.1.1). ∎

D.2 Application of Product Formula

THEOREM D.2
Let $a(t)$ be a right-continuous increasing process. Then

$$\int_0^t a^{n-1}(s-)da(s) \le \frac{a^n(t) - a^n(0)}{n} \le \int_0^t a^{n-1}(s)da(s) . \qquad \text{(D.2.1)}$$

PROOF The product formula can be written in the form

$$d(fg) = f_- dg + g df .$$

Let $f = a^{n-1}$ and $g = a$. Then

$$d\left(a^n\right) = a_-^{n-1} da + a\left(da^{n-1}\right) = a_-^{n-1} da + a\left[a_-^{n-2} da + a\left(da^{n-2}\right)\right]$$

$$= a_-^{n-1} da + a a_-^{n-2} da + a^2 d\left(a^{n-2}\right) .$$

Repeated appliction leads to the relation

$$d\left(a^n\right) = \left[a_-^{n-1} + a a_-^{n-2} + \cdots + a^{n-1}\right] da .$$

Since a is increasing, $a(t-) \le a(t)$ for all $t \ge 0$. Hence,

$$d\left(a^n\right) \le n a^{n-1} da .$$

This proves the right-side inequality in (D.2.1). Similar arguments can be used to prove the left-side inequality in (D.2.1). ∎

D.3 Exponential Formula

THEOREM D.3
Let $a(t)$ be a right-continuous increasing function with $a(0) = 0$ and let $u(t)$ be such that

$$\int_0^t |u(s)|\, da(s) < \infty,\ t \geq 0\,.$$

Then

$$x(t) = x(0) + \int_0^t x(s-)u(s)\, da(s) \tag{D.3.1}$$

admits a unique locally bounded ($\sup_{0\leq s\leq t} |x(s)| < \infty, t \geq 0$) solution given by

$$x(t) = x(0)\Pi_{0<s\leq t}(1 + u(s)\Delta a(s)) \exp\left(\int_0^t u(s) da^c(s)\right) \tag{D.3.2}$$

where $\Delta a(t) = a(t) - a(t-)$ and $a^c(t)$ is the continuous component of $a(t)$, that is, $a^c(t) = a(t) - \sum_{0\leq s\leq t} \Delta a(s)$.

PROOF [9] Let

$$f(t) = x(0)\Pi_{0<s\leq t}(1 + u(s)\Delta a(s))\,, \tag{D.3.3}$$

and

$$g(t) = \exp\left(\int_0^t u(s) da^c(s)\right) \tag{D.3.4}$$

where we define

$$\Pi_{0<s\leq t}(1 + u(s)\Delta a(s)) = 1 \text{ if } \Delta a(s) = 0 \text{ for all } s \in (0, t]\,.$$

Let $x(t) = f(t)g(t)$. Then, by the product formula,

$$x(t) = x(0) + \int_0^t f(s-)dg(s) + \int_0^t g(s)df(s)$$

$$= x(0) + \int_0^t f(s-)\, g(s)\, u(s)\, da^c(s)$$

$$+ \sum_{0<s\leq t} g(s)f(s-)u(s)\Delta a(s)\,.$$

Note that $\int_0^t g(s)df(s) = \sum_{0<s\leq t} g(s)\Delta f(s)$ and $\Delta f(s) = f(s-)u(s)\Delta a(s)$. But $f(s) - f(s-) = f(s-)\, u(s)\Delta a(s)$, which implies that $f(s) = f(s-)[1 + u(s)\Delta a(s)]$.

Hence

$$x(t) = x(0) + \int_0^t x(s-)u(s)da(s) \tag{D.3.5}$$

and $x(\cdot)$ is a solution of (D.3.1).

Suppose $x(t)$ and $y(t)$ are two locally bounded solutions of (D.3.1). Define

$$z(t) = x(t) - y(t)\,,$$

$$M(t) = \sup_{0\leq s\leq t} |x(t) - y(t)|\,,$$

$$\alpha(t) = \int_0^t |u(s)|da(s)\,.$$

Then, for fixed t, and $0 \leq s \leq t$,

$$|z(s)| \leq \int_0^s |z(v-)|\, |u(v)|da(v) \qquad \text{(since } x \text{ and } y \text{ are solution of (D.3.1))}$$

$$\le M(t)\alpha(s)$$

and hence,

$$|z(s)| \le \int_0^s |z(s-)|\,|u(v)|\,da(v)$$

$$\le M(t)\int_0^s \alpha(v-)d\alpha(v)$$

$$\le \frac{M(t)}{2}\alpha^2(s) \qquad \text{(by (D.2.1))}.$$

Repeating the process, we have

$$|z(s)| \le \frac{M(t)}{2}\int_0^s \alpha^2(v-)d\alpha(v) \le \frac{M(t)}{3!}\alpha^3(s)$$

and in general

$$|z(s)| \le \frac{M(t)}{n!}\alpha^n(s).$$

Let $n \to \infty$. Then we have $z(s) = 0$ for $0 \le s \le t$. Since t is arbitrary, the result follows. ∎

THEOREM D.4 **(Another version of the product formula)**
Suppose $f(t)$, $g(t)$ are RCLL functions of bounded variation with $f(0-) = 0$ and $g(0-) = 0$. Then

$$f(t)g(t) = \int_{[0,t]} f(s-)dg(s) + \int_{[0,t]} g(s)df(s).$$

If F is continuously differentiable, then

$$F(f(t)) = F(0) + \int_{[0,t]} F'(f(s-))df(s)$$

$$+ \sum_{0 \le s \le t} \{F(f(s)) - F(f(s-)) - F'(f(s-))\Delta f(s)\}$$

$$= \int_0^t F'\left(f(s))df^c(s) + \sum_{0 \le s \le t} (F(f(s)) - F(f(s-))\right) .$$

(Here f^c is the continuous component of the function of bounded variation f. Note that $\Delta f(0) = f(0) - f(0-) = f(0)$).

COROLLARY D.1
If $f(t)$ is RCLL and of bounded variation, then there is at most one locally bounded solution of

$$x(t) - x(0-) = \int_{[0,t]} x(s-)df(s), \; 0 \le t < \infty .$$

REMARK D.5 Discussion in this section is based on [1], [3], and [9]. ∎

Appendix E

A Useful Lemma

LEMMA E.1

[2] Let $F : [0, \infty) \to R$ be a regular right-continuous nondecreasing function and $\zeta : [0, \infty) \to R^k$ be a regular left-continuous vector-valued function. Suppose M_0 is a positive semidefinite matrix. Define

$$M(t) = M_0 + \int_{(0,t)} \zeta(s)\zeta'(s)dF(s) . \tag{E.1.1}$$

Suppose that $M(t)$ is nonsingular for some $t = \tau \geq 0$ (here A' denotes transpose of A). Then

$$\int_{\tau}^{t} \zeta'(s)M^{-1}(s)\zeta(s)dF(s) = \int_{\tau}^{t} \frac{1}{|M(s)|}d|M(s)| \tag{E.1.2}$$

for all $t \geq \tau$ where $|M(s)| = \det(M(s))$.

PROOF It is clear that $M(t)$ is nonsingular for $t \geq \tau$. Let us first consider the case when $\zeta(s)$ is a step function of the form

$$\zeta(s) = \sum_{i=1}^{n} \alpha_i \chi_{(s_{i-1},s_i]}(s), \ \tau < s \leq t \tag{E.1.3}$$

where $\tau = s_0 < s_1 < \cdots < s_n = t$. Let

$$\Delta_h^- F(s) = F(s+h-) - F(s-),\ h > 0 \tag{E.1.4}$$

and

$$\Delta F(s) = F(s) - F(s-)\,. \tag{E.1.5}$$

Then

$$M(s+h) = M(s) + \alpha_i \alpha_i' \Delta_h^- F(s),\ s_{i-1} < s < s+h \le s_i \tag{E.1.6}$$

and

$$M(s_i+) = M(s_i) + \alpha_i \alpha_i' \Delta F(s_i)\ . \tag{E.1.7}$$

Since $A = B + WW'$ implies $W'A^{-1}W = (|A| - |B|)/|A|$ for any k-dimensional vector W and matrices A and B of order k with $|A| \neq 0$ (cf. [7]), it follows that (choosing $A = -M(s),\ B = -M(s+h)$)

$$\zeta'(s)M(s)^{-1}\zeta(s)\Delta_h^- F(s) = (|M|(s+h) - |M(s)|)/|M(s)| \tag{E.1.8}$$

and

$$\zeta'(s_i)M(s_i)^{-1}\zeta(s_i)\,\Delta F(s_i) = (|M(s_i+) - M(s_i)|)\,/\,|M(s_i)| \tag{E.1.9}$$

using the fact that $|-M| = (-1)^n|M|$. Evaluating Equation (E.1.8) at $s = s_{i-1} + jh_N,\ h_N = (s_i - s_{i-1})/N\ ,\ j = 1, \ldots, N$, summing over j, and taking limit as $N \to \infty$, we obtain that

$$\int_{s_{i-1}}^{s_i} \zeta'(s)M(s)^{-1}\zeta(s)dF(s) = \int_{s_{i-1}}^{s_i} \frac{1}{|M(s)|} d(|M(s)| \tag{E.1.10}$$

(note that $M(s)$ is left-continuous). Summing (E.1.10) over i, we obtain the result (E.1.2) for step functions.

Let $\zeta(s)$ be any left-continuous function with right limits. Then there exists a sequence of step functions ζ_n of the type (E.1.3) such that

$$|\zeta_n - \zeta|_{[0,t]} = \sup_{0 \le s \le t} \|\zeta_n(s) - \zeta(s)\| \to 0\,. \tag{E.1.11}$$

Then

$$\int_s^t \zeta_n'(s) M_n^{-1}(s) \zeta_n(s) dF(s) = \int_s^t \frac{d\,(|M_n(s)|)}{|M_n(s)|} \tag{E.1.12}$$

for all n and since

$$|M_n - M|_{[\tau,t]} + \left| M_n^{-1} - M^{-1} \right|_{[\tau,t]} \to 0 \,, \tag{E.1.13}$$

the expression on the left-hand side of (E.1.12) converges to the expression on the left-hand side of (E.1.2). Let $m(s) = |M(s)|$ and $m_n(s) = |M_n(s)|$. Observe that

$$\left| \int_\tau^t \frac{dm_n}{m_n} - \int_\tau^t \frac{dm}{m} \right| \le \int_\tau^t \left| \frac{1}{m_n} - \frac{1}{m} \right| dm_n + \left| \int_\tau^t \frac{dm_n}{m} - \int_\tau^t \frac{dm}{m} \right| . \tag{E.1.14}$$

The first term of the right-hand side of (E.1.14) converges to zero by (E.1.13). As far as the second term is concerned, the processes m_n and m have the same jump times as F and the jumps of m_n converge to those of m from (E.1.13). Let m_n^c and m^c denote the continuous part of m_n and m, respectively. Let $\Delta m_s^n = m_n(s) - m_n(s-)$. Then

$$\left| \int_\tau^t \frac{dm_n}{m} - \int_\tau^t \frac{dm}{m} \right| \le \left| \int_\tau^t \frac{dm_n^c}{m} - \int_\tau^t \frac{dm^c}{m} \right| + \sum_{\tau < s \le t} \frac{1}{m} \left| \Delta m_s^n - \Delta m_s \right| . \tag{E.1.15}$$

The first term on the right-hand side of (E.1.15) converges to zero, since the set of discontinuity points of $\frac{1}{m}$ has m^c-measure zero, and the second term goes to zero by the bounded convergence theorem. This proves that the right-hand side of (E.1.12) converges to the right hand side of (E.1.2). ∎

COROLLARY E.1
Let $M^+(s) = M(s+)$. Then, under the assumptions of Lemma E.1,

$$\int_{\tau}^{t} \zeta'(s)M^{-1}(s)\zeta(s)dF(s) = O(\log \lambda_{\max}(M^+(t))) \,. \tag{E.1.16}$$

PROOF Let $m^+(s) = |M^+(s)|$ and $c(s) = \inf\{r \ : \ m^+(r) > s\}$. Then, by [8], it follows that

$$\begin{aligned}
\int_{\tau}^{t} \frac{d|M(s)|}{|M(s)|} &= \int_{\tau}^{t} \frac{dm^+(s)}{m^+(s-)} \\
&= \int_{m^+(\tau)}^{m^+(t)} \frac{ds}{m^+(c(s)-)} \\
&\le \int_{m^+(\tau)}^{m^+(t)} \frac{ds}{s} \\
&= \text{constant } + \log m^+(t) \\
&= O\left(\log \lambda_{\max}\left(M^+(t)\right)\right)
\end{aligned} \tag{E.1.17}$$

since $m^+(t) \le (\lambda_{\max}(M^+(t)))^k$. ∎

Appendix F

Contiguity

For each $n \geq 1$, consider a measurable space $(\Omega^{(n)}, \mathcal{F}^{(n)})$ endowed with probability measures $P^{(n)}$ and $P'^{(n)}$.

DEFINITION F.1 *The sequence $\{P^{(n)}\}$ is said to be contiguous to the sequence $\{P'^{(n)}\}$ if for all sequences $A^{(n)} \in \mathcal{F}^{(n)}$ such that $P^{(n)}(A^{(n)}) \to 0$ as $n \to \infty$, we have $P'^{(n)}(A^{(n)}) \to 0$ as $n \to \infty$. In such a case we denote $(P')^{(n)} \lhd (P^{(n)})$.*

For a detailed discussion of this concept and related concept of separability, see [11, p. 96].

Let

$$Q^{(n)} = \frac{P^{(n)} + P'^{(n)}}{2}, \quad Z^{(n)} = \frac{dP'^{(n)}}{dQ^{(n)}} / \frac{dP^{(n)}}{dQ^{(n)}}$$

with the usual convention. Then $Z^{(n)}$ is called the *extention* density of $P'^{(n)}$ with respect to $P^{(n)}$. If $P'^{(n)} << P^{(n)}$, then $Z^{(n)}$ is the usual density.

Le Cam's first lemma: *Suppose that*

$$\mathcal{L}\left(Z^{(n)} \middle| P^{(n)}\right) \xrightarrow{\mathcal{L}} Z as n \to \infty .$$

Then $(P'^{(n)}) \lhd (P^{(n)})$ if and only if $E(Z) = 1$.

Le Cam's third lemma: *(a) The following are equivalent:*

(i) $\mathcal{L}(Z^{(n)}|P^{(n)})$ converges weakly to a probability measure η on R_+ and $(P'^{(n)}) \lhd (P^{(n)})$;

(ii) $\mathcal{L}(Z^{(n)}|P'^{(n)})$ converges weakly to a probability measure η' on R_+.

(b) Suppose (i) or (ii) holds. Then $\eta'(dx) = x\ \eta(dx))$. *Further, if* $X^{(n)}$ *is another random variable such that* $\mathcal{L}((Z^{(n)}, X^{(n)})|P^{(n)})$ *converges weakly to a probability measure* $\bar{\eta}$, *then the* $\mathcal{L}((Z^{(n)}, X^{(n)})|P'^{(n)})$ *converges weakly to* $\bar{\eta}'$ *where* $\bar{\eta}'(dx, dy) = x\ \ \bar{\eta}(dx, dy)$.

For proofs, see [4, p. 253] or [11, p. 96].

References

[1] Bremaud, P. (1981) *Point Processes and Queues: Martingale Dynamics*, Springer, Berlin.

[2] Christopeit, N. (1986) Quasi-least-squares estimation in semimartingale regression models, *Stochastics*, **16**, 255–278.

[3] Elliott, R.J. (1982) *Stochastic Calculus and Applications*, Springer, New York.

[4] Jacod, J. and Shiryayev, A.N. (1987) *Limit Theorems for Stochastic Processes*, Springer, Heidelberg.

[5] Karandikar, R.L. (1983) Interchanging the order of stochastic integration and ordinary differentiation, *Sankhya Ser. A*, **45**, 120–124.

[6] Kunita, H. (1990) *Stochastic Flows and Stochastic Differential Equations*, Cambridge University Press, Cambridge.

[7] Lai, T.L. and Wei, C.Z. (1982) Least squares estimation in stochastic regression models with applications to identification and control of dynamical systems, *Ann. Statist.*, **10**, 154–166.

[8] Liptser, R.S. and Shiryayev, A.N. (1977) *Statistics of Random Processes: General Theory*, Springer, New York.

[9] Liptser, R.S. and Shiryayev, A.N. (1978) *Statistics of Random Processes: Applications*, Springer, New York.

[10] Métivier, M. (1982) *Semimartingales*, Walter de Gryuter, Berlin.

[11] Prakasa Rao, B.L.S. (1987) *Asymptotic Theory of Statistical Inference,* Wiley, New York.

[12] Sorensen, M. (1986) On sequential maximum likelihood estimation for exponential families of stochastic processes, *Internat. Statist. Rev.,* **54**, 191–210.

[13] Stroock, D.W. (1982) *Lectures on "Topics in Stochastic Differential Equations,"* Narosa, New Delhi.

[14] Stroock, D. and Varadhan, S.R.S. (1979) *Multidimensional Diffusion Processes,* Springer, Berlin.

Appendix G

Notes

Note: The numbering of references here corresponds to the list of references at the end of each chapter.

CHAPTER 1: Semimartingales

The concept of a semimartingale has been found to be of major interest in stochastic modeling, as it includes several types of processes such as point processes, diffusion processes, diffusion processes with jumps etc. A semimartingale essentially is the sum of a local martingale and a process that is of bounded variation. As the notion of a semimartingale and its properties are not widely discussed in the books on statistical inference and are not widely known to the statisticians and modelers, we give an extensive review of the theory of semimartingales giving proofs of the results occasionally. A brief introduction was earlier given in Prakasa Rao [53]. The discussion here is based on Elliott [12], Bremaud [6], Liptser and Shiryayev [44, 45], Kallianpur [33], Jacod and Shiryayev [30] and other related books. Section 1.2 deals with a discussion on the general concept of a stochastic process and stochastic integration with respect to a process of bounded variation. Properties of martingales are also discussed in this section. Doob–Meyer decomposition for a class of super martingales along with its implications for the square of a square integrable martingale are studied in Section 1.3. Stochastic integration with respect to a Wiener process and Ito stochastic differential equations are introduced in Section 1.4. The concepts of quadratic characteristic and quadratic variation for a martingale are defined and central limit theorems for martingales are investigated in this section. Local martingales were introduced in Section 1.5 and stochastic integration with respect to a local martingale as well as some limit theorems for local martingales such as strong law of large numbers and central limit theorem are discussed in this section. The concept of a semimartingale was defined and stochastic integration with respect to a semimartingale was discussed in Section 1.6. Stochastic differential equations with a semimartingale as the *noise* component are studied in this section. The notion of local characteristics for a semimartingale are introduced. Section 1.7 contains a discussion on the Girsanov's theorem for semimartingales. Limit theorems for semimartingales are studied in Section 1.8. Diffusion-type processes and some of their properties are given in Section 1.9 and point processes and some of their properties are discussed in Section 1.10 along with some examples.

CHAPTER 2: Exponential Families of Stochastic Processes

The material in this chapter is based on Kuchler and Sorensen [6, 8]. They have studied exponential families of Markov processes in Kuchler and Sorensen [7]. For classical theory of exponential families, see Barndorff-Nielsen [1].

CHAPTER 3: Asymptotic Likelihood Theory

This chapter discusses asymptotic likelihood theory due to Barndorff–Nielsen and Sorensen [3] and Kuchler and Sorensen [13]. Various concepts of the information quantities are introduced in Barndorff–Nielsen and Sorensen [3] and for earlier discussion on extending the concept of information in the context of stochastic processes, see Basawa and Prakasa Rao [4] and Basawa and Scott [5]. The discussion on asymptototic likelihood theory for exponential families is from Kuchler and Sorensen [13]. Section 3.4 contains results due to Barndorff–Nielsen and Sorensen [3] on the asymptotic likelihood theory for general processes.

CHAPTER 4: Asymptotic Likelihood Theory for Diffusion Processes with Jumps

This chapter deals with likelihood methods for inference for parameters that determine the drift and the jump mechanism of a diffusion process. It is assumed that a continuously observed sample path is available. Our discussion is based on Sorensen [25]. The case of diffusion processes without the jump component was discussed in Basawa and Prakasa Rao [4], Kutoyants [17], Liptser and Shiryayev [19], and more recently in Prakasa Rao [22].

CHAPTER 5: Quasi Likelihood and Semimartingales

The notion of quasilikelihood has been introduced by Wedderburn [24]. There are problems in statistical inference for stochastic processes where the likelihood functions involved are not easily computable and the maximum likelihood estimators of the parameters involved are difficult to calculate. In such cases and in several other cases, the quasi likelihood approach may be employed for estimation of the parameters. Godambe [4] introduced the concept of estimating functions in the context of discrete time stochastic processes. Section 5.1 is based on Godambe [4]. Godambe and Heyde [5] studied quasi likelihood and optimal estimation for continuous time stochastic processes. The concept of the optimal estimating function introduced in Section 5.2 is from Godambe and Heyde [5]. The discussion in Sections 5.3 and 5.4 on asymptotic properties of quasi likelihood estimators for a class of semimartingales is based on Sorensen [20, 21]. Section 5.4 deals with estimation for partially specified counting processes through the quasi likelihood approach following Greenwood and Wefelmeyer [7].

CHAPTER 6: Local Asymptotic Behavior of Semimartingale Experiments

The notion of local asymptotic normality for statistical experiments in the classical case of independent and identically distributed observations was introduced by Le Cam [14] and was later extended to the concept of local asymptotic mixed normality in the context of stochastic processes by Jeganathan [12]. Examples of families of measures generated by stochastic processes that form either locally asymptotic normal or mixed normal families are discussed in Basawa and Prakasa Rao [2], Basawa and Scott [3], and Kutoyants [13]. Our discussion extending the notion of local asymptotic mixed normality to semimartingales in Section 6.1 is based on Luschgy [17]. The concept of local asymptotic quadraticity was introduced by Luschgy [18]. Discussion in the Section 6.2 is from Luschgy [18]. Taraskin [23] studied the concept of local asymptotic infinite divisibility generalizing the notion of local asymptotic normality. Results in Section 6.3 are based on Taraskin [23]. Greenwood and Wefelmeyer [8] discussed the concept of local asymptotic normality in the case of an infinite dimensional parameter and applied their results to the problems of efficient estimation of an intensity function for some classes of counting processes. The discussion in the Sections 6.4 and 6.5 is based on Greenwood and Wefelmeyer [8]. Section 6.6 contains some exercises.

CHAPTER 7: Likelihood and Asymptotic Efficiency

Most of the the results in this chapter are based on a series of papers by Greenwood and Wefelmeyer extending the concept of local asymptotic normality to the infinite dimensional parameter for efficient estimation of functionals of the parameter. Sections 7.1 and 7.2 are based on partially and fully specified semimartingales and efficiency by Greenwood and Wefelmeyer [7]. The discussion on partial likelihood and asymptotic efficiency in the Section 7.3 is from Jacod [11] and Greenwood [4]. For related discussion, see Jacod [12]. Efficient estimation for partially specified filtered models is discussed in Greenwood and Wefelmeyer [8], and the results in the Section 7.4 are from Greenwood and Wefelmeyer [8].

CHAPTER 8: Inference for Counting Processes

Point processes occur widely in modeling of many phenomena in physical, technical, economic, biological sciences, and many other areas (Cox and Lewis [16] and Lewis [37]). Parametric inference for such processes has been discussed in Kutoyants [32] and more recently in Karr [29]. Statistical inference for counting processes in general is extensively discussed in Andersen et al. [6] and Fleming and Harrington [18]. Most of the results in the Sections 8.1 and 8.2 are based on Kutoyants [32]. The results on Bayesian inference for counting processes are due to Hjort [22]. The discussion on M-estimation for nonhomogeneous Poisson processes is based on Yoshida and Hayashi [58]. Section 8.3 contains results due to Svensson [55] on semiparametric inference for counting processess. The discussion on the estimation of the intensity function by the kernel method for nonhomogeneous Poisson processes is due to Kutoyants [32, 33, 34]. Results on the estimation of the intensity function by the kernel method for counting procesess are due to Ramlau–Hansen [52]. Results on the estimation by the method of sieves are due to Karr [28] and on the estimation by the method of penalty functions are due to Antoniadis [8]. The discussion on the method of martingale estimators follows Karr [29]. Results on the estimation by the maximum likelihood method are from Van de Geer [57]. The discussion on inference for additive–multiplicative hazard models follows Lin and Ying [39].

CHAPTER 9: Inference for Semimartingale Regression Models

Semimartingale regression models discussed in the Section 9.1 include discrete time models with stochastic regressors as well as continuous time diffusion models as special cases. The material in this section is based on Christopeit [5]. The maximum likelihood approach for the problem of estimation of parameters, in a system modeled by a scalar linear stochastic differential equation excited by Gaussian and Poisson noises, discussed in Section 9.2 is due to Brodeau and Le Breton [4]. Le Breton [11] discusses parameter estimation and input design for linear dynamical systems. Section 9.3 contains material on the estimation of the parameters of semimartingale regression model by the method of sieves following McKeague [14]. This model includes continuous time diffusion processes, point processes, as well as processes with both the diffusion process and the point process components. McKeague and Utikal [16] discuss the estimation of doubly cumulative hazard function in nonlinear semimartingale regression models. Section 9.4 is based on their work.

CHAPTER 10: Applications to Stochastic Modeling

A large number of applications of diffusion processes for stochastic modeling are discussed in Karlin and Taylor [10]. Applications of counting processes to survival analysis are given in Andersen et al. [1], Fleming and Harrington [4] and Karr [11]. Some recent applications to mathematical finance, forest management and market policy are discussed in Prakasa Rao [14]. Material in Section 10.2 is from Lo [12]. The introduction to applications to modeling of neuron movement in a nervous system in Section 10.3 is based on Wegman and Habib [15].

APPENDIX A–F

For a discussion on the Doléans measure in the Appendix A, see Métivier [10]. Kunita [6] gives a proof of the Burkholder's inequality for martingales discussed in the Appendix A as an application of the Ito's formula. Appendix B contains results due to Karandikar [5] on Fubini-type theorems for stochastic integrals and a result from Liptser and Shiryayev [8]. Sufficient conditions for the differentiability of an Ito stochastic integral are also discussed in Appendix B. The discussion in the Appendix C on the fundamental identity of sequential analysis follow from the general theory of sequential analysis, for instance, see Sorensen [12]. The material on Stieltjes–Lebesgue calculus in the Appendix D follows discussions in Bremaud [1], Elliott [3] and Liptser and Shiryayev [9]. Results in Appendix E are due to Christopeit [2]. Appendix F contains results due to Le Cam following Prakasa Rao [11] and Jacod and Shiryayev [4].

Index